THE ENGLISH LEGAL PROCESS

THE ENGLISH LEGAL PROCESS

Eleventh Edition

TERENCE INGMAN LLM, PhD

Formerly of the Newcastle Law School, University of Newcastle upon Tyne

OXFORD

UNIVERSITY PRESS

OXFORD
UNIVERSITY PRESS

Great Clarendon Street, Oxford OX2 6DP

Oxford University Press is a department of the University of Oxford.
It furthers the University's objective of excellence in research, scholarship,
and education by publishing worldwide in

Oxford New York

Auckland Cape Town Dar es Salaam Hong Kong Karachi
Kuala Lumpur Madrid Melbourne Mexico City Nairobi
New Delhi Shanghai Taipei Toronto

With offices in

Argentina Austria Brazil Chile Czech Republic France Greece
Guatemala Hungary Italy Japan Poland Portugal Singapore
South Korea Switzerland Thailand Turkey Ukraine Vietnam

Oxford is a registered trade mark of Oxford University Press
in the UK and in certain other countries

Published in the United States
by Oxford University Press Inc., New York

© Terence Ingman, 2006

First published 1983	Sixth edition 1996
Second edition 1987	Reprinted 1997
Reprinted 1989	Seventh edition 1998
Third edition 1990	Reprinted 1999
Reprinted 1991	Eighth edition 2000
Fifth edition 1994	Ninth edition 2002
Reprinted 1995	Tenth edition 2004
	Eleventh edition 2006
	Reprinted 2007

British Library Cataloguing in Publication Data

Data available

Library of Congress Cataloging in Publication Data

Data available

Typeset by RefineCatch Limited, Bungay, Suffolk
Printed in Great Britain
on acid-free paper by
Ashford Colour Press Ltd, Gosport, Hampshire

ISBN 978–0–19–929038–3

3 5 7 9 10 8 6 4 2

Outline Contents

Contents

Preface

The text of this new edition has been fully revised to take account of major developments in the legal process since publication of the tenth edition two years ago. The order of chapters has been rearranged so that the material on legislation and precedent appears earlier in the book. The material on private law remedies against public authorities, previously contained in Chapter 10, is now to be found in Chapter 9, while Chapter 10 itself has been renamed 'Remedies in Public Law'.

Among the new statutes noted are the relevant parts of the Civil Partnership Act 2004, the Employment Relations Act 2004, the Gender Recognition Act 2004, the Prevention of Terrorism Act 2005, and the Constitutional Reform Act 2005.

The troubled legislative history of the latter statute is well documented. It eventually received the Royal Assent in its amended form in March 2005, although many of its provisions are still not yet in force. The office of Lord Chancellor was, in the end, rescued from abolition if not substantial reform. The authority of the holder of that post to sit as a judge has been removed, and his role in the judicial appointments process has been significantly reduced (para 1.1.6). The Judicial Appointments Commission began work in April 2006 (ibid.).

In anticipation of the creation of the new Supreme Court of the United Kingdom, the present senior Lord of Appeal in Ordinary has been authorised to make the first Supreme Court Rules governing its practice and procedure. The government has announced that the Court will be housed in a redesigned and refurbished Middlesex Guildhall, a Grade II listed building. The work, costing some £30m, was originally expected to be completed by late 2008. This has already been put back to October 2009. It is thought that the Court will not come into existence until its premises are ready for occupation.

The statutory regulations governing conditional fee agreements have been revoked. Agreements made on or after 1 November 2005 are now regulated instead by a Law Society Code (para 1.9.4).

The provision of public funding for representation in the criminal courts continues to lurch from one crisis to another. It is expected that the Criminal Defence Service Bill will be passed into law during the course of 2006. The Bill is part of an effort to make savings of £35m on the criminal legal aid budget and thereby make more money available for civil cases. It will reintroduce means-testing in the magistrates' courts before the grant of public funding for representation, and will transfer from the courts to the Legal Services Commission the power to grant representation (para 1.9.7). The government believes that the means-testing will be 'simple and swift to administer'. The signs are not good. This form of means-testing has existed before but was abolished at the inception of the Criminal Defence Service in 2001—largely because the cost of administering it outweighed the savings to be made. It remains to be seen whether it will be any more financially effective this time around.

The provisions of the Criminal Justice Act 2003 changing the double jeopardy rule and allowing retrials in respect of a number of serious offences came into force on 4 April 2005. The first application to the Court of Appeal for an order to quash an acquittal and to authorise a retrial has been sanctioned by the Director of Public Prosecutions (para 7.5.3). The provisions of the Criminal Justice Act 2003 relating to the general right of appeal of the prosecution were also brought into force on the same day (para 7.5.4.2).

Recent judgments of the Court of Appeal and the High Court incorporated in the new edition include decisions on the relevance of public policy considerations in the interpretation of statutes (*R (Hicks)* v *Secretary of State for the Home Department*), modification of the forfeiture rule (*Dalton* v *Latham*), the meaning of a 'new or newly discovered fact' for the purposes of claiming statutory compensation for a miscarriage of justice (*R (Murphy)* v *Secretary of State for the Home Department*), increasing sentences for contempt of court (*Lomas* v *Parle*), exemplary damages in the law of tort (*Design Progression Ltd* v *Thurloe Properties Ltd*), the use of McKenzie friends in family proceedings (*In Re O (Children) (Hearing in Private: Assistance)*), the meaning of a 'matter relating to trial on indictment' (*R (Snelgrove)* v *Crown Court at Woolwich*), and bias in public law (*R* v *Abdroikov*; leave to appeal to the House of Lords has been granted in this case).

In addition, account is taken of important pronouncements by the House of Lords on the recovery of a success fee from an unsuccessful defendant under a conditional fee agreement (*Campbell* v *MGN Ltd (No. 2)*), the validity of legislation (*R (Jackson)* v *Attorney-General*), the use of headings and side-notes as aids to statutory construction (*R* v *Montila*), the limits of purposive construction (*R* v *Z (Attorney-General for Northern Ireland's Reference)*; *R* v *Bentham*), the incompatibility of English law with the European Convention on Human Rights (*Ghaidan* v *Godin-Mendoza; A* v *Secretary of State for the Home Department*), prospective overruling in the House of Lords (*In re Spectrum Plus Ltd*), how a judge should deal with allegations of impropriety made by a juror (*R* v *Smith*), and contempt of court committed by a juror (*Attorney-General* v *Scotcher*).

Extracts from *The Law Reports* and *The Weekly Law Reports* are reproduced with the permission of The Incorporated Council of Law Reporting for England and Wales, and those from *The All England Law Reports* are reproduced by permission of Reed Elsevier (UK) Ltd, trading as Lexis Nexis Butterworths. Crown Copyright material (including Acts, statutory instruments, and *Judicial Statistics 2004*, Cm 6565, 2005) is reproduced with the permission of the Controller of HMSO and the Queen's Printer for Scotland.

The assistance received from my contacts at Oxford University Press is again gratefully acknowledged. Once more, special thanks are due to Annie Henderson for her encouragement and support.

Terry Ingman
Crowthorne
Berkshire
March 2006

Preface to the First Edition

In writing *The English Legal Process* it was not my intention to provide a comprehensive survey of the English legal system. Rather, my aim was to produce a topical account of some of the more important institutions and practices which form part of our legal process.

There appeared to be a need for a textbook, as opposed to a source-book, which attempted to deal with the substantive law on those matters in an informative and unsophisticated manner, and which, at the same time, did not ignore their relationship to the 'real world' and avoided the impression that our legal process was without blemish. It is for these latter reasons that areas of current controversy or difficulty are touched upon. Court delays, bail, contempt of court, miscarriages of justice, the vetting and 'nobbling' of juries, and the continuing struggle of the ordinary citizen, occasionally frustrated by the courts, to challenge abuse of power on the part of central government and other public authorities: these are some of the areas where the legal process is malfunctioning. Although in places I have ventured to be critical, I did not perceive that as my principal function. I have tried to avoid 'knocking' for its own sake. I have knocked 'the system' only where, in my view, it is in need of it.

The English Legal Process is intended mainly for first-year students at universities, polytechnics and colleges who are studying the English legal system. It is hoped that they will find it a comprehensible treatment of the subjects with which it deals and that the material contained in it will be rendered more accessible and digestible to them by the use of frequent cross-referencing and relatively short, numbered paragraphs.

The subject of remedies is often ignored or, at best, treated only cursorily in works on the legal system. Yet it is an integral and important part of the civil legal process. The inclusion in this book of chapters on remedies is designed to familiarise students at an initial stage in their career with the major forms of redress available in both private and public law. It is believed that their inclusion is further justifiable on the ground that an early introduction to remedies may prove helpful to the study and understanding of other legal subjects, such as contract, tort, administrative law, land law and equity. Students who have advanced beyond the first year of their course may find the book useful for that reason.

The law-making process is never static. Because the book covers so many diverse topics, keeping the text up to date in the face of incessant legislative and judicial outpourings has been a particularly difficult task. Many parts of the manuscript had to be rewritten at short notice, sometimes more than once, in order to take account of the latest changes in the law effected by statute or judicial decision.

Among the new statutes to be considered were two important consolidating measures, the Magistrates' Courts Act 1980 and the Supreme Court Act 1981, and others which introduced important rules into our law, such as the Contempt of Court Act 1981, the Criminal Justice (Amendment) Act 1981, the Administration of Justice Act 1982, the Criminal Justice Act 1982 and the Forfeiture Act 1982.

As is to be expected, the courts have been even more productive than Parliament. Many important new cases, too numerous to mention here, have had to be accommodated in the text. The House of Lords has been unusually active over the past two or three years. In the first quarter of 1983 alone some 15 House of Lords' decisions affecting the text have been reported.

I acknowledge with gratitude my indebtedness to those in the Law Faculty at Newcastle upon Tyne who gave me their time during the preparation of this book. Bill Elliott, Mick Rowell, Ashley Wilton, John Clark, Ian Dawson and Joanna Gray, a former student, were all imposed upon. With characteristic generosity and good nature they supplied innumerable criticisms and suggestions which rescued me from many an error. Responsibility for those errors and other shortcomings which remain is, of course, mine alone. My thanks are also due to my secretary, Christine Markham, who worked on a disgraceful manuscript with such commendable intelligence, speed, accuracy and cheerfulness, to Alistair MacQueen and Heather Saward for their assistance at every step in the publishing process, and to Derek French, who edited the manuscript so conscientiously and skilfully and succeeded in making sense out of some of my more Delphic pronouncements.

I have endeavoured to state the law as at the end of March 1983. There have been developments since then which it was not possible to include in the text. Some of them may be mentioned here.

All the provisions of the Criminal Justice Act 1982 (so far as it affects England and Wales) come into force on various dates from January to May 1983.

Sir John Donaldson MR, in an effort to facilitate further the disposal of appeals in the Court of Appeal, civil division, has given guidance on the submission of 'skeleton arguments' by counsel (see para 1.3.5 and *The Times*, 13 April 1983).

The Lord Chancellor has taken the unusual step of dismissing two lay magistrates— apparently after a complaint from the estranged wife of one of them. The two justices, who became good friends and intended to marry, had refused to resign from the bench (see para 1.8.1).

The Coroners' Society is pressing the Home Secretary to introduce legislation to reverse the decision of the Court of Appeal in *R v West Yorkshire Coroner (ex parte Smith)* [1982] 3 All ER 1098, which requires a coroner to hold an inquest into certain deaths occurring abroad if the body is brought back to England (see para 2.5.3.2).

The wisdom of conferring on magistrates' courts the power to punish summarily a contempt committed in the face of the court came into question when a man was sent to gaol by a stipendiary magistrate, under s. 12(1) of the Contempt of Court Act 1981, for whispering in the public gallery (see para 5.4.4).

In a White Paper, the government has rejected the suggestion that there should be an independent review body to examine alleged miscarriages of justice. It is envisaged, however, that the Home Secretary will in future be prepared more readily to exercise his power to refer cases to the Court of Appeal, criminal division, under s. 17 of the Criminal Appeal Act 1968, and that the court will be prepared to make greater use of its powers to admit fresh evidence and order a retrial (see para 6.3.6 and *Miscarriages of Justice: Government Reply to the Sixth Report from the Home Affairs Committee, Session 1981 to 1982* Cmnd 8856, 1983).

Terry Ingman
Newcastle upon Tyne
April 1983

Table of Cases

Table of Statutes

European legislation

International legislation

Table of Secondary Legislation

European secondary legislation

Table of Reports

1

Courts of normal jurisdiction

1.1 Introduction

1.1.1 Civil courts and criminal courts

Civil courts exist in order to resolve disputes between private citizens or between a citizen and the state. These disputes may involve such matters as breach of contract, liability for harm in the law of tort, rights in property, marital status, or the wrongful exercise of power by some public authority. In such cases one party is seeking to obtain from the court some private remedy against the other.

Criminal courts exist in order to hear and determine accusations against persons that they have broken the criminal law. In most cases the accusation is made by someone representing the state. On a finding of guilt, the criminal courts have power to inflict punishment, commonly in the form of a fine or imprisonment.

In England and Wales there is no rigid line of demarcation between civil and criminal courts since almost all the courts exercise both types of jurisdiction. Exceptionally, the county courts are purely *civil* courts and exercise exclusively one type of jurisdiction.

1.1.2 Superior courts and inferior courts

Some courts are *superior* courts and others are *inferior* courts. Superior courts, like the House of Lords, the Court of Appeal, the High Court and the Crown Court, have unlimited jurisdiction and deal with the more important and difficult cases. Inferior courts, such as the county courts and the magistrates' courts, have limited jurisdiction and hear the less important and less difficult cases.

Inferior courts, but not superior courts, are subject to the supervisory prerogative jurisdiction of the High Court, which is discussed in Chapter 10. For the unusual position of the Crown Court, which is a superior court and yet is sometimes subject to supervision by the High Court, see para 10.5.2.3.

1.1.3 Courts of record and courts not of record

Another distinction is that between *courts of record* and *courts not of record*. A court of record is one whose proceedings are kept as a permanent record in the Public Record Office. A court of record may be a superior court or an inferior court. The House of Lords, the Court of Appeal, the High Court, the Crown Court, and the Employment

Appeal Tribunal are superior courts of record. The county courts and the coroners' courts are inferior courts of record. Magistrates' courts are courts not of record.

A superior court of record has power to punish all forms of contempt of court (see Chapter 4), while an inferior court of record can generally only punish contempt committed in the face of the court. The distinction between a court of record and a court not of record has little meaning now that the magistrates' courts have power themselves to punish contempt committed in the face of the court (Contempt of Court Act 1981, s. 12; see para 8.4.4).

The coroners' courts are inferior courts of record and have jurisdiction, therefore, to punish contempt committed in their face (*R* v *West Yorkshire Coroner (ex parte Smith) (No. 2)* [1985] 1 All ER 100, DC). Other contempts of the coroners' courts, such as prejudicial comment in the press, are not punishable by those courts themselves but may be punished on an application to the Queen's Bench Division of the High Court (*R* v *Davies* [1906] 1 KB 32, DC, and see para 8.4.5).

Consistory courts and courts martial are courts not of record and do not, therefore, possess any inherent power to punish for contempt of court. The Queen's Bench Divisional Court may, however, punish contempts on their behalf (see para 8.4.5).

1.1.4 Courts and tribunals

The distinction between a court and a tribunal is sometimes blurred. While every court is a tribunal, not every tribunal is a court. The nomenclature used is not a sure guide. Thus, the Employment Appeal *Tribunal* is a court (see para 2.3) while a local valuation *court* was a tribunal only.

The distinction between a court and a tribunal assumes particular importance in relation to punishment for contempt of court. The Queen's Bench Divisional Court has jurisdiction to punish a contempt committed in connection with the proceedings of an 'inferior court' (Rules of the Supreme Court (RSC) 1965, Ord. 52), but it has no general power to punish contempt of a tribunal. It does, however, have power to punish a contempt of a tribunal of inquiry established under the Inquiries Act 2005 (formerly the Tribunals and Inquiries (Evidence) Act 1921; see para 8.4.5).

In *Attorney-General* v *British Broadcasting Corporation* [1980] 3 All ER 161, HL, the House of Lords held that a local valuation court was not an inferior court because, although it was called a 'court', its functions were essentially administrative rather than judicial. It followed that the Divisional Court had no power to punish an alleged contempt of a local valuation court arising out of a BBC television programme about the religious sect known as the Exclusive Brethren.

In *General Medical Council* v *British Broadcasting Corporation* [1998] 3 All ER 426, CA, it was held that the Professional Conduct Committee of the General Medical Council is not a court. Although the Committee's function is a judicial one, it does not exercise the judicial power of the state. Its role is to exercise the self-regulatory power of the medical profession to maintain standards of professional conduct.

On the other hand, it has been held by the House of Lords that a mental health review tribunal *is* a court for the purposes of the law of contempt. Its functions are essentially judicial rather than administrative, and, in discharging them, it is exercising the judicial power of the state (*Pickering* v *Liverpool Daily Post and Echo Newspapers plc* [1991] 1 All ER 622, HL). Similarly, in *Peach Grey & Co.* v *Sommers* [1995] 2 All ER 513, DC, it was held that an employment tribunal is a court since it has many of the characteristics of a court of law and discharges judicial, as opposed to administrative, functions.

The opinion of their Lordships in *Attorney-General* v *British Broadcasting Corporation* [1980] 3 All ER 161, HL, was that a court of law is a court established to exercise the judicial power of the state. This definition has since been adopted by Parliament in s. 19 of the Contempt of Court Act 1981. It is explained in the following words of Lord Scarman in *Attorney-General* v *British Broadcasting Corporation*, above (at pp. 181–2):

I would identify a court in (or 'of') law, i.e., a court of judicature, as a body established by law to exercise, either generally or subject to defined limits, the judicial power of the state. In this context judicial power is to be contrasted with legislative and executive (i.e. administrative) power. If the body under review is established for a purely legislative or administrative purpose, it is part of the legislative or administrative system of the state, even though it has to perform duties which are judicial in character. . . . [T]he judicial power of the state exercised through judges appointed by the state remains an independent, and recognisably separate, function of government. Unless a body exercising judicial functions can be demonstrated to be part of this judicial system, it is not, in my judgment, a court in law.

Attorney-General v *British Broadcasting Corporation*, above, was applied by the Judicial Committee of the Privy Council in *Badry* v *Director of Public Prosecutions of Mauritius* [1982] 3 All ER 973, PC, in which it was said that the former case had plainly established that, in the absence of any statutory provision to the contrary, the common law relating to contempt of court applied by definition only to courts of justice properly so-called and to the judges of such courts of justice. It followed in the *Badry* case that the appellant's comments against a judge of the Supreme Court of Mauritius, including the statement that 'We must tear off his trousers in this country', did not constitute a contempt of court. The comments were 'vulgar, scurrilous, abusive and lacking in respect' ([1982] 3 All ER 973 *per* Lord Hailsham LC at p. 980) but they were directed against the judge as a commissioner or tribunal holding an inquiry into allegations of fraud and corruption and not in his capacity as a judge of the Supreme Court.

1.1.5 Qualifications for appointment as a professional judge: Courts and Legal Services Act 1990

The Courts and Legal Services Act 1990 achieved what were then the most radical reforms in court jurisdiction and procedure since the nineteenth century, and, in relation to the legal profession, the most fundamental changes ever made.' Of particular importance here are the changes made by the Act to the qualifications required for appointment as a judge. One effect of these changes is that *solicitors* are qualified for many more judicial posts than hitherto. Eligibility no longer depends solely on whether candidates happen to be barristers (or solicitors). Instead, candidates are eligible if they possess qualifications which are based on *rights of audience* in the courts granted by authorised bodies. This move also reflected the government's policy of stimulating competition by an extension of rights of audience and a widening of the choice of persons qualified to provide advocacy services.

A list of the qualifications for judicial appointment based on rights of audience is set out in s. 71(3) of the Act:

(a) a 'Supreme Court qualification' (a right of audience in relation to all proceedings in the Supreme Court);

(b) a 'High Court qualification' (a right of audience in relation to all proceedings in the High Court);

(c) a 'general qualification' (a right of audience in relation to any class of proceedings in any part of the Supreme Court, or all proceedings in county courts or magistrates' courts);

(d) a 'Crown Court qualification' (a right of audience in relation to all proceedings in the Crown Court);

(e) a 'county court qualification' (a right of audience in relation to all proceedings in county courts);

(f) a 'magistrates' court qualification' (a right of audience in relation to all proceedings in magistrates' courts).

1.1.6 Appointment of professional judges: Constitutional Reform Act 2005

Before the Constitutional Reform Act 2005, the Lord Chancellor had a major role in the selection and appointment of judges. He had sole responsibility for appointing district judges and deputy district judges to the county courts. His recommendation was necessary before the appointment by the Queen of High Court judges, circuit judges, recorders, and district judges (magistrates' courts). Even in the case of appointments made by the Queen on the advice of the Prime Minister (Lords of Appeal in Ordinary in the House of Lords, the Heads of Division in the High Court, and the Master of the Rolls and Lords Justices of Appeal in the Court of Appeal), it was the practice that the Prime Minister sought the advice of the Lord Chancellor before making a recommendation.

The original intention of the government (expressed in the Constitutional Reform Bill) was to abolish the post of Lord Chancellor completely, but this plan was thwarted when, in December 2004, the House of Lords as part of Parliament voted to keep the Lord Chancellor as a member of that House. The office of Lord Chancellor, which dates back some 1,400 years, was thus saved from abolition but not significant reform.

Under the Constitutional Reform Act 2005, the functions of the Lord Chancellor in relation to the *judiciary* have, in the main, been transferred to the Lord Chief Justice. The latter has become President of the Courts of England and Wales, Head of Criminal Justice, and Head of the Judiciary of England and Wales in place of the Lord Chancellor, with responsibility for the welfare, training, and deployment of judges, and for representing their views to ministers and Parliament. The holder of the new post of President of the Queen's Bench Division presides over that Division in place of the Lord Chief Justice.

The office of Vice-Chancellor of the Chancery Division of the High Court has been replaced by the new post of Chancellor of the High Court, who becomes the president of the Chancery Division in place of the Lord Chancellor.

As President of the Courts of England and Wales, the Lord Chief Justice is entitled to sit in the Court of Appeal, the High Court, the Crown Court, the county courts, and the magistrates' courts (Constitutional Reform Act 2005, s. 7). The Lord Chancellor is no longer head of the judiciary or eligible to sit as a judge in any court.

The qualifications for appointment as Lord Chancellor have been changed so that a person can only be recommended for appointment if he appears to the Prime Minister to be qualified by experience as a minister of the Crown, or as a Member of either House of Parliament, or as a qualifying legal practitioner, or as a university law teacher, or qualified by other experience which the Prime Minister considers to be relevant (ibid., s. 2).

With effect from April 2006, the Constitutional Reform Act has severely curtailed the Lord Chancellor's involvement in the choice and appointment of judges. Selections are now made by the Judicial Appointments Commission created by s. 61 of the Act, and appointments hitherto made by the Lord Chancellor acting alone (see above) are made by the Queen on his advice after selection by the Commission. He does, however, continue to be solely responsible for appointing lay justices to the magistrates' courts, but they are not professional judges and he is now under a duty of consultation before making these appointments (Courts Act 2003, s. 10(2A), as inserted by the Constitutional Reform Act 2005).

The Judicial Appointments Commission has 15 members—a lay chairman, five other lay members, five judicial members, two legal professionals, a tribunal member, and a lay magistrate (Constitutional Reform Act 2005, sch. 12). The Commission (and its selection panels) is under a duty to make selections for judicial appointment solely on merit and to ensure that persons selected are of good character. In addition, the Commission is under a duty to have regard to the need to encourage diversity in the range of persons available for selection (ibid., ss. 63–64).

The Act makes special provision for the appointment of judges to the proposed new Supreme Court of the United Kingdom. The first judges will be the 12 existing Lords of Appeal in Ordinary (ibid., s. 24). Thereafter when a vacancy arises, the Lord Chancellor must appoint an *ad hoc* selection commission of five members (including one member of the Judicial Appointments Commission) to be chaired by the President of the Supreme Court. The selection commission will choose one person for each vacancy and report the selection to the Lord Chancellor. There are provisions for accepting or rejecting the selection or requiring the selection commission to reconsider its choice (ibid., ss. 26–31).

In choosing candidates, selection commissions are under a duty to ensure that the judges of the Court will, between them, have knowledge of, and experience of practice in, the law of each part of the United Kingdom (ibid., s. 27(8)). This preserves the convention applicable presently to the House of Lords that there should usually be at least two Scottish Law Lords and one from Northern Ireland.

Appointments to the office of Justice of the Supreme Court will be made by the

Queen on the recommendation of the Prime Minister, but he can only recommend the person who was successful in the selection process as notified to him by the Lord Chancellor (ibid., ss. 23, 26(3)).

The 2005 Act also makes special provision for the appointment of the Lord Chief Justice, the Heads of Division (i.e., the Master of the Rolls, the President of the Family Division, the Chancellor of the High Court, and the President of the Queen's Bench Division), and judges of the Court of Appeal.

Selection of a candidate is made by a panel set up by the Commission. The selection panel of four members, comprising two senior judges and two members of the Commission, chooses one person for each vacancy and reports its selection to the Lord Chancellor. There are provisions for accepting or rejecting the selection or requiring the panel to reconsider its choice (ibid., ss. 67–84).

Appointments to the office of Lord Chief Justice, Master of the Rolls, President of the Family Division, Chancellor of the High Court, President of the Queen's Bench Division, and judge of the Court of Appeal are made by the Queen on the advice of the Prime Minister after the Commission has made a recommendation to the Lord Chancellor.

For a wide range of other judicial appointments, including puisne (i.e., lesser or assistant) judge of the High Court, circuit judge, district judge, and recorder, the Commission makes recommendations *direct* to the Lord Chancellor. One candidate is recommended by the Commission for each vacancy (ibid., ss. 85–94). Appointments are then made by the Queen on the recommendation of the Lord Chancellor following the recommendations from the Commission.

1.2 House of Lords

1.2.1 Introduction

In cases not involving European Community legislation, the House of Lords is the highest court in the United Kingdom. (For the relationship between the House of Lords, European Community legislation, and the Court of Justice of the European Communities, see para 2.1.1.) Judicial decisions of the House of Lords can be overruled only by statute or by a refusal of the House to follow them in later cases. (For judicial precedent in the House of Lords, see para 5.3.2.2.)

The House of Lords as a court will be abolished and replaced by a new Supreme Court of the United Kingdom when the relevant provisions of the Constitutional Reform Act 2005 are brought into force (see para 1.2.4 below).

1.2.2 Composition

The House of Lords as a court is composed of between seven and 12 Lords of Appeal in Ordinary together with any peer who holds or has held high judicial office, such as a retired Court of Appeal judge. The Lords of Appeal in Ordinary are made life peers (Appellate Jurisdiction Act 1876, s. 6), and are often called 'Law Lords' for short. One of them is designated as Senior Lord of Appeal in Ordinary.

Qualifications for appointment as a Lord of Appeal in Ordinary are:

(a) The holding of high judicial office for two years (Appellate Jurisdiction Act 1876, s. 6). 'High judicial office' means 'The office of Lord Chancellor . . . or of Judge of one of Her Majesty's superior courts of Great Britain and Ireland'. (Appellate Jurisdiction Act 1876, s. 25; the reference to 'superior courts of Ireland' is now a reference to the High Court in Northern Ireland and the Court of Appeal in Northern Ireland—ibid., as amended by the Judicature (Northern Ireland) Act 1978.)

(b) Possession of a 15-year Supreme Court qualification within the meaning of s. 71 of the Courts and Legal Services Act 1990 (para 1.1.5 above) (Appellate Jurisdiction Act 1876, s. 6, as amended by the Courts and Legal Services Act 1990).

(c) Practice for 15 years as an advocate in Scotland, or as a solicitor entitled to appear in the Court of Session and the High Court of Justiciary (ibid.).

(d) Practice for 15 years as a member of the Bar of Northern Ireland (ibid.).

1.2.3 Jurisdiction

The House of Lords has little *original* jurisdiction left. What remains includes the trial of disputed peerage claims and breaches of Parliamentary privilege in relation to itself.

The jurisdiction of the House is now almost exclusively *appellate*. It is the final appeal court for England and Wales and Northern Ireland in both civil and criminal cases, and for Scotland in civil cases only. Final Scottish criminal appeals are heard in Edinburgh in the High Court of Justiciary.

Most appeals heard by the House come from the Court of Appeal in England. An appeal can be taken from the Court of Appeal to the House only by leave of either court (Administration of Justice (Appeals) Act 1934, s. 1(1)). In criminal cases, there is the additional requirement that the appeal must involve a point of law of general public importance (Criminal Appeal Act 1968, s. 33(2)).

Every appeal to the House must be heard by at least *three* of the judges mentioned in para 1.2.2 (Appellate Jurisdiction Act 1876, s. 5). In practice, *five* of them usually sit together to form a court. Sometimes more than five may sit to hear an appeal which involves particularly difficult or important points. Thus, *seven* sat in the important civil case of *Cassell & Co. Ltd* v *Broome* [1972] AC 1027, HL, and in the important criminal case of *DPP* v *Majewski* [1977] AC 443, HL. More recently, a court of seven was assembled to hear the appeal in *Pepper* v *Hart* [1993] 1 All ER 42, HL, to rehear Senator Pinochet's attempt to avoid extradition to Spain on charges of crimes against humanity in *R* v *Bow Street Metropolitan Stipendiary Magistrate (ex parte Pinochet Ugarte) (Amnesty International intervening) (No. 3)* [1999] 2 All ER 97, HL, to reconsider its earlier decision in the case of *Rondel* v *Worsley* [1969] 1 AC 191, HL, in *Arthur J.S. Hall & Co.* v *Simons* [2000] 3 All ER 673, HL, and to review *McFarlane* v *Tayside Health Board* [2000] 2 AC 59, HL, in *Rees* v *Darlington Memorial Hospital NHS Trust* [2003] 3 WLR 1091, HL. The House of Lords sat as a court of *nine* towards the end of 2004 to hear the highly important case of *A* v *Secretary of State for the Home Department* [2005] 2 WLR 87, HL.

Each judge may deliver his own separate judgment, which in the House of Lords is called a 'speech' or 'opinion'. The hearing of the appeal is not a retrial. No oral evidence is given; the judges read the documents in the case and listen to counsel's arguments. The majority decision prevails. If for any reason the number of judges sitting is even, and the House is equally divided, the appeal is dismissed.

Although the House of Lords has an inherent jurisdiction to rehear an appeal, it is reluctant to do so except where a party has been the victim of an unfair procedure through no fault of his own. An appeal will not be reopened simply because the earlier decision is subsequently thought to be wrong. (See, e.g., *R* v *Bow Street Metropolitan Stipendiary Magistrate (ex parte Pinochet Ugarte) (No. 2)* [1999] 1 All ER 577, HL; and see further, para 7.1.5.)

1.2.4 Abolition of the House of Lords as a court: Constitutional Reform Act 2005

In addition to modifying the role of the Lord Chancellor, establishing an independent Judicial Appointments Commission with the function of recruiting and selecting judges (see para 1.1.6 above), and providing a statutory guarantee of judicial independence (see below), the Constitutional Reform Act 2005 will also ultimately abolish the House of Lords as a court. The House has existed as a professional court for over 130 years, but when the relevant provisions of the Act are brought into force it will be replaced by a new 'Supreme Court of the United Kingdom'. This is unlikely to be before suitable accommodation for the new Court is ready.

The Supreme Court of the United Kingdom will be a superior court of record, and will assume the appellate jurisdiction of the House of Lords together with the devolution jurisdiction of the Judicial Committee of the Privy Council (Constitutional Reform Act 2005, s. 40 and sch. 9). It will be the final appeal court for England and Wales and Northern Ireland in both civil and criminal cases, and for Scotland in civil cases. Unlike the United States Supreme Court, it will not have power to strike down legislation.

To avoid confusion, the existing Supreme Court of England and Wales (comprising the Court of Appeal, the High Court, and the Crown Court) will be renamed the 'Senior Courts of England and Wales', and the Supreme Court Act 1981 will be retitled the 'Senior Courts Act 1981' (ibid., s. 59(1) and sch. 11).

The Court will have 12 judges. There will be a President and a Deputy President, with the other judges of the Court to be known as 'Justices of the Supreme Court'. The first judges will be the 12 existing Lords of Appeal in Ordinary, the senior of whom becomes the President and the second senior the Deputy President. Thereafter, appointments are to be made in accordance with rules laid down in the 2005 Act (see para 1.1.6 above).

Eligibility for appointment as a judge of the Court will be the same as the present eligibility for appointment as a Lord of Appeal in Ordinary (see para 1.2.2 above).

The judges of the Court will hold office during good behaviour, subject to a power of removal by the Queen on an address presented to her by both Houses of Parliament

(Constitutional Reform Act 2005, s. 33). Provision is made for resignation, retirement, and medical retirement (ibid., ss. 35–36). The permanent membership of the Court may be supplemented on an *ad hoc* basis by 'acting judges' and judges who have ceased to hold high judicial office drawn from a 'supplementary panel' (ibid., ss. 38–39).

Like the House of Lords, the Court will be able to sit in panels. In general, the Court will be properly constituted for the purpose of hearing any appeal if it comprises an uneven number of at least three judges and at least half of them are permanent members of the Court. It will be possible for the President to direct a higher quorum for certain classes of proceedings and for specific proceedings (ibid., s. 42).

Once established, the practice and procedure of the Court will be governed by Supreme Court Rules made by the President (ibid., ss. 45–46). Prior to the creation of the Court and the office of President, the senior Lord of Appeal in Ordinary (who will become the first President anyway) has been given authority to make the first Rules (Constitutional Reform Act 2005 (Temporary Modifications) Order 2006, SI 2006, No. 227).

It is anticipated that the Court will, first, adopt the same approach to its own precedents as the House of Lords (see para 5.3.2.2), and, secondly, regard itself as normally bound by previous decisions of the House which it replaces.

A decision made by the Court in an appeal coming from one of the three jurisdictions within the United Kingdom will be binding within that jurisdiction, but only of persuasive effect in the others (Constitutional Reform Act 2005, s. 41). This is currently the position with decisions of the House of Lords, and the Act does not change it. Only decisions of the Court on devolution issues will be binding in all legal proceedings, although they will not be binding on the Court itself (ibid.).

The independence of the Court (and of all other courts) will be guaranteed by s. 3 of the Act, which imposes an obligation on ministers (including the Lord Chancellor), and others with responsibility for the administration of justice, to uphold the continued independence of the judiciary. In particular, ministers must not seek to influence particular judicial decisions through any special access to the judges, and the Lord Chancellor must 'have regard to' the need to defend judicial independence and the need for the judiciary to 'have the support necessary to enable them to exercise their functions'.

1.3 Court of Appeal

1.3.1 Introduction

The Court of Appeal was created by the Judicature Act 1873. Under the Act as originally drafted the Court of Appeal was to have been the ultimate appeal court since the jurisdiction of the House of Lords was to be abolished. In the end, the House of Lords was reprieved and the Court of Appeal was set up to hear civil appeals only. It now also hears criminal appeals after the Criminal Appeal Act 1966 created its criminal division to replace the Court of Criminal Appeal (see now the Supreme Court Act 1981, s. 3(1)).

The Court of Appeal, together with the High Court and the Crown Court, is part of the Supreme Court of England and Wales (Supreme Court Act 1981, s. 1(1)). The president of the Supreme Court of England and Wales is the Lord Chief Justice in his capacity as President of the Courts of England and Wales (Constitutional Reform Act 2005, s. 7). The Supreme Court of England and Wales will be renamed the 'Senior Courts of England and Wales' when the Supreme Court of the United Kingdom comes into existence.

The House of Lords, although for most purposes the 'supreme' court in the United Kingdom, is not part of the Supreme Court of England and Wales. This was probably a pure oversight caused originally by the uncertainty surrounding the future of the House between 1873 and 1876. The Employment Appeal Tribunal is also not part of the Supreme Court of England and Wales despite the fact that High Court judges sit in it and that appeals from its decisions lie to the Court of Appeal.

1.3.2 Composition

The Court of Appeal is composed of certain *ex officio* judges and up to 37 'ordinary' judges. The *ex officio* judges are any person who was Lord Chancellor before 12 June 2003 and is willing to sit, any Lord of Appeal in Ordinary willing to sit, the Lord Chief Justice, the Master of the Rolls, the President of the Queen's Bench Division of the High Court, the President of the Family Division of the High Court, and the Chancellor of the High Court (Supreme Court Act 1981, s. 2(1)–(2), as amended). An 'ordinary judge' of the Court of Appeal is styled 'Lord Justice of Appeal' or 'Lady Justice of Appeal' (ibid., s. 2(3), as substituted by the Courts Act 2003).

In addition, any former Court of Appeal or High Court judge, and any present High Court judge, may be requested to sit in the Court of Appeal (ibid., s. 9(1)). A former judge can refuse the request; a judge still in office cannot do so (ibid., s. 9(3)).

The 'ordinary' judges of the Court of Appeal are the Lords Justices and Lady Justices of Appeal appointed in accordance with the Constitutional Reform Act 2005 (see para 1.1.6 above). To be qualified for appointment as a Lord or Lady Justice of Appeal a person must either have a ten-year High Court qualification within the meaning of s. 71 of the Courts and Legal Services Act 1990 (para 1.1.5 above) or be a High Court judge (Supreme Court Act 1981, s. 10(3)(b), as amended by the Courts and Legal Services Act 1990). Thus it is possible for a person to be appointed as a Lord or Lady Justice direct from practice without any previous judicial experience. A good example is Sir Wilfred Greene, who was appointed a Lord Justice of Appeal in 1935 after a highly successful career of 27 years at the Bar. He later became Master of the Rolls (1937–49) and a Lord of Appeal in Ordinary (1949–50).

The president of the *civil* division of the Court of Appeal is the Master of the Rolls, who is also Head of Civil Justice. The president of the *criminal* division is the Lord Chief Justice (Supreme Court Act 1981, s. 3(2)). Each division also has a vice-president. Any number of courts of either division may sit at the same time, subject, of course, to the availability of judicial manpower (ibid., s. 3(5)).

Since January 1995, it has been possible for circuit judges to sit on appeals from the Crown Court in the criminal division of the Court of Appeal at the invitation of the Lord Chief Justice (Supreme Court Act, s. 9, as amended by the Criminal Justice and

Public Order Act 1994). Circuit judges were, however, initially forbidden to take part in appeals against a conviction before, or a sentence passed by, a High Court judge at the Crown Court. This restriction was removed by the Courts Act 2003.

1.3.3 Jurisdiction

1.3.3.1 Civil division

The jurisdiction of the Court of Appeal, civil division, is mainly to hear appeals in civil cases from all three divisions of the High Court and from the county courts. (For restrictions on the right to appeal, and for appeals to the Court of Appeal from other courts and tribunals, see para 7.1.) Since 1970, it has been possible in certain circumstances to 'leapfrog' the Court of Appeal and to take a civil appeal direct from the High Court to the House of Lords (para 7.1.4.3).

The Court of Appeal, civil division, is properly constituted for the purpose of exercising any of its jurisdiction if it consists of *one or more* judges (Supreme Court Act 1981, s. 54(2), as substituted by the Access to Justice Act 1999). In practice, most appeals are heard by *three* judges sitting together. However, appeals from the county courts, and appeals from the High Court in proceedings which could have been brought in a county court, can be heard by *two* judges (Court of Appeal (Civil Division) Order 1982, SI 1982, No. 543).

In addition, by s. 54(3)–(4A) of the Supreme Court Act 1981 (as substituted by the Access to Justice Act 1999), the Master of the Rolls is empowered to:

(a) give directions (with the concurrence of the Lord Chancellor) about the minimum number of judges required to sit in any *description of proceedings*;

(b) decide how many judges should sit in any *particular proceedings*, or designate any Lord or Lady Justice of Appeal to do this for him;

(c) give directions about what is to happen where proceedings in a case are partly heard and a judge is unable to continue.

If an appeal has been heard by two judges and they are equally divided, the case must be reargued before an uneven number of judges not less than three before any appeal can be taken to the House of Lords (Supreme Court Act 1981, s. 54(5)). This provision also applies to appeals heard in the criminal division (ibid., s. 55(5)).

Although it seems self-evident in the interests of justice and the preservation of public confidence in the bench, it is expressly provided that a judge is not allowed to hear an appeal from one of his own decisions (ibid., s. 56). This provision, too, applies equally to appeals in the criminal division (ibid.). It ousts the common law assumption that there is no inherent objection in allowing a superior court judge to hear an appeal from one of his own decisions.

In *Taylor* v *Lawrence* [2002] 3 WLR 640, CA, a Court of Appeal comprising five members (including the Lord Chief Justice and the Master of the Rolls) held that the Court of Appeal has power to *reopen* an appeal after it has given a final judgment and after that judgment has been drawn up if it is clearly established that a significant injustice has probably occurred and that there is no effective alternative remedy. Where the alternative remedy is an appeal to the House of Lords, the Court of Appeal will give permission to reopen the appeal only if it is satisfied that the House will not give permission to appeal.

Taylor v *Lawrence* was applied in *Seray-Wurie* v *Hackney London Borough Council* [2003] 1 WLR 257, CA, in which the Court of Appeal held that the *High Court*, when sitting as an appeal court, has a similar jurisdiction to reopen its decisions in exceptional circumstances in order to avoid a real injustice.

1.3.3.2 Criminal division

The jurisdiction of the Court of Appeal, criminal division, is to hear appeals in criminal cases from persons convicted at the Crown Court. An appeal lies against conviction and/or sentence, but only with leave of the Court of Appeal or if the trial judge certifies that the case is fit for appeal (see further, para 7.2.2).

Although they should not, strictly, be called 'appeals', the criminal division also hears cases referred to it by the Attorney-General and the Criminal Cases Review Commission (paras 7.3, 7.4, and 7.5.7).

The Court of Appeal has no jurisdiction to hear appeals in criminal cases from the High Court, including a divisional court thereof (Supreme Court Act 1981, s. 18(1)(a); see further, para 7.1.4.2). Such appeals will normally lie direct to the House of Lords (para 7.2.1.3). An exception is that an appeal lies to the Court of Appeal from the decision of a single judge of the High Court in a case of criminal contempt of court (other than a decision on appeal to him) (para 7.1.4.2).

An appeal in the criminal division will normally be heard by *three* judges. In particular, the court must consist of an uneven number of judges not less than three to hear the following matters (Supreme Court Act 1981, s. 55(4), as amended):

(a) an appeal against conviction;

(b) an appeal against a verdict of not guilty by reason of insanity;

(c) an appeal against a jury's finding of unfitness to plead by reason of a disability;

(d) an application for leave to appeal to the House of Lords;

(e) a reference by the Attorney-General of an unduly lenient sentence (see para 7.4).

An appeal by the defendant against *sentence only* can be heard by *two* judges. Where the two judges fail to agree, the case will be reargued before an uneven number of judges, normally three (ibid., s. 55(5)).

1.3.4 Procedure

The Court of Appeal can sit anywhere in England and Wales (Supreme Court Act 1981, s. 57(1)), although it usually sits in London. The criminal division sat for the first time outside London when, in November 1979, it sat for a week in Cardiff to hear appeals from the Crown Court in Wales. It has also sat in Manchester, Liverpool, Birmingham, and Bristol.

The hearing of an appeal by the Court of Appeal is not a retrial. The appeal is determined by the judges after reading the documents in the case and hearing counsel's arguments.

In the civil division where the court consists of an uneven number of judges, the majority decision prevails. Each judge is entitled to deliver his own separate judgment.

In the criminal division, normally only one judgment is delivered, either by the presiding judge or by another member of the court on his direction. Separate judgments can only be delivered in the criminal division where the presiding judge states that in his opinion the case involves a question of law on which it is convenient that separate judgments should be pronounced (ibid., s. 59). This rule is designed to prevent the regular recording of dissenting judgments on the criminal law where certainty is of special importance.

1.4 High Court of Justice

1.4.1 Introduction

The High Court of Justice, a superior court of record, was created as part of the Supreme Court of England and Wales by the Judicature Act 1873. For administrative convenience, it comprises three divisions: the Chancery Division, the Queen's Bench Division, and the Family Division. The three divisions are not, however, separate courts (see the Supreme Court Act 1981, s. 5(5)).

Originally five divisions were created, corresponding to some old courts which were at the same time abolished. The original five were the Chancery Division, the Queen's Bench Division, the Common Pleas Division, the Exchequer Division, and the Probate, Divorce and Admiralty (PDA) Division. In 1880, the Common Pleas and Exchequer Divisions were abolished and their jurisdiction taken over by the Queen's Bench Division. In 1971, under the provisions of the Administration of Justice Act 1970, the PDA Division was renamed the Family Division and its jurisdiction redistributed among the three divisions.

The number of divisions may be increased or reduced by Order in Council on a joint recommendation of the Lord Chancellor, the Lord Chief Justice, the Master of the Rolls, the President of the Queen's Bench Division, the President of the Family Division, and the Chancellor of the High Court (Supreme Court Act 1981, s. 7(1)–(2), as amended by the Constitutional Reform Act 2005).

Judges of the High Court sit mainly in London but it is possible for sittings of the High Court to be held at any place in England and Wales (ibid., s. 71(1)). It is the Lord Chancellor, in consultation with the Lord Chief Justice, who directs where sittings outside London shall take place (ibid., s. 71(2), as amended by the Constitutional Reform Act 2005). Standing directions authorise High Court civil cases to be tried at the following places: Birmingham, Bristol, Caernarvon, Cardiff, Carlisle, Chelmsford, Chester, Exeter, Leeds, Lewes, Lincoln, Liverpool, Manchester, Middlesbrough, Mold, Newcastle upon Tyne, Norwich, Nottingham, Oxford, Preston, Sheffield, Stafford, Swansea, Truro, Warwick, and Winchester. Sittings at places other than these can be authorised on an *ad hoc* basis. For example, in *St Edmundsbury & Ipswich Diocesan Board of Finance* v *Clark* [1973] Ch 323, Megarry J was given permission to sit in the Chancery Division at Iken in Suffolk so as to take the evidence of an 84-year-old witness who was in poor health (see also *Tito* v *Waddell (No. 2)* [1977] 3 All ER 129, para 1.4.4.4 below).

The High Court is principally a civil court. It does, however, exercise some important criminal jurisdiction. Its civil jurisdiction is virtually unlimited.

1.4.2 Composition

The High Court consists of the Lord Chief Justice, the President of the Queen's Bench Division, the President of the Family Division, the Chancellor of the High Court, the Senior Presiding Judge, the Vice-President of the Queen's Bench Division and a number of puisne judges (Supreme Court Act 1981, s. 4(1), as amended). The puisne judges are styled 'Justices of the High Court'. Their maximum number is 108. This number may be increased by Order in Council on the recommendation of the Lord Chancellor (ibid., s. 4(4)–(4A)).

To be qualified for appointment as a puisne judge of the High Court a person must either have a ten-year High Court qualification within the meaning of s. 71 of the Courts and Legal Services Act 1990 (para 1.1.5 above), or have been a circuit judge for at least two years (ibid., s. 10(3)(c), as amended by the Courts and Legal Services Act 1990). Puisne judges are appointed in accordance with the Constitutional Reform Act 2005 (see para 1.1.6 above).

Although assigned initially to a particular division dependent on the volume of business before the court, a puisne judge can sit in any division as and when required. He can be transferred to another division by the Lord Chief Justice after consulting the Lord Chancellor, but only with his consent and that of the senior judge of his present division (Supreme Court Act 1981, s. 5(2), as amended by the Constitutional Reform Act 2005).

In addition, judges and former judges of the Court of Appeal, former puisne judges of the High Court, circuit judges and recorders may be requested by the Lord Chief Justice to sit as additional judges of the High Court (Supreme Court Act 1981, s. 9(1)–(3), as amended). Such requests are rare, although it is not unusual for a circuit judge to be asked to take High Court work in the provinces or for a recorder to deal with urgent High Court applications in matrimonial and child cases.

The Lord Chief Justice in consultation with the Lord Chancellor may also appoint deputy judges of the High Court on a temporary basis in order to facilitate the disposal of business. The persons appointed must be qualified for appointment as puisne judges. A deputy judge has all the powers of a puisne judge of the High Court (Supreme Court Act 1981, s. 9(4)–(5), as amended). In September 1997, *solicitors* were appointed for the first time to sit as deputy High Court judges. The two appointees were Dr L.A. Collins QC and Mr A.L. Marriott QC. Earlier in the same year Dr Collins and Mr Marriott had become the first solicitors to be appointed Queen's Counsel. The first solicitor to be appointed as a full-time High Court judge was Sachs J, who was promoted from circuit judge in 1993. In 2000, Dr Collins became the first solicitor to be appointed as a full-time High Court judge *direct from practice*. He was assigned to the Chancery Division.

1.4.3 Queen's Bench Division

The Chancery Division is the senior division of the High Court but here it is proposed to deal first with the Queen's Bench Division because of the volume, diversity and importance of its work.

1.4.3.1 Composition

The Queen's Bench Division is headed by a President. He is assisted by a vice-president and a number of puisne judges. Over half of the total number of puisne judges of the High Court are attached to the Queen's Bench Division. This judicial strength reflects both the volume of business in the division and the fact that Queen's Bench judges spend a certain amount of time away from London on circuit in the provinces trying High Court civil actions and, as judges of the Crown Court, criminal cases.

1.4.3.2 Jurisdiction

The Queen's Bench Division acts mainly as a civil court where a single judge tries, at first instance, such cases as breach of contract and actions in tort. These cases are invariably tried by the judge sitting alone since juries are very rare in civil cases (para 6.6). Most actions are either settled or abandoned and only one per cent result in a trial.

1.4.3.3 The Commercial Court, the Admiralty Court, and the Technology and Construction Court

Within the Queen's Bench Division there exist the Commercial Court and the Admiralty Court (Supreme Court Act 1981, s. 6(1)). It also administers the Technology and Construction Court (formerly the Official Referees' Court).

The Commercial Court, which was officially made part of the Queen's Bench Division in 1971, hears cases which have been entered on a special commercial list kept in London, Liverpool and Manchester (ibid., s. 62(3)). The commercial list has existed since 1895, although it is only since 1971 that the cases on the list have been dealt with in a separate Commercial Court officially recognised by statute. These cases include banking and insurance disputes and the construction of mercantile documents such as negotiable instruments and charterparties. The judges of the Commercial Court are those puisne judges of the High Court nominated by the Lord Chief Justice after consulting the Lord Chancellor (ibid., s. 6(2), as amended). They are usually puisne judges of the Queen's Bench Division with special knowledge and experience of commercial affairs.

The advantages of the Commercial Court to the business community lie in the guidance provided by a substantial body of case law and in the court's simplified, speedier procedure. Thus, by consent of the parties, the strict rules of evidence may be relaxed so as to admit evidence which would normally be inadmissible. Moreover, the case may be decided on the documentary evidence alone, thereby saving the time and money involved in calling witnesses to give oral evidence.

In appropriate cases (for example, where the costs of litigation are likely to be out of all proportion to the amount at stake), parties are encouraged to consider the use of alternative dispute resolution, such as mediation, conciliation and arbitration, as an additional means of settling disputes (see *Practice Note (Commercial Court: alternative dispute resolution)* [1994] 1 All ER 34).

Although the Commercial Court is frequently used by foreign litigants for the resolution of international commercial disputes, the Court is a national or domestic court of England and Wales and is *not* an 'international' court (*Amin Rasheed Shipping Corpn. v Kuwait Insurance Co.* [1983] 2 All ER 884, HL). The very popularity of the Commercial Court has made it the victim of its own success (*Barclay's Bank plc v Bemister* [1989] 1 All ER 10, CA, *per* Sir John Donaldson MR at p. 12).

The Admiralty Court became part of the Queen's Bench Division on the demise of the Probate, Divorce and Admiralty Division in 1971. Admiralty jurisdiction includes the trial of claims for damages, loss of life or personal injury arising out of a collision between ships, or as a result of any defect in a ship, claims to the possession or ownership of a ship, claims for loss of or damage to goods carried in a ship, claims for salvage and towage in respect of a ship or aircraft, and claims for wages by a master or member of the crew of a ship (Supreme Court Act 1981, s. 20, as amended). The jurisdiction over claims for salvage in respect of ships is exercisable only where the ship in question was in *tidal* waters. It does not extend to ships in peril in *non-tidal* waters (*The Goring* [1988] 1 All ER 641, HL, disallowing a claim for remuneration for salvage services rendered to a small passenger vessel adrift in the River Thames).

The Admiralty Court, sitting as a prize court, also exercises the prize jurisdiction of the High Court (ibid., ss. 20(1)(d), 27 and 62(2)). Prize relates to ships and aircraft, and the goods carried therein, seized by the armed forces of a belligerent. It should be noted that an appeal from the High Court in a prize case lies to the Judicial Committee of the Privy Council and not to the Court of Appeal (ibid., s. 16(2)).

The judges of the Admiralty Court are those puisne judges of the High Court nominated by the Lord Chief Justice after consulting the Lord Chancellor (ibid., s. 6(2), as amended). A judge of the Admiralty Court often sits with lay nautical assessors, who are chosen from acting Masters of Trinity House and who keep the judge advised on questions of seamanship and navigation (ibid., s. 70).

In 1993, with a view to improving Admiralty Court practice and harmonising it with that of the Commercial Court, the appropriate parts of the *Guide to Commercial Court Practice* were extended to the Admiralty Court (see *Practice Note (Admiralty Court Practice)* [1993] 2 All ER 671). The two courts remain, however, separate and distinct. A new edition of the *Guide* was published in March 2002 and is now entitled the *Admiralty and Commercial Courts Guide*. At the same time, the courts became subject to CPR, Parts 58 (claims in the Commercial Court), 61 (claims in the Admiralty Court), and 62 (arbitration claims), together with their associated practice directions (see *Practice Statement (Admiralty and Commercial Courts: Procedure)* (2002) *The Times*, 2 April).

The Technology and Construction Court hears cases involving prolonged examination of technical issues arising from building and engineering disputes, computer litigation, sale of goods, valuation disputes, landlord and tenant (especially dilapidations), and torts relating to the occupation of land. It also deals with questions arising from arbitrations in building and engineering disputes.

1.4.3.4 The election court

Two judges of the Queen's Bench Division, sitting as an election court, have jurisdiction to pronounce on the validity of disputed Parliamentary elections. The matter is tried in open court without a jury. The judgment of the election court takes the form of a report to the Speaker which the House of Commons is bound to act on (Representation of the People Act 1983, ss. 120–126, as amended, s. 144). With leave of the election court, an appeal lies on a question of law to the Court of Appeal, whose decision is 'final and conclusive' (ibid, s. 157, as amended).

1.4.3.5 Divisional Court of the Queen's Bench Division and the Administrative Court

Two or more judges sitting together may constitute a Divisional Court of the Queen's Bench Division (Supreme Court Act 1981, s. 66(1) and (3)). The judges will usually be from the Queen's Bench Division, although it is provided that judges from the other two divisions are qualified to sit (ibid., s. 66(4)). The Divisional Court is presided over by the senior judge present (ibid., ss. 66(5) and 13).

Historically, the functions of the Queen's Bench Divisional Court are:

(a) To supervise public authorities, inferior courts and tribunals by entertaining applications for judicial review (Chapter 10).

(b) To hear applications for the writ of habeas corpus from persons who allege that they are being unlawfully detained. Note, however, that an application for a writ of habeas corpus made by a parent or guardian in relation to the custody of a minor is heard in the Family Division and not in the Queen's Bench Division (Supreme Court Act 1981, sch. 1).

(c) To hear criminal appeals on points of law by way of case stated direct from the magistrates' courts or via the Crown Court (Chapter 7).

(d) To hear applications to punish contempts committed in inferior courts (Chapter 8).

Cases in categories (b), (c) and (d) are these days more likely to be heard by a single judge than a Divisional Court by virtue of provisions in the Access to Justice Act 1999 designed to achieve a more economical use of judicial manpower.

Claims for judicial review (category (a) above) are now dealt with in the Administrative Court, which took over from the 'Crown Office List', and was established as part of the Queen's Bench Division, on 2 October 2000 (*Practice Direction (QBD: Administrative Court: Establishment)* [2000] 1 WLR 1654). Permission to claim judicial review is granted (or refused) by a single judge. If permission is granted, the substantive claim for judicial review may be heard by the Divisional Court if it concerns a *criminal* matter (where a person's liberty may be at stake). If, however, it concerns a *civil* matter it will normally be heard by a single judge unless it is unusually difficult and the court directs it to go before the Divisional Court.

1.4.4 **Chancery Division**

1.4.4.1 Composition

The president of the Chancery Division is the Chancellor of the High Court. He is assisted by a number of puisne judges.

1.4.4.2 Jurisdiction: original

The Chancery Division is purely a civil court, being the successor to the old Court of Chancery which was chiefly concerned with the administration of equity. Its jurisdiction covers those matters formerly dealt with by the Court of Chancery together with other matters assigned to it by various statutes. By s. 61(1) of, and sch. 1 to, the Supreme Court Act 1981, the jurisdiction covers all causes and matters relating to:

(a) the sale, exchange or partition of land, or the raising of charges on land;

(b) the redemption or foreclosure of mortgages;

(c) the execution of trusts;

(d) the administration of the estates of deceased persons;

(e) bankruptcy;

(f) the dissolution of partnerships or the taking of partnership or other accounts;

(g) the rectification, setting aside or cancellation of deeds or other written instruments;

(h) contentious probate business, including both the validity and the interpretation of wills;

(i) patents, trade marks, registered designs, copyright or design right;

(j) the appointment of a guardian of a minor's estate;

(k) company law.

1.4.4.3 Jurisdiction: appellate

In addition to the *first instance* or original jurisdiction just set out, the Chancery Division has some *appellate* jurisdiction as follows.

(a) A single judge may hear appeals from decisions of General or Special Commissioners on matters of taxation (Taxes Management Act 1970, ss. 56 and 56A), and appeals in insolvency cases from the county courts (Insolvency Act 1986, s. 375 and sch. 1 to the Supreme Court Act 1981).

(b) A Divisional Court of the Chancery Division, consisting of at least two judges, may hear appeals from decisions of the Adjudicator to HM Land Registry on land registration matters (Land Registration Act 2002, s. 111; RSC, Ord. 93, r. 10).

(c) When sitting as the Patents Court, one or more judges may hear appeals from certain decisions of the Comptroller-General of Patents, Designs and Trade Marks (Patents Act 1977, s. 97(1) and (2)). The Patents Court was first established as part of the Chancery Division in 1978, under the Patents Act 1977, to take over the functions of the Patents Appeal Tribunal. The Patents Court is part of the Chancery Division now by virtue of s. 6(1) of the Supreme Court Act 1981. The judges of the Patents Court are those puisne judges of the High Court nominated by the Lord Chief Justice after consulting the Lord Chancellor (Supreme Court Act 1981, s. 6(2), as amended). The Patents Court may be assisted by scientific advisers (ibid., s. 70(3)).

1.4.4.4 The hearing of High Court Chancery proceedings

The highly specialised work of the Chancery Division tends to be concentrated mainly in London. However, since the High Court can sit at any place in England and Wales, there is provision for High Court Chancery proceedings to be dealt with in the provinces by two High Court judges, one of whom is the Vice-Chancellor of the County Palatine of Lancaster. They sit at Birmingham, Bristol, Cardiff, Liverpool, Manchester, Preston, Leeds and Newcastle upon Tyne. There are also some specialist *circuit judges* exercising general Chancery jurisdiction in the provinces. They sit as High Court judges, having been requested to do so by the Lord Chief Justice under s. 9 of the Supreme Court Act 1981 (para 1.4.2 above).

On one celebrated occasion Megarry J took the Chancery Division beyond even the provinces. In *Tito* v *Waddell* [1975] 3 All ER 997, he held that he had power to conduct a 'view' outside the jurisdiction of the court altogether (i.e., outside England and Wales). Having visited Ocean Island and Rabi in the western Pacific, he produced, in *Tito* v *Waddell (No. 2)* [1977] 3 All ER 129, a judgment of some 180 pages (complete with table of contents) which took him more than four days to read out in court and which brought an end to litigation which had already occupied the court for 206 days. The report of *Tito* v *Waddell (No. 2)*, which involved the citation of over 200 cases, occupies 195 pages in the *All England Law Reports*.

1.4.5 Family Division

1.4.5.1 Composition

This Division is presided over by the President of the Family Division, who is also Head of Family Justice. The President is assisted by a number of puisne judges.

1.4.5.2 Jurisdiction

The jurisdiction of the Family Division is as follows (Supreme Court Act 1981, s. 61(1) and sch. 1, as amended).

(a) All matrimonial causes and matters (whether at first instance or on appeal).

(b) All causes and matters (whether at first instance or on appeal) relating to:

    (i)   legitimacy;

    (ii) the exercise of the inherent jurisdiction of the High Court with respect to minors, the maintenance of minors and any proceedings under the Children Act 1989, except proceedings solely for the appointment of a guardian of a minor's estate (which are dealt with in the Chancery Division);

    (iii) adoption;

      (iv) non-contentious or common-form probate business (contentious or solemn-form probate business is taken in the Chancery Division).

(c) Applications for consent to the marriage of a minor or for a declaration under s. 27B(5) of the Marriage Act 1949.

(d) Proceedings on appeal by persons who have been punished for disobeying an order, other than for the payment of money, made by a magistrates' court in matrimonial proceedings, or proceedings under part IV of the Family Law Act 1996 (family homes and domestic violence), or with respect to the guardianship of a minor (see the Magistrates' Courts Act 1980, s. 63(3)).

(e) Proceedings under the Children Act 1989.

(f) Proceedings under:

      (i)   part IV of the Family Law Act 1996;

      (ii)   the Child Abduction and Custody Act 1985;

      (iii) the Family Law Act 1986 (including declarations of status);

      (iv)   s. 30 of the Human Fertilisation and Embryology Act 1990.

(g) All proceedings relating to a debit or credit under s. 29(1) or s. 49(1) of the Welfare Reform and Pensions Act 1999.

(h) Proceedings for the purpose of enforcing an order made in any of the above proceedings.

(i) Proceedings under the Child Support Act 1991.

(j) All proceedings under ss. 6 and 8 of the Gender Recognition Act 2004.

(k) All civil partnership causes and matters (whether at first instance or on appeal).

(l) Applications for consent to the formation of a civil partnership by a minor or for a declaration under para 7 of sch. 1 to the Civil Partnership Act 2004.

(m) Applications under s. 58 of the Civil Partnership Act 2004 (declarations relating to civil partnerships).

1.4.5.3 Divisional Court of the Family Division

An appeal lies to the High Court on family matters, such as financial provision orders, from decisions of the magistrates' courts made under the Domestic Proceedings and Magistrates' Courts Act 1978 (ibid., s. 29), from their decisions to make or refuse to make any order under the Children Act 1989 or the Adoption and Children Act 2002 (Children Act 1989, s. 94, as amended), and from their decisions to make or refuse to make any order under part IV of the Family Law Act 1996, except for a decision to decline jurisdiction in any proceedings under part IV (Family Law Act 1996, s. 61).

    An appeal under the 1978 Act will usually be heard by a Divisional Court consisting of two or more judges. If the appeal is concerned only with the amount of any periodical or lump sum payment ordered to be made it will be heard by a single judge unless the President directs otherwise (Family Proceedings Rules 1991, SI 1991, No. 1247, r. 8.2).

An appeal under the Children Act 1989 will usually be heard by a single judge unless the President directs otherwise (ibid., r. 4.22). The hearing will normally be held in open court at the nearest convenient High Court centre (see *Practice Direction* [1992] 1 All ER 864).

1.4.6 Delay in the High Court

Although by reason of changes made in 1982 it is possible for High Court judges to sit in vacation to hear non-urgent cases as well as the usual emergency applications (see now *Practice Direction* to CPR, Part 39), it remains the position that the problems caused by delays in the High Court are exacerbated by the long vacation during August and September. While the Crown Court and the county courts operate fairly normally during this period, High Court judges do not sit regularly in court.

Of course, it is not true to say that High Court judges do no work at all during the long vacation. In *Libyan Arab Foreign Bank* v *Bankers Trust Co.* [1989] 3 All ER 252, Staughton J, delivering in the Commercial Court on 2 September 1987 a reserved judgment which occupies 34 pages in the *All England Law Reports*, concluded with the following postscript (at p. 288):

In August of this year there were 20 working days. Fourteen of them were entirely consumed in the preparation of this judgment. In those circumstances it is a shade disappointing to read in the press and elsewhere that High Court judges do no work at all in August or September and have excessively long holidays.

It seems undeniable, however, that the long vacation does lead to under-utilisation of valuable judicial time when there is a need for increased judicial productivity in the High Court.

In March 1994, the Lord Chancellor appointed Lord Woolf (then a Lord of Appeal, later Master of the Rolls and Lord Chief Justice) to inquire into the problems of cost, delay and complexity associated with the civil justice system. Lord Woolf's final report, published in July 1996, contained some 300 recommendations for a 'new landscape' for civil litigation (*Access to Justice: final report to the Lord Chancellor on the civil justice system in England and Wales*, HMSO, July 1996). Many of these recommendations have been implemented.

The Civil Procedure Act 1997 facilitated the introduction of Lord Woolf's proposed civil justice reforms by initiating new arrangements for making rules of court. The Supreme Court Rule Committee and the County Court Rule Committee were replaced by a single committee, the Civil Procedure Rule Committee, which has power to make Civil Procedure Rules (CPR) 'with a view to securing that the civil justice system is accessible, fair and efficient' (s. 1(3)), and which must try to produce rules which are 'both simple and simply expressed' (s. 2(7)).

The Civil Procedure Act 1997 also created an advisory body, the Civil Justice Council, for the purposes of considering how to make the civil justice system more accessible, fair and efficient, advising the Lord Chancellor and the judiciary on the development of the system, referring proposals for change in the system to the Lord Chancellor and the Civil Procedure Rule Committee, and making proposals for research (s. 6).

Most of the Civil Procedure Rules 1998 (SI 1998, No. 3132), as amended, came into force on 26 April 1999. They regulate practice and procedure in the civil division of the Court of Appeal, in the High Court, and in the county courts. No longer are the three courts governed by separate procedural codes. The Civil Procedure Rules comprise 76 Parts, together with associated practice directions, and eight pre-action protocols. Part 1 emphasises that the Rules are a new procedural code with the over-riding objective of enabling the court to deal with cases justly. In furtherance of the overriding objective, courts are required actively to manage cases.

Civil proceedings are started when the court issues a claim form (formerly a 'writ' or 'summons') at the request of the claimant (formerly the 'plaintiff'). Proceedings (whether for damages or for a specified sum) cannot be started in the High Court unless the value of the claim is more than £15,000, or, where the proceedings include a claim for damages for personal injuries, £50,000 or more (*Practice Direction* to CPR, Part 7).

Cases are allocated under CPR, Part 26, by procedural judges to one of three tracks—the small claims track, the fast track, or the multi-track—taking into account a number of factors, including the financial value and the complexity of the claim.

The *small claims track* is the normal track for:

(a) any claim for personal injuries which has a financial value of not more than £5,000 where the claim for damages for personal injuries is not more than £1,000;

(b) any claim which includes a claim by a tenant of residential premises against a landlord for repairs or other work to the premises where the estimated cost of the repairs or other work is not more than £1,000 and the financial value of any other claim for damages is not more than £1,000; and

(c) any other claim which has a financial value of not more than £5,000.

Cases generally suitable for the small claims track include consumer disputes, acci-dent claims, disputes about the ownership of goods, and most disputes between a landlord and tenant other than those for possession (*Practice Direction* to CPR, Part 26). A claim for a remedy for harassment or unlawful eviction relating to residential premises will not be allocated to the small claims track whatever the financial value of the claim. A case involving a disputed allegation of dishonesty will not usually be regarded as suitable for the small claims track (ibid.).

The *fast track* is the normal track for any claim for which the small claims track is not the normal track and which has a financial value of not more than £15,000.

The *multi-track* is the normal track for any claim for which the small claims track or the fast track is not the normal track.

The working of the Civil Procedure Rules is kept under constant review.

1.5 **Crown Court**

1.5.1 **Introduction**

Before the Courts Act 1971 the major criminal courts were the Assizes (which also heard civil cases) and the Quarter Sessions. One of the main criticisms of these courts was that they did not sit continuously. The Commissioners of Assize went on circuit to the counties only two or three times a year while the Quarter Sessions sat four times a year. The Assize circuits were outdated and were not based on the large centres of population thrown up by the industrial revolution.

As much as one-quarter of valuable judicial time was spent simply on travelling from one Assize town on the circuit to the next. The system of criminal justice was severely strained by the increasing crime rate of the twentieth century. Accused persons spent long periods in jail awaiting trial. At the Assizes, civil cases tended to be neglected in order to get through as much of the criminal list as possible. A major defect of the Quarter Sessions was that the existence of a large number of part-time judges made it difficult to achieve any consistency in sentencing.

1.5.2 **Courts Act 1971**

In 1967, a Royal Commission on Assizes and Quarter Sessions was appointed under the chairmanship of Lord Beeching. It reported in 1969 and made many recommendations for the improvement of the court system in England and Wales (*Report of the Royal Commission on Assizes and Quarter Sessions*, Cmnd 4153, 1969). Most of these recommendations were implemented with unusual speed by the Courts Act 1971.

The principal changes brought about by the Courts Act 1971 may be summarised as follows.

(a) The Assizes and Quarter Sessions were abolished (Courts Act 1971, ss. 1(2)–(3). In their place was established a single Crown Court as part of the Supreme Court of England and Wales (ibid., ss. 4(1) and 1(1); see now s. 1(1) of the Supreme Court Act 1981).

(b) The judges of the Crown Court are High Court judges, circuit judges, recorders, district judges (magistrates' courts), and justices of the peace (Courts Act 1971, s. 4(2); see now s. 8(1) of the Supreme Court Act 1981, as amended by the Courts Act 2003).

(c) The High Court can sit at any place in England and Wales designated by the Lord Chancellor (after consulting the Lord Chief Justice) to take civil work (Courts Act 1971, s. 2; see now s. 71 of the Supreme Court Act 1981, as amended by the Constitutional Reform Act 2005). The list of civil cases is no longer linked to the criminal list, although the High Court, when on circuit, often sits in the old Assize court buildings.

(d) The Court of Chancery of the County Palatine of Lancaster and the Court of

Chancery of the County Palatine of Durham and Sadberge were merged with the High Court (Courts Act 1971, s. 41). However, the office of Vice-Chancellor of the County Palatine of Lancaster was expressly preserved by s. 44(1) (see para 1.4.4.4 above).

(e) Certain ancient local courts of record which still heard some civil cases were abolished (ibid., s. 43). They were the Tolzey and Pie Poudre Courts of the City and County of Bristol, the Liverpool Court of Passage, the Norwich Guildhall Court, and the Court of Record for the Hundred of Salford.

(f) In the place of individual local authorities, the Lord Chancellor was given overall responsibility for the court system, including the provision of administrative staff (ibid., s. 27; provisions now contained in the Courts Act 2003, s. 2).

(g) The Lord Chancellor was made responsible for the summoning and empanelling of jurors in both civil and criminal cases in the High Court, Crown Court and county courts (ibid., ss. 31–32; provisions now contained in the Juries Act 1974, ss. 2 and 5).

1.5.3 Circuits, court centres, and the classification of offences

The Courts Act 1971, and the administrative arrangements made thereunder by the Lord Chancellor, produced a smoother and more unified court system in the provinces. England and Wales are divided into six circuits each with a circuit administrator, headquarters and staff. In addition, each circuit has at least two Presiding Judges from the High Court to assist with administration. A Senior Presiding Judge for England and Wales is appointed from among the Lords and Lady Justices of Appeal. Appointments to the post of Presiding Judge and Senior Presiding Judge are made by the Lord Chief Justice with the agreement of the Lord Chancellor. (These arrangements, which have existed on an informal basis since 1970, were put on a statutory footing by s. 72 of the Courts and Legal Services Act 1990.)

The Crown Court is not a local court. It is a single court which can sit anywhere in England and Wales (Supreme Court Act 1981, s. 78(1)), although sittings are usually confined within each circuit to the towns designated as court centres. When the Crown Court sits in the City of London it is known as the Central Criminal Court (often referred to as the Old Bailey), thus preserving the name of a court first established in 1834 (see Supreme Court Act 1981, s. 8(3), re-enacting the Courts Act 1971, s. 4(7)).

The six circuits with (in parentheses) some of their first-tier centres are: (i) Midland and Oxford (Birmingham, Nottingham); (ii) North-Eastern (Leeds, Newcastle upon Tyne, Sheffield); (iii) Northern (Liverpool, Manchester); (iv) South-Eastern (London, Norwich); (v) Wales and Chester (Cardiff, Chester, Swansea); (vi) Western (Bristol, Exeter, Winchester). The full list of first-tier centres appears in para 1.4.1 above.

The court centres, of which there are presently some 90, are designated as such by the Lord Chancellor after consulting the Lord Chief Justice (Supreme Court Act 1981, s. 78(3), as amended by the Constitutional Reform Act 2005, formerly s. 4(6) of the Courts Act 1971). They are either first-tier, second-tier or third-tier centres. At first-tier centres, High Court judges try both civil and criminal cases while circuit judges and recorders try criminal cases only. Second-tier centres deal with criminal cases only but are served by High Court judges, circuit judges and recorders. At third-tier centres, circuit judges and recorders try criminal cases only.

For the purposes of trial in the Crown Court, criminal offences are classified according to gravity and complexity into three classes (see now *Practice Direction (Criminal Proceedings: Consolidation), Part III* [2002] 1 WLR 2870, as amended by *Practice Direction (Crown Court: Classification and Allocation of Business)* [2005] 1 WLR 2215; these directions were given by the Lord Chief Justice, with the concurrence of the Lord Chancellor, under s. 75(1) of the Supreme Court Act 1981, formerly s. 4(5) of the Courts Act 1971, and replaced directions first given in 1971).

Class 1 offences, including murder, manslaughter, abortion, and offences under the Official Secrets Acts, are triable by a High Court judge, or by a circuit judge or deputy High Court judge or deputy circuit judge if he is authorised by the Lord Chief Justice to try murder cases and the case has been released by a Presiding Judge for trial by such a judge.

Class 2 offences, including rape and other sexual offences, are tried by a High Court judge, or by a circuit judge or deputy High Court judge or deputy circuit judge or a recorder if he is authorised by the Lord Chief Justice to try class 2 offences.

Class 3 offences, covering all offences not falling into classes 1 and 2 (such as kidnapping, causing death by dangerous driving, grievous bodily harm, and robbery), can be tried by a High Court judge, or by a circuit judge or a deputy circuit judge or a recorder in accordance with guidance given by a Presiding Judge.

1.5.4 Composition of the Crown Court

The judges of the Crown Court are High Court judges, circuit judges, recorders, district judges (magistrates' courts), and justices of the peace (Supreme Court Act 1981, s. 8(1), as amended by the Courts Act 2003). The High Court judges are drawn mainly from the Queen's Bench Division.

Circuit judges are full-time permanent judges. They are appointed in accordance with the Constitutional Reform Act 2005 (see para 1.1.6 above). To be qualified for appointment as a circuit judge a person must:

(a) have a ten-year Crown Court or ten-year county court qualification within the meaning of s. 71 of the Courts and Legal Services Act 1990 (para 1.1.5 above);

(b) be a recorder; or

(c) have held a full-time appointment for at least three years in one of the offices listed in Part 1A of sch. 2 to the Courts Act 1971 (Courts Act 1971, s. 16(3), as amended by the Courts and Legal Services Act 1990). Part 1A of sch. 2 to the Courts Act 1971 was inserted by the 1990 Act, and the appointments mentioned include Social Security Commissioner, chairman of an employment tribunal, coroner, district judge, and district judge (magistrates' courts).

The many new posts of circuit judge created when the Crown Court was originally set up were, in the main, filled automatically from the ranks of existing county court judges (see Courts Act 1971, s. 16(5) and sch. 2). Before recommending any person for appointment as a circuit judge, the Lord Chancellor must take steps to satisfy himself that that person's health is satisfactory (Courts Act 1971, s. 16(4)).

A circuit judge also sits as a county court judge to hear civil cases (County Courts Act 1984, s. 5(1)). He may also be requested by the Lord Chancellor to sit as an *additional* judge of the High Court under the provisions of the Supreme Court Act 1981, s. 9(1)–(2).

A circuit judge retires at the age of 70 (Courts Act 1971, s. 17(1), as amended by the Judicial Pensions and Retirement Act 1993). A circuit judge may be removed from office by the Lord Chancellor (with the agreement of the Lord Chief Justice) on the ground of incapacity or misbehaviour (ibid., s. 17(4), as amended by the Constitutional Reform Act 2005).

Recorders are also appointed in accordance with the Constitutional Reform Act 2005. But, unlike circuit judges, recorders are part-time judges appointed for a specified period (not less than five years). The period may be extended from time to time by the Lord Chancellor, although not beyond the age of 75. The appointment of a recorder must state how often and for how long during the specified period he will be required to sit as a judge. His appointment can be terminated by the Lord Chancellor (with the agreement of the Lord Chief Justice) for failing to make himself available as agreed, and for incapacity or misbehaviour (ibid., s. 21(6), as substituted by the Constitutional Reform Act 2005). A recorder may also sit as a judge of the county court to hear civil cases (County Courts Act 1984, s. 5(3), as amended).

To be qualified for appointment as a recorder a person must have a ten-year Crown Court or a ten-year county court qualification within the meaning of s. 71 of the Courts and Legal Services Act 1990 (para 1.1.5 above) (Courts Act 1971, s. 21(2), as amended by the Courts and Legal Services Act 1990).

District judges (magistrates' courts) (para 1.7.2 below) were made judges of the Crown Court for certain purposes by the Courts Act 2003.

Justices of the peace (in effect, lay magistrates) may only sit as judges of the Crown Court with a High Court judge, circuit judge or recorder. The number of justices of the peace hearing any particular case must not exceed four (Supreme Court Act 1981, s. 8(1)(c)). When the Crown Court is hearing appeals, the presence of between two and four justices along with a High Court judge, circuit judge or recorder is mandatory (ibid., s. 74(1)). In addition, justices *may* sit with a circuit judge or recorder when trying cases on indictment, except for cases listed for plea of not guilty (ibid., ss. 8(1) and 75(2); *Practice Direction (Criminal Proceedings: Consolidation), Part IV* [2002] 1 WLR 2870). When sitting with justices, the judge must consult them on the question of sentence (*R* v *Newby* (1984) 6 Cr App R(S) 148, CA), and on any defence submission that there is no case to answer (*R* v *Southwark Crown Court (ex parte Mitchell)* [1994] COD 15, DC).

To add to the usual judges of the Crown Court already mentioned, judges and former judges of the Court of Appeal and former puisne judges of the High Court may be requested by the Lord Chief Justice to sit in the Crown Court on an *ad hoc* basis (Supreme Court Act 1981, s. 9(1)–(2), as amended). Such requests are rare. The request can be refused in the case of each category of additional judge mentioned (ibid., s. 9(3)).

The Lord Chief Justice (with the agreement of the Lord Chancellor) can appoint deputy circuit judges, for periods of at least five years, in order to facilitate the disposal of business in the Crown Court or a county court. The persons appointed as deputy circuit judges must previously have held office as judges of the Court of Appeal or High Court or as circuit judges (Courts Act 1971, s. 24(1)(a), as amended by the Constitutional Reform Act 2005). A deputy circuit judge has all the powers of a circuit judge (Courts Act 1971, s. 24(2)).

In April 2000, the Lord Chancellor introduced new conditions of service for part-time judicial office-holders, including deputy circuit judges, recorders, retired judges of the Court of Appeal, and retired puisne judges of the High Court. (The new conditions also apply to chairmen and lay member of tribunals: see paras 3.2.2 and 3.3.2.)

The new conditions are intended to guarantee the independence of part-time judges from the dismissal and non-renewal powers of the executive in the light of art. 6 of the European Convention on Human Rights (right to a fair trial) and the Scottish case of *Starrs* v *Procurator Fiscal, Linlithgow* 2000 SLT 42, in which it was held by the High Court of Justiciary that a judge (in this case, a temporary sheriff) who had no security of tenure, and whose appointment was subject to annual renewal by the executive, was not 'independent' within the meaning of art. 6.

These part-time appointments are now for a period of not less than five years, subject to the relevant upper age limit. So long as it is administratively possible, the offer of a minimum number of sitting days will be guaranteed. Appointments which are renewable will normally be renewed automatically, except for limited and specified grounds. Removal from office is likewise only on limited and specific grounds.

Subject to statutory provision, the specified grounds for *non-renewal* are:

(a) misbehaviour;

(b) incapacity;

(c) persistent failure (without good reason) to comply with sitting requirements;

(d) failure to comply with training requirements;

(e) sustained failure to observe the standards reasonably expected from a holder of such office;

(f) a reduction in numbers because of changes in operational requirements;

(g) a structural change to enable recruitment of new appointees.

Decisions not to renew on grounds (f) and (g) are taken on a 'first in, first out' basis, and the decision to use such grounds (and the extent to which they will be used) is made by the Lord Chancellor with the concurrence of the Lord Chief Justice.

The grounds for *removal* are those in (a)–(e), above.

1.5.5 Jurisdiction of the Crown Court

The jurisdiction of the Crown Court is as follows.

(a) It has exclusive jurisdiction in relation to trial on indictment for offences wherever committed (Supreme Court Act 1981, s. 46), although, as noted in para 1.5.3 above, offences are divided into three classes for the purposes of the distribution of work among the judges. Where the accused pleads not guilty, trial on indictment in the Crown Court takes place before judge and jury.

(b) It hears appeals by persons convicted summarily in the magistrates' courts. This, as well as heads (c) and (d) below, was part of the jurisdiction of the old Quarter Sessions which was given to the Crown Court by the Courts Act 1971 and is preserved by the Supreme Court Act 1981, s. 45(2).

In the course of hearing an appeal, the Crown Court may correct any error in the order or judgment incorporating the decision which is being appealed against (Supreme Court Act 1981, s. 48(1)). At the conclusion of the hearing, the Crown Court has power to confirm, reverse or vary any part of the decision under appeal or to remit the case with its opinion thereon to the magistrates' court which made the decision (ibid., s. 48(2)). If the appeal is decided against the accused, the Crown Court has power to impose any sentence which the magistrates' court could have imposed. This power is wide enough to allow the imposition of a punishment which is *more severe*, as well as one which is less severe, than that actually inflicted by the magistrates' court (ibid., s. 48(4)). The Crown Court cannot increase the sentence in a case which has been referred to it under s. 11 of the Criminal Appeal Act 1995 by the Criminal Cases Review Commission (Criminal Appeal Act 1995, s. 11(6)).

(c) To sentence persons committed for sentence following conviction in the magistrates' courts (para 1.7.6.2.4 below).

(d) It has a limited civil jurisdiction which allows it to hear, in particular, licensing appeals from the magistrates' courts, and appeals against decisions of chief officers of police in firearm and shotgun certificate cases (Firearms Act 1968, s. 44, as substituted by the Firearms (Amendment) Act 1997).

1.5.6 Delay in the Crown Court

The benefits to the legal process achieved by the reorganisation of the criminal courts in the early 1970s are undoubted. But the problem of delay, which plagues the criminal justice system in England and Wales, remains unresolved. A person committed to the Crown Court to be tried must wait (sometimes in custody) an average of 15.5 weeks for the trial to begin. The worst figures are for the South Eastern circuit (outside London), where the average waiting time is 17.2 weeks. The shortest waiting time of 10.5 weeks is on the Wales and Chester circuit (*Judicial Statistics 2004*, Cm 6565, 2005, p. 94).

In 1961, an official committee looking into the business of the criminal courts recommended that the waiting time should not exceed eight weeks on average (*Report of the Interdepartmental Committee on the Business of the Criminal Courts*, Cmnd 1289, 1961). In 1969, the Royal Commission on Assizes and Quarter Sessions noted that this target had not been achieved and that the problem was particularly acute in London (*Report*, Cmnd 4153, 1969). The Courts Act 1971 came into force on 1 January 1972, and for that year and the following three years there was a progressive reduction in the waiting time. In 1976, there was an upturn. Since 1979, however, there has been another gradual reduction—from 17.6 weeks in 1979 to 15.5 weeks in 2004.

The *actual* waiting times compare unfavourably with the *theoretical* maximum period of eight weeks prescribed under s. 77 of the Supreme Court Act 1981 by r. 39.1 of the Criminal Procedure Rules 2005 (SI 2005, No. 384). The reason for the discrepancy between the actual and the theoretical periods is that a judge of the Crown Court has power to order that a trial need not begin within the eight-week period. Moreover, s. 77 is directory and not mandatory so that failure to comply with it does not nullify the trial (*R* v *Urbanowski* [1976] 1 All ER 679, CA; *R* v *Spring Hill Prison Governor (ex parte Sohi)* [1988] 1 All ER 424, DC).

The principal causes of delay in the Crown Court are the large number of defendants being committed for trial, the small number of judges available to try them, and the slowness of trial by jury, which, in turn, is due in no small measure to the increasing prolixity of counsel and judges (*R* v *Lawrence* [1981] 1 All ER 974, HL, *per* Lord Hailsham LC at p. 975). To cope with the bulk of the work there are only 624 full-time circuit judges and 1,417 part-time recorders.

The long delay between committal and trial has been condemned judicially as 'nothing short of a disgrace to our legal system' (*R* v *Lawrence*, above, *per* Lord Diplock at p. 979). As well as introducing anxiety and uncertainty into the life of the defendant, such delay may amount to a denial of justice since the recollection of witnesses on both sides dims as the months pass (ibid., *per* Lord Hailsham LC at p. 975).

In an attempt to ensure that cases are as well prepared for trial as possible, and that sufficient information is available for a trial date to be fixed, 'plea and directions hearings' were introduced in almost all Crown Court cases in July 1995 (the details are now contained in *Practice Direction (Criminal Proceedings: Consolidation), Part IV* [2002] 1 WLR 2870). At a plea and directions hearing, the defendant's plea is taken. If the plea is one of guilty, the judge is expected to sentence the defendant at this stage wherever possible. If the defendant pleads not guilty, the prosecution and defence are expected to assist the judge to identify the key issues of law and fact in readiness for the ensuing trial before a jury.

In 1999, the government published a consultation paper, *Transforming the Crown Court* (Lord Chancellor's Department, September 1999). The paper promised the most radical changes to the Crown Court in almost 30 years with the aims of cutting delays and cancellations, improving performance, and delivering enhanced levels of service to victims of crime, witnesses and jurors.

In December of the same year, the Lord Chancellor established a review of the criminal courts. The remit of the review, under the chairmanship of a Court of Appeal judge, Auld LJ, was to examine the practices, procedures and rules of evidence of the criminal courts at every level, with a view to ensuring that justice is delivered fairly. Regard was to be had to streamlining processes, increasing efficiency, the interests of all parties (including victims and witnesses) and the promotion of public confidence in the rule of law. Auld LJ's review was intended to be complementary to the changes already signalled in *Transforming the Crown Court* (above) and to the work done by Lord Woolf on the *civil* justice system between 1994 and 1996 (para 1.4.6 above).

The review was published in October 2001 (*Review of the Criminal Courts of*

England and Wales, Lord Chancellor's Department, 2001), and was followed by a government White Paper, *Justice For All* (Cm 5563, 2002). While the government rejected the central recommendation for a unified Criminal Court to replace the Crown Court and the magistrates' courts, it did accept many of the review's proposals and these are being implemented by the Criminal Justice Act 2003.

1.6 County courts

1.6.1 Introduction

The common law courts at Westminster were unable to provide cheap justice in small cases. Between 1822 and 1827 an average of 90,000 causes were entered each year in the common law courts, of which 30,000 were for claims of £20 or less. The cost of recovering £20 in a defended action far exceeded the amount of the claim.

The modern county courts were established by the County Courts Act 1846 in order to hear small civil cases at first instance. Their jurisdiction originally was over most personal actions (especially in contract and tort) involving not more than £20. That financial limit was raised progressively from £20 to £50 (1850), £100 (1903), £200 (1937), £400 (1955), £500 (1966), £750 (1969), £1,000 (1974), £2,000 (1977), and £5,000 (1981) until it was finally abolished in the wake of reforms introduced by the Courts and Legal Services Act 1990.

In its final report in 1988, the civil justice review body made a number of recommendations for change in the county courts (*Report of the Review Body on Civil Justice*, Cm 394, 1988). Many of these recommendations were subsequently implemented.

1.6.2 Composition

In England and Wales there are about 220 county courts sitting locally. The regular judges are 624 circuit judges and 416 district judges (formerly known as 'registrars'). The post of 'county court judge' disappeared when, under the provisions of the Courts Act 1971, all existing county court judges became circuit judges (Courts Act 1971, ss. 16 and 20, and sch. 2).

All circuit judges are competent to sit in any county court, although the Lord Chief Justice, after consulting the Lord Chancellor, must specifically assign one or more circuit judges to each county court district (County Courts Act 1984, s. 5(1), as amended by the Constitutional Reform Act 2005).

A district judge is appointed in accordance with the Constitutional Reform Act 2005 (para 1.1.6 above) from persons who have a seven-year general qualification within the meaning of s. 71 of the Courts and Legal Services Act 1990 (para 1.1.5 above) (County Courts Act 1984, s. 9, as amended by the Courts and Legal Services Act 1990). An appeal can be taken from the district judge's decision to the circuit judge. Apart from this right of appeal, the county courts have no appellate jurisdiction; they are civil courts of first instance.

In addition to the regular judges, every judge of the Court of Appeal and of the High Court and every recorder is, *ex officio*, competent to sit in any county court. If he consents to do so, he may sit at such times and on such occasions as the Lord Chief Justice considers desirable after consulting the Lord Chancellor (County Courts Act 1984, s. 5(3), as amended by the Constitutional Reform Act 2005). It would, however, be extremely rare to find a Court of Appeal or High Court judge sitting in a county court.

1.6.3 Jurisdiction

1.6.3.1 Introduction

The jurisdiction of the county courts is limited in three ways. First, as noted in para 1.6.3.2 below, there is sometimes a *financial* limit beyond which the county courts have no jurisdiction to hear cases. Such cases would have to be taken to the High Court, which has unlimited jurisdiction. There is, however, an exception which allows the parties to agree to give jurisdiction to the county court where the case should normally have gone to the Queen's Bench Division of the High Court (County Courts Act 1984, s. 18).

Secondly, there is a *geographical* limitation. The claimant is not allowed to pick and choose his forum. The court can deal with a case at any place that it considers appropriate (CPR, r. 2.7). Where the defendant is an individual, and the claim is for a specified amount of money and was commenced otherwise than in the defendant's home court, the proceedings will automatically be transferred by the court to the defendant's home court when a defence is filed (CPR, r. 26.2). The 'defendant's home court' means the county court for the district in which the defendant resides or carries on business (CPR, r. 2.3).

Thirdly, there are limitations on the powers of the county courts to grant remedies. They cannot grant the prerogative remedies of mandatory, quashing, and prohibiting orders (County Courts Act 1984, s. 38(3), as substituted by the Courts and Legal Services Act 1990, reinforcing s. 1(10) of the 1990 Act which provides that the county courts are not to be given jurisdiction to hear applications for judicial review).

They cannot grant search orders (*Anton Piller* orders) at all. Subject to certain exceptions (notably when dealing with family proceedings or when seeking to preserve the subject matter of the dispute), they cannot grant freezing injunctions (*Mareva* injunctions). They cannot revoke or vary a search order made by a High Court judge. They cannot *revoke* a freezing injunction granted by a High Court judge; they can *vary* it but only if all the parties agree on the terms of the variation. (These limitations on the granting of search and freezing relief, which do not apply to a Court of Appeal or High Court judge sitting in a county court, or to the judge of the patents county court at the Central London County Court (para 1.6.3.3 below), are contained in the County Court Remedies Regulations 1991, SI 1991, No. 1222, made under the County Courts Act 1984, s. 38, as substituted by the Courts and Legal Services Act 1990.)

Where a county court is unable to grant the appropriate remedy, application will have to be made instead to the High Court.

It is somewhat anomalous that the same circuit judge who, when sitting in a *county court*, has limited powers in relation to freezing injunctions and search orders has the full range of powers if he happens to be sitting in the *High Court*. If he has been appointed on a temporary basis as an additional or deputy judge of the High Court under s. 9 of the Supreme Court Act 1981 (para 1.4.2 above), he has all the powers of such a judge to grant, revoke or vary freezing injunctions and search orders. In March 2002, the Lord Chancellor published a consultation paper indicating the government's wish to extend the jurisdiction of these particular circuit judges so as to enable them to grant, revoke or vary *freezing injunctions* when sitting in a county court (*Freezing Injunctions and Search Orders in Civil Proceedings*, Lord Chancellor's Department, March 2002). If the proposal is ever adopted, the County Court Remedies Regulations 1991 will need to be amended. The government has no present intention to empower these circuit judges to deal with *search orders* in a county court, nor to extend the power to deal with freezing injunctions to all circuit judges.

The county courts are entirely creatures of statute. It follows that their jurisdiction is entirely statutory, conferred either by the consolidating County Courts Act 1984 or by particular statutes of which there are many. They are extremely busy courts. In 2004, 1.6 million proceedings were commenced in the county courts as opposed to a combined total of 49,000 in the Queen's Bench Division and the Chancery Division of the High Court, although only a small fraction of proceedings ever come to trial (*Judicial Statistics 2004*, Cm 6565, 2005, pp. 29, 38, 51). When compared with the High Court, the county courts provide a speedy and inexpensive forum for the trial of civil cases.

1.6.3.2 General jurisdiction under the County Courts Act 1984

(a) Claims founded on contract or tort, or for money due under a statute (County Courts Act 1984, ss. 15 and 16; High Court and County Courts Jurisdiction Order 1991, SI 1991, No. 724, as amended).

By s. 1 of the Courts and Legal Services Act 1990, the Lord Chancellor was empowered to confer jurisdiction on the county courts in relation to proceedings in which the High Court has jurisdiction, and *vice versa*. He was further empowered by s. 1 to allocate business between the High Court and the county courts according to criteria which include the following:

(i)   the value of a claim;

(ii) the nature of the proceedings;

(iii) the parties to the proceedings;

(iv) the degree of complexity likely to be involved in any aspect of the proceedings; and

(v)   the importance of any question likely to be raised by, or in the course of, the proceedings.

These provisions gave effect to proposals made by the civil justice review body for the more rational distribution of civil cases between the High Court and the county courts and for the easier transfer of cases up and down the system. It was anticipated that the allocation of civil business to the Queen's Bench Division of the High Court would in future reflect not so much the amount at stake (although this would still be a relevant factor) but the likely complexity and importance of particular proceedings.

The Lord Chancellor exercised his powers in the High Court and County Courts Jurisdiction Order 1991 (SI 1991, No. 724, as amended). This provides that claims which include a claim for damages for personal injuries (such as might arise in contract and tort) are to be commenced in a county court unless the value of the claim is £50,000 or more. Claims of which the value is £50,000 or more are to be tried in the High Court unless, in accordance with the criteria laid down for the transfer of claims (see below), they are more suitable for trial in a county court.

The county courts have no original jurisdiction to entertain claims for the tort of defamation (libel and slander) (County Courts Act 1984, s. 15(2)(c)). A county court may be given such jurisdiction, however, if the parties agree to it under s. 18 of the County Courts Act 1984, or the High Court transfers the case under s. 40 (as substituted by the Courts and Legal Services Act 1990).

In addition to claims for defamation, many other proceedings commenced in the High Court are transferable to a county court under s. 40 of the County Courts Act 1984. The section gives the High Court power to order a transfer either of its own motion or on the application of a party to the proceedings. The effect of a transfer is to free High Court judges for other, more important, work and to expedite hearings.

Any transfer will be to a county court considered appropriate by the High Court having regard to the convenience of the parties and the state of business in the courts concerned (s. 40(4)). Section 40 does not apply to family proceedings (s. 40(9)). These proceedings may, however, be transferred under the Matrimonial and Family Proceedings Act 1984 (para 1.6.3.3 below).

Guidance has been given on the appropriateness of transfers under s. 40. When considering whether to transfer proceedings to a county court the High Court must have regard to the following criteria under CPR, r. 30.3:

(i)   the financial value of the claim and the amount in dispute, if different;

(ii) whether it would be more convenient or fair for hearings (including the trial) to be held in some other court;

(iii) the availability of a judge specialising in the type of claim in question;

(iv) the complexity of the facts, legal issues, remedies or procedures involved;

(v)   the importance of the outcome of the claim to the public in general;

(vi) the facilities available at the court where the claim is being dealt with and whether they may be inadequate because of any disabilities of a party or potential witness;

(vii) whether the making of a declaration of incompatibility under s. 4 of the Human Rights Act 1998 has arisen or may arise; and

(viii) in the case of civil proceedings by or against the Crown, the location of the relevant government department or officers of the Crown and, where appropriate, any relevant public interest that the matter should be tried in London.

In addition, it is provided by the *Practice Direction* to CPR Part 29 that certain types of proceedings are particularly suitable for trial in the High Court and, therefore, should not normally be transferred to a county court. These are cases involving:

    (i)   professional negligence;

    (ii) fatal accidents;

    (iii) allegations of fraud or undue influence;

    (iv) defamation;

    (v) malicious prosecution or false imprisonment;

    (vi) claims against the police;

    (vii) contentious probate claims.

(b) Claims for the recovery of land, or in which the title to land comes into question, whatever the value of the land (County Courts Act 1984, s. 21).

(c) Equity matters, such as trusts, mortgages and dissolution of partnerships, where the amount of the fund or the value of the property involved does not exceed £30,000 (ibid., s. 23).

(d) Applications for financial-provision orders under s. 2 of the Inheritance (Provision for Family and Dependants) Act 1975, whatever the value of the deceased's estate (ibid., s. 25).

(e) Contentious probate proceedings concerning an application for a grant or revocation of probate or letters of administration where the net value of the estate at the time of death was less than £30,000 (ibid., s. 32, as substituted by the Administration of Justice Act 1985).

(f) Certain county courts designated by the Lord Chancellor have jurisdiction in specified admiralty proceedings. The financial limit in each case is £5,000, except in a claim for salvage where the value of the property saved must not exceed £15,000. These limits may be increased in individual cases by agreement of the parties (ibid., ss. 26–27).

1.6.3.3 Special jurisdiction under other statutes

(a) Certain county courts designated by the Lord Chancellor may hear insolvency cases and may deal with all matters relating to companies with paid-up capital not exceeding £120,000 (Insolvency Act 1986, ss. 117 and 373).

(b) The Central London County Court has been designated by the Lord Chancellor as a patents county court with the same jurisdiction as the High Court to deal with patent and design matters, with the exception that it cannot hear appeals from the Comptroller-General of Patents, Designs and Trade Marks (Patents County Court (Designation and Jurisdiction) Order 1994, SI 1994, No. 1609, made under the authority of s. 287(1) of the Copyright, Designs and Patents Act 1988).

(c) Certain county courts designated by the Lord Chancellor as divorce county courts have jurisdiction to hear petitions for divorce, nullity of marriage or judicial separation (Matrimonial and Family Proceedings Act 1984, s. 33).

(d) Certain county courts designated by the Lord Chancellor as civil partnership proceedings county courts have jurisdiction to hear civil partnership causes (i.e., actions for the dissolution or annulment of a civil partnership or for the legal separation of civil partners) (Matrimonial and Family Proceedings Act 1984, s. 36(A), as inserted by the Civil Partnership Act 2004).

(e) County courts have jurisdiction to hear such family proceedings as are transferred to them under s. 38 of the Matrimonial and Family Proceedings Act 1984, as amended by the Matrimonial Proceedings (Transfers) Act 1988, the Children Act 1989, and the Civil Partnership Act 2004.

These family proceedings may include wardship proceedings *except* applications for an order that a minor be made, or cease to be, a ward of court. County courts are thus able to deal with ancillary issues arising in wardship cases, such as residence, contact and education, but the High Court alone retains jurisdiction to decide whether a minor should be a ward of court or not.

(f) All county courts have jurisdiction to hear proceedings under the Children Act 1989. The allocation of these proceedings, and other family proceedings, as between judges in county courts is governed by a *Practice Direction* [1991] 4 All ER 764.

(g) All county courts have jurisdiction to grant orders against molestation and to exclude a spouse, civil partner, or a cohabitant from the home whether any other proceedings are pending or not (Family Law Act 1996, part IV, as amended by the Civil Partnership Act 2004; the High Court and the magistrates' courts also have jurisdiction under this statute).

(h) Certain county courts designated by the Lord Chancellor may hear civil actions brought by individuals in respect of alleged acts of unlawful racial discrimination or harassment otherwise than in the field of employment (Race Relations Act 1976, ss. 57 and 67, as amended).

The judge has power to award damages, grant injunctions and make declarations in such cases (ibid., s. 57(2)). Any damages awarded may include compensation for injury to feelings whether or not they include compensation under any other head (ibid., s. 57(4)). Damages for injury to feelings should normally not be of an exaggerated amount but should be more than a merely nominal sum (*Alexander* v *Home Office* [1988] 2 All ER 118, CA, where, on appeal, damages of £50 were increased to £500).

In addition, the Commission for Racial Equality may apply to a designated county court for an injunction to restrain *persistent* racial discrimination or harassment (ibid., s. 62, as amended).

The circuit judge will normally sit with two assessors, appointed by the Home Secretary for their special knowledge and experience of problems connected with race relations. If the parties consent, the judge may sit without assessors (ibid., s. 67(4)). The function of assessors is explained in *Ahmed* v *Governing Body of the University of Oxford* [2003] 1 WLR 995, CA.

Complaints about unlawful racial discrimination or harassment in *employment* are dealt with by the employment tribunals and not by the county courts (ibid., s. 54, as amended).

(i) All county courts have jurisdiction to hear claims from individuals alleging unlawful sexual discrimination or harassment in the fields of education and the provision of goods, facilities, services and premises (Sex Discrimination Act 1975, s. 66(1) and (2), as amended).

Damages, injunctions and declarations are the available remedies (Sex Discrimination Act 1975, s. 66(2)). Complaints of unlawful sexual discrimination or harassment in *employment* are heard by the employment tribunals and not by the county courts (ibid., s. 63(1), as amended).

(j) All county courts have jurisdiction to hear claims from individuals that they have been the victims of unlawful disability discrimination in the provision of goods, facilities and services (Disability Discrimination Act 1995, s. 25).

Any damages awarded may include compensation for injury to feelings whether or not they include compensation under any other head (ibid., s. 25(2)). Complaints of unlawful disability discrimination or harassment in *employment* are heard by the employment tribunals and not by the county courts (ibid., s.17A (formerly s. 8)).

1.6.4 Small claims in the county courts

1.6.4.1 Introduction

A scheme for the hearing of small claims using the existing county courts was introduced in 1973 by the Administration of Justice Act of that year (see now County Courts Act 1984, s. 64, as amended by the Courts and Legal Services Act 1990). The original 1973 financial limit of £75 has been raised progressively to £100 (1974), £200 (1977), £500 (1981), £1,000 (1991), £3,000 (1996), and £5,000 (1999).

A claim which has a financial value of £5,000 or less is dealt with on the small claims track (CPR, r. 26.6). This track is 'intended to provide a proportionate procedure by which most straightforward claims with a financial value of not more than £5,000 can be decided, without the need for substantial pre-hearing preparation and the formalities of a traditional trial, and without incurring large legal costs' (*Practice Direction* to CPR, Part 26). The court has power to reallocate a claim to a different track later (CPR, r. 26.10).

The small claims track is also the appropriate one for:

(a) a claim for personal injuries having a financial value of not more than £5,000 where the claim for damages for personal injuries is not more than £1,000; and

(b) a claim which includes a claim by a tenant of residential premises against his landlord for repairs (or other work) to the premises where the estimated cost of the repairs (or other work) is not more than £1,000 and the financial value of any other claim for damages is not more than £1,000 (ibid., r. 26.6).

In addition, a claim with a value above these financial limits can be allocated to the small claims track if the parties consent (ibid., r. 26.7). However, such a claim will not be allocated to the small claims track, even where the parties are agreeable, unless the court is satisfied that it is suitable for that track (*Practice Direction* to CPR, Part 26).

Most small claims are about debt, goods, and services. In 2004, 59 per cent of claimants using the small claims procedure were individual consumers and 41 per cent were local traders and professional firms (*Judicial Statistics 2004*, Cm 6565, 2005, p. 55).

Small claims in the county courts have been the subject of several official reports over the last two decades. Generally, these have revealed a fairly high level of satisfaction among litigants, including those who lost their cases. Neither costs, nor the procedure involved, seem to be a deterrent to would-be suitors. Historically, the major defect in dealing with small claims has been the lack of effective enforcement procedures against recalcitrant defendants.

In a report published in 2005, the Constitutional Affairs Select Committee of the House of Commons concluded that the small claims procedure generally works well and provides an important means for the pursuit of claims in an informal atmosphere at low cost and reasonable speed (*The Courts: small claims*, First Report of Session 2005–06, HC 519, December 2005). Nevertheless, a number of deficiencies were identified. First, the lack of proper IT facilities in the county courts was criticised because reliance on a paper-intensive system slows down proceedings and leads to unnecessary waste. Secondly, enforcement mechanisms were described as inadequate and in need of improvement. Thirdly, the claims limits in personal injury and housing disrepair cases (see above) were said to be in need of review, and it was suggested that these limits could usefully be raised to £2,500.

The government responded to the Committee's report by reaffirming its commitment to the small claims procedure and its determination to make it better through improved information leaflets, IT systems, and enforcement mechanisms. It was said that *all* case track limits are under consideration (*The Courts: small claims*, Government Response to the Constitutional Affairs Select Committee's Report, Cm 6754, February 2006).

1.6.4.2 Hearings

The *final* hearing of a small claim (at which the claim is determined) may be preceded by a *preliminary* hearing. However, a preliminary hearing can only be held where:

(a) the judge considers that special directions are needed to ensure a fair hearing, and it appears necessary for a party to attend at court to ensure that he understands what he must do to comply with the special directions; or

(b) to enable the judge to dispose of the claim on the basis that one or other of the parties has no real prospect of success at a final hearing; or

(c) to enable the judge to strike out a statement of case or part of a statement of case on the basis that the statement, or the part to be struck out, discloses no reasonable grounds for bringing or defending the claim (CPR, r. 27.6).

Furthermore, the judge, when considering whether to hold a preliminary hearing, must have regard to the desirability of limiting the expense to the parties of attending court (ibid.).

At or after the preliminary hearing, any appropriate directions are given, the date of the final hearing is fixed (if it has not been fixed already), the parties are given at least 21 days' notice of that date unless they agree to accept less notice, and they are informed of the amount of time allowed for the final hearing. If all the parties agree, the preliminary hearing can be treated as the final hearing of the claim (ibid.).

Most small claims will be determined after a hearing. If, however, the parties consent to the claim being dealt with by the judge on the basis of the statements and documents filed with the court a hearing is not necessary (CPR, r. 27.10).

Most hearings are conducted by the district judge, although the circuit judge also has jurisdiction to conduct them. There were 46,100 small claims hearings in the county courts in 2004, the vast majority of them conducted by the district judge (*Judicial Statistics 2004*, Cm 6565, 2005, p. 54). In what follows, the judge is assumed to be the district judge.

The hearing can be held at the court (usually in the judge's room, but it may be in a courtroom), or at any other place convenient to the parties. The general rule is that the hearing is to be in public (CPR, r. 27.2, applying r. 39.2; *Practice Direction* to CPR, Part 27). It may be in private if the parties agree (*Practice Direction* to CPR, Part 27), or if publicity would defeat the object of the hearing, or if it involves matters relating to national security, or if it involves confidential information (including information relating to personal financial matters) and publicity would damage that confidentiality, or if a private hearing is necessary to protect the interests of any child or patient, or if the judge considers it to be necessary in the interests of justice (CPR, r. 39.2). If the hearing takes place elsewhere than at the court, for example, at the home or business premises of a party, it will not be in public (*Practice Direction* to CPR, Part 27).

The hearing is an informal one at which the strict rules of evidence do not apply, evidence need not be taken on oath, and the judge can limit cross-examination (CPR, r. 27.8). The judge can adopt any method of proceeding which he considers to be fair (ibid.). In particular, he can question any witness himself before allowing any other person to do so; refuse to allow cross-examination of any witness until all the witnesses have given evidence in chief; and limit cross-examination of a witness to a fixed time, or to a particular subject or issue, or both (*Practice Direction* to CPR, Part 27).

An expert cannot give evidence (whether written or oral) at a hearing without the permission of the judge (CPR, r. 27.5). In any event, expert evidence is restricted to what is reasonably required to resolve the proceedings. The expert's duty is to help the court on the matters within his expertise, and this duty overrides any obligation to the person from whom he has received instructions or by whom he is paid (ibid., r. 27.2, applying rr. 35.1 and 35.3).

The judge must give reasons for his decision (ibid., r. 27.8). They can be given as briefly and simply as the nature of the case allows. They will normally be given orally at the hearing, or, failing that, they can be given later either in writing or at a hearing fixed for that purpose (*Practice Direction* to CPR, Part 27). Where the judge decides the case without a hearing, or a party has given written notice that he will not attend the hearing and does not attend it, the judge will prepare a note of his reasons and the court will send a copy to each party (ibid.).

The judge has power to grant interim injunctions (CPR, r. 27.2, applying r. 25.1), but not search orders (*Anton Piller* orders) and freezing injunctions (*Mareva* injunctions) (*Practice Direction* to CPR, Part 25). He also has power to grant any final remedy which could be granted if the proceedings were on the fast track or the multi-track, such as injunctions, damages and specific performance (CPR, r. 27.3).

1.6.4.3 Representation

At the hearing of a small claim, representation of a party by a lawyer (which, for this purpose, means a barrister, a solicitor, or a legal executive employed by a solicitor), or by a lay representative (see further below), is permitted (*Practice Direction* to CPR, Part 27), but must be paid for by the party personally even if he wins the case. This is because of the general rule in small claims cases that a party is not to be ordered to pay the other party's costs except in the limited circumstances permitted by the Civil Procedure Rules (see para 1.6.4.4 below). A corporate party can be represented by a lawyer, by any of its officers or employees, or by a lay representative (*Practice Direction* to CPR, Part 27; *Avinue Ltd* v *Sunrule Ltd* [2004] 1 WLR 634, CA, para 1.6.4.6 below).

By virtue of s. 11 of the Courts and Legal Services Act 1990, the Lord Chancellor was empowered by order to provide that there must be no restriction on the persons who may exercise rights of audience in relation to certain proceedings (including small claims hearings) in a county court. This provision implemented the civil justice review body's recommendation that in small claims cases a party should have a statutory right to be represented by a lay representative of his choice.

The Lord Chancellor's power was first exercised in an Order made in 1992. This was revoked and replaced by the Lay Representatives (Rights of Audience) Order 1999 (SI 1999, No. 1225), which provides that any person can exercise rights of audience as a lay representative at small claims hearings, except:

(a) where his client does not attend the hearing;

(b) at any stage after judgment; or

(c) on any appeal brought against the district judge's decision.

Notwithstanding these restrictions, the court, exercising its general discretion to hear anybody, can hear a lay representative even in circumstances excluded by the 1999 Order (*Practice Direction* to CPR, Part 27).

The conduct of lay representatives is governed by other provisions of s. 11 of the Courts and Legal Services Act 1990. The county court may refuse to hear a person who otherwise has a right of audience under the section if he is behaving in an unruly manner in any proceedings (s. 11(4)). A person exercising a right of audience under s. 11 may be disqualified by order of the judge from exercising that right in any county court if the judge has reason to believe that that person has intentionally misled the court, or otherwise demonstrated his unsuitability to exercise that right, whether in the current or any other proceedings (s. 11(6)). The judge must give his reasons for making the order, which can be appealed against to the Court of Appeal (s. 11(7)–(8)). An order may be revoked at any time by any judge of a county court (s. 11(9)).

1.6.4.4 Costs

In small claims cases, a party will not normally be ordered to pay the other party's costs. However, pursuant to CPR, r. 27.14, as amended, and *Practice Direction* to CPR, Part 27, some small costs can be awarded by the judge, representing, for example:

(a) the fixed costs attributable to issuing the claim;

(b) any court fees paid by another party;

(c) the reasonable travel and subsistence expenses of a party or witness;

(d) a sum not exceeding the amount prescribed in the *Practice Direction* to CPR, Part 27, for loss of earnings or loss of leave of a party or witness;

(e) a sum not exceeding the amount prescribed in that *Practice Direction* in respect of an expert's fees.

In addition, further costs can be awarded against a party who has behaved unreasonably. Although a party's rejection of an offer in settlement does not of itself amount to behaving unreasonably, the court can take it into account when applying the unreasonableness test (CPR, r. 27.14, as amended).

The general rule that a party will not be ordered to pay the other party's costs is designed to discourage the employment of lawyers at small claims hearings (*Hobbs v Marlowe* [1978] AC 16, HL, *per* Lord Elwyn-Jones LC at p. 32).

1.6.4.5 Rehearings

A party who did not attend, and was not represented at, the hearing of the claim, and who did not give the requisite notice to the court, can apply to have the decision set aside and the claim re-heard. This is not, technically, an appeal. The judge can only set aside the decision if the applicant had a good reason for not attending or being represented at the hearing or giving notice to the court, and has a reasonable prospect of success at the hearing (CPR, r. 27.11).

1.6.4.6 Appeals

The only grounds for appeal are that the judge's decision is:

(a) wrong, or

(b) unjust because of a serious procedural or other irregularity in the proceedings (CPR, r. 52.11).

A 'serious irregularity' consisting of misconduct on the part of the judge is very difficult to establish (*Starmer* v *Bradbury* (1994) *The Times*, 11 April, CA, where the claimant's allegations that the district judge appeared partial and lost control of the proceedings were dismissed).

An appeal from a decision of a district judge is dealt with by a circuit judge (Access to Justice Act 1999 (Destination of Appeals) Order 2000, SI 2000, No. 1071, art. 3). When hearing an appeal, the circuit judge has all the powers of the district judge, and he also has power to affirm, set aside or vary the decision, or order a new hearing (CPR, r. 52.10). When the circuit judge allows an appeal, he will, if possible, dispose of the case at the same time without referring it back to the district judge or ordering a new hearing (*Practice Direction* to CPR, Part 27).

Permission to appeal is not required at present for an appeal (a 'first' appeal) from a district judge to a circuit judge against a decision made in the small claims track.

COURTS OF NORMAL JURISDICTION   41

According to the government in a consultation paper published in 2000, this is a confusing anomaly and small claims appeals should be brought within the strict appeal rules contained in CPR, Part 52 (*Small Claims Appeals: Proposed New Procedures*, Lord Chancellor's Department, June 2000). No action has so far been taken on this proposal. If it is ever implemented, then permission to appeal will be required in all small claims cases.

Any 'second' appeal lies from the circuit judge to the Court of Appeal (Access to Justice Act 1999 (Destination of Appeals) Order 2000, art. 5). Such an appeal requires the permission of the Court of Appeal itself (CPR, r. 52.13). By the Access to Justice Act 1999, s. 55(1), the Court of Appeal cannot give permission unless:

(a) the appeal would raise an important point of principle or practice, or

(b) there is some other compelling reason for the appeal to be heard.

In *Avinue Ltd v Sunrule Ltd* [2004] 1 WLR 634, CA, the Court of Appeal, first, granted leave to appeal on the ground that the case raised an important point of principle or practice, and, secondly, decided that the defendants had not had a fair trial by reason of a serious procedural or other irregularity in the proceedings.

At the hearing, a director of the defendants told the district judge that he was Greek and did not have a good understanding of English, but the district judge refused to allow the defendants to be represented by a lay representative of their choice. The reason given was that, under the *Practice Direction* to CPR, Part 27, a company, while it can be represented by an officer or employee, cannot be represented by any other lay person. The defendants ended up being represented by the director. They lost the case and their appeal was dismissed by the circuit judge, who also refused permission to appeal.

The Court of Appeal held that the district judge had misinterpreted the *Practice Direction*, and that a corporate party can be represented by a lay representative on the same conditions as an individual litigant (para 1.6.4.3 above).

1.7 Magistrates' courts

1.7.1 Introduction

The office of magistrate or justice of the peace dates back to 1195 when Richard I appointed 'keepers of the peace' to pursue, arrest and punish those who offended against the King's peace. In 1363, a statute provided that the justices should meet at least four times a year in each county. These meetings, which existed in addition to the petty sessions of local magistrates, became known as the Quarter Sessions and lasted until finally abolished by the Courts Act 1971.

Originally the magistrates had many administrative as well as judicial duties. They were responsible for such local government activities as the poor law, highways and bridges, and weights and measures. Responsibility for the administration of local government was transferred from the justices to elected local authorities in the nineteenth century.

The title, 'justice of the peace', covers both lay magistrates and professional district judges (magistrates' courts), who are *ex officio* justices of the peace by virtue of s. 25 of the Courts Act 2003.

1.7.2 District judges (magistrates' courts)

Most of what follows is concerned with magistrates' courts composed of lay justices (often referred to as 'lay magistrates').

However, it should be noted that in addition to lay justices there are, in London and some of the larger provincial towns, some 129 district judges (magistrates' courts) who sit alone and exercise the same jurisdiction as a bench of lay magistrates (Courts Act 2003, s. 26). They are also judges of the Crown Court for certain purposes (ibid., s. 65 and sch. 4). They are full-time, paid judges, appointed from persons who have a seven-year general qualification within the meaning of s. 71 of the Courts and Legal Services Act 1990 (para 1.1.5 above) (Courts Act 2003, s. 22). They are appointed in accordance with the Constitutional Reform Act 2005 (para 1.1.6 above).

District judges (magistrates' courts) can be removed from office by the Lord Chancellor, with the agreement of the Lord Chief Justice, on the ground of incapacity or misbehaviour (Courts Act 2003, s. 22, as amended by the Constitutional Reform Act 2005). By virtue of the Judicial Pensions and Retirement Act 1993, as amended by the Constitutional Reform Act 2005, they normally retire at the age of 70, but, if it is desirable in the public interest, they may be authorised on an individual basis by the Lord Chief Justice to continue in office for a year at a time up to the age of 75.

The Lord Chancellor can appoint deputy district judges (magistrates' courts) on a temporary basis in order to facilitate the disposal of business. The persons appointed as deputies must be qualified for appointment as district judges (magistrates' courts). Deputy district judges (magistrates' courts) have all the judicial powers of district judges (magistrates' courts). They may be removed from office by the Lord Chancellor, with the agreement of the Lord Chief Justice, for incapacity or misbehaviour (Courts Act 2003, s. 24, as amended by the Constitutional Reform Act 2005). There are some 167 deputy district judges (magistrates' courts) in post.

District judges (magistrates' courts) were formerly called 'stipendiary magistrates'. They were renamed by the Access to Justice Act 1999. That Act also abolished the office of 'metropolitan stipendiary magistrate' formerly held by stipendiary magistrates appointed to sit in the Inner London area. The resultant merger of the provincial and metropolitan professional magistracy into a unified judicial body was intended to increase speed and efficiency in the magistrates' courts.

High Court and Crown Court judges have the powers of a district judge (magistrates' courts) in relation to criminal cases and family proceedings in the magistrates' courts (Courts Act 2003, s. 66).

1.7.3 Lay magistrates: appointment and training

Lay magistrates are appointed on behalf of the Queen by the Lord Chancellor (Courts Act 2003, s. 10(1)).

Each county has a local advisory committee, usually under the chairmanship of the

Lord-Lieutenant of the county with the rest of its membership remaining secret to discourage lobbying. Names of potential lay magistrates are put forward to the local advisory committee by interested bodies, such as political parties, trade unions, and chambers of commerce. In addition, the committee may, if it wishes, advertise for candidates to put themselves forward. Candidates may be interviewed. The details of those found to be acceptable by the committee are sent to the Lord Chancellor, who makes the final decision on appointments.

In appointing lay magistrates, the Lord Chancellor is under a duty of consultation (Courts Act 2003, s. 10(2A), as inserted by the Constitutional Reform Act 2005). In particular, this will involve consulting local advisory committees.

There is no formal limit on the number of lay magistrates that can be appointed. At present there are some 28,253 magistrates (51 per cent men and 49 per cent women) in England and Wales dispensing justice in approximately 700 magistrates' courts. Lay magistrates are unqualified and unpaid, although they are expected to attend courses of instruction (Courts Act 2003, s. 19), and they are paid allowances for travel, subsistence and loss of earnings (ibid., s. 15).

Local commission and petty sessions areas, and local magistrates' courts committees, were abolished, and the former geographical limitation on a lay magistrate's jurisdiction was removed, by the Courts Act 2003. There is now only one commission of the peace for England and Wales. The Lord Chief Justice is under a statutory duty to assign every lay magistrate to one or more of the 'local justice areas' into which the two countries are divided. The statute makes it clear that such assignment does not limit jurisdiction to that area (ibid., s. 10, as amended by the Constitutional Reform Act 2005). Thus, lay magistrates have, in theory, a national jurisdiction, although it is not envisaged that they will be required to sit outside their normal place of sitting either on a regular basis or without their consent.

The Lord Chief Justice, in consultation with the Lord Chancellor, has a statutory responsibility for the training, development and appraisal of lay magistrates (ibid., s. 19, as amended by the Constitutional Reform Act 2005). The training available puts emphasis on learning through the experience of actually sitting in court. The object is to allow magistrates to demonstrate that they have acquired the knowledge and skills necessary for their work. Attendance at courses plays a part in the training process, but emphasis is placed on learning by sitting as a magistrate—supported and assisted by specially selected colleagues trained to act as monitors. The training scheme is also designed to emphasise equality of treatment for all who appear in the magistrates' courts regardless of race, creed, colour, ethnic origin, gender, religion, disability, or sexual orientation.

1.7.4 Lay magistrates: retirement, the supplemental list, resignation, and removal from office

Lay magistrates do not officially 'retire'. At the age of 70, their names are entered in 'the supplemental list' as recognition of their service (Courts Act 2003, s. 13(1)). Those who have not reached the age of 70 can request to be placed in the list (ibid., s. 13(4)). In addition, the Lord Chancellor, with the agreement of the Lord Chief Justice, can direct a name to be entered in the supplemental list on the ground of incapacity (ibid., s. 13(5), as amended by the Constitutional Reform Act 2005). Lay magistrates whose names are in the supplemental list are no longer qualified to act as justices of the peace (ibid., s. 12).

Lay magistrates can resign at any time (ibid., s. 11(1)).

They can be removed from office by the Lord Chancellor, with the agreement of the Lord Chief Justice, on behalf of the Queen for incapacity, misbehaviour, persistent failure to meet prescribed standards of competence, or for declining or neglecting to perform their functions (ibid., s. 11(2), as amended by the Constitutional Reform Act 2005).

1.7.5 Justices' clerks

1.7.5.1 Appointment

Lay magistrates are assisted by justices' clerks appointed by the Lord Chancellor (and assigned to one or more local justice areas) from persons who have a five-year magistrates' court qualification within the meaning of s. 71 of the Courts and Legal Services Act 1990, or who are barristers or solicitors who have served for not less than five years as an assistant to a justices' clerk, or who have previously been a justices' clerk (Courts Act 2003, ss. 2 and 27).

The Lord Chancellor also has power to appoint assistants to justices' clerks ('assistant clerks': ibid.).

1.7.5.2 Functions of the justices' clerk

The main duty of a justices' clerk is to advise the lay magistrates on matters of law (including procedure and practice), the range of penalties available, and any relevant decisions of the superior courts or other guidelines, and he may tender such advice even when not specifically requested if he thinks he should do so (see Courts Act 2003, s. 28, and *Practice Direction (Criminal Proceedings: Consolidation), Part V* [2002] 1 WLR 2870). The magistrates are not bound by law to accept their clerk's advice since they are independent and are responsible for making their own decisions and fixing their own sentences.

In certain circumstances, and in accordance with rules made by the Lord Chancellor in agreement with the Lord Chief Justice, the justices' clerk can exercise the powers of a single lay magistrate (Courts Act 2003, s. 28, as amended by the Constitutional Reform Act 2005).

The independence of the justices' clerk is guaranteed by a statutory provision to the effect that, in the exercise of his functions, he is not subject to the direction of the Lord Chancellor or any other person (ibid., s. 29(1)). An assistant clerk is not subject to the direction of any person other than the justices' clerk (ibid., s. 29(2)).

It has famously been said to be of 'fundamental importance that justice should not only be done but manifestly and undoubtedly be seen to be done' (*R v Sussex Justices (ex parte McCarthy)* [1924] 1 KB 256, DC, *per* Lord Hewart CJ at p. 259). In accordance with this principle, the justices' clerk should not *instruct* the magistrates what their actual decisions should be, or what sentence to impose, and, further, should not even *appear* to be taking part in decisions. To this end, the justices' clerk should not normally retire with the magistrates when they consider their verdict. But he may legitimately be called in by them to give advice on matters of law (including procedure and practice). He should not, however, be a party to findings of fact.

Any request to the clerk to accompany them while they retire should be made by the magistrates clearly and in open court (*R* v *Eccles Justices (ex parte Fitzpatrick)* (1989) 89 Cr App R 324, DC; *Clark (Procurator Fiscal, Kirkcaldy)* v *Kelly* [2003] 2 WLR 1586, PC (Scotland)). It is recognised as good practice that the clerk should return to the court before the magistrates rather than stay with them throughout the whole of their retirement, although so to stay is not regarded as improper in itself.

1.7.6 Functions of magistrates' courts

1.7.6.1 Introduction

The functions of magistrates' courts include the following:

    (a) to exercise summary criminal jurisdiction;

    (b) to determine the mode of trial of offences triable either way;

    (c) to exercise some civil jurisdiction.

1.7.6.2 Summary criminal jurisdiction

1.7.6.2.1 *Classification of offences* The Criminal Law Act 1977 provided for three classes of offence for the purpose of mode of trial.

    (a) *Offences triable only summarily.* All summary offences are created by statute and most of them are of a relatively minor nature. There are hundreds of summary offences, and they are triable only in a magistrates' court.

    (b) *Offences triable only on indictment.* These are the most serious crimes and include treason, murder, manslaughter and rape. They are triable only in the Crown Court.

    (c) *Offences triable either way.* These are the offences listed in sch. 1 to the Magistrates' Courts Act 1980, or which are made triable either way by individual statutes. They are triable either summarily in the magistrates' court or on indictment in the Crown Court.

Magistrates try mainly *summary* offences; i.e., offences which are triable without a jury, such as most motoring offences. The case must be heard by at least two, and not more than three, lay justices or a district judge (magistrates' court) (Magistrates' Courts Act 1980, s. 121(1); Courts Act 2003, s. 26(1)). Magistrates may try a case in the absence of the accused but they cannot sentence him to imprisonment or detention in his absence (Magistrates' Courts Act 1980, s. 11(1) and (3), as amended).

Following conviction for a summary offence, the magistrates may, in general, impose a prison sentence not longer than the period specified in the statute creating the offence, or 12 months, whichever is the shorter (increased from six months by the Criminal Justice Act 2003, s. 154). If the defendant has been convicted of two or more summary offences the magistrates may impose consecutive prison sentences, but these must not exceed 65 weeks in the aggregate (Magistrates' Courts Act 1980, s. 133(1), as amended by the Criminal Justice Act 2003). Many summary offences are not punishable by imprisonment at all.

The maximum fine for a summary offence is such sum as may be specified in the statute creating the offence, or £1,000, whichever is the higher (Magistrates' Courts Act 1980, s. 34(3), as amended).

Before deciding on the amount of a fine to be imposed on an offender, the magistrates must inquire into the defendant's financial circumstances and, when fixing the amount, must take these into account (so far as they are known, or appear, to the court) together with the other circumstances of the case. The amount of any fine imposed must be such as in the opinion of the court reflects the seriousness of the offence (Criminal Justice Act 2003, s. 164).

Other sentences available to the court for dealing with adult offenders include absolute discharge, conditional discharge, compensation order, and community order. Conditions can be attached to a community order. They include unpaid work, curfew (including electronic monitoring), drug rehabilitation, alcohol treatment, and supervision.

The magistrates' courts also have jurisdiction to try offences *triable either way*; i.e. offences which are triable *either* summarily by magistrates or on indictment in the Crown Court. Following summary conviction of the defendant for an offence triable either way, the magistrates' powers of punishment are limited to a maximum of 12 months' imprisonment for any one offence (Magistrates' Courts Act 1980, s. 32(1), as amended by the Criminal Justice Act 2003), and a maximum fine of £5,000 (Magistrates' Courts Act 1980, s. 32(9), as amended). If the magistrates convict the defendant summarily of two or more offences triable either way, they may impose consecutive prison sentences of up to 65 weeks in the aggregate (ibid., s. 133(1), as amended by the Criminal Justice Act 2003).

1.7.6.2.2 *Procedure for determining the mode of trial of offences triable either way* A procedure for determining the mode of trial of offences triable either way is set out in the Magistrates' Courts Act 1980, ss. 17A–26, as amended by the Criminal Justice Act 2003. The allocation jurisdiction may be exercised by a single justice (Magistrates' Courts Act 1980, s. 17E, as inserted by, and s. 18(5), as substituted by, the Criminal Justice Act 2003).

The first step in the procedure is to ascertain whether (if the offence were to proceed to trial) the accused would plead guilty or not guilty (Magistrates' Courts Act 1980, ss. 17A–17C, as inserted by the Criminal Procedure and Investigations Act 1996). The charge is read to the accused and the court explains to him in ordinary language that:

(a) he may indicate his plea;

(b) if he indicates that he would plead guilty the court will try him summarily; and

(c) he may be committed for sentence to the Crown Court under s. 3 of the Powers of Criminal Courts (Sentencing) Act 2000, as amended by the Criminal Justice Act 2003 (para 1.7.6.2.4 below).

The court then asks the accused about his plea, and, if he indicates that he would plead guilty, proceeds to try him summarily.

If the accused indicates that he would plead not guilty, or fails to indicate any plea at all, the court does not at that point proceed to try him summarily but will instead reach a decision on 'allocation'; i.e., whether the offence is more suitable for summary trial or trial on indictment.

Before making its decision, the court must give the prosecution an opportunity to inform it of the accused's previous convictions (if any), and must give both sides an opportunity to make representations about mode of trial. In making its decision, the court must consider whether the sentencing powers of a magistrates' court would be adequate for the offence, any representations about mode of trial made by the prosecution or the accused, and any allocation guidelines (or revised allocation guidelines) issued as definitive guidelines under s. 170 of the Criminal Justice Act 2003 (Magistrates' Courts Act 1980, s. 19(1)–(3), as substituted by the Criminal Justice Act 2003).

If, after considering the relevant matters, the court decides that *summary* trial is more suitable it must explain to the accused in ordinary language that:

(a) it appears to be more suitable for him to be tried summarily for the offence;

(b) he can either consent to be so tried or, if he wishes, be tried on indictment; and

(c) in the case of a specified offence (within the meaning of s. 224 of the Criminal Justice Act 2003), that if he is tried summarily and is convicted by the court, he may be committed for sentence to the Crown Court under s. 3A of the Powers of Criminal Courts (Sentencing) Act 2000. (Magistrates' Courts Act 1980, s. 20(1)–(2), as substituted by the Criminal Justice Act 2003.)

At this stage, the accused can request an indication ('an indication of sentence') of whether a custodial or non-custodial sentence would be more likely to be imposed if he pleaded guilty and were to be tried summarily. The court may, but is not bound to, give such an indication. If an indication of sentence *is* given, the court must ask the accused whether he wishes to reconsider his plea indication. If he does, he must be asked whether (if the offence were to proceed to trial) he would plead guilty or not guilty. If he indicates that he would plead guilty, the court will proceed to sentence him (ibid., s. 20(3)–(7), as substituted by the Criminal Justice Act 2003). He cannot be given a custodial sentence unless one was 'indicated' (ibid., s. 20A(1), as inserted by the Criminal Justice Act 2003).

If the court does not give an indication of sentence, or if the accused declines to reconsider his plea indication or does not indicate that he would plead guilty, the court must ask him whether he consents to be tried summarily or wishes to be tried on indictment. If he consents to be tried summarily, the magistrates proceed with a summary trial. If he does not consent to be tried summarily, he is sent to the Crown Court for trial (ibid., s. 20(8)–(9), as substituted by the Criminal Justice Act 2003).

If, after considering the relevant matters, the court decides that trial *on indictment* is more suitable, the court must tell the accused that its decision is that it is more suitable for him to be tried on indictment. The accused is then sent to the Crown Court for trial (ibid., s. 21, as substituted by the Criminal Justice Act 2003).

Normally the accused must be present in court during the proceedings for determining the mode of trial of an offence triable either way (ibid., s. 18(2)). But if the accused is represented by a legal representative, his presence is not required if he consents to the proceedings being conducted in his absence and the court is satisfied that there is a good reason for so proceeding (ibid., s. 23(1)). This would cover, for example, the illness of the accused. If the accused is *not* represented, proceedings cannot continue in his absence and would need to be adjourned.

In addition, the court may proceed in the absence of the accused, and without his consent, where it is not practicable to proceed in his presence because of his disorderly conduct before the court (ibid., s. 18(3)).

Where the accused is legitimately absent, the procedure for determining mode of trial outlined above is followed with appropriate modifications. In particular, the accused's consent to summary trial may be signified by the legal representative (if any) representing him (ibid., s. 18(3) and s. 23(2)–(5), as amended by the Criminal Justice Act 2003).

Where the court has allocated an either-way offence for summary trial, the prosecution can apply for the offence to be tried on indictment instead. The application must be made before the summary trial begins, and the court can only grant the application if satisfied that the sentence which a magistrates' court would have power to impose for the offence would be inadequate. If the application is granted, the accused is sent to the Crown Court for trial (ibid., s. 25(2)–(2B) and (2D), as substituted by the Criminal Justice Act 2003).

The procedure outlined above for determining the mode of trial of an offence triable either way applies where the accused is aged 18 or over. In the case of *children* (aged ten or more but under 14) and *young persons* (aged 14 or more but under 18), a similar procedure is available in some circumstances (ibid., s. 24, as amended by, and ss. 24A–24D, as inserted by, the Criminal Justice Act 2003).

1.7.6.2.3 *Restrictions on reports of allocation and sending proceedings* Printed and broadcast reports of allocation and sending proceedings are required to be limited to the following matters (Crime and Disorder Act 1998, s. 52A, as inserted by the Criminal Justice Act 2003):

(a) the identity of the court and the name of the justice or justices;

(b) the name, age, home address and occupation of the accused;

(c) the offence or offences, or a summary of them, with which the accused is charged;

(d) the names of counsel and solicitors engaged in the case;

(e) where the proceedings are adjourned, the date and place to which they are adjourned;

(f) any arrangements for bail;

(g) whether a right to representation funded by the Legal Services Commission as part of the Criminal Defence Service was granted to the accused. (Crime and Disorder Act 1998, s. 52A(7).)

It is a summary offence to report or broadcast any information not listed above. On conviction, the proprietors, editors and publishers of newspapers or periodicals, and the BBC, independent television companies, and their senior programme executives, are liable to a fine. Proceedings for this offence cannot be instituted without the consent of the Attorney-General (ibid., s. 52B(1)–(3), as inserted by the Criminal Justice Act 2003).

The magistrates' court has power to lift the reporting restrictions. But the court must not do this if the accused, or one of the accused, objects, unless it is satisfied, after hearing the representations of the accused, that it is in the interests of justice to do so (ibid., s. 52A(2)–(4)).

If the restrictions are lifted, reporting in the media ceases to be confined to the matters specified above. Full reporting is possible, including details of the evidence, subject to the understanding that official secrets must not be divulged and subject to what is said below about the Contempt of Court Act 1981. Even if the reporting restrictions are not lifted, the evidence may be published at the end of any trial. By this time, however, the evidence will have lost much of its news value.

The reporting restrictions outlined above operate *in addition to* any other statutory limitations on reports of court proceedings (ibid., s. 52A(10)). For example, by s. 4(2) of the Contempt of Court Act 1981 a court has a discretion to order that the publication of any report of legal proceedings held in public be postponed for such period as the court thinks necessary. But a court, including a magistrates' court, can only make such an order 'where it appears necessary for avoiding a substantial risk of prejudice to the administration of justice' in the proceedings before it or in any other proceedings which are pending or imminent (see further, para 8.2.2.4.5).

1.7.6.2.4 *Committal to the Crown Court for sentence* As seen in para 1.7.6.2.1 above, there are statutory limits on the punishments that may be imposed by a magistrates' court for *offences triable either way*. There is a maximum sentence of imprisonment of 12 months, though the court may impose consecutive sentences of up to 65 weeks in the aggregate in the case of two or more offences triable either way. The maximum fine that can be inflicted is £5,000.

Magistrates' sentencing powers were increased to these levels by amendments to the law made in the Criminal Justice Act 2003 so that fewer cases will have to go to the Crown Court. It is present government policy that the magistrates' courts should take more responsibility for sentencing the defendants they convict. To this end, the circumstances in which they can commit defendants to the Crown Court for sentence because they feel that their own sentencing powers are inadequate have been made more limited than hitherto.

For example, the magistrates can, in specified circumstances, commit adult and young defendants to the Crown Court for sentence if of the opinion that the offence, or the combination of the offence and one or more offences associated with it, is so serious that the Crown Court should have the power to pass sentence (Powers of Criminal Courts (Sentencing) Act 2000, s. 3, as substituted by, and s. 3B, as inserted by, the Criminal Justice Act 2003). In certain other specified circumstances the magistrates *must* commit *dangerous* adult and young defendants to the Crown Court for sentence (Powers of Criminal Courts (Sentencing) Act 2000, ss. 3A and 3C, as inserted by the Criminal Justice Act 2003).

On a committal for sentence, the Crown Court will inquire into the circumstances of the case in the same way as if the defendant had been convicted there. The Crown Court may impose any sentence up to the maximum prescribed for that particular offence if it were tried on indictment (ibid., s. 5, as substituted by, and s. 5A, as inserted by, the Criminal Justice Act 2003).

1.7.6.2.5 *Youth courts* A child (aged ten or over but under 14) or young person (aged 14 or over but under 18) will normally be tried in a youth court, even when charged with an indictable offence (Magistrates' Courts Act 1980, s. 24(1), as substituted by the Criminal Justice Act 2003; but see below). A youth court is a magistrates' court sitting for that purpose and composed of lay justices or a district judge (magistrates' courts) authorised to sit by the Lord Chancellor (Children and Young Persons Act 1933, s. 45, as substituted by the Courts Act 2003). Unless the case is being heard by a district judge (magistrates' courts) sitting alone, there must be three justices on the bench and at least one of them must be a woman and at least one must be a man (Youth Courts (Constitution) Rules 1954 (SI 1954, No. 1711), as amended).

The procedure in a youth court is aimed at avoiding both the criminal environment of the magistrates' court and excessive publicity. Among the persons allowed to be present are the justices, court officers, the parties and their legal representatives, witnesses, representatives of news gathering or reporting organisations, and other persons authorised by the court to be present (Children and Young Persons Act 1933, s. 47(2), as amended). The attendance at court of a parent or guardian may be required in any case, and is mandatory in the case of a child or young person under the age of 16, unless it would be unreasonable to require such attendance in the circumstances (ibid., s. 34A, as inserted by the Criminal Justice Act 1991).

The name or photograph of any child or young person appearing in the case must not be published in any newspaper or broadcast without the authority of the court (Children and Young Persons Act 1933, s. 49, as substituted by the Criminal Justice and Public Order Act 1994). The words 'conviction' and 'sentence' must not be used in connection with children and young persons (ibid., s. 59, as amended).

Some of the punishments available to a youth court are different from those that may be imposed on adults by the magistrates' court. In the case of a *child*, there is a maximum permitted fine of £250 (Powers of Criminal Courts (Sentencing) Act 2000, s. 135(2)). A child may be made the subject of a supervision order, compensation order, attendance centre order, curfew order, anti-social behaviour order, or detention and training order (if not less than 12). In the case of a *young person*, the maximum permitted fine is £1,000 (ibid., s. 135(1)). He may be made the subject of a youth community order (if aged 16), compensation order, anti-social behaviour order, or detention and training order.

The maximum permitted fines may be raised (or lowered) by the Home Secretary by statutory instrument (Magistrates' Courts Act 1980, s. 143, as amended).

There are some exceptional cases where a child or young person will not be tried summarily in a youth court but will be sent for trial to the Crown Court.

First, where the offence charged is *homicide* (Crime and Disorder Act 1998, s. 51A, as inserted by the Criminal Justice Act 2003).

Secondly, in the case of some *firearms* offences where (by s. 51A of the Firearms Act

1968, as inserted by the Criminal Justice Act 2003) a minimum sentence of three years' detention must be passed unless the court is of the opinion that there are exceptional circumstances relating to the offence or to the offender which justify its not doing so (Crime and Disorder Act 1998, s. 51A, as inserted by the Criminal Justice Act 2003).

Thirdly, where a *child or young person* is charged with an offence for which the penalty in the case of a person aged 18 or over is 14 years' imprisonment or more, or with certain sexual offences, or (if aged at least 14 but under 18) with causing death by dangerous driving or causing death by careless driving while under the influence of drink or drugs, and the magistrates' court considers that if he is found guilty it ought to be possible for him to be sentenced by the Crown Court (ibid.).

Fourthly, where the offence is a specified violent offence (including manslaughter and certain explosive and firearms offences) or a specified sexual offence (including rape) (ibid.).

Fifthly, where a *child or young person* is charged jointly with an adult, and the magistrates' court considers it necessary in the interests of justice to commit them both for trial to the Crown Court (Crime and Disorder Act 1998, s. 51, as substituted by the Criminal Justice Act 2003).

No *criminal* proceedings at all can be brought against a child under the age of ten because of the irrebuttable presumption that he is incapable of committing a crime (Children and Young Persons Act 1933, s. 50, as amended by the Children and Young Persons Act 1963). Ten is the so-called age of criminal responsibility in English law. A child under that age who gets into trouble and who is, for example, beyond parental control may be brought before a family proceedings court (see para 1.7.6.5.2 below) in *care* proceedings under the Children Act 1989, which are *civil*, and in which he may be made the subject of, *inter alia*, a care order or a supervision order involving the local authority. Alternatively, the child may be made the subject of a child safety order under the Crime and Disorder Act 1998.

In criminal proceedings against a child aged between ten and 14, it was formerly the case that the prosecution had to prove beyond reasonable doubt not only that he committed the offence but also that he knew that what he was doing was seriously wrong as opposed to merely naughty or mischievous. This rebuttable common law presumption had attracted much criticism in recent years, and in 1994 an attempt was made to get rid of it by judicial decision in *C v Director of Public Prosecutions* [1995] 2 All ER 43, HL. Here the Queen's Bench Divisional Court, departing from some of its own earlier decisions, had held that the presumption was no longer part of English law. However, the Divisional Court was reversed on appeal to the House of Lords, although their Lordships also expressed dissatisfaction with the presumption and called on Parliament to review it. It was finally abolished by the Crime and Disorder Act 1998.

The presumption that a boy under the age of 14 is incapable of sexual intercourse was abolished by the Sexual Offences Act 1993.

The Children and Young Persons Act 1969 provided for a completely new way of treating children and young persons. The proposed procedure was aimed broadly at the replacement of criminal proceedings by care proceedings. However, the relevant provisions were never brought into force and after more than 20 years they were repealed by the Criminal Justice Act 1991.

1.7.6.3 Civil jurisdiction

1.7.6.3.1 *Introduction* The magistrates' courts have jurisdiction to hear family proceedings, particularly under the Domestic Proceedings and Magistrates' Courts Act 1978, the Children Act 1989, and part IV of the Family Law Act 1996 (family homes and domestic violence). They have powers over the recovery of council tax and charges for water, gas and electricity. As licensing justices, magistrates are responsible for granting, revoking and renewing licences for the sale of liquor.

1.7.6.3.2 *Family proceedings courts*

1.7.6.3.2.1 *Introduction* Perhaps the most important and time-consuming aspect on the civil side is the jurisdiction, first conferred by s. 4 of the Matrimonial Causes Act 1878, over family proceedings. When a magistrates' court is dealing with family proceedings it is called a 'family proceedings court' (Magistrates' Courts Act 1980, s. 67(1), as substituted by the Courts Act 2003).

The term 'family proceedings' includes adoption proceedings (Magistrates' Courts Act 1980, s. 65(1)), which, before 1979, were dealt with, somewhat anomalously, in the juvenile court (now called the youth court). 'Family proceedings' also include applications for residence, contact and care orders under the Children Act 1989, proceedings under part 1 of the Domestic Proceedings and Magistrates' Courts Act 1978 (Magistrates' Courts Act 1980, s. 65(1), as amended), proceedings under part IV of the Family Law Act 1996, and applications for child safety orders in respect of children below the age of ten under the Crime and Disorder Act 1998.

Proceedings under part I of the Domestic Proceedings and Magistrates' Courts Act 1978 enable the magistrates to order financial provision, which may take the form of periodical payments and/or a lump sum (not exceeding £1,000), to be made by a spouse in cases of failure to provide reasonable maintenance for the other spouse or any child of the family. Part IV of the Family Law Act 1996 confers jurisdiction on the magistrates to make occupation orders and non-molestation orders in cases of domestic violence.

1.7.6.3.2.2 *Composition of family proceedings courts* Lay justices and district judges (magistrates' courts) cannot sit in a family proceedings court unless they are authorised to do so by the Lord Chancellor (Magistrates' Courts Act 1980, s. 67(2)–(3), as substituted by the Courts Act 2003).

Under s. 66(1) of the Magistrates' Courts Act 1980, as substituted by the Access to Justice Act 1999, a family proceedings court must be composed of:

(a) two or three lay justices; or

(b) a district judge (magistrates' courts) as chairman sitting with one or two lay justices; or

(c) a district judge (magistrates' courts) sitting alone if it is not practicable for him to sit with one or two lay justices.

Except where the court is composed of a district judge (magistrates' courts) sitting alone, it must, so far as practicable, include both a man and a woman (Magistrates' Courts Act 1980, s. 66(2), as substituted by the Access to Justice Act 1999).

1.7.6.3.2.3 Restrictions on attendance at family proceedings The hearing and determination of family proceedings must be kept separate from other business dealt with by the magistrates' court so far as this is consistent with efficiency (Magistrates' Courts Act 1980, s. 69(1)). When family proceedings (other than *adoption* proceedings) are being heard the only persons entitled to be present are:

(a) officers of the court;

(b) the parties, their legal representatives, witnesses, and other persons directly concerned in the case;

(c) representatives of newspapers or news agencies;

(d) any other person who the court may in its discretion permit to be present, but permission must not be refused to 'a person who appears to the court to have adequate grounds for attendance' (ibid., s. 69(2)).

In the case of *adoption* proceedings, only persons in categories (a) and (b) above are entitled to be present; those in categories (c) and (d) are excluded (ibid., s. 69(3)).

Although in *any* family proceedings witnesses are generally entitled to be present, the court still retains its discretion to exclude them until they are called for examination (ibid., s. 69(6)). When hearing *any* family proceedings which involve indecent evidence the court may exclude everyone during the taking of the evidence except officers of the court, the parties and their legal representatives, and other persons directly concerned in the case. But the magistrates can only do this if they think it is necessary in the interests of the administration of justice or of public decency (ibid., s. 69(4)).

1.7.6.3.2.4 Restrictions on newspaper and broadcast reports of family proceedings The only information that may be published is as follows (Magistrates' Courts Act 1980, s. 71(1A)):

(a) the names, addresses, and occupations of the parties and witnesses;

(b) the grounds of the application, and a concise statement of the charges, defences, and counter-charges in support of which evidence has been given;

(c) submissions on any points of law arising in the proceedings, and the decision of the court on the submissions;

(d) the decision of the court and any observations made by the court in giving it.

In the case of *adoption* proceedings the information that may lawfully be published is limited to categories (c) and (d) above; information in categories (a) and (b) must not be published (ibid., s. 71(2), as amended).

Breach of these reporting restrictions is a summary offence punishable by a fine. A prosecution may be started only with the consent of the Attorney-General. The restrictions do not forbid the publication of otherwise prohibited particulars in a newspaper or periodical of a technical nature which is bona fide intended for circulation among members of the legal or medical professions (ibid., s. 71(3)–(5)).

1.7.7. **Future of magistrates' courts**

It has traditionally been a criticism of appointments to the lay magistracy that they are political in character and made by the Lord Chancellor from a limited social class. This problem occupied a Royal Commission in 1910 and again in 1948 (*Report of the Royal Commission on Justices of the Peace*, Cd 5250, 1910; *Report of the Royal Commission on Justices of the Peace*, Cmd 7463, 1948). The position has undoubtedly improved following the efforts of the Labour government which took office in 1964. Some of the mystique surrounding magisterial appointments has disappeared and newspaper advertisements inviting people to put themselves forward for consideration can sometimes be seen.

The Lord Chancellor has listed six key qualities defining personal fitness and suitability for appointment: good character, understanding and communication, social awareness, maturity and sound temperament, sound judgment, and commitment and reliability.

Since the Employment Protection Act 1975, it has been theoretically less difficult for working people to secure time off work in order to perform their duties as magistrates. An employer is bound by statute (currently the Employment Rights Act 1996) to permit an employee who is a lay magistrate to take time off during working hours to enable him to perform his duties (Employment Rights Act 1996, s. 50(1)). How much time off the employee is entitled to will depend on what is reasonable in the circumstances, bearing in mind particularly how much time off the employee has already been allowed and the effect of the employee's absence on the employer's business (ibid., s. 50(4)). There is no express statutory duty on the employer to pay the employee during the time he takes off to act as a magistrate. An employee who is not allowed time off (or sufficient time off) can complain to an employment tribunal, which can award him such compensation as is just and equitable in the circumstances (ibid., s. 51(1), (3)–(4)).

It is a common criticism of lay justices that they do not know the law. But this is to misunderstand their role in the judicial process. The duty to act judicially and impartially with an appreciation of the basics of procedure and evidence is, in their case, far more important than the acquisition of a detailed knowledge of substantive law (see *Report of the Royal Commission on Justices of the Peace*, Cmd 7463, 1948, para 89). For this, they are entitled to ask for the advice of their professional clerk.

It is obviously preferable for the magistrates to get their decisions right in the first place, but when, inevitably, mistakes are made they can be rectified by an appeal to a higher court. There is an appeal to the Crown Court on points of law and fact, and a further appeal from there to the Queen's Bench Division of the High Court on a point of law only. Alternatively, on a point of law, an appeal can be taken direct from the magistrates' court to the High Court by way of 'case stated' (para 7.2.1.2). Any excess of jurisdiction, or violation of the principles of natural justice, on the part of the magistrates can be cured by an application for judicial review made to the Administrative Court, which may quash the offending decision.

Lack of uniformity in sentencing is a constant source of concern. Precise uniformity is impossible to achieve unless Parliament were to lay down fixed penalties for each offence. This would be most undesirable as it would exclude that flexibility which the court needs in order to arrive at the most appropriate penalty in the light of any mitigating circumstances. Variations in sentences for the same offence are inevitable as long as there is a system in which Parliament lays down a *maximum* but *no minimum* penalty. There are many thousands of lay magistrates exercising the same jurisdiction throughout England and Wales, and local and individual attitudes towards particular offences are bound to be reflected in the sentences imposed for them.

Magistrates can seek the advice of their clerk on the range of sentences available for particular offences. In addition, sentencing guidelines, which are intended to assist in achieving consistency in sentencing, are issued by the Sentencing Guidelines Council. Every court is under a duty to take into account any relevant guidelines when sentencing a defendant (Criminal Justice Act 2003, s. 172). If the court departs from any relevant guidelines issued by the Council it must give reasons for doing so (ibid., s. 174(2)). The court when passing sentence must in almost every case state in open court its reasons for deciding on that particular sentence, and must also explain to the defendant in ordinary language the effect of the sentence (ibid., s. 174(1)).

It is sometimes said that magistrates are too old. It is probably true that many candidates will not put themselves forward for appointment until the passing of those years which involve the heaviest family commitments. Some, no doubt, wait until retirement approaches. While problems may arise when advancing years are associated with mental infirmity, or lack of touch with the outside world, or political or social prejudice, these are failings which are capable of disturbing the judgment and behaviour of persons of all ages.

It is, in any event, the Lord Chancellor's general policy not to appoint anyone over the age of 65 (or under 27). At the age of 70, magistrates are removed to the 'supplemental list' (Courts Act 2003, s. 13(1)), and are then no longer qualified to act as justices of the peace (ibid., s. 12). A lay magistrate of any age may be removed to the supplemental list on the ground of incapacity (ibid., s. 13(5)). The ultimate sanction is removal from office by the Lord Chancellor, with the agreement of the Lord Chief Justice, for incompetence or neglect of duty (ibid., s. 11(2)), although this is a course resorted to only rarely.

Perhaps a greater potential problem is the extent to which a magistrates' court may be dominated by its clerk and/or the police. The justices' clerk is there to advise and assist, not to dictate or even to be seen taking an over-active part in the decision-making process (para 1.7.5.2 above). The expression 'police court' is an unfortunate misnomer, a relic of the days when the magistrates' court was often situated in the same building as, or next door to, the local police station. But the suspicion remains in some quarters that, on a conflict of evidence, the magistrates are too ready to accept the 'official', police version. Police officers are frequent witnesses and may come to be on reasonably familiar terms with the bench.

Some magistrates' courts exposed themselves to a certain amount of criticism during the miners' strike of 1984–85. The criticisms were made mainly in relation to the imposition of wide bail conditions and the appearance of 'conveyor-belt justice'. The dispensing of 'group justice' (i.e., putting into the dock together defendants who have been arrested on different occasions or at different places) was condemned by the Lord Chief Justice (*R* v *Mansfield Justices (ex parte Sharkey)* [1985] 1 All ER 193, DC, *per* Lord Lane CJ at p. 203).

In another context, the practice among some magistrates of claiming anonymity has been declared unlawful (*R* v *Felixstowe Justices (ex parte Leigh)* [1987] 1 All ER 551, DC). Open justice, it was said, demands that the names of those who sit in judgment should be known. It is an 'occupational hazard', which magistrates and other judges should accept, that they will occasionally be subjected to criticism, and even vilification, and that they will be pestered by persons who bear some grievance (ibid., *per* Watkins LJ at pp. 560–1).

Other problems have arisen over the extent to which a magistrates' court may sit in private or protect the identity of a defendant or witness. The Magistrates' Courts Act 1980, s. 121(4) (as substituted by the Courts Act 2003), provides that, subject to the provisions of any enactment to the contrary, a magistrates' court 'must sit in open court' if it is trying a summary offence, trying summarily an either way offence, imposing imprisonment, hearing a complaint, or holding an inquiry into the means of an offender.

That provision notwithstanding, it seems that the court has an *inherent* jurisdiction to exclude the public and the press, and to sit in private, if the administration of justice so requires. Such a course should, however, be regarded as exceptional, and there must be a compelling reason for deciding to sit in private (*R* v *Governor of Lewes Prison (ex parte Doyle)* [1917] 2 KB 254, DC; *R* v *Malvern Justices (ex parte Evans)* [1988] 1 All ER 371, DC, *per* Watkins LJ at p. 378).

Similarly, a magistrates' court has inherent jurisdiction to take steps to protect the identity of a defendant or witness, but only in those rare cases where disclosure of that identity would interfere with the administration of justice (*Attorney-General* v *Leveller Magazine Ltd* [1979] AC 440, HL; *R* v *Evesham Justices (ex parte McDonagh)* [1988] 1 All ER 371, DC, in which the magistrates' decision to prohibit publication of the defendant's home address was declared unlawful on the ground that, being concerned with a desire to spare the defendant from harassment from his former wife, it had nothing to do with the administration of justice).

In view of the fact that some 97 per cent of criminal cases begin and end in the magistrates' courts, it would be manifestly impracticable to abolish them without providing for their replacement by whatever system was considered to be more desirable. The appointment of large numbers of district judges (magistrates' courts) would bring the advantage of full-time professionalism and mitigate further the problem of lack of uniformity in sentencing. However, even assuming that a sufficient supply of persons was available with the requisite knowledge of the law and with other essential judicial attributes, the financial cost would be enormous. Such a move would also seriously undermine the involvement of the layman in the administration of justice.

In 2000, a research report commissioned by the government concluded that, in the main, the magistrates' courts, whether constituted by lay magistrates or district judges (magistrates' courts), work well and command general confidence (*The Judiciary in the Magistrates' Courts*, Home Office and Lord Chancellor's Department, December 2000).

Not surprisingly, the research found that district judges, because they have legal knowledge and experience and sit full-time and alone, are significantly faster and otherwise more efficient than lay magistrates, who need to confer with each other and often take the advice of their clerk. Evidence was found that district judges are more likely to remand defendants in custody and to sentence more heavily than their lay colleagues.

Auld LJ's Review, published in October 2001, relied heavily on this piece of research in recommending that lay magistrates and district judges should continue to exercise summary jurisdiction (*Review of the Criminal Courts of England and Wales*, Lord Chancellor's Department, 2001). Auld LJ reported that he had detected no 'hidden agenda' whereby district judges and justices' clerks would be allowed to squeeze lay magistrates out of the system.

Addressing the annual general meeting of the Magistrates' Association in October 2001, the then Lord Chancellor, Lord Irvine, once again reassured lay magistrates that the government remained committed to retaining them as 'a cornerstone' of the criminal justice system. This commitment was reflected in the White Paper of July 2002, *Justice For All* (Cm 5563, 2002), in which the government rejected Auld LJ's recommendation for a unified Criminal Court and announced its decision to keep the magistrates' courts as separate courts. New arrangements for the administration of the magistrates' courts are being implemented by the Courts Act 2003, which will repeal and replace the Justices of the Peace Act 1997.

The government's confidence in, and commitment to, the lay magistracy was most recently expressed by the Secretary of State for Constitutional Affairs and Lord Chancellor, Lord Falconer of Thoroton, in a speech to the annual conference of the Magistrates' Association in November 2005. He even went so far as to say, in response to representations received, that his department would stop using the term 'lay magistrate' because, he said, it does not properly reflect the qualities expected of, and the 'professional' task performed by, modern day magistrates. He also announced a relaxation of the rules governing the use of the initials 'JP' (justice of the peace) after their name in letterheads and elsewhere. It remains the case, however, that any attempt to use the initials to obtain a personal benefit could be construed as misconduct.

It may be proper to conclude that, in spite of the criticisms made of them and the uncertainty to which they have been subjected in recent years, magistrates' courts seem assured of an active part in the judicial process for a long time ahead if men and women of the right quality and motivation can be recruited to the bench.

1.8 Judicial misconduct, retirement, and immunity from suit

1.8.1 Judicial misconduct and retirement

Before the Revolutionary Settlement of 1701 the judges were dismissible at the King's pleasure. Now, as a direct result of the Act of Settlement (1701), judges of the House of Lords, Court of Appeal and High Court hold office during good behaviour, subject to a power of removal by Her Majesty on an address presented to her by both Houses of Parliament (Appellate Jurisdiction Act 1876, s. 6 (House of Lords); Supreme Court Act 1981, s. 11(3) (Court of Appeal and High Court)). Judges of the new Supreme

Court of the United Kingdom will hold office on the same conditions (Constitutional Reform Act 2005, s. 33).

The Lord Chancellor was an exception. As a government minister, he was (and remains) dismissible at any time by the Queen on the advice of the Prime Minister. In the days when he could act as a judge (prior to the Constitutional Reform Act 2005), he was expressly excluded from the security of tenure normally associated with superior court judges.

Save for the rare dismissal of a Lord Chancellor occasioned by political or other exigency, no judge of a superior court in England has been removed from office since the Act of Settlement of 1701. The Conservative Lord Chancellor, Viscount Kilmuir, resigned in 1962 at the instance of the Prime Minister, Mr Macmillan, during the major Cabinet reshuffle known as the 'night of the long knives'. Lord Macclesfield LC was dismissed for corruption in 1725, a fate which had befallen Francis Bacon (Lord St Albans) a century earlier. In 1865, Lord Westbury LC, while not actually dismissed, resigned after allegations of nepotism and abuse of patronage had been made against him. (For the cases of Bacon, Macclesfield and Westbury, see Gibb, *Judicial Corruption in the United Kingdom*, 1957, pp. 1–51.)

An *Irish* judge, Sir Jonah Barrington, was removed from office in 1830. He was found to have misappropriated £700 paid into the Admiralty Court of Ireland. Having fled to France, an address was presented by Parliament to the Crown and he was dismissed from office (see Gibb, op. cit., pp. 64–72).

English judges have generally shown themselves to be above corruption and, on the whole, impeccably behaved. It must be repeated, however, that even if it were possible to define with any precision 'good' or 'bad' behaviour, it is still difficult to get rid of an English judge of the High Court and above. The Lord Chancellor had no legal power to dismiss, although, as head of the judiciary, he could privately suggest resignation in the guise of voluntary early retirement. Thus, Hallett J was persuaded to resign by Viscount Kilmuir LC following serious criticism by the Court of Appeal in *Jones v National Coal Board* [1957] 2 QB 55, CA, that he interrupted counsel too much and asked too many questions. Harman J resigned in April 1998 following criticism by the Court of Appeal of his conduct of a case in which he had delayed giving judgment for 20 months after the conclusion of the hearing (see *The Times*, 14 February 1998 and *Goose* v *Wilson Sandford & Co.* (1998) *The Times*, 19 February, CA). What is needed for *compulsory* removal is nothing short of a joint address presented to the Queen by both Houses of Parliament.

The Lord Chancellor is no longer Head of the Judiciary. By s. 7 of the Constitutional Reform Act 2005, this office is now held by the Lord Chief Justice. Without limiting what he can do *informally*, the Lord Chief Justice's recently acquired statutory disciplinary powers include giving *formal* advice, or a *formal* warning or reprimand, to a judge in any court except the Supreme Court of the United Kingdom (Constitutional Reform Act 2005, ss. 108–109).

Lords of Appeal in Ordinary retire at the age of 70, but they can continue to hear appeals (as former Law Lords) in the House of Lords and the Judicial Committee of the Privy Council until reaching the age of 75. Thereafter, they are disqualified from hearing appeals and from membership of the Judicial Committee (Judicial Pensions and Retirement Act 1993).

The retiring age for a judge of the Court of Appeal or High Court is 70 (Supreme Court Act 1981, s. 11(2), as amended by the Judicial Pensions and Retirement Act 1993). A judge may, by giving written notice to the Lord Chancellor, resign his office at any time before reaching retirement age (Supreme Court Act 1981, s. 11(7)), although resignation other than for reasons of health or old-age is rare.

Lord Devlin caused controversy when he resigned as a Lord of Appeal in Ordinary in 1964 at the early age of 58. He explained many years later that he had found life as a Law Lord 'utterly boring' (see *The Times*, 11 June 1985; he died in 1992 at the age of 85). There was more controversy in 1970 when Fisher J resigned from the High Court bench in order to go into the City at a greatly enhanced salary. Lord Denning had already decided to retire as Master of the Rolls in 1982 at the age of 83 before it became clear that he could not carry on much longer following the discovery of a possible libel in his book, *What Next in the Law*. He tells his own story of his retirement in *The Closing Chapter* (1983, Section One). He had previously maintained that he would die in office, being quoted more than once as saying that he had 'all the Christian virtues except resignation'. He was able to sit on the bench to such an advanced age because he was appointed before a statutory retirement age was introduced for judges in 1959. He died in March 1999 at the age of 100.

There is provision for vacating the office of a Court of Appeal or High Court judge who is so ill as to be incapable of both the performance of his duties and resignation. The procedure requires the Lord Chancellor, acting on a medical certificate and with the concurrence of senior judges, to make a written declaration that the office has been vacated. The judge in question is then deemed to have resigned (Supreme Court Act 1981, s. 11(8) and (9)).

Judges of the new Supreme Court of the United Kingdom will retire at 70, and there will be similar provision for resignation and medical retirement (Constitutional Reform Act 2005, ss. 35–36).

Circuit judges and recorders do not enjoy the same security of tenure as their senior judicial colleagues. Although the Crown Court is part of the Supreme Court of England and Wales, circuit judges and recorders are not judges of that Court for the purposes of appointment, tenure of office and precedence (Supreme Court Act 1981, s. 151(4)). Their position remains governed by the Courts Act 1971.

Circuit judges can be removed by the Lord Chancellor, with the agreement of the Lord Chief Justice but without the intervention of Parliament, for incapacity or misbehaviour (Courts Act 1971, s. 17, as amended by the Constitutional Reform Act 2005). A circuit judge was removed in 1983 after pleading guilty to, and being fined on, charges of evading customs duty on cigarettes and whisky. In 1985, another was allegedly threatened with dismissal unless he abided by the so-called 'Kilmuir rules', which provided guidance to judges about participation in journalism and appearances on radio and television. (These 'rules' were abolished by a later Lord Chancellor, Lord Mackay LC, in 1987.) The same judge was 'seriously rebuked' in 1990 after describing the Lord Chief Justice as an 'ancient dinosaur living in the wrong age'.

A circuit judge was rebuked in 1992 for kissing a court usher in his chambers. The same judge was rebuked in 1997 for using a racially offensive expression during the course of a trial.

In 1999, a circuit judge was severely reprimanded, but not dismissed, following conviction for a drink-driving offence. Another circuit judge resigned in the same year. This judge, who had been suspended on full pay since November 1995, was tried in March 1998 for conspiracy to commit offences under the Theft Act 1978 while in practice as a solicitor. The jury failed to reach a verdict, and in October 1998 the Attorney General imposed a permanent stay on any further prosecution by entering a *nolle prosequi* ('unwilling to prosecute'). The judge's resignation pre-empted any decision to remove him from office. It was known that the Lord Chancellor was considering whether the evidence disclosed in the judge's trial would justify removing him for misbehaviour.

A circuit judge was reprimanded in 2002 for having fallen asleep during the closing stages of a rape trial.

Recorders, who are part-time judges, can be removed by the Lord Chancellor (with the agreement of the Lord Chief Justice) for incapacity, misbehaviour, or for failing to sit as often as agreed (Courts Act 1971, s. 21(6), as substituted by the Constitutional Reform Act 2005). Another sanction available against recorders is not to extend their appointment on the expiry of the specified period.

District judges can be removed from office by the Lord Chancellor, with the agreement of the Lord Chief Justice but without the intervention of Parliament, for incapacity or misbehaviour (County Courts Act 1984, s. 11, as amended by the Constitutional Reform Act 2005). A district judge of the Principal Registry of the Family Division who had been convicted of a drink-driving offence was 'severely reprimanded' (but not dismissed) in 1989. Another district judge was reprimanded for the same reason in 1993. The same judge was charged with a similar offence in 1997. However, before the case was heard it became apparent that he was no longer able to perform his judicial duties on account of alcohol dependence and depression, and he was allowed to retire from the bench on grounds of ill-health. At his subsequent trial, he was convicted and sent to prison for 28 days.

A district judge (magistrates' courts) is dismissible by the Lord Chancellor (with the agreement of the Lord Chief Justice) on the ground of incapacity or misbehaviour (Courts Act 2003, s. 22(5), as amended by the Constitutional Reform Act 2005).

Lay magistrates can be removed from office by the Lord Chancellor, with the agreement of the Lord Chief Justice, on behalf of the Queen for incapacity, misbehaviour, incompetence, or neglect of duty (ibid., s. 11(2), as amended by the Constitutional Reform Act 2005). Prior to the Courts Act 2003, they were removable by the Lord Chancellor without showing cause. One was removed in 1997 after the Lord Chancellor got to hear of a 'mooning' incident in which she was photographed exposing her bottom during an argument (see *The Times*, 18 September 1997, under the headline 'JP who exposed bottom loses her seat').

There is no official retiring age for lay magistrates but they are put on the 'supplemental list' at the age of 70 (Courts Act 2003, s. 13(1)), and thereupon they cease to be qualified to act as justices of the peace (ibid., s. 12). They can request to be put on the supplemental list before they reach the age of 70 (ibid., s. 13(4)). Alternatively, their names can be ordered to be placed there by the Lord Chancellor (with the agreement of the Lord Chief Justice) on the ground of incapacity (ibid., s. 13(5), as amended by the Constitutional Reform Act 2005).

1.8.2 Immunity from suit

Judicial immunity from suit means that a judge of a *superior* court cannot be sued for damages for acts done in his judicial capacity in good faith, even though he has acted mistakenly or in ignorance of his powers (see, e.g., *Sirros* v *Moore* [1975] QB 118, CA, in which a civil suit against a circuit judge for false imprisonment arising out of mistaken action in the Crown Court was unsuccessful). Immunity from suit enables a judge to perform his duties with complete independence and freedom from fear of repercussions. The judge is only liable in a civil action for damages if he was not acting judicially, knowing that he had no jurisdiction to do what he did.

Although he cannot succeed in a civil action for damages, the hapless litigant who is damaged in some way at the hands of the judiciary is not totally without a remedy. He may, for example, appeal against the judge's decision to a higher court. If he has been imprisoned he may try to secure his release by taking habeas corpus proceedings.

Prior to 1991, magistrates, as judges of *inferior* courts, had a more limited form of immunity from suit. They could, for example, be civilly liable for acts done without, or in excess of, jurisdiction even though sitting judicially at the time. The opportunity was taken in the Courts and Legal Services Act 1990 to amend the law so as to give magistrates the same immunity from suit as that enjoyed by judges of superior courts. The law on this matter is now contained in the Courts Act 2003.

A justice of the peace (whether a lay magistrate or a professional district judge (magistrates' courts)) is immune from civil action for anything done or omitted to be done while executing his duty as a magistrate *and* acting within his jurisdiction, even though he has acted mistakenly or in ignorance of his powers (Courts Act 2003, s. 31(1)). If he goes outside his jurisdiction while purporting to execute his duty as a magistrate he will be liable to be sued, but only if it is proved that he acted *in bad faith*, such as where he knew he had no jurisdiction to do what he did (ibid., s. 32(1)).

Furthermore, a justice of the peace is immune from paying *costs* in any proceedings in respect of any act or omission in the execution (or purported execution) of his duty as a justice (ibid., s. 34(1)). The immunity from the payment of costs does not extend to any proceedings in which the justice is himself being tried for an offence or in which it is proved that he acted in bad faith (ibid., s. 34(3)).

Justices' clerks and assistant clerks in the magistrates' courts enjoy immunities on similar terms (ibid., ss. 31(2), 32(2), and 34(2)–(3)).

It should be noted that other participants in the legal process may have immunity from suit in respect of things said or done by them as participants. This immunity extends to the parties to a case, witnesses, advocates and jurors. Thus, a witness or advocate cannot be sued in slander for anything in his remarks that may be defamatory. In *Munster* v *Lamb* (1883) 11 QBD 588, CA, for example, a solicitor was held not liable in slander for things said in court about a third party while representing a person on a criminal charge.

Immunity from suit does not, however, give the participants complete freedom to say and do as they please. A witness who tells lies may fall foul of the crime of perjury, and an advocate whose conduct falls below professionally accepted standards may be disciplined by his professional body. Moreover, every participant in court proceedings, with the exception of the judge, is subject to the law of contempt of court.

1.9 Access to justice

1.9.1 Introduction: background to the Access to Justice Act 1999

It is axiomatic that making available the elaborate structure of courts outlined in the preceding paragraphs is useless in itself unless those courts are accessible, regardless of financial status, to those with meritorious cases.

To many people, litigation, or even the obtaining of legal advice, is impossible unless the cost is subsidised in some way. The state has accepted responsibility in this area for several decades now, although the financial limits imposed on eligibility for assistance have never been over-generous. There is always a danger that justice may be denied to those who fall outside these limits and who, contrary to the presumption, cannot afford to proceed out of their own pockets.

State financial assistance is not available at all before the vast majority of tribunals. Many view this omission as an ironic denial of justice given that Parliament created many of these tribunals for the protection and assistance of ordinary citizens who are then denied the public funding with which to assert their rights adequately. Historically, the exclusion of tribunals has been justified on the ground that their procedures are straightforward enough to allow people to represent themselves.

A scheme for the provision of legal aid and advice to those who could not otherwise afford to employ a lawyer to assist with advice, or in the bringing or defending of proceedings, was introduced by the Legal Aid and Advice Act 1949. The Act of 1949 was repealed and replaced by the Legal Aid Act 1974, which, as subsequently amended, governed the scheme until 1 April 1989 when it, in turn, was repealed and replaced by the Legal Aid Act 1988.

It was a feature of the revised scheme introduced by the 1988 Act that the framework provided in the Act was to be supplemented by detailed regulations made by the Lord Chancellor, who had the very wide power to make such regulations as appeared to him necessary or desirable for giving effect to the Act or for preventing abuses of it. Responsibility for the administration of legal aid was transferred from the Law Society (where it had rested since 1949) to the Legal Aid Board established by the Act, and the Board was given power, acting under the direction of the Lord Chancellor, to provide advice, assistance and representation by means of contracts with, or grants or loans to, other persons or bodies, such as Citizens' Advice Bureaux, law centres and designated firms of solicitors.

In spite of the new arrangements, public expenditure on legal aid and advice continued to spiral out of control. The overall net cost rose from £650m in 1990 to £1,604m in 1998. Much of the blame for this was placed on lawyers, whose fees accounted for some 90 per cent of the legal aid bill. It also became evident that more money was being paid out each year to help fewer people.

From 1992, successive governments indicated a determination to bring expenditure within reasonable limits. The Conservative Lord Chancellor, Lord Mackay, attacked the problem of escalating costs in 1993 by making new, more stringent, regulations on financial eligibility, and by imposing standard fees for the remuneration of solicitors performing criminal legal aid work in the magistrates' courts. Further progress was made by his introduction of conditional fee agreements in 1995.

The funding of civil litigation on a contingency basis was suggested in the Green Paper, *Contingency Fees* (Cm 572, 1989) published in 1989. The US contingency fee model, under which the lawyer claims a percentage of a client's damages if he wins a case but receives no fee if he loses, was rejected in favour of a system which does not give the lawyer a stake in the client's damages but instead allows him, if the case is won, to receive his usual fee together with a success fee.

Following the Green Paper, conditional fee agreements were authorised by s. 58 of the Courts and Legal Services Act 1990, which came into force in July 1993 and provided for such agreements to be allowed in proceedings specified by the Lord Chancellor, who was also given power to prescribe the maximum permitted percentage success fee. Section 58 prohibited the use of conditional fee agreements in criminal proceedings and in most family proceedings.

In 1995, after a lengthy period of consultation with senior judges, the legal profession, consumer bodies and other interested organisations, the maximum percentage success fee was fixed at 100 per cent, and three types of proceedings were prescribed in which conditional fee agreements were to be available, namely, personal injury cases, insolvency proceedings, and cases before the European Court of Human Rights.

The Conservative government's further proposals to deal with the funding crisis in legal aid provision were overtaken when the Labour government elected in May 1997 decided to pursue its own, more radical, agenda under the guidance of Lord Irvine LC. This involved removing most civil cases from the scope of legal aid, the substitution of public funding by private funding in the form of conditional fee agreements, and the investigation of other alternative funding mechanisms for paying lawyers' bills, such as the extended use of legal expenses insurance and of membership litigation schemes operated by legal expenses insurers, trade unions, and motoring organisations.

A Consultation Paper of March 1998, *Access to Justice with Conditional Fees*, was followed by the extension of conditional fee agreements to all civil cases, except family cases, in July 1998. However, most of the government's agenda was implemented in the Access to Justice Act 1999, which effected the most significant changes in the funding of court proceedings since the legal aid and advice scheme was introduced in 1949. That scheme, as subsequently revised, was replaced by a Community Legal Service and a Criminal Defence Service authorised by the 1999 Act.

On 1 April 2000, the Legal Services Commission (created by the Access to Justice Act 1999, s. 1), the Community Legal Service (ibid., s. 4), and the Funding Code (ibid., s. 5) were established, and the scope of conditional fee agreements (ibid., s. 27) was extended again. The Criminal Defence Service was established on 2 April 2001.

The Legal Services Commission replaced the Legal Aid Board and is responsible for developing and maintaining the Community Legal Service and the Criminal Defence Service (ibid., s. 12). The Commission allocates funds to priority areas through contracts, and works in partnership with funders of legal services, including local authorities.

The Commission has set priorities at both national and local levels to ensure that resources are spent on the areas of greatest need (ibid., s. 6). The Commission has also introduced contracting for legal services (ibid.). Contracts are placed with 'quality-assured' providers only. These include firms of solicitors, law centres, advice bureaux, and other non-profit-making agencies. Community Legal Service Partnerships have been formed between local authorities, charities, and advice providers.

The previous civil and family legal aid budget was replaced by the Community Legal Service Fund, the size of which is fixed by the Secretary of State. The Fund has to be administered by the Commission in a way which obtains the best possible value-for-money for taxpayers. All applications for funding must meet the tests set out in the Funding Code (approved by Parliament), which replaced the former civil legal aid merits test.

The Code sets less stringent tests for higher priority cases and more stringent tests for lower priority cases, such as money claims. It is thus harder to get funding for lower priority cases (and impossible if the case is unmeritorious) than under the previous legal aid scheme. Strict cost-benefit tests must be satisfied to ensure that funding is justified, and the Code also allows the Commission to refuse funding where there are alternative ways of resolving a dispute or funding litigation.

Public expenditure on access to justice continues to be a source of anxiety for the government. In 2004, total net expenditure on publicly-funded legal services was £2,059m, broadly the same as the previous year's figure (*Judicial Statistics 2004*, Cm 6565, 2005, p. 144).

1.9.2 Services excluded from the Community Legal Service

Except as noted below, the Commission is not allowed to fund certain services as part of the Community Legal Service (Access to Justice Act 1999, s. 6(6) and sch. 2, as amended). Those excluded services are, *inter alia*:

(1) services consisting of the provision of help (beyond the provision of general information about the law and the legal system and the availability of legal services) in relation to—

  (a) allegations of personal injury or death, other than allegations relating to clinical negligence;

  (b) allegations of negligently caused damage to property;

  (c) conveyancing;

  (d) boundary disputes;

  (e) the making of wills;

  (f) trust law;

  (g) the creation of lasting powers of attorney under the Mental Capacity Act 2005;

  (h) the making of advance decisions under the Mental Capacity Act 2005;

  (i) defamation or malicious falsehood;

  (j) company or partnership law;

  (k) other matters arising out of the carrying on of a business, or

(l)   attending an interview conducted on behalf of the Secretary of State with a view to his reaching a decision on a claim for asylum.

(2)   Advocacy in any proceedings *except*—

(a)   proceedings in—

(i)   the House of Lords in its judicial capacity;

(ii)   the Supreme Court of the United Kingdom;

(iii)   the Judicial Committee of the Privy Council in the exercise of its jurisdiction under the Government of Wales Act 1998, the Scotland Act 1998, or the Northern Ireland Act 1998;

(iv)   the Court of Appeal;

(v)   the High Court;

(vi)   any county court;

(vii) the Employment Appeal Tribunal;

(viii) any mental health review tribunal;

(ix)   the Asylum and Immigration Tribunal;

(x)   the Special Immigration Appeals Commission, or

(xi)   the Proscribed Organisations Appeal Commission;

(b)   proceedings in the Crown Court—

(i)   for the variation or discharge of an order under s. 5 of the Protection from Harassment Act 1997;

(ii)   which relate to an order under s. 10 of the Crime and Disorder Act 1998;

(iii)   which relate to an order under para 6 of sch. 1 to the Anti-terrorism, Crime and Security Act 2001; or

(iv)   which relate to specified provisions of the Proceeds of Crime Act 2002;

(c)   proceedings in a magistrates' court—

(i)   for or in relation to an order under part I of the Domestic Proceedings and Magistrates' Courts Act 1978 (financial provision) or sch. 6 to the Civil Partnership Act 2004;

(ii)   under the Children Act 1989;

(iii)   under s. 30 of the Human Fertilisation and Embryology Act 1990;

(iv)   under s. 20 of the Child Support Act 1991;

(v)   under part IV of the Family Law Act 1996 (family homes and domestic violence);

(vi)   for the variation or discharge of an order under s. 5 of the Protection from Harassment Act 1997;

    (vii) under ss. 8 or 11 of the Crime and Disorder Act 1998;

  (viii) for an order or direction under paras 3, 5, 6, 9 or 10 of sch. 1 to the Anti-terrorism, Crime and Security Act 2001; or

    (ix) for an order or direction under ss. 295, 297, 298, 301 or 302 of the Proceeds of Crime Act 2002.

The Secretary of State is empowered to amend the list of excluded services by adding new services to it, or by omitting or varying any services (Access to Justice Act 1999, s. 6(7), as amended). He also has power to direct the Commission to fund any excluded service or, if requested by the Commission, to authorise its funding in an individual case (ibid., s. 6(8), as amended).

    Acting under the power conferred by s. 6(8), it was directed in April 2001 that the Commission could fund Investigative and Litigation Support in relation to non-clinical personal injury cases (which are otherwise excluded services). These are cases which involve high investigative or overall costs or a significantly wider public interest.

    The same direction authorised the Commission to fund *conveyancing* services as part of Legal Help where this is necessary to give effect to a court order made in proceedings which were funded by the Commission, or to give effect to an agreement reached, with the help of funded services, to settle or avoid family proceedings. In addition, the Commission was authorised to fund Legal Help in relation to *the making of wills* where the client is:

  (a) aged 70 or over; or

  (b) a disabled person within the meaning of the Disability Discrimination Act 1995; or

  (c) the parent of such a disabled person who wishes to provide for that person in a will; or

  (d) the parent of a minor who is living with the client but not with the other parent, and the client wishes to appoint a guardian for the minor in a will.

The Commission was also authorised to fund Legal Representation, including excluded services, in any of the following types of case:

  (a) judicial review proceedings, except where they arise out of the carrying on of the client's business unless the proceedings concern the serious wrongdoing, abuse of position or power, or significant breach of human rights by a public authority;

  (b) proceedings against public authorities concerning serious wrongdoing, abuse of position or power, or significant breach of human rights;

  (c) housing cases, except proceedings relating to business tenancies or otherwise arising out of the carrying on of the client's business unless possession of the client's home is in issue;

  (d) family proceedings;

  (e) professional negligence proceedings, except where the alleged negligence relates to services provided to the client's business;

(f) personal insolvency proceedings;

(g) public interest cases, except where the proceedings arise out of the carrying on of the client's business;

(h) hearings at which the liberty of the client is in issue.

By another direction made under s. 6(8) in April 2001, the Commission was authorised to fund Legal Help, Help at Court, and Legal Representation in all proceedings before the *Protection of Children Act Tribunal* (established by the Protection of Children Act 1999). Such Help and Representation were also extended to all proceedings (including cases arising out of the running of a business) before the *General and Special Commissioners of Income Tax* and the *VAT and Duties Tribunal,* provided that it is in the interests of justice for the client to be legally represented, *and* the proceedings concern penalties which the courts have declared to be criminal in European Court of Human Rights terms or where an appellant reasonably seeks to argue that the penalties under consideration by the tribunal are criminal in European Court of Human Rights terms.

Appeals from the Protection of Children Act Tribunal, the General and Special Commissioners of Income Tax, and the VAT and Duties Tribunal to the High Court, the Court of Appeal, and the House of Lords can also be funded in the same specified circumstances.

It will be seen that most *personal injury* cases are outside the scope of the Community Legal Service. They are funded instead by conditional fee agreements (para 1.9.4 below). Cases of clinical negligence, and other personal injury not caused by negligence (e.g., claims against the police for assault and battery), are within the scope of the Service.

The prohibition on the public funding of proceedings in *defamation* (i.e., libel and slander), which existed under the former legal aid scheme, is continued. Indeed, it has been extended to proceedings for the analogous, but separate, tort of *malicious falsehood* (thus overruling the public funding aspect of the decision in *Joyce* v *Sengupta* [1993] 1 All ER 897, CA).

Criminal proceedings in the magistrates' courts and Crown Court (which are excluded services) are funded under the Criminal Defence Service.

1.9.3 The Community Legal Service

1.9.3.1 Levels of service

The Funding Code (Part 1—Criteria) provides that the Commission will fund the following levels of service as part of the Community Legal Service:

(a) Legal Help;

(b) Help at Court;

(c) General Family Help;

(d) Legal Representation (either Investigative Help or Full Representation);

(e) Help with Mediation;

(f) Family Mediation;

(g) other services authorised by the Secretary of State from time to time.

'Legal Help' includes advice on how the law applies in particular circumstances.

'Help at Court' covers help and advocacy for a client in relation to a particular hearing without formally acting as a legal representative.

'General Family Help' is help in relation to a family dispute, including assistance in resolving the dispute by negotiation or otherwise.

'Legal Representation' covers legal representation for a party to proceedings (including litigation and advocacy services) or for a person who is contemplating taking proceedings.

'Help with Mediation' authorises help and advice in support of Family Mediation, including help in drawing up any agreement reached in mediation.

'Family Mediation' covers mediation of a family dispute, including an assessment of whether mediation appears suitable to the dispute.

1.9.3.2 Standard criteria: financial eligibility

The standard criteria are set out in the Funding Code (Part 1—Criteria).

An application for funding for any level of service *will* be refused, *inter alia*, if:

(a) it relates to law other than the law of England and Wales, except where this is permitted by or under s. 19 of the Access to Justice Act 1999; or

(b) the client is not financially eligible.

In addition, an application *may* be refused if it appears unreasonable to grant funding in the light of the client's conduct in connection with this or any other application or in connection with any proceedings.

The financial eligibility conditions are contained in the Community Legal Service (Financial) Regulations 2000 (SI 2000, No. 516), as amended, made under s. 7 of the Access to Justice Act 1999.

Under these Regulations, the following services, *inter alia*, are available without reference to the client's financial resources and without payment of any contribution:

(a) services consisting exclusively of the provision of general information about the law and legal system and the availability of legal services;

(b) initial legal advice consisting of such amount of Legal Help and Help at Court as is authorised under a contract to be provided without reference to the client's financial resources;

(c) Legal Representation in certain proceedings under the Children Act 1989;

(d) Legal Representation in proceedings before a mental health review tribunal under the Mental Health Act 1983, where the client's case or application to the tribunal is, or is to be, the subject of the proceedings.

A client is eligible for:

(a) Legal Representation before (i) the Asylum and Immigration Appeal Tribunal, and (ii) the High Court in respect of an application under s. 103A of the Nationality, Immigration and Asylum Act 2002 if his monthly disposable income does not exceed £649, and his disposable capital does not exceed £8,000;

(b) Legal Help, Help at Court, and Family Mediation if his monthly disposable income does not exceed £649 and his disposable capital does not exceed £8,000; a client who is eligible for Family Mediation is also eligible for Help with Mediation in relation to family mediation;

(c) Legal Representation in respect of family proceedings before a magistrates' court, other than proceedings under the Children Act 1989 or Part IV of the Family Law Act 1996, if his monthly disposable income does not exceed £649, and his disposable capital does not exceed £8,000;

(d) Legal Representation (other than as provided for in (a) and (c) above) and General Family Help if his disposable monthly income does not exceed £649, but he may be refused such services where:

(i) his disposable capital exceeds £8,000; and

(ii) it appears to the assessing authority that the probable cost of the funded services would not exceed the contribution payable by him.

A client whose gross monthly income exceeds £2,350 is not eligible for any funded services.

Those in receipt of certain state benefits, like income support, income-based job-seeker's allowance, or guarantee state pension credit, are automatically eligible since they are deemed to have disposable incomes and disposable capital which do not exceed the limits specified.

No contribution is payable in respect of services provided under (a) and (b), above. In the case of (c) and (d), above, a contribution is payable where the client's monthly disposable income exceeds £279 and his disposable capital exceeds £3,000. Contributions are payable to the assessing authority, which also has the responsibility for calculating them.

A service supplier must not provide any funded service to the client prior to the assessment of resources unless in accordance with Funding Code procedures, or they are authorised to do so by the Legal Services Commission in a contract. When making an application, the client must provide the assessing authority with the information necessary to enable it to determine whether he satisfies the financial eligibility conditions and to calculate his disposable income and disposable capital.

Disposable income is what is left of the client's weekly income after deducting income tax, national insurance contributions, prescribed allowances in respect of his spouse or partner, children and other dependants, and a reasonable amount in respect of regular maintenance payments for a former partner, a child, or a relative who is not a member of his household.

The client's *disposable capital* ignores the amount or value of the subject matter of the dispute to which the application relates, the first £100,000 of the value of his interest in his house, and (unless the circumstances are exceptional having regard in particular to the quantity or value of the items concerned), his furniture, clothing and tools of his trade. His disposable capital is what is left of his cash, savings, investments, and other valuable items after deducting prescribed allowances in respect of his spouse or partner, children and other dependants.

Gross income means total income from all sources before deductions are made.

In calculating the disposable income and disposable capital of the client, the resources of his partner are treated as his resources unless he has a contrary interest in the dispute in respect of which the application is made.

Except where eligibility is being assessed under (d), above, where the client is a child the resources of a parent, guardian or any other person who is responsible for maintaining him, or who usually contributes substantially to his maintenance, are treated as his resources, unless, having regard to all the circumstances, including the age and resources of the child and any conflict of interest, it appears inequitable to do so.

If it appears to the assessing authority that the person concerned, with intent to reduce the amount of his disposable income or disposable capital, has:

(a) directly or indirectly deprived himself of any resources; or

(b) transferred any resources to another person; or

(c) converted any part of his resources into resources which under the Regulations are to be wholly or partly disregarded,

the resources in question are still treated as part of his resources.

The client must inform the assessing authority immediately of any change in his financial circumstances of which he is, or should reasonably be, aware which has occurred since assessment of his resources, and which might affect the terms on which he was assessed as eligible to receive funded services.

1.9.3.3 Specific criteria

Under the Funding Code (Part 1—Criteria), applications for particular levels of service must satisfy both the standard criteria which apply to all levels of service (para 1.9.3.2 above) *and* certain specific criteria.

1.9.3.3.1 *Legal Help* Help may only be provided if:

(a) there is sufficient benefit to the client to justify work or further work being done; and

(b) it is reasonable for funding to come from the Community Legal Service Fund having regard to any other potential sources of funding.

1.9.3.3.2 *Help at Court* The two specific criteria for Legal Help (above) apply. In addition, Help at Court may only be provided if:

(a) advocacy is appropriate and will be of real benefit to the client; and

(b) Legal Representation is not more appropriate to the proceedings given their contested nature or the nature of the hearing.

1.9.3.3.3 *Legal Representation* An application *will* be refused if the case has been, or is likely to be, allocated to the small-claims track (see para 1.6.4.1 above).

An application *may* be refused if:

(a) alternative funding is available (e.g., through insurance, but not by means of a conditional fee agreement); or

(b) there are alternatives to litigation which should be tried first; or

(c) other levels of service appear more appropriate, such as Legal Help or Help at Court; or

(d) it appears unreasonable to fund representation.

In addition, *Full Representation* will be refused if the case is suitable for a conditional fee agreement and the client is likely to be able to avail himself of one. It will also be refused if prospects of success are 'unclear' (not fitting into any other category because further investigation is needed), 'borderline' (not better than 50 per cent) and the case has no significant wider public interest, or 'poor' (clearly less than 50 per cent so that the claim is likely to fail).

If the claim is primarily for damages and has no significant wider public interest, Full Representation will be refused unless:

(a) prospects of success are 'very good' (80 per cent or more) and likely damages exceed likely costs;

(b) prospects of success are 'good' (60–80 per cent) and likely damages exceed likely costs by a ratio of 2:1; or

(c) prospects of success are 'moderate' (50–60 per cent) and likely damages exceed likely costs by a ratio of 4:1.

If the claim is not primarily for damages and has no significant wider public interest, Full Representation will be refused unless the likely benefits from the proceedings justify the likely costs, so that a reasonable private-paying client would be prepared to litigate.

1.9.3.3.4 *General Family Help, Help with Mediation, Family Mediation, and Legal Representation in family proceedings*   General Family Help will be refused unless the benefits to the client justify the likely costs in the sense that a reasonable private paying client would be prepared to proceed at his own expense. It may be refused if either mediation or Legal Help is more appropriate.

Help with Mediation may only be provided if:

(a) the client is participating, or has reached an agreement or settlement, in family mediation and is in need of the services provided by Help with Mediation; and

(b) there is sufficient benefit to the client to justify work or further work being done.

It may be refused if Legal Help is more appropriate.

Legal Representation in family proceedings is available only as Full Representation; Investigative Help is not available. Moreover, the specific criterion relating to the small-claims track, and the specific criteria for Full Representation (para 1.9.3.3.3 above) concerning conditional fee agreements and prospects of success, do not apply to applications for Legal Representation in family proceedings.

In private law domestic violence cases in which an injunction, committal order, or other order protecting a person from harm, is sought, Legal Representation will be refused if:

(a) the prospects of obtaining the order are 'poor' (clearly less than 50 per cent so that the application is likely to fail); or

(b) the likely costs are out of proportion to the likely benefits.

In family proceedings concerning financial provision, Legal Representation *will* be refused if:

(a) in accordance with the Funding Code (Part 2—Procedures), the case must first be referred to a mediator in order to decide whether Mediation is suitable; or

(b) if prospects of success are—

 (i) 'borderline' (not better than 50 per cent) or 'unclear' (not fitting into any other category because further investigation is needed), unless the case has overwhelming importance for the client or a significant wider public interest; or

 (ii) 'poor' (clearly less than 50 per cent so that the application is likely to fail); or

(c) the likely benefits for the client do not justify the likely costs.

It *may* be refused if:

(a) mediation is more appropriate; or

(b) reasonable attempts have not been made to resolve the dispute (by negotiation or otherwise) without recourse to legal proceedings; or

(c) it is reasonable for the proceedings to be funded privately.

1.9.3.3.5 *Special cases* The Funding Code (Part 1—Criteria) lays down additional criteria applicable in cases of emergency and in special cases, such as very expensive cases, judicial review and other claims against public authorities, clinical negligence, housing, mental health, and immigration.

1.9.3.4 The statutory charge

It may happen that the amount of costs (if any) recovered for a successfully funded client, together with his own contribution (if any), is not sufficient to reimburse the Community Legal Service Fund in respect of the sums actually expended by the Legal Services Commission in funding the services for the client. In this event, the Commission has a first charge (known as 'the statutory charge') on any property recovered or preserved by him in any proceedings or in any compromise or settlement of any dispute (Access to Justice Act 1999, s. 10(7)). This means, for example, that any damages awarded in proceedings may be swallowed up by the Community Legal Service Fund before they are ever received by the funded claimant.

The Community Legal Service (Financial) Regulations 2000 (SI 2000, No. 516), as amended, lay down detailed rules about the statutory charge. Maintenance payments and (unless the circumstances are exceptional having regard in particular to the quantity or value of the items concerned), the client's furniture, clothing, and tools of his trade are exempt from the charge. However, the former exemption covering the first £3,000 of any lump sum, or of the value of any property (e.g., the matrimonial home), recovered by the client in certain family proceedings, was removed in April 2005. There are provisions for postponement of the charge, the payment of interest, and for substitution of the charged property.

In the case of Legal Help or Help at Court, the statutory charge may sometimes be in favour of the supplier of the service. The supplier, with the authority of the Commission, can waive either all or part of the amount of the statutory charge where its enforcement would cause grave hardship or distress to the client or would be unreasonably difficult because of the nature of the property.

The Commission itself can, if it considers it equitable to do so, waive some or all of the amount of the statutory charge in cases of 'wider public interest', which means the potential of proceedings to produce real benefits for individuals other than the client (apart from any general benefits which normally flow from proceedings of the type in question).

1.9.3.5 Costs against a funded client

By virtue of s. 11(1) of the Access to Justice Act 1999, except in prescribed circumstances, costs ordered against a client in relation to proceedings funded for him must not exceed an amount which is reasonable for him to pay having regard to all the circumstances, including:

(a) the financial resources of all the parties; and

(b) their conduct in connection with the dispute.

Detailed rules for assessing financial resources are contained in the Community Legal Service (Costs) Regulations 2000 (SI 2000, No. 441), as amended.

The limit, imposed by s. 11(1), on costs awardable against a funded client is referred to as 'cost protection' in the Community Legal Service (Cost Protection) Regulations 2000 (SI 2000, No. 824), as amended, which prescribe the circumstances in which cost protection does not apply.

1.9.3.6 Costs of a successful non-funded party

Section 11(4) of the Access to Justice Act 1999, and regulations made thereunder in the Community Legal Service (Cost Protection) Regulations 2000, above, provide for the recovery of costs against the Legal Services Commission in certain circumstances.

A non-funded party appearing in proceedings against a funded party may, at the discretion of the court, be awarded all or part of his costs from the Community Legal Service Fund (other than any costs that the client is required to pay under a s. 11(1) costs order), provided that:

(a) the non-funded party wins the case; and

(b) cost protection applies; and

(c) as regards costs incurred in a court of first instance, the proceedings were begun by the funded party, the non-funded party is an individual, and the court is satisfied that the non-funded party will suffer financial hardship unless the costs order is made; and

(d) in any case, the court is satisfied that it is just and equitable in the circumstances that the costs should be paid out of public funds (for what is 'just and equitable' in this context, see *Hanning* v *Maitland (No. 2)* [1970] 1 QB 580, CA).

In determining whether conditions (c) and (d) are satisfied, the court is to have regard to the resources of the non-funded party and of his partner. However, the partner's resources are not taken into account where the partner has a contrary interest in the funded proceedings.

In relation to proceedings in a county court, the High Court, or the Court of Appeal, 'the court' responsible for awarding costs to a non-funded party means a costs judge or district judge. In proceedings in a magistrates' court, it is a single justice or the justices' clerk. In the Employment Appeal Tribunal, it means the registrar of the Tribunal. In the House of Lords, it is the Clerk to the Parliaments.

1.9.4 Conditional fee agreements

Section 58(2) of the Courts and Legal Services Act 1990 (as substituted by the Access to Justice Act 1999) defines a conditional fee agreement as:

an agreement with a person providing advocacy or litigation services which provides for his fees and expenses, or any part of them, to be payable only in specified circumstances.

If such an agreement satisfies all the statutory conditions, it is not unenforceable by reason only of its being a conditional fee agreement (ibid., s. 58(1)). The agreement must be in writing and must state the percentage success fee, if any (ibid., s. 58(3), (4)).

In essence, a conditional fee agreement is one by which a legal representative supplying advocacy or litigation services agrees with his client that fees and expenses are to be payable on a 'no win, no fee' basis. The legal representative usually agrees that he will be paid nothing if the case is lost, but that he will take a 'success fee' (or percentage 'uplift' on his normal fee) if the case is won. This arrangement has the advantage for the client that he does not have to worry about paying his own legal representative if he loses the case. He may, however, still be liable to pay his successful opponent's costs, although this possibility can be insured against.

Legal expenses insurance policies are either 'before the event' policies (typically contained in existing buildings, contents, and motor insurance policies, or provided as a benefit of trade union membership) or 'after the event' policies (where insurance is only taken out after a claim has arisen to cover the risk of losing a case and having to pay the opponent's costs). After the event policies are often arranged by solicitors and claims management companies. The premium payable is based on the risks involved in taking on the case and the likely costs.

Criminal proceedings (with one exception) and family proceedings are excluded from the scope of enforceable conditional fee agreements (ibid., s. 58A, as inserted by the Access to Justice Act 1999). Criminal proceedings under s. 82 of the Environmental Protection Act 1990 for an order that a statutory nuisance be put right are included. However, the agreement in such a case cannot provide for a success fee (ibid., and Conditional Fee Agreements Order 2000 (SI 2000, No. 823)).

All civil proceedings, except family proceedings, are thus within the scope of enforceable conditional fee agreements which may provide for a success fee. The maximum percentage success fee is fixed at 100 per cent of the normal fee (Conditional Fee Agreements Order 2000).

In addition to s. 58 of the Courts and Legal Services Act 1990, conditional fee agreements were governed for several years by statutory regulations made thereunder. However, responses to two consultation papers (*Simplifying Conditional Fee Agreements*, Department for Constitutional Affairs, June 2003, and *Making Simple Conditional Fee Agreements A Reality*, Department for Constitutional Affairs, June 2004) convinced the government that the regulations were ineffective, unnecessary, and too complex, and that the arrangements for conditional fee agreements should be simplified. It was noted that the provisions governing client care and the content of agreements were already covered by the Professional Rules of the Law Society.

As a result, and with a view to providing simplification, greater transparency, and better protection for clients, the existing regulations were revoked and not replaced. Instead, any conditional fee agreement and collective conditional fee agreement (below) entered into on or after 1 November 2005 is now governed by the Law Society's Solicitors' Costs Information and Client Care Code 1999, as amended. Sections 58 and 58A of the Courts and Legal Services Act 1990 (above), and s. 29 of the Access to Justice Act 1999 (below), remain in force.

The Code specifies that a client should be given 'the best information possible' about likely overall costs and that costs information should not be inaccurate or misleading. The solicitor is urged to explain the circumstances in which the client may be liable for costs (both his own and the other party's), the client's right to have costs assessed if the solicitor intends to seek payment of costs from the client, and any interest the solicitor may have in recommending a particular insurance policy or other funding.

The Law Society's model conditional fee agreement sets out what is, and what is not, covered by the agreement, explains liability for costs, specifies details of the success fee, and provides for signature by the solicitor and the client. The model agreement is accompanied by model information for clients.

Subject to rules of court, the successful party's success fee is recoverable from the losing party (Courts and Legal Services Act 1990, s. 58A, as inserted by the Access to Justice Act 1999), as is the premium paid under an insurance policy (Access to Justice Act 1999, s. 29; see also CPR, r. 44.3A, and *Practice Direction* to CPR, Part 44). The House of Lords has held that the success fee is recoverable even though the successful party could have afforded to litigate without the assistance of a conditional fee agreement (*Campbell* v *MGN Ltd* (*No. 2*) [2005] 1 WLR 3394, HL, which was concerned with a success fee of almost £280,000).

The aim of the conditional fee legislation is to widen access to justice, particularly for those who fall outside the financial eligibility limits for public funding and who are deterred from going to court by the costs of litigation. There is, however, no means-test, and it follows that conditional fee agreements are available to clients who are wealthy enough not to need them (*Campbell* v *MGN Ltd* (*No. 2*), above). Under a conditional fee agreement, the costs risk is transferred from the client (or the tax-payer) to the legal representative, who is given an incentive to take on cases by means of the success fee which is intended to be proportionate to the risk of failure. In practice, it seems that the average percentage success fee charged by legal representatives is 43 per cent. In addition, it appears that they voluntarily limit the success fee to no more than 25 per cent of any damages recovered by the client.

Collective conditional fee agreements are available for the purpose of funding class actions. A collective conditional fee agreement is a conditional fee agreement with a legal representative which does not refer to specific proceedings, but provides for fees to be payable on a common basis in relation to a class of proceedings, whether or not the person liable to pay the fees under the agreement is the client of the legal representative.

1.9.5 The Criminal Defence Service

1.9.5.1 Introduction

The Criminal Defence Service was established on 2 April 2001. It replaced the former criminal legal aid scheme, and includes within its scope the public funding of advice and assistance and of representation in criminal and contempt proceedings. The Service is governed by the Access to Justice Act 1999, as amended, and regulations made thereunder.

Section 12(1) of the 1999 Act requires the Legal Services Commission to 'establish, maintain and develop' a Criminal Defence Service 'for the purpose of securing that individuals involved in criminal investigations or criminal proceedings have access to such advice, assistance and representation as the interests of justice require'.

1.9.5.2 Meaning of 'criminal proceedings'

'Criminal proceedings' include criminal trials, appeals and sentencing hearings, proceedings for dealing with a person under the Extradition Act 2003, binding over proceedings, appeals on behalf of a convicted person who has died, proceedings for contempt committed in the face of any court (ibid., s. 12(2)(a)–(f)), and such other proceedings before any court or other body as may be prescribed by the Secretary of State (ibid., s. 12(2)(g)).

In the Criminal Defence Service (General) (No. 2) Regulations 2001 (SI 2001, No. 1437), as amended, the following proceedings are among those prescribed as 'criminal proceedings' for the purposes of s. 12(2)(g):

(a) civil proceedings in a magistrates' court arising from failure to pay a sum due or to obey an order of that court where such failure carries the risk of imprisonment;

(b) proceedings under ss. 1, 1D, and 4 of the Crime and Disorder Act 1998 relating to anti-social behaviour orders;

(c) proceedings under ss. 1G and 1H of the 1998 Act relating to intervention orders, in which an application for an anti-social behaviour order has been made;

(d) proceedings under s. 8(1)(b) of the 1998 Act relating to parenting orders made where an anti-social behaviour order or a sex offender order is made in respect of a child;

(e) proceedings under s. 8(1)(c) of the 1998 Act relating to parenting orders made on the conviction of a child;

(f) proceedings under s. 9(5) of the 1998 Act to discharge or vary a parenting order made as mentioned in (d) or (e) above;

(g) proceedings under s. 10 of the 1998 Act to appeal against a parenting order made as mentioned in (d) or (e) above;

(h) proceedings under ss. 14B, 14D, 14G, 14H, 21B, and 21D of the Football Spectators Act 1989 (banning orders and references to a court);

(i) proceedings under s. 137 of the Financial Services and Markets Act 2000 to appeal against a decision of the Financial Services and Markets Tribunal;

(j) proceedings under ss. 2, 5, and 6 of the Anti-social Behaviour Act 2003 relating to closure orders; and

(k) proceedings under s. 5A of the Protection from Harassment Act 1997 relating to restraining orders on acquittal.

The proceedings under (a) to (k) above would normally be classified as *civil* proceedings. Prescribing them as *criminal* means, first, that individuals involved in them have a right to representation provided that the interests of justice test set out in sch. 3 to the Access to Justice Act 1999 is satisfied, and, secondly, that representation in the proceedings cannot be funded as part of the Community Legal Service. In practice, it will probably be rare for representation to be granted in these proceedings since the fact that advocacy assistance is also available for them will usually indicate that it is not in the interests of justice for a representation order to be made.

In addition, the following proceedings are to be regarded as *incidental to* the criminal proceedings from which they arise:

(a) in the Crown Court, following committal for sentence by a magistrates' court;

(b) to quash an acquittal under the Criminal Procedure and Investigations Act 1996; and

(c) for confiscation and forfeiture in connection with criminal proceedings under RSC Ord. 115 (contained in sch. 1 to the Civil Procedure Rules 1998).

Because these proceedings are deemed to be incidental to existing criminal proceedings, no separate application for representation is required. They are covered by any existing magistrates' court or Crown Court representation order.

Applications for judicial review or habeas corpus relating to any criminal investigations or proceedings are *not* to be regarded as incidental to such criminal investigations or proceedings. It follows that an application to have these proceedings publicly funded would need to be made under the Community Legal Service and not the Criminal Defence Service.

1.9.5.3 Suppliers of services

The Legal Services Commission is empowered to secure advice, assistance and representation services through contracts with legal representatives in private practice. Alternatively, these services are available through salaried public defenders employed directly by the Commission or by non-profit-making organisations (Access to Justice Act 1999, ss. 12–14).

All contractors have to meet quality-assurance standards. Salaried defenders employed by the Commission, or by any bodies established by the Commission, are subject to a code of conduct prepared by the Commission. The code includes duties to avoid discrimination, to protect the interests of the individuals for whom services are provided, to the court, to avoid conflicts of interest, of confidentiality, and to act in accordance with professional rules (ibid., s. 16, and Criminal Defence Service Code of Conduct 2001).

1.9.5.4 Funding

The overall funding of the Criminal Defence Service is the responsibility of the Secretary of State. The Legal Services Commission is responsible for the day-to-day funding of the various services available. The Secretary of State is required to pay to the Commission the cost of advice, assistance and representation funded by the Commission (Access to Justice Act 1999, s. 18(1), as amended). Thus, the Criminal Defence Service (like the criminal legal aid scheme before it, but unlike the Community Legal Service Fund) is demand-led. For its part, the Commission must aim to obtain the best possible value for money when funding services as part of the Criminal Defence Service (ibid., s. 18(3)).

Solicitors acting under the duty solicitor schemes (mentioned in para 1.9.6.1 below) are remunerated by the Legal Services Commission in accordance with the Criminal Defence Service (Funding) Order 2001 (SI 2001, No. 855), as amended, and the General Criminal Contract Remuneration Rates, as amended, which cover preparation, travel and waiting, routine letters and telephone calls, and (where appropriate) advocacy. Under the police station scheme, they are entitled to payment for stand-by duty not resulting in advice or assistance, and there are enhanced rates of payment for work done during unsocial hours.

Remuneration for most representation in criminal proceedings in the Crown Court, Court of Appeal, and House of Lords (para 1.9.7.3 below) is paid in accordance with detailed rules laid down in the Criminal Defence Service (Funding) Order 2001 (SI 2001, No. 855), as amended.

In the following paragraphs, the various services available under the umbrella of the Criminal Defence Service are considered, namely, advice and assistance, representation in criminal proceedings, and representation in contempt proceedings.

1.9.6 Advice and assistance

1.9.6.1 Extent

Advice and assistance to persons appearing before magistrates' courts used to be provided under voluntary duty solicitor schemes. In 1982, these schemes were put on a statutory footing following a recommendation of the Royal Commission on Legal Services (*Report*, Cmnd 7648, 1979). They became fully operational during 1984 and 1985. From 1 January 1986, the duty solicitor schemes were extended to cover the provision of advice and assistance to persons who had been arrested and held in custody in police stations, and persons who were voluntarily at police stations helping the police with their enquiries. The schemes have been retained as part of the Criminal Defence Service and are operated under the control of the Legal Services Commission (see below).

By s. 13(1) of the Access to Justice Act 1999, as amended by the Criminal Defence Service (Advice and Assistance) Act 2001, the Commission is required to provide such advice and assistance (including advocacy assistance) as it considers appropriate:

(a) for individuals who are arrested and held in custody at a police station or other premises; and

(b) in prescribed circumstances, for individuals who:

   (i) do not fall within (a) above but who are involved in investigations which may lead to criminal proceedings;

   (ii) are before a court or other body in such proceedings; or

   (iii) have been the subject of such proceedings.

The full extent of the advice and assistance available under the Criminal Defence Service is set out in the Criminal Defence Service (General) (No. 2) Regulations 2001 (SI 2001, No. 1437), as amended, by which the Commission is under a duty to fund appropriate advice and assistance (including advocacy assistance) for any individual who:

(a) is the subject of an investigation which may lead to criminal proceedings;

(b) is the subject of criminal proceedings;

(c) requires advice and assistance regarding his appeal or potential appeal against the outcome of any criminal proceedings or an application to vary a sentence;

(d) requires advice and assistance regarding his sentence;

(e) requires advice and assistance regarding his application or potential application to the Criminal Cases Review Commission;

(f) requires advice and assistance regarding his treatment or discipline in prison (other than in respect of actual or contemplated proceedings regarding personal injury, death or damage to property);

(g) is the subject of proceedings before the Parole Board;

(h) requires advice and assistance regarding representations to the Home Office in relation to a mandatory life sentence or other parole review;

(i) is a witness in criminal proceedings and requires advice regarding self-incrimination;

(j) is a volunteer; or

(k) is detained under sch. 7 to the Terrorism Act 2000.

A 'volunteer' is defined as a person who, for the purpose of assisting with an investigation, attends voluntarily at a police station or a customs office, or at any other place where a constable or customs officer is present, or accompanies a constable or customs officer to a police station or a customs office or any such other place, without having been arrested.

'Advice and assistance' covers help from a solicitor in the form of general advice on matters such as bail and whether to plead guilty or not guilty, writing letters, negotiating, getting a barrister's opinion, and preparing a written case. It does not include representation in court. For that, advocacy assistance (see below) or a representation order (para 1.9.7 below) is required.

'Advocacy assistance' covers help from a solicitor in preparing a case, and initial representation (including a first application for bail) in certain proceedings in the magistrates' court and the Crown Court. It also extends to representation for prisoners facing disciplinary charges before the prison authorities, and for those serving discretionary and automatic life sentences, and those detained at Her Majesty's Pleasure, whose cases are referred to the Parole Board. Representation for those who have failed to pay a fine or obey an order of the magistrates' court and are at risk of imprisonment is also covered.

In compliance with its statutory duty to provide advice and assistance, the Legal Services Commission, acting under s. 3(4) of the Access to Justice Act 1999 (which authorises it to 'make such arrangements as it considers appropriate for the discharge of its functions, including the delegation of any of its functions'), has made the Duty Solicitor Arrangements 2001, as amended.

These Arrangements provide for two duty solicitor schemes—the police station duty solicitor scheme and the magistrates' court duty solicitor scheme—with the primary objective of ensuring that individuals requiring advice and assistance (including advocacy assistance) at a police station or a magistrates' court, and who choose not to, or are unable to, obtain such help from their own solicitor, can have access to the services of a duty solicitor.

Under the Duty Solicitor Arrangements, England and Wales are divided into regions and, within each region, local schemes are in operation. Duty solicitors are normally expected to serve on both schemes. There are detailed rules for membership and continued membership of the schemes. The Commission has power to suspend or remove from a scheme duty solicitors who, *inter alia*, unreasonably fail to attend a police station, send a representative to the police station when they should have attended personally, unreasonably fail to carry out a duty or fail to comply with the Arrangements, face an outstanding criminal charge or have been convicted of a criminal offence, or do not demonstrate the necessary level of competence.

1.9.6.2 Eligibility

There is no merits test for advice and assistance.

Under the Criminal Defence Service (General) (No. 2) Regulations 2001 (SI 2001, No. 1437), as amended, the following advice and assistance is available free of charge regardless of the individual's financial resources and no contribution is payable:

(a) all advice and assistance provided to an individual who is arrested and held in custody at a police station or other premises;

(b) all advocacy assistance before a magistrates' court or the Crown Court;

(c) all advice and assistance provided by a court duty solicitor in accordance with his contract with the Commission;

(d) all advice and assistance provided to a volunteer during his period of voluntary attendance;

(e) all advice and assistance provided to an individual being interviewed in connection with a serious service offence; and

(f) all advice and assistance provided to an individual who is the subject of an identification procedure carried out by means of video recordings.

A 'serious service offence' ((e) above) is defined as an offence under the Army Act 1955, the Air Force Act 1955, or the Naval Discipline Act 1957 which cannot be dealt with summarily.

There is a means test for cases not falling within (a) to (f) above. An individual is only eligible for advocacy assistance in connection with his treatment or discipline in prison (other than in respect of actual or contemplated proceedings regarding personal injury, death or damage to property), or in connection with proceedings before the Parole Board, if his weekly disposable income does not exceed £194 and his disposable capital does not exceed £3,000.

In all other cases not falling within (a) to (e) above, an individual is eligible for advice and assistance if his weekly disposable income does not exceed £92 and his disposable capital does not exceed £1,000. If an individual is eligible for advice and assistance, no contribution is payable.

Those individuals who receive specified state benefits, such as income support, income-based jobseeker's allowance, or guarantee state pension credit, are automatically eligible since they are deemed to have disposable incomes and disposable capital which do not exceed the prescribed limits.

Disposable income and disposable capital are assessed by the solicitor supplying the service in accordance with sch. 1 to the Criminal Defence Service (General) (No. 2) Regulations 2001 (SI 2001, No. 1437), as amended. The assessment rules with regard to disposable income, disposable capital, the treatment of the resources of a partner or of the parent or guardian of a child, and the deliberate dissipation of resources are similar to those applicable under the Community Legal Service (para 1.9.3.2 above).

1.9.7 Representation

1.9.7.1 Method of application

Applications for representation are normally made to the court itself rather than to the Legal Services Commission (except as noted below). The rules governing applications for representation orders are contained in the Criminal Defence Service (General) (No. 2) Regulations 2001 (SI 2001, No. 1437), as amended.

An application for a representation order in respect of proceedings in a magistrates' court or the Crown Court is made either in writing on a prescribed form to the court or the appropriate officer, or orally to the court, which may, if it wishes, refer it to the appropriate officer for determination. The 'appropriate officer' is the justices' clerk in a magistrates' court or the court manager in the Crown Court. The term 'appropriate officer' includes an officer designated by him to act on his behalf.

It is usual for the appropriate officer to be entrusted with making decisions about representation. The appropriate officer or the court may grant or refuse the first application. If the application is refused, the appropriate officer must provide the applicant with written reasons for the refusal and details of the appeal process (para 1.9.7.2 below).

Representation for proceedings in the Crown Court may, alternatively, be granted by the magistrates' court which convicted the defendant, or sent him for trial, or committed him for sentence.

The vast majority of defendants appearing at the Crown Court for trial or for sentence are represented at public expense. In 2004, the proportions were 94.5 per cent and 79 per cent, respectively (*Judicial Statistics 2004*, Cm 6565, 2005, p. 143).

An application for a representation order in respect of proceedings in the Court of Appeal or the House of Lords can be made orally or in writing to the Court of Appeal, or a judge of the court, or in writing to the appropriate officer. The 'appropriate officer' is the registrar of criminal appeals in the case of the criminal division of the Court of Appeal or the head of the Civil Appeals Office in the case of the civil division (or an officer designated by either to act on his behalf).

Where an application is made to the court, it can refer it to a judge or the appropriate officer for determination. Where an application is made to a judge, he can refer it to the appropriate officer.

The appropriate officer cannot refuse the application; he must either grant it or refer it to the court or a judge of the court.

The House of Lords cannot grant representation orders in any proceedings.

An application for a representation order in respect of some proceedings must be made to the Legal Services Commission and not to the court. They comprise the proceedings which have been prescribed as criminal for the purposes of s. 12(2)(g) of the Access to Justice Act 1999 (see para 1.9.5.2 above), with the exception of proceedings under s. 137 of the Financial Services and Markets Act 2000. If the application is refused, the Commission must provide the applicant with written reasons for the refusal and details of the appeal process (para 1.9.7.2 below).

1.9.7.2 Appeals against refusals to grant representation orders

The rules governing appeals against refusals are made under the authority of the Access to Justice Act 1999, s. 14 and sch. 3. They are contained in the Criminal Defence Service (Representation Order Appeals) Regulations 2001 (SI 2001, No. 1168), as amended. Appeals are made to the body which refused the application, and are made by way of a renewed application.

If representation is refused for proceedings in a magistrates' court or the Crown Court, whether by the appropriate officer or by the court itself, the applicant has a right to renew his application either orally or in writing to the same court, or in writing to the appropriate officer. If the application is renewed to the appropriate officer, he cannot refuse it. He must either grant it or refer it to the court (or to a district judge (magistrates' court), or a single justice of the peace, in the case of a magistrates' court). If the application is renewed to the court, the court can grant or refuse it.

If representation is refused by the Court of Appeal or the appropriate officer, the applicant can make a renewed application either orally or in writing to the court, or in writing to the appropriate officer. If made to the appropriate officer, he cannot refuse it. He must either grant the order or refer the renewed application. He can refer it either to a judge of the court, who can grant but not refuse the order, or to the court, which can grant the order or refuse the application. The court must give reasons for the refusal of any application. The reasons must be given in writing if the application was made in writing.

If representation is refused by the Legal Services Commission, the applicant can make a renewed application in writing to the Funding Review Committee of the Commission, which can grant the order or refuse the application. The Commission must give written reasons for the refusal of any application.

1.9.7.3 Extent

In the Crown Court, a representation order normally extends to preparation of the defendant's case by a solicitor and representation in court by an advocate. 'Advocate' is defined in the Criminal Defence Service (General) (No. 2) Regulations 2001 (SI 2001, No. 1437), as amended, as 'a barrister, or a solicitor who has obtained a higher courts advocacy qualification in accordance with regulations and rules of conduct of the Law Society'.

In proceedings before a magistrates' court, the order can only cover representation by an advocate in the case of:

(a) an indictable offence, including an either-way offence; or

(b) proceedings under s. 9 of, or para 6 of sch. 1 to, the Extradition Act 1989

where the court is of the opinion that, because of circumstances which make the proceedings unusually grave or difficult, representation by both a solicitor and an advocate would be desirable.

The court can grant an order for representation by an advocate alone:

(a) in any proceedings referred to in s. 12(2)(f) of the Access to Justice Act 1999 (proceedings for contempt committed in the face of any court);

(b) for an appeal to the Court of Appeal or the Courts-Martial Appeal Court; or

(c) in cases of urgency where it appears to the court that there is no time to instruct a solicitor:

(i) for an appeal to the Crown Court; or

(ii) in proceedings in which a person is committed to, or appears before, the Crown Court for trial or sentence.

There are detailed rules about the use of Queen's Counsel, the use of more than one advocate, and the withdrawal of representation by the court or the Legal Services Commission. If a representation order is withdrawn, the defendant can appeal against the withdrawal on one occasion to the body which withdrew the order.

By s. 15(1) of the Access to Justice Act 1999, an individual granted representation can select any representative who is willing to act for him. However, s. 15(2) further provides that regulations may limit this freedom of choice in cases of prescribed descriptions. The Criminal Defence Service (General) (No. 2) Regulations 2001

(SI 2001, No. 1437), as amended, lay down that for certain specified proceedings a representative must be chosen from among those who are authorised to provide representation under a crime franchise contract with the Legal Services Commission, or who are public defenders employed by the Commission to provide representation.

The specified proceedings are:

(a) any criminal proceedings in a magistrates' court;

(b) any proceedings prescribed as criminal taking place in the Crown Court (para 1.9.5.2 above);

(c) any appeal by way of case stated from a magistrates' court; and

(d) any proceedings which are preliminary or incidental to proceedings mentioned in (a) to (c) above.

Proceedings for contempt committed in the face of the court are expressly excluded from this requirement, with the result that a person involved in such proceedings can choose any legal representative.

In very high cost cases (including very high cost *fraud* cases), the Legal Services Commission can require representation to be provided under a very high cost case contract. It can cease to fund representation by the defendant's existing representative if the representative or the Commission do not wish to enter into such a contract because, for example, they cannot agree terms. The defendant is then free to choose a different representative who is willing to enter into a very high cost case contract. Additionally, in a very high cost *fraud* case the Commission can cease to fund representation if the defendant's solicitor representatives are not members of the Serious Fraud Panel, which is a panel of solicitors approved by the Commission as suitable to deal with very high cost fraud cases.

The Criminal Defence Service (General) (No. 2) Regulations 2001 (SI 2001, No. 1437), as amended, define a 'very high cost case' as a case which, if it proceeds to trial, would be likely to last for 41 days or longer. The Commission decides whether this definition is satisfied in any particular case. A 'very high cost fraud case' is defined as a very high cost case in which the defendant is charged with an offence which is 'primarily or substantially founded on allegations of fraud or other serious financial impropriety, or involves complex financial transactions'.

A solicitor who has conduct of a very high cost case is obliged to notify the Commission in writing as soon as is practicable. A solicitor whose failure to do this without good reason causes a loss to public funds can be refused payment of his costs up to the extent of that loss (Criminal Defence Service (General) (No. 2) Regulations 2001 (SI 2001, No. 1437), as amended).

1.9.7.4 Eligibility

1.9.7.4.1 *Merits test* The criminal trial courts do not have an unfettered discretion to grant representation. Whether a right to representation should be granted is determined according to the interests of justice (Access to Justice Act 1999, s. 14 and sch. 3).

It was accepted by successive Lord Chancellors before 1988 that the question whether or not it was in the interests of justice to grant representation was to be determined in accordance with some non-statutory criteria suggested in the *Report of the Departmental Committee on Legal Aid in Criminal Proceedings* (Cmnd 2934, 1966). These so-called 'Widgery criteria' (named after the chairman of the Committee) were eventually put down in statutory form in the Legal Aid Act 1988. They are now contained in para 5 of sch. 3 to the Access to Justice Act 1999.

The criteria are as follows:

(a) whether the individual would he likely to lose his liberty or livelihood or suffer serious damage to his reputation;

(b) whether the proceedings may involve consideration of a substantial question of law;

(c) whether the individual may be unable to understand the proceedings or to state his own case;

(d) whether the proceedings may involve the tracing, interviewing or expert cross-examination of witnesses on behalf of the individual; and

(e) whether it is in the interests of another person that the individual be represented.

This list of criteria is not exhaustive since sch. 3, para 5, says that they are factors which *must* be taken into account. Other factors not mentioned in para 5 may also be relevant. The Secretary of State is empowered to amend the list by adding new factors or by varying any factor.

1.9.7.4.2 *Recovery of defence costs orders*   The practice of means-testing every defendant, which was a feature of the former criminal legal aid scheme, has been abolished (but note the Criminal Defence Service Bill, below). The courts no longer conduct a means test before deciding whether to grant representation, which is available to all, regardless of means and without contribution, who satisfy the 'interests of justice' test (Access to Justice Act 1999, ss. 14(1) and 17(1), and sch. 3).

However, s. 17(2) of the 1999 Act, and regulations made thereunder in the Criminal Defence Service (Recovery of Defence Costs Orders) Regulations 2001 (SI 2001, No. 856), as amended, introduced the recovery of defence costs order (RDCO) under which funded defendants can be compelled, at the conclusion of the trial, to pay towards the cost of their defence when they can afford to do so.

A magistrates' court cannot make an RDCO, but a Crown Court judge can. Under the RDCO the defendant will be ordered to pay some or all of the cost of his defence, which can include the cost of any earlier representation in a magistrates' court. An RDCO can provide for payment to be made forthwith, or in specified instalments.

The defendant can be required to provide financial information. If he fails to do so, the court will make an RDCO for the full costs incurred under the representation order.

The judge or the appropriate officer can request the Legal Services Commission to investigate the defendant's financial resources in order to assist the judge. The defendant's assets can be frozen during the investigation. The Commission will investigate the defendant's resources in detail and prepare a report for the court. If the defendant's means are complicated, the case can be referred to the Special Investigations Unit of the Commission.

In calculating the defendant's financial resources, the following are left out of account (unless the circumstances are exceptional):

(a) the first £3,000 of capital available to him;

(b) the first £100,000 of equity in his principal residence; and

(c) his income, where the court or the Commission are satisfied that his gross annual income does not exceed £25,250.

These exceptions apart, the amount or value of every source of income, and every resource of a capital nature, can be taken into account, and the financial resources of the defendant's partner are treated as those of the defendant.

An RDCO cannot be made against a funded defendant who:

(a) has appeared in a magistrates' court only;

(b) is committed for sentence to the Crown Court;

(c) is appealing against sentence to the Crown Court; or

(d) has been acquitted, other than in exceptional circumstances.

In deciding whether to make an RDCO under the exception in (d) above, the judge must consider whether it is reasonable in all the circumstances of the case to make one. In all other cases not falling within (a) to (d) above, an RDCO *must* be made.

The Criminal Defence Service Bill provides for the reintroduction in the magistrates' courts of means-testing before representation can be granted, and for the Legal Services Commission to take over from the courts the power to grant representation. The Bill is a reflection of the government's intention to make savings on the criminal legal aid budget and to make more funds available for civil cases.

1.9.8 Representation in contempt proceedings

Public funding for representation in proceedings for contempt committed, or alleged to have been committed, in the face of the court, is included within the scope of the Criminal Defence Service (Access to Justice Act 1999, s. 12(2)(f)). This is a potentially useful measure for helping to prevent the injustice that might otherwise occur in the case of an unrepresented defendant in what are usually very hasty proceedings. The court's decision whether to grant representation is based on the 'interests of justice' (ibid., sch. 3, para 5).

A defendant involved in such proceedings can select any legal representative. Choice is not limited to representatives who have contracted with the Legal Services Commission or are public defenders employed by the Commission.

2

Courts of special jurisdiction

2.1 Court of Justice of the European Communities (the European Court)

2.1.1 Introduction

By the European Communities Act 1972, the United Kingdom became a member of the European Communities on 1 January 1973. The Communities are:

(a) the European Coal and Steel Community established in 1951 by the Treaty of Paris (the ECSC Treaty);

(b) the European Economic Community established in 1957 by the Treaty of Rome (the EEC Treaty);

(c) the European Atomic Energy Community established in 1957 by the second Treaty of Rome (the Euratom Treaty).

As from November 1993, the European Economic Community and the EEC Treaty were renamed, respectively, the 'European Community' and the 'EC Treaty' by the Treaty on European Union (TEU, signed at Maastricht in 1992). At the same time the TEU created the 'European Union', a completely new body based on the European Communities.

With effect from May 1999, the TEU and the Treaties establishing the Communities were amended by the Treaty of Amsterdam (signed at Amsterdam in 1997). A number of articles of the EC Treaty were repealed, and the rest were renumbered.

Since the accession of the United Kingdom, the Court of Justice of the European Communities has stood above the House of Lords as the ultimate court in disputes with a European flavour. In internal, domestic cases the House of Lords remains the final appeal court in the United Kingdom.

The Court of Justice of the European Communities was first established in 1954 as the Court of Justice of the European Coal and Steel Community. It was renamed in 1958. The Court, which sits in Luxembourg, is often referred to simply as 'the European Court', but it is not to be confused with the European Court of Human Rights, which sits in Strasbourg, nor with the International Court of Justice, which sits at The Hague.

Attached to, but independent of, the European Court is the *Court of First Instance of the European Communities.* This was established by the Single European Act, signed in February 1986, which added a new art. 225 to the EC Treaty providing for a Court of

First Instance with jurisdiction to decide at first instance certain types of case brought by natural or legal persons, particularly actions brought by private individuals and cases relating to unfair competition between businesses.

The Court of First Instance began work in 1989 and was created in order to relieve pressure on the European Court, which was becoming overburdened with cases, and to offer ordinary citizens and businesses better legal protection under Community law.

A new article, art. 225a, was inserted in the EC Treaty by the Treaty of Nice, which was signed at Nice in 2001. Article 225a authorised *judicial panels* to be attached to the Court of First Instance in an effort to alleviate the growing burden on both it and the European Court. The function of judicial panels is to decide at first instance certain classes of case in specific areas. A right of appeal lies to the Court of First Instance, normally on points of law only (ibid.).

2.1.2 Composition

The European Court consists of one judge per Member State chosen from among persons whose independence is beyond doubt and who are eligible for appointment to the highest judicial posts in their own countries, or who are jurists of recognised competence (EC Treaty, art. 221). They are appointed by the governments of the Member States. Their appointment is for six years initially, and reappointment is permissible. The Court has a President, who is appointed from among themselves by the judges. The presidency is held for three years (ibid., art. 223).

The Court is assisted in its work by eight advocates-general, who must have the same qualifications for appointment as the judges. The advocate-general assigned to a particular case delivers an opinion, in open court and with complete impartiality and independence, in which he indicates the issues raised and the reasoned conclusions he has reached (ibid., arts. 222–223). The opinion, while not binding on the Court, is taken into account (and carries great weight) when the Court is considering its decision.

It is the practice of the European Court to deliver only one judgment so that it is never known whether there were any dissentients among the judges. The judgment often consists of a series of terse propositions from which it is difficult to extract any *ratio decidendi*. (For the meaning of *ratio decidendi* and the role of judicial precedent in the European Court, see paras 5.2 and 5.3.2.1.) For this reason, the usually fuller opinion of the advocate-general can be of considerable assistance in understanding what the Court has decided.

The Court can sit in chambers of three or five judges, or as a Grand Chamber of 13 judges, or in plenary session (where *all* the judges of the Court take part). It can sit in plenary session when a Member State or a Community institution which is a party to the proceedings so requests, or in particularly complex or important cases. Most cases are dealt with by chambers of three or five judges. For the sake of efficiency in a greatly enlarged European Union, the Court is more likely to sit as a Grand Chamber than in plenary session.

The Court of First Instance consists of at least one judge per Member State chosen from persons whose independence is beyond doubt and who possess the ability required for appointment to high judicial office (ibid., art. 224). They are appointed by the governments of the Member States. Appointment is for six years initially, and reappointment is permitted. The judges elect the President of the Court from among themselves for a term of three years. There are no advocates-general, but the duties of advocate-general can be performed in a limited number of cases by a judge designated for that purpose. The Court of First Instance can sit in chambers of three or five judges. It can sit in plenary session in certain particularly important cases.

Judicial panels attached to the Court of First Instance consist of members chosen from persons whose independence is beyond doubt and who possess the ability required for appointment to judicial office. They are appointed by the Council (ibid., art. 225a, as inserted by the Treaty of Nice).

2.1.3 Jurisdiction

2.1.3.1 Introduction

The function of the European Court is to ensure that Community law is interpreted and applied in the same way in each Member State. The vast majority of cases heard by the European Court are brought by Member States, or by the Community institutions, or are referred to it by national courts. It has only limited power to deal with cases brought by individual citizens. There are four most commonly exercised pieces of jurisdiction.

First, proceedings can be initiated before it by the Commission or a Member State if it is alleged that another Member State is failing to fulfil its obligations under Community law. It can impose a financial penalty on a Member State which has not complied with its judgment following a finding that the Member State is in breach of its Treaty obligations (EC Treaty, arts. 226–228).

Secondly, it has power to review the legality of, and annul, certain acts of the Community institutions if they were not adopted correctly or were not authorised by the Community Treaties (ibid., arts. 230–231).

Thirdly, it can deal with complaints from Member States and Community institutions that the European Parliament, the Council, or the Commission has failed to make certain decisions as required by the Treaties (ibid., arts. 232–233).

Fourthly, it has jurisdiction to give 'preliminary rulings'. It is the responsibility of the courts in each Member State to see that Community law is properly applied in that country. If a national court is doubtful about the interpretation or validity of Community law, and in order to minimise the risk that courts in different countries might interpret Community law in different ways, the preliminary ruling procedure enables it to ask the European Court for advice (ibid., art. 234). The function of the European Court under art. 234 is limited to giving rulings on the interpretation of the Treaty, and on the interpretation and validity of secondary Community legislation. It can neither rule on the consistency of national law with Community law nor interpret national law (see *Costa v ENEL* [1964] CMLR 425, CJEC). The final decision in a referred case is made by the *national* court in accordance with the European Court's preliminary ruling. The preliminary ruling procedure is dealt with in more detail in para 2.1.3.2 below.

With the exception of actions and proceedings assigned to a judicial panel and those reserved in the Statute for the European Court, the Court of First Instance exercises the first instance jurisdiction of the European Court in disputes between the Communities and its officials and other staff, and in actions brought against a Community institution by private individuals and businesses, including actions for declaration of failure to act, annulment, and damages (ibid., art. 225).

Originally, references for a preliminary ruling under the EC Treaty, art. 234, were expressly excluded from the jurisdiction of the Court of First Instance. Now, however, as a result of changes made by the Treaty of Nice, the Court of First Instance has jurisdiction to decide questions referred for a preliminary ruling under art. 234 in specific areas laid down by the Statute of the European Court. Decisions of the Court of First Instance on such references can exceptionally be reviewed by the European Court where there is a serious risk of the unity or consistency of Community law being affected (ibid., art. 225). If the Court of First Instance considers that a reference for a preliminary ruling requires a decision of principle likely to affect the unity or consistency of Community law, it can, instead of deciding the matter itself, refer the case to the European Court (ibid.).

There is a right of appeal to the European Court from a decision of the Court of First Instance, but only on points of law (ibid.). If an appeal is successful, the European Court must quash the decision and either give final judgment itself or refer the case back to the Court of First Instance for final judgment.

2.1.3.2 References under article 234 of the EC Treaty

2.1.3.2.1 *Introduction* The European Court has jurisdiction to give preliminary rulings on references made to it by United Kingdom courts under art. 234 of the EC Treaty. It has similar jurisdiction under art. 41 of the ECSC Treaty and art. 150 of the Euratom Treaty. Article 234 of the EC Treaty (as amended by TEU) is here set out in full:

The Court of Justice shall have jurisdiction to give preliminary rulings concerning:

(a) the interpretation of this treaty;

(b) the validity and interpretation of acts of the institutions of the Community and of the European Central Bank;

(c) the interpretation of the statutes of bodies established by an act of the Council, where those statutes so provide.

Where such a question is raised before any court or tribunal of a Member State, that court or tribunal may, if it considers that a decision on the question is necessary to enable it to give judgment, request the Court of Justice to give a ruling thereon.

Where any such question is raised in a case pending before a court or tribunal of a Member State, against whose decisions there is no judicial remedy under national law, that court or tribunal shall bring the matter before the Court of Justice.

2.1.3.2.2 *The meaning of article 234* Article 234 is designed to secure the uniform judicial interpretation of European Community law (*Sociale Verzekeringsbank v H.J. Van der Vecht* [1968] CMLR 151, 161, CJEC; *H.P. Bulmer Ltd v J. Bollinger SA* [1974] Ch 401, CA, *per* Lord Denning MR at p. 425). A reference for a preliminary ruling may be made by a national court without any application from the parties and, although a party may seek to persuade the court to make a reference, the decision belongs to the national court. Similarly, it has been held that the Court of Appeal, on its own initiative and without the agreement of the parties, has power to *withdraw* a reference it has already made under art. 234. However, it is reluctant to do so unless it is clear that the reference will serve no useful purpose (*Royscot Leasing Ltd* v *Commissioners of Customs and Excise* [1999] 1 CMLR 903; (1998) *The Times*, 23 November, CA).

If a national court makes a reference under art. 234, the national proceedings will be stayed pending receipt of the preliminary ruling from the European Court. In the meantime, it is the responsibility of the national court to protect the parties' Community law rights by granting interim relief if that is required in the circumstances (*Amministrazione delle Finanze dello Stato* v *Simmenthal SpA* [1978] 3 CMLR 263, CJEC). The effectiveness of the reference system established by art. 234 would be undermined if the national court were not able to award interim relief until after receipt of the European Court's preliminary ruling (*Factortame Ltd* v *Secretary of State for Transport (No. 2)* [1991] 1 All ER 70, CJEC).

Where a rule of national law is the sole impediment to the granting of interim relief in such cases, Community law insists that it is the obligation of the national court to set aside that rule of national law (ibid.). In the case of an English court this may involve defying the hitherto accepted common law by subjecting the Crown to an interlocutory injunction and/or by disapplying a statute of the United Kingdom Parliament (*Factortame Ltd* v *Secretary of State for Transport (No. 2)*, above, where the House of Lords, in response to a preliminary ruling from the European Court, granted an interlocutory injunction against the Secretary of State for Transport which, in effect, disapplied certain statutory regulations, made under the Merchant Shipping Act 1988, pending final judgment by the European Court on the merits of the case; see further, para 4.7.2).

It will be observed that under the second paragraph of art. 234 there is a discretion whether or not to refer the question to the European Court (the 'court or tribunal *may* . . .'). Under the third paragraph, however, there is a positive duty to refer the question (the 'court or tribunal *shall* . . .').

In relation to the English legal system, the third paragraph of art. 234 certainly applies to the House of Lords since, under English law (the 'national law'), there is no appeal ('judicial remedy') from decisions of the House. The first case to be referred to the European Court by the House of Lords was a criminal case, *R* v *Henn* [1980] 2 All ER 166, CJEC and HL.

It is still an open question whether the Court of Appeal, like the House of Lords, is *bound* to refer a question in those cases where it is effectively the final appellate court. It must be remembered that no appeal at all is possible from the Court of Appeal to the House of Lords in probate proceedings which emanated from a county court or in insolvency proceedings which emanated from the Chancery Division (para 7.1.5). Moreover, in all other cases, an appeal from the Court of Appeal to the House of Lords lies only with permission of either court.

The third paragraph of art. 234 is open to two interpretations. First, it only makes a reference compulsory in the case of the highest judicial body of a Member State. This is the construction favoured by Lord Denning MR (*H.P. Bulmer Ltd* v *J. Bollinger*

SA [1974] Ch 401, CA, at pp. 420–1; the other members of the court, Stamp and Stephenson LJJ, did not commit themselves on the question). Secondly, it imposes an obligation to refer on any national court, not necessarily the highest in the land, whose decision is (or may be) final in a particular case. This wider interpretation is the one favoured by the European Court itself and seems to have received the greater support of the two (*Costa* v *ENEL* [1964] CMLR 425, CJEC).

The wider interpretation was preferred in *SA Magnavision NV* v *General Optical Council (No. 2)* [1987] 2 CMLR 262, DC, where the Queen's Bench Divisional Court was prepared to assume that it was a final court in a criminal case for the purposes of the third paragraph of art. 234, having prevented a further appeal to the House of Lords by refusing to certify under s. 1 of the Administration of Justice Act 1960 (para 7.2.1.3) that the case before it involved a point of law of general public importance. (See further on this case, para 2.1.3.2.4 below.)

The *Magnavision* approach was approved in *Chiron Corp* v *Murex Diagnostics Ltd* [1995] All ER (EC) 88, CA, in which it was said that the Court of Appeal is a court of last resort for the purposes of the third paragraph of art. 234 in those cases where there is no right of appeal against its decision to the House of Lords. Lord Denning's statement in *H.P. Bulmer Ltd* v *J. Bollinger SA* ([1974] Ch 401, CA, at p. 420) that 'short of the House of Lords, no other English court is bound to refer a question to the European Court at Luxembourg' was said to be mistaken ([1995] All ER (EC) 88 *per* Balcombe LJ at p. 93). Where, however, there is a right of appeal to the House of Lords but the Court of Appeal refuses leave to appeal, the court of last resort is the House of Lords and not the Court of Appeal (*Chiron Corp* v *Murex Diagnostics Ltd*, above). This approach is supported by the decision of the European Court in the case of *Criminal Proceedings against Lyckeskog* [2003] 1 WLR 9, CJEC.

2.1.3.2.3 *Exercise of the discretion under the second paragraph of article 234* The second paragraph of art. 234 applies to any United Kingdom court or tribunal below the House of Lords where the decision in a particular case is not final. Thus, in the exercise of the discretion, references to the European Court have been made by the Court of Appeal (*Macarthys Ltd* v *Smith* [1979] 3 All ER 325, CA—the first ever reference by the Court of Appeal); the High Court (*Van Duyn* v *Home Office* [1975] Ch 358—the first ever reference by any English court); the Employment Appeal Tribunal (*Burton* v *British Railways Board* [1982] 3 All ER 537, CJEC); and even by a magistrates' court (*R* v *Marlborough Street Stipendiary Magistrate (ex parte Bouchereau)* [1977] 3 All ER 365, DC; *R* v *Plymouth Justices (ex parte Rogers)* [1982] 2 All ER 175, DC).

Guidelines for the exercise of the discretion under art. 234 were suggested by Lord Denning MR in *H.P. Bulmer Ltd* v *J. Bollinger SA* [1974] Ch 401, CA, at pp. 420–5. It is important to appreciate, however, that these guidelines have been criticised on the ground that some of them are unduly restrictive and, if followed, will reduce the number of opportunities for asking the European Court to resolve doubts about important principles of Community law (Bridge (1975–6) 1 EL Rev, p. 20). The guidelines appear to have been accepted by Stephenson LJ in the *Bulmer* case (at p. 430), but not with any great enthusiasm. In addition, it may be argued that the Court of Appeal has no power to lay down any guidelines at all since the discretion under art. 234 is vested directly in national courts and should not be limited by a national appellate court (see *Bethell* v *Sabena* [1983] 3 CMLR 1, *per* Parker J at p. 4).

Lord Denning's guidelines are as follows.

(a) A reference to the European Court should only be made if a ruling of that court is necessary to enable the English court to give judgment in the case. 'Necessary' means that the ruling would be conclusive of the case. If other matters remained to be decided, then the ruling would not be 'necessary' within the meaning of art. 234. In a criminal trial on indictment, for example, this means that it would rarely be proper for the Crown Court judge to seek a preliminary ruling before the facts of the alleged offence have been established. It is generally preferable for the trial judge himself to decide the question of Community law that has arisen and to have his decision reviewed on appeal through the hierarchy of our national courts (*R* v *Henn* [1980] 2 All ER 166, HL).

(b) There is no need to refer a question which has already been decided by the European Court in a previous case. This guideline is open to objection. Since the European Court is not bound by its own previous decisions, it may be advisable for an English court to submit the same question where it considers that the previous decision of the European Court may have been wrong or where there are new factors which ought to be put before the Court.

(c) There is no need to refer a point which is reasonably clear and free from doubt. This is the so-called *acte clair* doctrine. Where it arises there is no need to interpret the law but only to apply it. It has been said that magistrates' courts in particular should exercise considerable caution before referring a question to Europe and should generally decide any question of Community law themselves. If their decision is wrong it can be appealed against to a higher English court, which is a more suitable forum to assess the appropriateness of a reference and to formulate the question to be referred (*R* v *Plymouth Justices (ex parte Rogers)* [1982] 2 All ER 175, DC, *per* Lord Lane CJ at p. 182).

(d) In deciding whether to make a reference, the English court must consider all the circumstances. Of particular consequence are factors such as not overloading the European Court with references, the length of time which may elapse before a ruling can be obtained, the difficulty and importance of the point to be referred, the expense involved in obtaining a ruling, and the wishes of the parties.

None of these matters is expressly mentioned in art. 234. Nevertheless in the *Bulmer* case itself, time and expense were important factors in the decision to refuse to refer the question to Europe. The view of the Court of Appeal was that the point was important, but not difficult, and should be decided by the English courts. By contrast, in *Macarthys Ltd* v *Smith* [1979] 3 All ER 325, CA, the Court of Appeal concluded that it would be less expensive and more convenient to send the case to the European Court than to the House of Lords (see especially *per* Lawton LJ, at p. 334).

It should be noted that the costs of a reference to the European Court are included as part of the proceedings before the English court, so that, for example, a representation order in criminal proceedings will cover an application for a ruling to the European Court (*R v Marlborough Street Stipendiary Magistrate (ex parte Bouchereau)* [1977] 3 All ER 365, DC).

(e) The discretion to refer the case or not belongs to the English court and not to the European Court. This means that the European Court cannot be asked to give guidance on how the discretion should be exercised and that a party cannot complain to the European Court about the exercise of the discretion one way or the other by the English court. The appropriate remedy for a party who wishes to challenge the decision to refer or not to refer is to appeal to a higher English court.

2.1.3.2.4 *References under the third paragraph of article 234* As explained in para 2.1.3.2.2 above, the third paragraph of art. 234 applies to the House of Lords as our highest national court. It may also apply to the Court of Appeal or the Queen's Bench Divisional Court in a case where that court is effectively the final appellate body.

Despite the clear wording of the third paragraph of art. 234, no court (not even the House of Lords) is always bound to refer points of Community law. In *CILFIT v Ministro della Sanita* [1983] 1 CMLR 472, CJEC, the European Court provided guidance for national courts falling within the third paragraph of art. 234.

First, the national court is not bound to refer a question to the European Court if the question is not relevant, i.e., if the answer to the question could have no influence on the outcome of the case. This is similar to Lord Denning's guideline at (a) in para 2.1.3.2.3 above.

Secondly, the national court can rely on previous rulings of the European Court and need not refer the same questions again. However, the national court is always entitled to make a fresh reference on the same point if it sees fit (cf Lord Denning's guideline at (b) in para 2.1.3.2.3 above).

Thirdly, the national court need not make a reference where the correct application of European Community law is so obvious as to leave no room for reasonable doubt. The European Court has thus approved the *acte clair* doctrine, although not in such wide terms as Lord Denning had expressed it in the *Bulmer* case (see para 2.1.3.2.3 above, at (c)). According to the European Court, the question whether or not there is reasonable doubt must be assessed by the national court having regard to the special characteristics of Community law, the peculiar problems presented by its interpretation, and the risk of causing conflicts of legal rulings within the European Union (*CILFIT v Ministro della Sanita*, above). Lord Denning's view of the *acte clair* doctrine must now be taken to have been modified in the light of this opinion from the European Court.

It has been held to be too late to make a reference to the European Court once judgment has been given by the national court because such a case can no longer be said to be 'pending' before the national court within the meaning of the third paragraph of art. 234, and any decision by the European Court on such a reference could not be described as a 'preliminary ruling' (*SA Magnavision NV v General Optical Council (No. 2)* [1987] 2 CMLR 262, DC; *Chiron Corp v Murex Diagnostics Ltd* [1995] All ER (EC) 88, CA).

In the *Magnavision* case, above, the Queen's Bench Divisional Court would not have made a reference even if it had not been too late to do so, since, applying the *acte clair* doctrine, the judges found the relevant point of Community law to be quite clear. In the *Chiron* case, above, it was pointed out by the Court of Appeal in approving the *Magnavision* decision that a preliminary ruling by the European Court under art. 234 should logically precede the judgment of the national referring court. Once, therefore, the Court of Appeal has given final judgment in a case it has no power to make a reference under art. 234.

2.2 Judicial Committee of the Privy Council

2.2.1 Composition

Before 1833, the Crown prerogative to hear appeals from courts in all the dominions was exercised through a lay committee of the whole Privy Council. Since the Judicial Committee Act of 1833, all such appeals have been heard by a special committee created by that Act and called the Judicial Committee of the Privy Council.

The Judicial Committee is composed of members of the Privy Council who hold or have held high judicial office in the United Kingdom (including past and present judges of the House of Lords and of the Court of Appeal), together with members of the Privy Council who hold or have held the office of chief justice or judge in a superior court in Australia or in any other superior court in the Commonwealth named by Order in Council (Judicial Committee Act 1833, s. 1, as substituted by the Constitutional Reform Act 2005; Judicial Committee Amendment Act 1895, s. 1). Among the superior Commonwealth courts 'named' are those in Barbados, Trinidad and Tobago, the Bahamas, Jamaica, and the Eastern Caribbean. South Africa was removed from the list in 1962.

2.2.2 Procedure

As a matter of convenience, the Judicial Committee sits in London. Appeals are heard at the bar of the Privy Council; the judges do not wear robes. The quorum is three, although Commonwealth appeals and devolution references are usually heard by a panel of five judges. In the past, the court was most commonly made up of Lords of Appeal in Ordinary. For this reason, decisions of the Judicial Committee carry strong persuasive authority but they are not absolutely binding on English courts. (For judicial precedent in relation to the Judicial Committee, see para 5.3.2.9.)

Technically, the Committee's decision in a case is not a judgment but *advice* in the form of a report or recommendation to Her Majesty (Judicial Committee Act 1833, s. 3). In practice, the advice is always followed and is implemented by an Order in Council. In theory, the advice to Her Majesty should be unanimous. For this reason, the practice was for only one judgment to be prepared so that it was never known publicly whether there had been any dissension among the judges. This is still the normal practice, although, since 1966, it has been possible for a member of the Committee to publish a dissenting judgment (Judicial Committee (Dissenting Opinions) Order, dated 4 March 1966). The first dissenting opinion given under the authority of the 1966 Order was delivered by Lord Morris in *National and Grindlays Bank Ltd* v *Dharamshi Vallabhji* [1967] 1 AC 207, PC, a case in which his Lordship dissented on a question of statutory interpretation.

2.2.3 Jurisdiction

The jurisdiction of the Judicial Committee of the Privy Council extends to Commonwealth appeals, appeals from the Disciplinary Committee of the Royal College of Veterinary Surgeons, prize appeals, ecclesiastical appeals, and certain other matters.

2.2.3.1 Commonwealth jurisdiction

The Judicial Committee hears appeals from courts in United Kingdom overseas territories and from the independent members and associate members of the Commonwealth, except for those independent Commonwealth countries which have stopped sending their final appeals to London either because special legislation has been passed or because Her Majesty has ceased to be Head of State there. Among the territories and independent countries whose final appeals are still heard by the Committee are Antigua and Barbuda, the Bahamas, Barbados, Bermuda, Jersey, Guernsey, the Falkland Islands, Gibraltar, Jamaica, the Isle of Man, and the Seychelles.

The jurisdiction of the Committee has been specially extended to the republics of Dominica, Trinidad and Tobago, Kiribati, and Mauritius, where the Queen is not Head of State though she is recognised as Head of the Commonwealth. The appeal from a republic either lies to the Head of State (who refers it to the Judicial Committee) or it is expressed as lying to the Committee itself and not to the Queen in Council. By agreement with the Sultan of Brunei, the Committee hears appeals in *civil* cases from the Brunei Court of Appeal. Its advice is given to the Sultan.

The jurisdiction has been ended in Aden, Australia, Botswana, Burma, Canada, Cyprus, Ghana, Guyana, Hong Kong, India, Kenya, Malaysia, Malta, Nigeria, Pakistan, Sierra Leone, Singapore, Sri Lanka, Tanzania, and Uganda, among others. Aden and Burma did not become members of the Commonwealth on independence. Pakistan left the Commonwealth in 1972; although she was readmitted as a member in 1989, the jurisdiction of the Committee was not revived. The same is true of South Africa, which was not a member of the Commonwealth between 1962 and 1994. The jurisdiction of the Committee in relation to Malaysia, an independent monarchy not owing allegiance to the Queen, was terminated at the end of 1984.

Appeals from Australia on *state* law ceased to lie to the Judicial Committee in 1986 (Australia Act 1986, s. 11), thus completing the process of withdrawal begun in 1968 when the Australian Parliament abolished appeals to the Committee on *federal* law. The final appeal court from all Australian courts, both state and federal, is now the High Court of Australia.

The right of appeal to the Committee from the republic of Singapore was abolished in 1994. The Committee was replaced as the final appellate court for Hong Kong by the Hong Kong Court of Final Appeal when that territory became a Special Administrative Region of China on 1 July 1997. Arrangements for two serving United Kingdom Law Lords to sit occasionally in the Hong Kong Court of Final Appeal as non-permanent judges were completed in January 1998 when Hong Kong's Provisional

Legislative Council endorsed the appointment of the first two nominees, Lord Nicholls and Lord Hoffmann. In October 2003, legislation was passed in New Zealand abolishing appeals to the Committee in respect of all cases heard by the Court of Appeal of New Zealand after the end of 2003.

A Caribbean Court of Justice is in the process of being established. This will take over the Committee's appellate jurisdiction in respect of most of the Commonwealth countries in the Caribbean.

Most of the appeals heard by the Judicial Committee are in *civil* cases. The Committee in particular does not sit as a court of appeal against *criminal sentence*, and will not grant special leave to appeal in criminal cases unless there has been a substantial injustice, such as where the appellant has been denied a fair trial because of a violation of the principles of natural justice (*Ibrahim* v *R* [1914] AC 599, PC; *Practice Note* (1932) 48 TLR 300; *Badry* v *Director of Public Prosecutions of Mauritius* [1982] 3 All ER 973, PC).

The Judicial Committee has re-emphasised that unless some other matter is specifically referred to it under s. 4 of the Judicial Committee Act 1833 (para 2.2.3.5 below) its Commonwealth jurisdiction is as an appellate court only. Accordingly, it cannot, for example, act as a court of first instance to adjudicate on a question of the constitutionality of a mandatory sentence of death where that question has not yet been determined by the appellate courts of the country concerned (*Walker* v *R* [1993] 4 All ER 789, PC, *per* Lord Griffiths at p. 791).

The continuing influence of the Judicial Committee within the Commonwealth is illustrated by the case of *Pratt* v *Attorney-General for Jamaica* [1993] 4 All ER 769, PC, in which it was held that the state's failure to carry out a death sentence as swiftly as practicable after sentence amounted to 'inhuman or degrading punishment or other treatment' contrary to the Jamaican constitution.

The Judicial Committee commuted from death to life imprisonment the sentences of two men, thus rescuing them from 'death row' where they had been for almost 15 years. The fate of 105 other prisoners who had been awaiting execution for more than five years (23 of them for longer than 10 years) also depended on the outcome of the appeal. *Pratt* was applied in *Guerra* v *Baptiste* [1995] 4 All ER 583, PC, an appeal from Trinidad and Tobago, and considered in *Henfield* v *Attorney-General of the Commonwealth of the Bahamas* [1996] 3 WLR 1079, PC, in both of which the death sentence was commuted.

In *Bowe* v *R* [2006] UKPC 9, PC, the Judicial Committee went so far as to decide that a provision in the constitution of the Bahamas which, on the face of it, imposed a *mandatory* death penalty for murder ('whoever commits murder shall be liable to suffer death') should be interpreted as imposing only a *discretionary* death penalty. It is well established in international human rights law that a mandatory death penalty amounts to 'inhuman or degrading punishment'. Since such punishment was expressly prohibited by another provision in the Bahamas constitution it followed that a mandatory death penalty was unconstitutional.

2.2.3.2 Disciplinary jurisdiction

For many years the Judicial Committee of the Privy Council heard appeals from persons in the United Kingdom whose names had been ordered by the appropriate disciplinary body to be erased from certain professional registers. This jurisdiction extended to health care professionals, such as medical practitioners, dentists, opticians, chiropodists, physiotherapists, and osteopaths. It also covered veterinary surgeons.

By virtue of the National Health Service Reform and Health Care Professions Act 2002, the relevant provisions of which came into force on 1 April 2003, appeals of this kind involving health care professionals no longer lie to the Judicial Committee. They go instead to the High Court. Those concerning veterinary surgeons still lie to the Judicial Committee.

2.2.3.3 Prize jurisdiction

The Judicial Committee hears appeals in prize cases from decisions of the High Court sitting as a Prize Court (Supreme Court Act 1981, s. 16(2)). (For the meaning of 'prize', see para 1.4.3.3.) This jurisdiction is rarely exercised.

2.2.3.4 Ecclesiastical jurisdiction

The wide jurisdiction once possessed by the Judicial Committee to hear appeals from ecclesiastical courts ended in 1963. However, it still hears appeals from decisions of the Arches Court of Canterbury and the Chancery Court of York given in causes of faculty not involving a matter of doctrine, ritual, or ceremonial (Ecclesiastical Jurisdiction Measure 1963), and appeals against pastoral schemes (Pastoral Measure 1983). The ecclesiastical jurisdiction is rarely exercised.

2.2.3.5 Other jurisdiction

The Judicial Committee has power to entertain an application for a declaration that a person purporting to be a member of the House of Commons is disqualified or has been disqualified at any time since his election (House of Commons Disqualification Act 1975, s. 7(1) and (2)). For the purpose of determining any preliminary issue of fact, the Committee may order the issue to be tried in the High Court, the Court of Session, or in the High Court in Northern Ireland, depending on the location of the constituency involved (ibid., s. 7(4)).

Her Majesty may make a special reference to the Judicial Committee asking it to hear or consider any matter whatsoever (Judicial Committee Act 1833, s. 4). The special reference need not necessarily be concerned with a judicial decision and, in the past, this procedure has been used in such diverse matters as the conduct and powers of colonial judges, the privileges of the Jersey Bar, boundary disputes between dominions, and the eligibility of a person to sit and vote in the House of Commons. The Committee advises the Queen on the outcome of the reference in the same way as on an appeal.

Under the Government of Wales Act 1998, the Scotland Act 1998, and the Northern Ireland Act 1998, 'devolution questions' may be referred to the Judicial Committee for determination. In relation to Scotland and Northern Ireland, for example, these include the question of whether any provision in a Bill would be within the legislative competence of the Scottish Parliament or the Northern Ireland Assembly, respectively. The Acts provide that the decision of the Judicial Committee on a devolution question is binding in all legal proceedings, except proceedings before the Judicial Committee itself.

When the relevant provisions of the Constitutional Reform Act 2005 come into force, the devolution jurisdiction of the Committee will be taken over by the new Supreme Court of the United Kingdom.

2.3 Employment Appeal Tribunal

2.3.1 Introduction

The Employment Appeal Tribunal was established in 1976 as the successor to the ill-fated National Industrial Relations Court (NIRC). NIRC was created under a Conservative government by the Industrial Relations Act 1971 and became unpopular with certain trade unions chiefly because of the exercise of its jurisdiction over 'unfair industrial practices'. This unpopularity led to a refusal of recognition and to open defiance of its orders on the part of some unions, resulting in the imprisonment of some members for contempt of court. The Labour Party committed itself to the repeal of the Industrial Relations Act 1971 and the abolition of NIRC. These objectives were achieved by the Trade Union and Labour Relations Act 1974 (since repealed).

NIRC was abolished on 31 July 1974 and, for a time, appeals in employment cases were heard by the Divisional Court of the Queen's Bench Division. On 30 March 1976, the Employment Appeal Tribunal came into existence. It was created as from that date by the Employment Protection Act 1975 (see now s. 20(1) of the Employment Tribunals Act 1996). Although called a 'tribunal' for psychological reasons in view of the trade union movement's distaste for courts (and especially NIRC), it is, in fact, a superior court of record (Employment Tribunals Act 1996, s. 20(3)). Thus, it has power to punish for contempt of court and is not subject to the supervisory jurisdiction of the Queen's Bench Division.

2.3.2 Composition

The Employment Appeal Tribunal consists of those High Court and Court of Appeal judges nominated by the Lord Chief Justice after consulting the Lord Chancellor, at least one judge of the Scottish Court of Session nominated by the Lord President of that Court, and lay members appointed by Her Majesty on the joint recommendation of the Lord Chancellor and the Secretary of State (Employment Tribunals Act 1996, s. 22(1), as amended by the Constitutional Reform Act 2005). The judicial members cannot be nominated unless they consent (Employment Tribunals Act 1996, s. 22(4)). One of the judicial members is appointed president by the Lord Chief Justice after consulting the Lord Chancellor and with the agreement of the Lord President of the Court of Session (ibid., s. 22(3), as amended by, and s. 22(3A) as inserted by, the Constitutional Reform Act 2005).

The lay members are persons with special knowledge or experience of industrial relations and are representatives of employers and of workers (ibid., s. 22(2)). There are no independent lay members. A lay member can be removed by the Lord Chancellor, after consultation with the Secretary of State, on the grounds of incapacity, misbehaviour, bankruptcy, or absence from sittings of the Tribunal for longer than six months without the permission of the president (ibid., s. 25(4)). Any such removal requires the agreement of the Lord Chief Justice or, in the case of Scottish lay members, the Lord President of the Court of Session (ibid., s. 25(5), (6), as inserted by the Constitutional Reform Act 2005). A lay member can resign at any time by written notice to the Lord Chancellor and the Secretary of State (Employment Tribunals Act 1996, s. 25(2)).

The Employment Appeal Tribunal has a central office in London but is authorised to sit at any place in Great Britain in any number of divisions concurrently (ibid., ss. 20(2) and 28(1)). An appeal must normally be heard by a judge with either two or four lay members who represent employers and workers in equal numbers. However, with the consent of the parties, an appeal can be heard by a judge sitting with one lay member or by a judge sitting with three lay members (ibid., s. 28(2)–(3)).

2.3.3 Jurisdiction

The jurisdiction of the Employment Appeal Tribunal is largely limited to the hearing of appeals. It has only a very limited original jurisdiction, and, unlike NIRC, it has no jurisdiction over such matters as 'unfair industrial practices'. The appeals come mainly from employment tribunals on questions of law arising in cases decided under various statutes. The matters covered include redundancy, equal pay, written particulars of contracts of employment, sex discrimination (in the field of employment only), racial discrimination (employment cases only), disability discrimination (employment cases only), unfair dismissal, unlawful deductions from wages, and unlawful refusal of employment, or of the services of an employment agency, on grounds related to trade union membership (Employment Tribunals Act 1996, s. 21). Cases involving sexual, racial, or disability discrimination in fields other than employment are heard at first instance by the county courts, with an appeal lying to the Court of Appeal.

2.3.4 Procedure

The procedure before the Employment Appeal Tribunal is relatively quick, informal, simple, and cheap. The strict rules of evidence do not apply. Any party may appear in person or be represented by counsel, or by a solicitor, or by a representative of a trade union or an employers' association, or by any other person he chooses (Employment Tribunals Act 1996, s. 29(1)).

The Tribunal can order the attendance and examination of witnesses and the discovery of documents. It has power to enforce its own orders, although contempt of court can only be punished by, or with the consent of, the judge (ibid., s. 36(4)).

Public funding under the Community Legal Service is available for representation before the Tribunal. There are some circumstances in which the Tribunal can order a party to pay the costs and expenses of another party, but it must have regard to a person's ability to pay when considering whether to make an order. If the conduct of the proceedings by a party's *representative* justifies it, the Tribunal can disallow all or part of the representative's costs or expenses, or order the representative to pay all or part of the costs or expenses incurred by a party (ibid., s. 34, as substituted by the Employment Act 2002).

2.3.5 Appeals

The decision of the Employment Appeal Tribunal is final on any question of *fact*, except in the case of committal for contempt of court. On a question of *law*, an appeal lies to the Court of Appeal or, in the case of Scottish proceedings, to the Court of Session. Leave to appeal must first be obtained from the Tribunal or from the Court of Appeal (or Court of Session in the case of Scottish proceedings) (Employment Tribunals Act 1996, s. 37). A further appeal lies, with leave, to the House of Lords.

2.4 Restrictive Practices Court

This Court is to be abolished, having outlived its usefulness. It was established by the Restrictive Trade Practices Act 1956 to determine whether restrictive agreements and information agreements relating to the supply of goods or services are valid, or whether they are contrary to the public interest and void. The composition and procedure of the Court are presently regulated by the Restrictive Practices Court Act 1976.

The Competition Act 1998 introduced new investigation and enforcement procedures into competition law, and established the Competition Commission. As amended by the Enterprise Act 2002, the 1998 Act also created an independent Competition Appeal Tribunal to hear appeals from certain decisions made by the Competition Commission, the Secretary of State, the Office of Fair Trading, and industry regulators.

The Competition Act 1998 has already repealed the Restrictive Trade Practices Act 1976 under which references were formerly made to the Restrictive Practices Court. From a day yet to be appointed, the Restrictive Practices Court Act 1976 will also be repealed, thus signalling the abolition of the Court. In the meantime, the Court will continue, under transitional provisions contained in the 1998 Act, to deal with references already made to it.

2.5 Coroners' courts

2.5.1 Introduction

Coroners were first appointed in the twelfth century in order to assist the sheriffs in their criminal jurisdiction. Coroners also kept watch on the sheriffs, whose honesty was suspected by the King in connection with royal revenues, such as fines and forfeitures, which were due from persons convicted of serious crimes ('Pleas of the Crown').

The coroners also kept a roll of local crimes to present to the royal judge on his next visit to the area, and held inquiries or 'inquests' into cases of unexplained death and 'deodand'. A deodand (literally, 'to be given to God', *deo dandum*) was an animate or inanimate object which had caused the death of a human being. It was forfeited to the King, or to the lord of the manor, to be put to pious uses in order to appease the wrath of God. The commonest deodands were said to be horses, oxen, carts, boats, mill-wheels, and cauldrons (Pollock and Maitland, *History of English Law*, 2nd ed., 1898, vol. 2, pp. 473–4). Deodand was abolished in 1846 after the claiming of a railway engine had demonstrated that the continued forfeiture of lethal objects could prove an expensive barrier to progress.

The investigation of unexplained deaths remains the chief function of the coroner's court, although the purpose is no longer to add to the royal revenues through fines but to satisfy the public conscience (see *R* v *West Yorkshire Coroner (ex parte Smith)* [1982] 3 All ER 1098, CA, *per* Donaldson LJ at p. 1108).

Most of the law relating to coroners' courts is contained in the Coroners Act 1988, which consolidated (with some amendments) legislation passed between 1844 and 1983.

2.5.2 The coroner

By law, coroners must be appointed by the appropriate local authority for each coroner's district in a metropolitan county or Greater London, for each non-metropolitan county, and for the City of London (Coroners Act 1988, s. 1(1)–(1A)). For appointment as a coroner, a person must either have a five year general qualification within the meaning of s. 71 of the Courts and Legal Services Act 1990, or be a legally qualified medical practitioner of at least five years' standing (ibid., s. 2(1), as amended by the Courts and Legal Services Act 1990).

A coroner may be removed from office by the Lord Chancellor with the agreement of the Lord Chief Justice for inability or misbehaviour in the discharge of his duty (Coroners Act 1988, s. 3(4), as substituted by the Constitutional Reform Act 2005). Alternatively, a coroner who has been convicted of the criminal offence of corruption, or of wilful neglect of his duty, may be ordered by the court of conviction to be removed from office and to be disqualified from acting as coroner (Coroners Act 1988, s. 3(6)).

The Lord Chief Justice and all the puisne judges of the High Court are *ex officio* coroners, although they are never called on to act in that capacity.

2.5.3 Jurisdiction

2.5.3.1 Treasure

At common law, treasure trove was coin, plate, or bullion, made of gold or silver, which had been hidden and of which the true owner was unknown. Any money or coin or other object not containing a substantial amount of gold or silver was not treasure trove. (This is Coke's definition (3 Co.Inst 132), preferred to all others by the Court of Appeal in *Attorney-General of the Duchy of Lancaster* v *G.E. Overton (Farms) Ltd* [1982] 1 All ER 524, CA.)

The Treasure Act 1996 replaced the common law rules on treasure trove with a statutory code. 'Treasure trove' was abolished and replaced with 'treasure', which is defined in s. 1 as including, *inter alia*, any found object (except a single coin) containing at least 10 per cent precious metal (i.e., gold or silver) and at least 300 years old. Coins are only treasure in two circumstances: (a) if there are at least two of them in the same find, and they are at least 300 years old and contain 10 per cent precious metal, or (b) there are at least ten of them and they are at least 300 years old.

In addition, the Secretary of State has power to alter the meaning of 'treasure' so as to (a) exclude from the definition any class of object which would otherwise be treasure, and (b) include within the definition any designated class of object which he considers to be of outstanding historical, archaeological or cultural importance (Treasure Act 1996, s. 2).

Treasure belongs, subject to any prior interests and rights, to any franchisee of the Crown (e.g., the Duchy of Cornwall, the Duchy of Lancaster and the City of London) or to the Crown itself (ibid., s. 4).

The coroner must be notified of any find if the finder believes, or has reasonable grounds for believing, that the found object is treasure (ibid., s. 8(1)). Failure to notify is a criminal offence (ibid., s. 8(3)). Any doubt about whether an object is treasure will be resolved by a coroner's inquest, which will be held without a jury unless the coroner orders otherwise (ibid., s. 7). Before an inquest the coroner must notify the British Museum or the National Museum of Wales (ibid., s. 9).

In the case of treasure which vests in the Crown and is to be transferred to a museum, the Secretary of State must decide whether a reward is to be paid by the museum and, if so, how much. The reward must not exceed the treasure's market value as determined by the Secretary of State. Payment of the reward is not enforceable against a museum or the Secretary of State (ibid., s. 10).

The Secretary of State is responsible for devising a code of practice, dealing with such matters as rewards and to whom treasure should be offered. The code may also include guidance for people who search for, or find, treasure (ibid., s. 11).

2.5.3.2 Deaths

The coroner has jurisdiction to inquire into violent or unnatural deaths, sudden deaths where the cause is unknown, and deaths in prison (Coroners Act 1988, s. 8(1)).

As long as the body is now lying within his district, the coroner has jurisdiction to hold an inquest into such a death which occurred outside England and Wales (*R v West Yorkshire Coroner (ex parte Smith)* [1982] 3 All ER 1098, CA). It is not obligatory for a coroner holding an inquest to view the body (ibid., s. 11(1)). The coroner has power to order the exhumation of the body of a person buried within his district (ibid., s. 23(1)). In the case of a sudden death from an unknown cause, the coroner may order a post-mortem examination of the body (ibid., s. 19(1)). If this shows that the death was from natural causes the coroner is not compelled to hold an inquest unless the deceased died in prison (ibid., s. 19(3)–(4), and see *R v Greater Manchester North District Coroner (ex parte Worch)* [1987] 3 All ER 661, CA).

A coroner has no jurisdiction to hold an inquest under s. 8 unless there is a body within his district. If, therefore, there is no body still in existence, resort must be had to s. 15 of the 1988 Act under which the Home Secretary can direct a coroner to hold an inquest notwithstanding the destruction of the body 'by fire or otherwise'.

2.5.4 **Procedure**

An inquest must be held in public except where the coroner considers that privacy is desirable in the interest of national security (Coroners Rules 1984 (SI 1984, No. 552), r. 17). This rule has two purposes. First, to allow interested parties and members of the public to hear the evidence of the circumstances of a death; secondly, to prevent the public from hearing evidence which affects national security. The coroner at an inquest which does not involve national security is not prevented by the rule from allowing a witness to give evidence out of public sight from behind a screen. Such an inquest is still 'held in public' (R v *Newcastle upon Tyne Coroner (ex parte A)* (1998) *The Times*, 19 January, where a member of the Northumbria Police armed response team obtained an order quashing the coroner's decision not to allow him to give evidence while screened from public sight).

The proceedings and evidence at an inquest are to be directed solely towards ascertaining who the deceased was; how, when, and where he came by his death; and the particulars required to be registered concerning the death (ibid., r. 36(1)). Neither the coroner nor the jury (if there is one) must express any opinion on any other matters (ibid., r. 36(2)).

In particular, no verdict should be framed in such a way as to appear to determine any question of criminal liability on the part of a named person, or any question of civil liability (ibid., r. 42). There appears to be a conflict between this prohibition and a verdict which states that the cause of death was aggravated, or contributed to, by 'lack of care'. This formula is increasingly used following the death in hospital or in prison of a person who, unable to look after himself, was being cared for by others and the death was due to starvation or exposure, or similar causes, brought about by the failure of his carers to look after him properly.

It has been suggested, however, that the apparent conflict is avoided if the verdict simply records that 'death was aggravated, or contributed to, by lack of care' and omits all reference to any particular hospital, prison, or person. Such a verdict is acceptable because it contains no suggestion that a legal duty of care has been broken giving rise to potential criminal liability for manslaughter or civil liability for negligence (R v *East Berkshire Coroner (ex parte Buckley)* (1992) *The Times*, 1 December, DC; R v *North Humberside and Scunthorpe Coroner (ex parte Jamieson)* [1994] 3 All ER 972, CA; R v *Surrey Coroner (ex parte Wright)* [1997] 1 All ER 823).

'Lack of care' signifies neglect, and neglect is only appropriate as part of a verdict where there is a direct causal link between the neglect and the death. Thus, while it is not neglect merely to present a prison inmate with opportunities to kill himself, it might be otherwise if a prison officer failed to intervene in a prisoner's obvious ongoing attempt at suicide (R v *North Humberside and Scunthorpe Coroner (ex parte Jamieson)*, above).

Rule 36(1), above, was examined by the House of Lords in the context of systemic neglect in R *(Middleton)* v *West Somerset Coroner* [2004] 2 WLR 800, HL, which arose out of the suicide by hanging of a prison inmate whose family alleged that he was a suicide risk and should have been placed on a suicide watch. The rule was found to be in conflict with art. 2 (right to life to be protected by law) of the European Convention on Human Rights. The requirement of r. 36 that an inquest be directed solely to ascertaining the identity of the deceased, and how, when, and where he came by his death was held to have been judicially interpreted in too restrictive a fashion because the word 'how' had been construed to mean only 'by what means'.

Article 2, as interpreted by the European Court of Human Rights, insists that an inquest jury should be allowed to express its conclusion on the central, factual issues in the case. The House decided that 'how' should be given the wider interpretation, 'by what means *and in what circumstances*'. This change will bring the inquest procedure into line with the requirement of art. 2 by enabling the jury to give its factual conclusion on the central issues, although it will remain the responsibility of the coroner to decide what *form* the verdict should take (short, narrative, or in response to questions put by him).

The House made it clear that the prohibitions against attributing liability and expressing extraneous opinions, imposed by rr. 42 and 36(2) respectively, must continue to be observed. Verdicts should avoid using expressions like 'neglect' or 'carelessness' since they are suggestive of civil liability. The House (through Lord Bingham [2004] 2 WLR 800 at [37] and [45]) recommended the use, in appropriate cases, of the following verdict: 'The deceased took his own life, in part because the risk of his doing so was not recognised and appropriate precautions were not taken to prevent him doing so'. This form of words violates neither of these two rules. (See also *R (Sacker)* v *West Yorkshire Coroner* [2004] 1 WLR 796, HL, decided on the same day as, and applying, *R (Middleton)* v *West Somerset Coroner* [2004] 2 WLR 800, HL, above.)

In most cases, the coroner has a *discretion* to summon a jury to the inquest (Coroners Act 1988, s. 8(4)). In a few cases, to be considered in para 2.5.6 below, the coroner *must* summon a jury. A coroner's jury consists of between seven and eleven persons (ibid., s. 8(2)). In order to sit on a coroner's jury a person must be qualified under the Juries Act 1974, s. 1, as substituted by the Criminal Justice Act 2003 (Coroners Act 1988, s. 9(1)).

Prospective jurors must in general be summoned formally in writing (Coroners Act 1988, s. 8(2)(a); Coroners Rules 1984, r. 45). There is an exception where a jury would be otherwise *incomplete*. Here, the coroner can require any person(s) in the vicinity of the place of the inquest to be summoned without written notice (for example, by informal oral communication with them) in order to make up the correct number of jurors required (Coroners Rules 1984, r. 48; this provision is analogous to s. 6 of the Juries Act 1974 (see para 6.4), which applies to the informal summoning of jurors to the Crown Court, High Court, and county courts). There is, however, no power to summon an *entire* jury by means of an informal oral communication (*R* v *Merseyside Coroner (ex parte Carr)* [1993] 4 All ER 65, DC, in which inquest proceedings were declared to be a nullity where the entire jury of nine had been obtained informally from a pool of jurors waiting at the nearby Crown Court).

A coroner has no *statutory* power to discharge a juror, except in a case where there is doubt as to the juror's capacity to act effectively by reason of physical disability or insufficient understanding of English (Coroners Rules 1984, r. 52). It is not certain whether he has a *common law* power to discharge a juror without having to start the inquest all over again with a fresh jury. It has been suggested, *obiter*, that a coroner has a common law power to discharge a juror and to continue the inquest provided that the number of jurors does not fall below the minimum number of seven (*R* v *Merseyside Coroner (ex parte Carr)*, above, *per* Neill LJ at p. 73).

At the inquest, witnesses attend and give evidence on oath. If necessary, their attendance can be compelled. The procedure is inquisitorial, unlike the procedure in the ordinary courts of law. Interested parties or their representatives are permitted to question witnesses. 'Representatives' include authorised advocates (as defined in s. 119(1) of the Courts and Legal Services Act 1990), and not just barristers and solicitors (Coroners Rules 1984, r. 20(1), as amended).

Within his limited role in the legal process, it is somewhat anomalous that a coroner has wider powers than those of a High Court judge. Unlike the latter, the coroner decides which witnesses to call and then takes an active part in their examination. If the evidence given at the inquest is likely to be of a technical nature, the coroner is entitled to invite an assessor to sit with him if he considers it necessary to do so. Under the control of the coroner, the assessor can examine witnesses who give technical evidence. However, the assessor must not be allowed to give expert evidence himself since that could create the impression that the assessor's evidence was accorded by the coroner greater weight than it deserved (*R v Surrey Coroner (ex parte Wright)* [1997] 1 All ER 823).

With the exception of the coroner's summing-up to the jury, no person is allowed to address the jury (or the coroner) on the *facts* (Coroners Rules 1984, r. 40). The coroner must, however, allow submissions on the *law* to be made to him, such as a submission (made in the absence of the jury) about what verdict ought or ought not to be left to the jury. A coroner's failure to do so may amount to a violation of natural justice leading to the quashing of the inquest by the High Court (*R v Southwark Coroner (ex parte Hicks)* [1987] 2 All ER 140, DC; *R v East Berkshire Coroner (ex parte Buckley)* (1992) *The Times*, 1 December, DC).

Where a jury has been used at the inquest, its verdict need not be unanimous as long as there are not more than two dissentients (Coroners Act 1988, s. 12(2)). The jury cannot return a verdict of murder, manslaughter, or infanticide against a named person (ibid., s. 11(6)). The appropriate verdict in such circumstances is that the deceased was killed unlawfully. The standard of proof required for a verdict of unlawful killing is the standard applicable in criminal cases, namely, satisfaction beyond reasonable doubt (*R v West London Coroner (ex parte Gray)* [1987] 2 All ER 129, DC; *R v Wolverhampton Coroner (ex parte McCurbin)* [1990] 2 All ER 759, CA).

If criminal proceedings are already pending in respect of the death, the inquest is adjourned until the outcome of those proceedings is known (ibid., s. 16(1), as amended by the Criminal Justice Act 2003). The coroner may then resume the adjourned inquest if he thinks there is sufficient cause to do so (ibid., s. 16(3)). However, the finding of the inquest must not be inconsistent with the outcome of the criminal proceedings (ibid., s. 16(7)). The decision whether to resume an adjourned inquest under s. 16(3) is clearly of a highly discretionary nature. The coroner, however, need only be satisfied that there is 'sufficient cause' for a resumption; it is not necessary that there should be exceptional circumstances (*R v Inner West London Coroner (ex parte Dallaglio)* [1994] 4 All ER 139, CA; see further on this case, para 2.5.5 below).

If, before the conclusion of an inquest, the coroner is informed by the Lord Chancellor that (a) a public inquiry conducted or chaired by a judge is being, or is to be, held into the circumstances of the death, and (b) the Lord Chancellor considers that the cause of death is likely to be adequately investigated by the inquiry, the coroner must adjourn the inquest unless there is an 'exceptional reason' not to do so (Coroners Act 1988, s. 17A(1), as inserted by the Access to Justice Act 1999). If the inquest is adjourned, the Lord Chancellor must send the coroner the findings of the public inquiry as soon as reasonably practicable after their publication (ibid., s. 17A(3)). The inquest must not be resumed by the coroner unless in his opinion there is an exceptional reason for doing so (ibid., s. 17A(4)). If the inquest is not resumed, the coroner must send to the registrar of deaths a signed certificate stating the findings of the public inquiry (ibid., s. 17A(6)). These provisions significantly reduce the possibility that concurrent inquests and public inquiries into the same deaths will produce inconsistent findings of fact.

In addition to unlawful killing, other verdicts suggested by the Coroners Rules 1984 include death from natural causes, death from dependence on drugs or non-dependent abuse of drugs, death by accident or misadventure, that the deceased killed himself, and an open verdict for use where the evidence does not fully disclose the cause of death.

In the case of a death caused by casual glue-sniffing, the appropriate verdict is death by solvent abuse, or a similar verdict, rather than the more offensive and uninformative one of 'death by drug abuse' (R v Inner South London Coroner (ex parte Kendall) [1989] 1 All ER 72, DC).

It has been held that there is no distinction between the two words, 'accident' and 'misadventure', and that in modern parlance the word 'accident' is to be preferred (R v Portsmouth Coroner (ex parte Anderson) [1988] 2 All ER 604, DC).

Suicide is never to be presumed. It must be proved by evidence and, if it is not, the appropriate verdict is an open one (R v Cardiff Coroner (ex parte Thomas) [1970] 3 All ER 469, DC; R v HM Coroner for the City of London (ex parte Barber) [1975] 3 All ER 538, DC). Suicide is not an appropriate verdict unless the death has ensued within a year and a day of a self-inflicted injury (R v Inner West London Coroner (ex parte De Luca) [1988] 3 All ER 414, DC).

2.5.5 Judicial review of coroner's proceedings

It is not possible to appeal against the finding of a coroner's inquest. The proceedings are, however, subject to judicial review in the High Court, which has an inherent common law power of supervision over inquests. On an application for judicial review, an inquest can be quashed by a quashing order for fraud, excess or refusal of jurisdiction, violation of the principles of natural justice, or for an error of law made within the coroner's jurisdiction. (For the distinction between an appeal and judicial review, see para 10.9.1.)

In R v Inner West London Coroner (ex parte Dallaglio) [1994] 4 All ER 139, CA, a coroner's decisions were quashed on the ground of his apparent bias. The case arose following the deaths in August 1989 of 51 people in a collision on the River Thames between the dredger Bowbelle and the passenger launch Marchioness. Inquests were opened but adjourned pending the outcome of criminal proceedings against the master of the dredger. In an official letter, the coroner described some relatives and survivors as 'mentally unwell', and, at a meeting with journalists, he referred to one bereaved mother as 'unhinged'. He later decided, first, not to accede to a request from some of the bereaved families to remove himself from the proceedings on the ground of apparent bias, and, secondly, not to resume the inquests. It was held that the coroner's unfortunate choice of words had given rise to a real possibility that he had unconsciously allowed himself to become biased against the relatives. His decisions were quashed and the matter was remitted to a different coroner for a fresh decision on whether there should be a resumption of the inquests. (A different coroner ordered the inquests to be resumed and, in April 1995, a coroner's jury brought in a verdict of unlawful killing. See *The Times*, 8 April 1995, and see further on this case, para 10.5.3.2.1.)

An alternative to judicial review is for the matter to be taken to the High Court, by or with the authority of the Attorney-General, in the form of a statutory application to quash the inquest under s. 13 of the Coroners Act 1988. Under this statutory procedure the grounds for interference are probably wider than those available to the court at common law, extending as they do to 'fraud, rejection of evidence, irregularity of proceedings, insufficiency of inquiry, the discovery of new facts or evidence, or otherwise' (ibid., s. 13(1)). The High Court has power to quash an inquest, and order a fresh one, on these grounds where it is necessary or desirable in the interests of justice to do so (ibid.).

Under the s. 13 procedure, it is not necessary to show a *probability* that a fresh inquest would produce a different verdict; a *possibility* of a different verdict is sufficient (*Re Rapier (deceased)* [1986] 3 All ER 726, DC, in which it was the coroner himself who, through the Attorney-General, made the successful statutory application to quash; but note *R* v *Inner London North District Coroner (ex parte Linnane) (No. 2)* [1991] COD 12, DC, and *R* v *West Berkshire Coroner (ex parte Thomas)* (1991) *The Times*, 25 April, DC, in both of which a fresh inquest was refused despite irregularities in the course of the original inquest).

If the High Court decides to interfere, its powers under s. 13 are limited to quashing the proceedings and ordering a fresh inquest. It has no power to substitute a verdict for one wrongly returned at the inquest, although such a power has been said to be desirable (*R* v *Birmingham and Solihull Coroner (ex parte Secretary of State for the Home Department)* (1990) *The Independent*, 2 August, DC, *per* Watkins LJ).

Although it is possible to seek to have an inquest quashed by combining an application under s. 13 of the Coroners Act 1988 with an application for judicial review, it has been held to be preferable to proceed under s. 13 in cases where the Attorney-General has given his authority under that section (*R* v *West Berkshire Coroner (ex parte Thomas)*, above).

Section 13 can also be used to compel the holding of an inquest in cases where the coroner refuses or neglects to hold one which ought to be held (ibid., s. 13(1)). In *Terry* v *East Sussex Coroner* [2001] 3 WLR 605, CA, it was decided that this provision does not confer a wide jurisdiction to order an inquest where the coroner was not at fault. The High Court should not declare itself satisfied that an inquest 'ought to be held' unless the coroner has misdirected himself in law or his factual conclusions are irrational.

2.5.6 Circumstances in which a coroner's jury is compulsory

There are some circumstances in which the summoning of a jury is compulsory. By s. 8(3) of the Coroners Act 1988, the coroner *must* have a jury in cases of:

(a) deaths in prison;

(b) deaths occurring in police custody, or resulting from an injury caused by a police officer in the purported execution of his duty;

(c) deaths caused by an accident, poisoning or disease, notice of which is required to be given to a government department under any statute or to a health and safety at work inspector;

(d) deaths occurring 'in circumstances the continuance or possible recurrence of which is prejudicial to the health or safety of the public'.

Category (b) was introduced in 1982 following public disquiet over, *inter alia*, the number of deaths occurring in police custody. 'Police custody' for this purpose does not necessarily involve direct physical control of a person. Thus, a man was held to have died 'in police custody', within the meaning of s. 8(3)(b) of the Coroners Act 1988, where he died in a hospital without a police guard. The deceased had become ill and had been moved to the hospital from the police station at which he had begun to serve a term of imprisonment because of overcrowding in the prisons. The court granted a mandatory order directing the coroner to summon a jury (*R v Inner London North District Coroner (ex parte Linnane)* [1989] 2 All ER 254, DC).

The deceased's son was concerned that any neglect in his father's care should be publicly exposed. When at the subsequent inquest the jury returned a verdict of death from natural causes, the coroner having refused to call as a witness a doctor consulted by the son and having withdrawn the issue of lack of care from the jury, the son applied under s. 13 of the Coroners Act 1988 (para 2.5.5 above) to have the inquest quashed. It was held that, although there had been 'rejection of evidence' and 'insufficiency of inquiry' on the part of the coroner within the meaning of s. 13, the application would be dismissed since it would be wrong to order a fresh inquest 21 months after the death and there was little evidence to show lack of care by the police (*R v Inner London North District Coroner (ex parte Linnane) (No. 2)* [1991] COD 12, DC).

The words in category (d) fell to be construed by the Court of Appeal in *R v Hammersmith Coroner (ex parte Peach)* [1980] 2 All ER 7, CA, a case which was decided before the addition of category (b) to the list. A teacher from New Zealand, Blair Peach, was killed in Southall while watching a demonstration which turned into a riot. It was alleged that the violent blow to the head which killed him was struck by a police officer. It was further alleged that the weapon used was much heavier than a police truncheon. Police inquiries unearthed a number of unauthorised weapons in the lockers of policemen who had been present at the demonstration.

The coroner refused to summon a jury to the inquest, and the deceased's family applied to the High Court for quashing and mandatory orders. The High Court refused the orders but, on appeal, the Court of Appeal granted a quashing order to quash the coroner's decision to sit without a jury and a mandatory order to compel him to have a jury when the inquest resumed.

The Court of Appeal held that the true construction of the words in category (d), above, was that the coroner must sit with a jury if the circumstances of the death were such that similar fatalities might recur and it was reasonable to expect that proper action ought to be taken by some responsible public body to prevent that happening. In the present case, there was reason to suspect that the death had resulted from the unauthorised use of a potentially lethal weapon by a police officer. In that situation the presence of a jury was required because, if that suspicion was confirmed, it would be reasonable to expect the police authority to take action to prevent it happening again. The possible recurrence of those circumstances would be 'prejudicial to the health or safety of the public' (see the other examples given by Lord Denning MR at [1980] 2 All ER 7, 10).

There need be no causative link between the deceased's death and the circumstances in which it occurred. Accordingly, a coroner must summon a jury if the deceased has died in circumstances the continuance or possible recurrence of which would be prejudicial to public health or safety even though they may not have caused the deceased's death (*R v Inner London North District Coroner (ex parte Linnane)* [1989] 2 All ER 254, DC).

If the circumstances of the case come within category (d) above, the coroner must summon a jury even though the death took place abroad. In the case of *In re Neal (coroner: jury)* (1995) *The Times*, 9 December, DC, the death occurred in a holiday apartment in Spain and was caused by carbon monoxide gas emanating from a water heater. The Surrey coroner sat without a jury and returned an open verdict. The deceased's father applied under s. 13 of the Coroners Act 1988 to have the inquest quashed and a fresh one ordered. In deciding that the coroner should have sat with a jury, the court said it was just as important that travellers to Spain should be protected from the dangers of gas heaters as those who stayed at home. However, the court, in the exercise of its discretion, declined to quash the inquest proceedings because the case was an old one, the death having occurred two years earlier, and in the circumstances an open verdict could not be impugned as irrational.

2.6 Public funding of proceedings

As regards proceedings before the Court of Justice of the European Communities, the Court itself can grant public funding. However, in the case of an application for a preliminary ruling under art. 234 it will only do so if public funding is not available under the domestic law of the country from which the reference emanated. This is because the reference is regarded as a step in the national proceedings, and the European Court will normally leave questions of costs and funding to be determined by the national court. Public funding under the Community Legal Service is available in preliminary ruling proceedings originating in courts in England and Wales.

Public funding is available for representation in proceedings in the Employment Appeal Tribunal (Access to Justice Act 1999, sch. 2).

Public funding is *not* available for representation in proceedings in the Judicial Committee of the Privy Council (except in relation to its jurisdiction under the Government of Wales Act 1998, the Scotland Act 1998, or the Northern Ireland Act 1998: ibid.).

Public funding is *not* available for representation in coroners' courts, except in the case of deaths in police or prison custody, or in other cases where there is a significant wider public interest in the client being represented.

2.7 Reform of coroners' courts

Government proposals for improving efficiency and increasing public confidence in the coroners' courts were published in March 2004 (*Reforming the Coroner and Death Certification Service: A Position Paper*, Cm 6159, 2004).

The proposals were made in response to a comprehensive review of the coroner service published by the Home Office in June 2003 (*Death Certification and Investigation in England, Wales and Northern Ireland: The Report of a Fundamental Review*, Cm 5831, 2003), and to one of Dame Janet Smith's reports into the activities of Dr Harold Shipman published in July 2003 (*Shipman Inquiry, Third Report: Death Certification and the Investigation of Deaths by Coroners*, Cm 5854, 2003).

Important points arising from the government's Position Paper include the following.

(a) Consideration is being given to the question of overall responsibility for the coroner service. Presently, it lies with the Home Office but the *Fundamental Review* recommended that it be moved to the Department for Constitutional Affairs (formerly the Lord Chancellor's Department).

(b) The coroner service for England and Wales should be a single, nationally managed, service supervised by a Chief Coroner and an advisory Coronial Council.

(c) It is envisaged that there would be full-time area coroners, supplemented by part-time coroners as case-loads and local circumstances require.

(d) The Chief Coroner would be responsible for the deployment of coroners, and for overseeing professional standards and complex inquests.

(e) Coroners should have legal qualifications, although existing medically-qualified coroners would be able to apply for posts in the proposed new structure.

(f) Coroners should be subject to a programme of mandatory professional development.

(g) A medical examiner employed by the coroner service should provide a second certification of death in all cases.

(h) Coroners should have increased powers to enter premises or seize documents when investigating the circumstances of a death.

(i) In the case of inquests of exceptional length and complexity, the coroner should be able to appoint a lawyer to act as counsel to the inquest.

(j) Inquests should continue to be mandatory in certain cases, including deaths in custody, multiple deaths after a disaster, workplace deaths, and deaths alleged to have been caused by agents of the state.

(k) The use of juries at inquests should continue on the present basis. The maximum number of jurors should be reduced to nine, although an inquest might continue provided there were no fewer than seven jurors.

(l) The use of narrative verdicts is preferred to the current short-form verdicts like 'accident' or 'misadventure', which were found to be used inconsistently and not always sufficiently explanatory. A narrative verdict is a short, descriptive, and non-judgmental, account of the facts leading to a death, and will provide a fuller explanation than is presently the case. Before reaching a final conclusion, the decision of the House of Lords in *R (Middleton)* v *West Somerset Coroner* [2004] 2 WLR 800 (para 2.5.4 above) will be taken into account.

(m) A Family Charter should be established, setting out exactly what the deceased's family can expect from the coroner service.

(n) It is intended to give further consideration to the proposal of the *Fundamental Review* that public funding for legal representation at inquest proceedings should be extended to families in all cases where a public authority is legally represented.

(o) In response to the recommendation of the *Fundamental Review* that the coroner's jurisdiction in respect of treasure should be abolished, options are being considered to ensure that historic finds can be investigated without undue diversion of resources away from inquisitions into deaths.

Responsibility for the coroner service was transferred from the Home Office to the Department for Constitutional Affairs in June 2005. In February 2006, the government announced its intention to reform the coroner system by legislation. Although there will be new national leadership and standards, the proposal to create a national coroner organisation has been dropped (*Coroners' Service Reform: Briefing Note*, Department for Constitutional Affairs, February 2006).

3

Tribunals

3.1 Introduction

Alongside the ordinary civil courts already described in Chapters 1 and 2 there exist many hundreds of tribunals dealing with a wide variety of disputes arising between the individual citizen and the state, or between citizen and citizen. The proliferation of tribunals has been a special feature of the development of judicial administration in England and Wales over recent decades. The number of tribunals and their importance have increased so significantly that it is no longer justifiable to regard tribunals merely as an appendage to the ordinary courts of law. They are an integral part of the ordinary legal process, a position which has been reached largely due to the implementation of the *Report of the Committee on Administrative Tribunals and Enquiries* (the Franks Committee), Cmnd 218, 1957. Most of the committee's recommendations were implemented by the Tribunals and Inquiries Act 1958 (since repealed and replaced by the Tribunals and Inquiries Act 1971, which, in turn, was repealed and replaced by the Tribunals and Inquiries Act 1992).

The distinction between a court and a tribunal was noted in para 1.1.4. Tribunals are regarded as inferior to the ordinary courts of law, even though for the most part they are independent in the exercise of their various jurisdictions. Because of this inferiority, and by reason of the fact that even tribunals described as 'administrative' are exercising *judicial* functions, they are subject to the supervisory prerogative jurisdiction of the High Court so that complaints of unlawful conduct on the part of tribunals may lead to the granting of remedies against them by the Administrative Court (see Chapter 10).

The workload of tribunals greatly exceeds that of the county courts. As with the county courts, most of the cases heard by tribunals are small cases. Even in a small case, however, the outcome is of great importance to the parties directly involved. Some matters heard by tribunals may involve substantial sums of money, as in the case of claims for redundancy payments or compensation for unfair dismissal determined by employment tribunals. Others may involve the deprivation of a person's freedom, as in the decision of a mental health review tribunal that a patient should not be discharged from compulsory detention.

In this chapter it is proposed to examine (a) administrative tribunals, which have been established by statute to resolve a myriad of specific disputes; (b) employment tribunals, an important species of administrative tribunal, which hear certain defined disputes arising out of the employment relationship; and (c) domestic tribunals, which are usually established privately in order to enforce discipline within a profession or trade union.

3.2 Administrative (or statutory) tribunals

3.2.1 Reasons for existence

Administrative tribunals are established, in the main, to resolve disputes between a private citizen and a central government department, such as claims to social security benefits, disputes which require the application of specialised knowledge or experience, such as the assessment of compensation following the compulsory purchase of land, and other disputes which by their nature or quantity are considered unsuitable for the ordinary courts, such as immigration appeals or fixing a fair rent for premises.

The main reasons for the creation of administrative tribunals may be identified as the relief of congestion in the ordinary courts of law, the provision of a speedier and cheaper procedure than that afforded by the ordinary courts, and the desire to have specific issues dealt with by persons with an intimate knowledge and experience of the problems involved. Thus, tax experts sit as Special Commissioners of income tax, surveyors sit on the Lands Tribunal, and medical practitioners sit on medical appeal tribunals to determine medical questions under the industrial injuries scheme.

3.2.2 Defects in the tribunal system

Several criticisms of administrative tribunals are common.

(a) Although representation (whether by a lawyer, trade union official, social worker, or other person) is allowed before the vast majority of tribunals, there remain a few tribunals where *legal representation* is not permitted.

(b) A more fundamental criticism is that, with a few exceptions, *public funding* is not available for representation in tribunal proceedings. (See further, para 3.5 below.)

(c) Many tribunals decide disputes to which the government minister who appointed the members of the tribunal is a party. This makes it difficult for a tribunal to achieve the *appearance* of impartiality, although there is no evidence of bias *in fact* towards a minister's case.

The Franks Committee (para 3.1 above) recommended that all chairmen of tribunals should be appointed by the Lord Chancellor, and that other members should be appointed by the Council on Tribunals. These proposals were not acceptable to the government of the day, although in many cases a compromise formula was reached under which the Lord Chancellor appointed a panel of chairmen from which the appropriate minister made his selection. Ordinary members were usually appointed by the appropriate minister, but quite often they were not dismissible without the consent of the Lord Chancellor.

Matters changed in 2000 when the terms of service of part-time judicial office-holders, including tribunal appointments at both chairman and lay member levels, were revised so as to guarantee their independence. Part-time tribunal appointments are now for a renewable fixed period, subject to the relevant upper age limit. Appointments are normally to be renewed automatically, except for limited and specified grounds, and removal from office is only on limited and specific grounds. The grounds for non-renewal and removal are set out in para 1.5.4.

The result of the changes is that while decisions not to renew appointments or to remove from office are still taken by the Secretary of State, he can only act with the agreement of the Lord Chief Justice and following an investigation conducted by a judge nominated by the Lord Chief Justice. Thus, although part-time tribunal members remain appointable by a government minister they cannot now be removed, or refused renewal by him, without the involvement of the senior full-time judiciary.

(d) There remain a few instances where tribunals are not obliged to give *reasons for their decisions*. However, the vast majority of tribunals are covered by the requirement of s. 10 of the Tribunals and Inquiries Act 1992 that reasons for decisions must be given if requested. The reasons, which can be furnished either in writing or orally, are taken to form part of the decision of the tribunal and, as such, to be incorporated in the record of the proceedings (ibid., s. 10(6)), even if given orally. This provision enables the Administrative Court, on an application for judicial review, to quash a tribunal's decision for error of law on the face of the record if any reason given is bad in law (para 10.5.3.3).

(e) Sometimes there is *no right of appeal* against a tribunal's decision, although an application for judicial review is not an 'appeal' and can therefore be made to the Administrative Court, on the grounds of error of law, violation of the principles of natural justice, or excess of jurisdiction, notwithstanding the fact that an 'appeal' may be precluded. (For the distinction between an appeal and judicial review, see para 10.9.1.)

Since administrative tribunals are statutory bodies, a right of appeal can only be conferred by statute. In fact, a right of appeal does exist from the decisions of most tribunals, although Parliament has not prescribed a uniform appellate procedure. In some instances, an appeal lies from one tribunal to another; in others, it lies from a tribunal to a minister. Or, again, an appeal may lie from a tribunal to a court of law.

3.2.3 Control of administrative tribunals

To a large extent the criticisms of administrative tribunals mentioned in para 3.2.2 are countered by other rules of law. Thus, tribunals are obliged to observe the principles of natural justice, namely, that no man may be a judge in his own cause (the rule against bias) and that both sides to the dispute must be heard and, moreover, given a hearing which is fair (see further, para 10.5.3.2).

Administrative tribunals are public statutory bodies inferior to the ordinary courts of law and, as such, they are subject to the supervision of the Administrative Court, which, on an application for judicial review of a tribunal's decision, may issue the prerogative remedies of quashing orders, mandatory orders, and prohibiting orders. It may do so for a variety of reasons, such as where there has been a violation of the principles of natural justice, or where the tribunal has refused to exercise its jurisdiction in a particular case, or has exceeded or threatened to exceed its jurisdiction, or where the tribunal has made an error of law. (For judicial review and the prerogative remedies, see Chapter 10.)

3.2.4 Examples of administrative tribunals

There are over 2,000 tribunals falling under the supervision of the Council on Tribunals. The enormous variation from one tribunal to another in terms of composition, jurisdiction, procedure, and appellate machinery makes it difficult to describe and compare tribunals. One common factor is that they were all created by Act of Parliament. Some examples of the more important administrative tribunals are the Asylum and Immigration Tribunal, employment tribunals, the Lands Tribunal, medical appeal tribunals, mental health review tribunals, valuation tribunals, rent assessment committees, the VAT and Duties Tribunal, General Commissioners of income tax, Special Commissioners of income tax, the Financial Services and Markets Tribunal, Social Security and Child Support Commissioners, social security appeal tribunals, the Transport Tribunal, and the traffic commissioners.

3.3 Employment tribunals

3.3.1 Introduction

The employment tribunals, which sit locally, were originally established as 'industrial tribunals' with a somewhat limited role by the Industrial Training Act 1964, since when their jurisdiction has been expanded considerably—notably under the employment legislation of the 1960s and 1970s. Their composition, jurisdiction and procedure are now regulated by the Employment Tribunals Act 1996 (as amended by the Employment Relations Act 1999) and regulations made thereunder by the Secretary of State.

'Industrial tribunals' were renamed 'employment tribunals' by the Employment Rights (Dispute Resolution) Act 1998 with effect from 1 August 1998.

3.3.2 Composition

When hearing a case, each employment tribunal is normally composed of a legally qualified chairman sitting with two lay members. However, with the consent of the parties a case can be heard by the chairman and one lay member. Some proceedings may be heard by the chairman sitting alone (Employment Tribunals Act 1996, s. 4, as amended by the Employment Rights (Dispute Resolution) Act 1998).

The chairman, who is appointed by the Lord Chancellor, must have a seven-year general qualification within the meaning of s. 71 of the Courts and Legal Services Act 1990 (Employment Tribunals (Constitution and Rules of Procedure) Regulations 2004 (SI 2004, No. 1861), reg. 8(3)). The lay members are taken from panels appointed

by the Secretary of State after consultation with appropriate organisations representing employers and employees respectively (ibid.).

In *Scanfuture UK Ltd* v *Secretary of State for Trade and Industry* [2001] IRLR 416,
EAT, it was held by the Employment Appeal Tribunal that the arrangements operating
in 1999 for the appointment of lay members of an employment tribunal contravened
art. 6 of the European Convention on Human Rights (right to a fair trial). Since those
members were appointed by the Secretary of State who was a party to the proceedings,
the employment tribunal was not 'independent and impartial'. Furthermore, the
defect was not cured by the existence of an appellate body to correct the mistakes of
the employment tribunal because the jurisdiction of the Employment Appeal Tribunal is not 'full', being limited to questions of *law* and not extending to questions
of *fact*.

By the time the *Scanfuture* case was decided, the Lord Chancellor had already
implemented new arrangements for appointment to part-time employment tribunal
posts at both chairman and lay member levels. The Employment Appeal Tribunal in
Scanfuture declared itself satisfied that, for the future, these arrangements provided
sufficient guarantees to exclude any legitimate doubt over the independence and
impartiality of employment tribunals.

The new arrangements are set out in para 3.2.2 above.

3.3.3 Procedure and appeals

An employment tribunal normally sits in public, except where it is in the interests of
national security or of confidentiality to sit in private (Employment Tribunals Act
1996, s. 10(2)–(5), as substituted by the Employment Relations Act 1999).

It has been held that the correct test for deciding whether a hearing is in public is
not whether any member of the public is prevented from attending it, but whether he
is able to attend if he wishes to do so. Thus, a hearing conducted in an office within a
secure area, and behind a locked door with a push-button coded lock, is not a public
hearing because the coded door lock is a physical barrier preventing all access to the
public and eliminating any chance of a member of the public 'dropping in' to see how
the tribunal proceedings are conducted. The fact that no one actually attempts to gain
access is irrelevant—the tribunal is sitting in private when it has no jurisdiction to do
so (*Storer* v *British Gas plc* [2000] 2 All ER 440, CA).

The proceedings are comparatively informal as the strict rules of evidence do not
apply (Employment Tribunals (Constitution and Rules of Procedure) Regulations
2004 (SI 2004, No. 1861), sch. 1, para 14). Any party may appear before an employment tribunal in person, or be represented by counsel or by a solicitor, or by a
representative of a trade union or an employers' association, or by any other person
whom he desires to represent him (Employment Tribunals Act 1996, s. 6(1)). This
right to representation is an unqualified right, and the tribunal has no power to stop a
party's chosen representative from acting or to order that party to represent himself
(*Bache* v *Essex County Council* [2000] 2 All ER 847, CA).

Although legal representation is permitted before an employment tribunal, public
funding is not available to pay for it. The tribunal has power to award costs against a
party if he or his representative has acted vexatiously, abusively, disruptively, or

otherwise unreasonably in relation to the proceedings (Employment Tribunals (Constitution and Rules of Procedure) Regulations 2004, above, sch. 1, para 40).

The decision of the tribunal may be unanimous or by majority. An appeal lies to the Employment Appeal Tribunal from most decisions of an employment tribunal, but the appeal can be taken only on questions of *law* as opposed to *fact* (Employment Tribunals Act 1996, s. 21).

3.3.4 Jurisdiction

The jurisdiction of the employment tribunals is as conferred on them by the Employment Tribunals Act 1996 and by other statutes passed before and after the 1996 Act (ibid., s. 2). It includes the following matters.

(a) Questions relating to written particulars of employment.

(b) Questions relating to redundancy payments.

(c) Disputes relating to equality clauses under the equal pay legislation.

(d) Complaints of unfair dismissal.

(e) Appeals against improvement and prohibition notices under the health and safety legislation.

(f) Complaints of unlawful sexual discrimination in employment.

(g) Complaints of unlawful racial discrimination in employment.

(h) Complaints relating to maternity pay; time off work for trade union activities, for public duties or to look for work or training; refusal to provide a written statement of reasons for dismissal; and failure to provide an itemised pay statement.

(i) Complaints of unlawful deductions from wages.

(j) Complaints of unreasonable exclusion or expulsion from a trade union where there exists a union membership agreement.

(k) Complaints from individuals of having been unjustifiably disciplined by a trade union.

(l) Complaints of unlawful refusal of employment, or of the services of an employment agency, on grounds related to trade union membership.

(m) Complaints of unlawful disability discrimination in employment.

(n) Complaints of discrimination or victimisation on grounds of religion or belief or of sexual orientation.

(o) The Lord Chancellor has power to confer jurisdiction on the employment tribunals to hear claims for damages for breach of the contract of employment, or of any other contract connected with employment, excluding claims for damages for personal injuries (Employment Tribunals Act 1996, s. 3, repeating a provision first enacted in 1971).

Under this rule, the employment tribunals have been given jurisdiction to adjudicate on contract claims made by both employees and employers which arise, or are outstanding, on the termination of the employee's employment. Excluded are claims for breach of a contractual term relating to intellectual property, restraint of trade, the provision of living accommodation, or the imposition of an obligation of confidence. There is a limit of £25,000 on the amount of compensation that can be ordered in these proceedings (Employment Tribunals Extension of Jurisdiction (England and Wales) Order 1994 (SI 1994, No. 1623)).

3.4 Domestic tribunals

3.4.1 Introduction

Most domestic tribunals are set up by private bodies for their own internal purposes. Examples are the disciplinary committees of a trade union, professional association, or university. These domestic tribunals differ from administrative tribunals in that the latter are set up by *statute* to decide matters of a more *public* nature. Some domestic tribunals, however, are created by statute, such as the bodies which have been established to regulate the professional conduct of medical practitioners, dentists, opticians, and veterinary surgeons.

3.4.2 Control by the ordinary courts of law

All domestic tribunals, however created, are subject to the control of the ordinary courts of law in two ways. First, a domestic tribunal must observe the principles of natural justice. If it fails to do so, its proceedings may be challenged in the courts. For example, in *R v Aston University Senate (ex parte Roffey)* [1969] 2 QB 538, DC, it was made clear that a student ought not to be expelled from his university (not even for failing his examinations twice) without first being given an opportunity of making representations in his own defence. On the facts of this particular case, however, relief was refused because of the applicant's delay in seeking it.

Secondly, a domestic tribunal, however created, must not exceed its jurisdiction as laid down in its rules or its statute of creation. For example, a trade union which acts *ultra vires* (beyond its powers) by expelling a member for some conduct not covered by its rules can be sued for a declaration (*Kelly v NATSOPA* (1915) 84 LJKB 2236, CA) that the expulsion was wrongful and for an injunction (*Lee v Showmen's Guild* [1952] 2 QB 329, CA) to prevent the union from acting on the wrongful expulsion. The expelled person may, in addition, be awarded damages for breach of the contract that exists between a member and his trade union, the basis of which is the union rules. If the wrongful expulsion has caused financial loss, as where the claimant is unable to find another job because of the operation of a closed shop, which happened in *Bonsor v Musicians' Union* [1956] AC 104, HL, the damages may be quite substantial.

In relation to trade union membership, there are now special statutory procedures which exist *in addition to* the common law position just discussed.

3.4.3 Remedies against domestic tribunals

It has been seen that all domestic tribunals are under an obligation to observe the principles of natural justice and to act within their powers. A breach of either obligation may lead to the intervention of the ordinary courts of law at the suit of an interested party. There is, however, a difference in the remedies available according to the type of tribunal involved. For unlawful conduct on the part of domestic tribunals exercising functions which are purely *private*, the appropriate remedies are the private law remedies of declarations, injunctions, and damages. The powers of such voluntary tribunals exist within the field of *private* contract law and the prerogative remedies of quashing orders, mandatory orders, and prohibiting orders, which operate in the field of *public* law, are not available against them. They are not public bodies performing public duties (*R* v *Criminal Injuries Compensation Board (ex parte Lain)* [1967] 2 QB 864, DC, *per* Lord Parker CJ at p. 882).

The legal position remains the same after the introduction in 1978 of a new procedure for claiming public law remedies, called an 'application for judicial review'. Thus, in *R* v *British Broadcasting Corporation (ex parte Lavelle)* [1983] 1 All ER 241 it was held that an application for judicial review was not available as a means of challenging the decision of a disciplinary tribunal set up by the BBC. The appropriate remedy against a tribunal of this character is to seek damages, or an injunction or declaration, in ordinary private law proceedings for breach of contract, although, if such remedy has already been inappropriately claimed in an application for judicial review, the court has a discretion to allow the proceedings to continue as if they had been begun as a private law action (CPR, r. 54.20; para 10.8.6.3). This is a sensible provision which obviates the necessity to refuse the application and compel the claimant to start the action all over again, with a consequent throwing away of legal costs, before a different court.

On the other hand, judicial review *is* available against domestic tribunals exercising functions of a *public* nature (see, e.g., *R* v *Panel on Takeovers and Mergers (ex parte Datafin plc)* [1987] 1 All ER 564, CA; para 10.8.2.4).

3.5 Public funding of proceedings

Public funding for representation is available in proceedings before the Asylum and Immigration Tribunal, mental health review tribunals, the Protection of Children Act Tribunal (established by the Protection of Children Act 1999), and (under certain conditions) the General and Special Commissioners of income tax and the VAT and Duties Tribunal. It is also available in the Employment Appeal Tribunal, but this is a court rather than a tribunal.

Research has shown that representation before tribunals significantly improves a claimant's chances of success. However, public funding is not available for most administrative tribunal proceedings. Thus, for example, although there is an unrestricted right to representation in such important tribunals as employment tribunals, social security appeal tribunals, Social Security Commissioners, and medical appeal tribunals, no public funding is available to assist with its cost. Persons involved in proceedings before such tribunals may, however, qualify for other forms of help under the Community Legal Service.

For access to justice and public funding, see para 1.9.

3.6 Reform of the tribunal system

In May 2000, the government commissioned an independent review of the tribunal system under the chairmanship of Sir Andrew Leggatt, a former Lord Justice of Appeal. The task of the review was to examine the administrative justice system as a whole in the light of its coherence, accessibility, and organisation. The report of the review team was published in 2001 as *Tribunals for Users—One System, One Service* (Lord Chancellor's Department, August 2001). It contained many recommendations for fundamental change.

A consultation paper based on the Leggatt report was issued in 2001 and, in March 2003, the government announced that there would be created, over a period of time, a new, unified Tribunals Service with responsibility for most central government tribunals. The aims are to increase accessibility to tribunals, raise customer service standards, and improve administration. Following a 'transitional year', the Tribunals Service is expected to be fully established by April 2006.

4

Legislation and statutory interpretation

4.1 Introduction

In the English legal system the law-making process is shared by two bodies. While Parliament claims to be the sole domestic law-maker, its powers are supplemented by the activities of the judiciary, which traditionally has adopted a much less overt role.

Parliament, consisting of the Queen (whose royal assent is necessary by constitutional convention before a Bill can become an Act), the House of Lords, and the House of Commons, passes legislation in the form of Acts of Parliament, alternatively called statutes. The interrelationship between legislation and the judiciary is of crucial importance, for the judges are the ultimate enforcers of the law, whether civil or criminal, and whether parliamentary in origin or judge-made. A judge will spend a significant proportion of his time on the interpretation of legislation.

Once a statute is in force it must be applied by the courts. The judges have 'judicial notice' of all public Acts whenever passed (Interpretation Act 1978, s. 3) and of all private Acts passed after 1850 (ibid., s. 22(1) and sch. 2, para 2). A private Act passed before 1851 must be pleaded and proved in evidence unless there is a deeming provision in it by which it is to be treated as a public Act and judicially noticed as such (1 Bl Com 85; *Greswolde* v *Kemp* (1842) Car & M 635).

'Judicial notice' is a convenient device whereby knowledge of certain matters is attributed to judges so that those matters do not need to be proved in evidence every time they arise in a particular case. (See *R* v *Simpson* [1983] 3 All ER 789, CA, at pp. 793–4.) If a judge makes a decision without reference to a relevant statute, either because he was ignorant of it or it was not cited to him by counsel, that in itself provides a ground for appeal. The judges also have judicial notice of the European Union Treaties, the *Official Journal* of the Union, and decisions and opinions of the European Court of Justice or any court attached thereto (European Communities Act 1972, s. 3(2), as amended).

If a statutory provision is ambiguous, the judge must *interpret* it before he can *apply* it. Statutory interpretation, which is by no means a precise art, involves the application of certain rules and presumptions (para 4.10 below). Before looking at the role of statutory interpretation in the judicial process it will be helpful to examine more closely parliamentary and delegated legislation, and the relationship between Parliament and the judiciary, especially in the light of the United Kingdom's membership of the European Union.

4.2 Advantages of statute law

Several advantages are claimed for statute law—usually at the expense of judge-made law. (The advantages of judge-made law are discussed in para 5.6.)

Since Parliament is supreme, a statute can abrogate any rule of law, whether contained in a previous statute or a previous case. The judges have no reciprocal power to abrogate a statutory provision (see, e.g., *British Railways Board* v *Pickin* [1974] AC 765, HL, para 4.6 below). There are, however, some circumstances arising from our membership of the European Union in which the judges may be expected to disapply the provisions of a statute of the United Kingdom Parliament in order to give priority and effect to directly applicable Community law (see para 4.7 below).

Normally when Parliament abolishes a rule of law it will only do so for the future so that previous transactions based on the old rule are not affected. Occasionally, Parliament passes legislation which has retrospective effect (para 4.10.3.6 below). When an English court overrules a decided case it usually does so with retrospective effect (para 5.3.1).

The constitutional doctrine of the separation of powers is to a greater extent satisfied by legislation than by judge-made law. Parliament *makes* law but does not *enforce* it, whereas judge-made law, or case law, is made by the very people who enforce it.

While statute law can be known in advance, case law can only be known at the same time that it is made. If there is no statutory provision covering a particular legal point, the parties may not know what their rights are until after the judge has decided the dispute between them.

Of course, much of the advantage claimed here for statute law is lost if the particular statutory provision is ambiguous—a state of affairs which is very common. A clear, unambiguous statute can undoubtedly prevent much litigation. The proper development of case law is too often dependent on the 'accidents of litigation'. The courts are not allowed for the most part to give opinions on hypothetical, as opposed to real, cases involving actual litigation between citizens. Thus, unsettled points in the common law tend to remain unresolved for a long time, either because a similar factual situation has not arisen or because the parties cannot afford, or do not wish, to litigate because of the costs involved. A statute could settle such doubtful points of law immediately, although perhaps a more practicable solution would be the establishment of a suitors' fund to allow points of law of general public importance to be taken to the highest court at the public expense.

It should be possible for statutory provisions to be expressed, not only in authoritative form, but clearly. This is not always achievable with judge-made law where it may often be necessary to separate the *ratio decidendi* from the *obiter dicta* in order to discover what a case actually decided (para 5.5).

It is said that Parliament is more in touch with the outside world than is the judiciary. It is certainly true that Parliament can more quickly turn public opinion and social policy into new law. Although, since 1966, the House of Lords as our highest court has not been absolutely bound to follow its own previous decisions (para 5.3.2.2.1), it still does not enjoy the freedom to make new law possessed by Parliament. The legislature is free to make law on any subject it thinks fit, while the courts are constrained by the facts of the cases before them and by the constitutional understanding that they must not trespass on the function of Parliament. (See the examples given in para 5.3.2.2.3.)

4.3 Consolidating statutes and codifying statutes

4.3.1 Consolidation

Consolidation is the re-enactment in one statute of some topic in the law, previously contained in several different statutes, but without changing the law. 'All consolidation Acts are designed to bring together in a more convenient, lucid and economical form a number of enactments related in subject-matter . . . previously scattered over the statute book' (*Farrell* v *Alexander* [1977] AC 59, HL, *per* Lord Simon at p. 82).

A consolidating statute is presumed not to change the law but only to re-enact it in a different place. It is therefore possible, when interpreting such a statute, to apply cases already decided on the meaning of the replaced Acts. In cases of difficulty or ambiguity, the Acts which have been consolidated may themselves be scrutinised (*Farrell* v *Alexander*, above; *R* v *Heron* [1982] 1 All ER 993, HL; *Cullen* v *Rogers* [1982] 2 All ER 570, HL).

In *Epping Forest District Council* v *Essex Rendering Ltd* [1983] 1 All ER 359, HL— not, one would have thought, a case involving great difficulty or ambiguity—the House of Lords considered the legislative history of the Public Health Act 1936, a consolidating measure, before deciding that the requirement of the written consent of a local authority under that Act to the establishment of an offensive trade was mandatory and not merely directory. The legislative precursors of the Public Health Act 1936 were so emphatic that the requirement was mandatory that it was not permissible to construe a consolidating measure in such a way as to change the nature of the requirement from mandatory to directory, even though the 1936 Act was expressed in slightly different language.

On the other hand, in *Sheldon* v *R.H.M. Outhwaite Ltd* [1995] 2 All ER 558, HL, the House declined to look at the Acts consolidated by the Limitation Act 1980 in construing s. 32(1)(b) of that Act because the wording of paragraph (b) was clear and involved no difficulty or ambiguity.

Only corrections and minor improvements can be made during the process of consolidation (Consolidation of Enactments (Procedure) Act 1949, s. 1). If a joint committee of both Houses of Parliament certifies that a Bill is a consolidating Bill (and does not change the law) it can expect a speedy passage through Parliament since full debate is largely unnecessary. (For the various types of consolidating statute, see *R* v *Heron* [1982] 1 All ER 993, HL, *per* Lord Scarman at p. 999.)

4.3.2 Codification

Codification is a restatement in one place of the law on a particular topic, which, if necessary, may also alter the law. It embraces not only previous statutes but also common law principles derived from previous cases. There was a flurry of codification towards the end of the nineteenth century when certain commercial law topics were set down in codified form in, for example, the Bills of Exchange Act 1882, the Partnership Act 1890, and the Sale of Goods Act 1893. In the twentieth century, the Theft Act 1968, based on the *Eighth Report of the Criminal Law Revision Committee: Theft and Related Offences* (Cmnd 2977, 1966), attempted to codify the law relating to theft.

These codifying statutes, which are proceeded with in a piecemeal fashion, are not 'codes' in the Continental sense because there is no attempt to put down in one place the whole of English law. They do, however, resemble the Continental type of code in that they put the law on a particular topic into a compact form which makes it easier to apply.

Codification of the law is a difficult and long process (see Diamond, 'Codification of the law of contract', (1968) 31 MLR 361). It is one of the objectives of the Law Commission, a full-time law reform body for England and Wales set up by the Law Commissions Act 1965. Codification of the law is expressly mentioned in s. 3(1) of the Act. One of the Law Commission's first tasks was to prepare a code of the law of contract, but the project soon ran into difficulties and had to be abandoned as a comprehensive exercise (see Diamond, op. cit., and Law Commission, *Eighth Annual Report* (for 1972–73)).

4.4 Preparation of legislation

The responsibility for initiating the vast majority of modern legislation rests with the government. Private Members' Bills are unlikely to become law because of shortage of parliamentary time.

The legislative proposal of a particular government department may be approved in principle in the cabinet and then handed to the Parliamentary draftsman ('Parliamentary Counsel to the Treasury') to be put into legal language in the form of a Bill. Sometimes the Bill is preceded by a White Paper containing the government's proposals for legislation.

Legal language is used because the English practice is to legislate by using precise words seeking to cover every conceivable situation and allowing of no loopholes. Complaint is frequently made that the use of legal language makes the legislation unintelligible to the layman.

There is sufficient evidence of this; indeed, some legislation is unintelligible to lawyers. Displays of judicial exasperation with legislative drafting are not uncommon. In *Davis* v *Johnson* [1979] AC 264, HL, Viscount Dilhorne (at p. 333) said of s. 1 of the Domestic Violence and Matrimonial Proceedings Act 1976 (now repealed and replaced by the Family Law Act 1996) that '[f]ew, if any, sections of a modern Act can have given rise to so much litigation in so short a time and to such a difference of opinion'. The Landlord and Tenant Act 1987 has been branded as 'ill-drafted, complicated and confused' (*Denetower Ltd* v *Toop* [1991] 3 All ER 661, CA, *per* Sir Nicolas Browne-Wilkinson V-C at p. 668), while the Dangerous Dogs Act 1991 has been said to bear 'all the hallmarks of an ill-thought out piece of legislation' (*R* v *Ealing Justices (ex parte Fanneran)* (1995) *The Times*, 9 December, DC, *per* Rougier J).

As long ago as 1975, the *Report of the Committee on the Preparation of Legislation* (Cmnd 6053) recommended that legislation should be arranged to suit the convenience of users, not legislators. It was said that the modest system of issuing explanatory notes with new Bills should be extended, and that statements of purpose and principle should be encouraged. It was recommended that there should be more parliamentary draftsmen, and more and quicker consolidation of statutory provisions. The practice of issuing explanatory notes with public Bills did not become a regular one until 1999 (see further, para 4.10.5.6 below).

It is important for democracy that the law should command the respect of the people who are expected to abide by 'the rules'. It is equally important that they should know what 'the rules' are. Legislation expressed in vague and cumbersome language is likely to bring the law into disrepute because it cannot be understood by those whose duty it is to observe it. Clarity and simplicity of expression should be the objectives of the draftsman.

Difficulties often arise where one statute is amended by a subsequent statute which simply refers to the previous statute without setting out the law as amended. Sometimes statutes are amended more than once in this way, so that it becomes increasingly difficult and time-consuming to discover what the relevant law is. It is necessary to read two or more statutes side by side. In the absence of regular consolidation, a preferable system of amendment would be for the amending provision to set out the amended law.

In *Merkur Island Shipping Corporation* v *Laughton* [1983] 2 All ER 189, HL, Lord Diplock made a plea for greater clarity and simplicity in legislative drafting. This was in a case where the House of Lords had to consult three Acts of Parliament, none of them intelligible by itself, in order to decide whether some secondary industrial action by employees was actionable in tort. His Lordship said (at pp. 198–9):

I see no reason for doubting that those on whom the responsibility for deciding whether and if so what industrial action shall be taken in any given circumstances wish to obey the law, even though it be a law which they themselves dislike and hope will be changed through the operation of this country's constitutional system of parliamentary democracy. But what the law is, particularly in the field of industrial relations, ought to be plain. It should be expressed in terms that can be easily understood by those who have to apply it even at shop floor level. . . . Absence of clarity is destructive of the rule of law; it is unfair to those who wish to preserve the rule of law; it encourages those who wish to undermine it. The statutory provisions which it became necessary to piece together into a coherent whole . . . are drafted in a manner which, having regard to their subject matter and the persons who will be called on to apply them, can in my view, only be characterised as most regrettably lacking in the requisite degree of clarity.

4.5 Commencement of an Act of Parliament

A Bill consists of clauses, subclauses, and paragraphs. After it has been through its Parliamentary stages and has received the royal assent it is an Act, consisting of sections, subsections, and paragraphs. The royal assent to a Bill has not been refused since Queen Anne refused to assent to the Scottish Militia Bill in 1707. There is no law which requires the royal assent to be given, although there is probably today a constitutional convention that it will not be withheld.

An Act comes into force on the date specified in the 'commencement section', if any, situated usually towards the end of the statute. If there is no commencement section, there may be an 'appointed day section', authorising a Secretary of State to implement the Act by means of an order made by statutory instrument. The Act may provide that different parts can be brought into effect on different dates.

An appointed day section gives the minister a discretion to bring the Act, or parts of it, into effect when he feels it is appropriate to do so. He cannot be compelled to implement the Act on any particular date. His discretion, however, is not completely unfettered. Parliament must be taken to intend that legislation will come into force at some time, and that its commencement will not depend entirely on ministerial whim. The minister will not be allowed to frustrate the intentions of Parliament by, for example, deciding not to implement the Act at all. Any such decision would be an abuse of power and unlawful.

Thus, in *R v Secretary of State for the Home Department (ex parte Fire Brigades Union)* [1995] 2 All ER 244, HL, a decision by the Home Secretary never to implement certain parts of the Criminal Justice Act 1988—which would have established a statutory Criminal Injuries Compensation Scheme in place of the non-statutory, prerogative-based scheme introduced in 1964—was held to be unlawful. (Subsequently, the relevant parts of the Criminal Justice Act 1988 were repealed by the Criminal Injuries Compensation Act 1995 without ever having been brought into force. By the same Act, the prerogative scheme, under which compensation was assessed on common law principles, was replaced by an inferior statutory, tariff-based scheme.)

If there is no commencement section and no appointed day section, the Act comes into force on the day it receives the royal assent, which nowadays is normally signified simply by being announced to the two Houses of Parliament (Royal Assent Act 1967). Personal assent by the Queen during a visit to Parliament is still possible (ibid.), although this has not been done since 1854.

An Act of Parliament is deemed to have been in force for the whole of the day of its commencement (Interpretation Act 1978, s. 4, derived from the Interpretation Act 1889 and the Acts of Parliament (Commencement) Act 1793). For example, in *Tomlinson v Bullock* (1879) 4 QBD 230, DC, the court was concerned with the Bastardy Laws Amendment Act 1872, which had laid down that any unmarried woman who had an illegitimate child 'after the passing of this Act' could apply for an affiliation order against the father. The royal assent to the Act was given on 10 August 1872, and the Act came into operation immediately. The applicant's illegitimate child was born on the same day. It was held that she could apply for an affiliation order. The actual hour of assent was irrelevant as the statute was deemed to have been in force for the whole of that day.

4.6 **Parliamentary sovereignty and the validity of statutes**

Because Parliament is supreme it is not possible for anyone to challenge the validity of a statute in the courts, even though it is unreasonable or its passage was produced by fraud or some other irregularity. There were suggestions down to the early seventeenth century that the judges could declare void an unreasonable statutory provision. In *Dr Bonham's Case* (1610) 8 Co Rep 114, Coke CJ said (at p. 118) that 'when an Act of Parliament is against common right and reason, or repugnant, or impossible to be performed, the common law will control it, and adjudge such act to be void'. (See also *Day* v *Savadge* (1615) Hob 85, *per* Hobart CJ at p. 97: 'Even an Act of Parliament, made against natural equity, as to make a man judge in his own cause, is void in itself'.)

However, Coke's statement was *obiter* and is not consistent with his support of parliamentary sovereignty expressed in the *Institutes* (4 Inst 36). If the judges were to arrogate to themselves the power to control legislation in the way suggested, conflict with Parliament would be inevitable.

In more recent times, Lord Denning MR once suggested, extra-judicially, that the courts may have power to strike down legislation on grounds of unconstitutionality along the lines of the function exercised by the Supreme Court of the USA (*The Misuse of Power: The Richard Dimbleby Lecture 1980*, reprinted in Denning, *What Next in The Law*, 1982, pp. 309–31).

Lord Denning's view is not only unorthodox but also bereft of authority. The true legal position is exemplified by decisions like that of the House of Lords in *British Railways Board* v *Pickin* [1974] AC 765, HL, where the respondent challenged the validity of a private Act alleging that British Rail had fraudulently concealed facts from Parliament. The respondent complained that the Act would deprive him of his land. It was held, applying in particular a dictum of Lord Campbell in *Edinburgh & Dalkeith Railway Co.* v *Wauchope* (1842) 8 CL & F 710, at p. 725, that the respondent could not challenge the validity of the statute. His claim was struck out as frivolous, vexatious, and an abuse of the process of the court. The attitude of the House of Lords was summarised thus by Lord Morris (at pp. 788–9):

The question of fundamental importance which arises is whether the court should entertain the proposition that an Act of Parliament can be so assailed in the courts that matters should proceed as though the Act or some part of it had never been passed. I consider that such doctrine would be dangerous and impermissible . . . When an enactment is passed there is finality unless and until it is amended or repealed by Parliament. In the courts there may be argument as to the correct interpretation of the enactment: there must be none as to whether it should be on the Statute Book at all.

(See also *R* (*Jackson*) v *Attorney-General* [2005] 3 WLR 733, HL, in which a challenge was mounted to the ban on fox hunting with dogs imposed by the Hunting Act 2004 enacted under the terms of the Parliament Act 1911 (which, as amended by the Parliament Act 1949, allows a statute to be enacted without the assent of the House of Lords in Parliament). The House of Lords in its judicial capacity, sitting nine-strong, held that both the 1949 Act and the 2004 Act are valid.)

4.7 Parliamentary sovereignty and the European Union

4.7.1 Introduction

Section 2(1) of the European Communities Act 1972 provides as follows:

All such rights, powers, liabilities, obligations and restrictions from time to time created or arising by or under the Treaties, and all such remedies and procedures from time to time provided for by or under the Treaties, as in accordance with the Treaties are without further enactment to be given legal effect or used in the United Kingdom shall be recognised and available in law, and be enforced, allowed and followed accordingly.

The result of s. 2(1) is that European Community law, whether arising from the treaties or from Community regulations, and whether such law has already been made or is to be made in the future, is to be directly applicable in the United Kingdom without the need for the United Kingdom Parliament to pass a statute each time. In addition, by s. 2(4), any 'enactment' (which is wide enough to cover a statutory instrument as well as a statute) passed or to be passed in the United Kingdom must be construed with directly applicable European Community law in mind.

Thus, the 1972 Act lays down that European Community law overrides existing domestic law whenever the two conflict and the former is directly applicable (*Duke* v *GEC Reliance Ltd* [1988] 1 All ER 626, HL; *Factortame Ltd* v *Secretary of State for Transport (No. 1)* [1989] 2 All ER 692, HL, *per* Lord Bridge at pp. 700–1; *Factortame Ltd* v *Secretary of State for Transport (No. 2)* [1991] 1 All ER 70, HL, *per* Lord Bridge at p. 108). It also lays down a presumption of interpretation that future United Kingdom statute law is to be read subject to European Community law. It is antici-pated thereby that Parliament will not pass legislation which is inconsistent with Community law. It is presumed that a United Kingdom statute is consistent with Community law unless and until it is declared by a court to be inconsistent (*Factortame Ltd* v *Secretary of State for Transport (No. 1)* [1989] 2 All ER 692, HL, *per* Lord Bridge at pp. 702–3).

Although the English courts will strive to achieve consistency with Community law where a United Kingdom statute has been passed in order to give effect to our Com-munity obligations (*Pickstone* v *Freemans plc* [1988] 2 All ER 803, HL; *Litster* v *Forth Dry Dock and Engineering Co. Ltd* [1989] 1 All ER 1134, HL, para 4.10.6 below), they will not deliberately misconstrue the meaning of a United Kingdom statute in an effort to enforce against an individual some provision of Community law which is not directly applicable and which, therefore, creates no directly enforceable rights between individuals (*Duke* v *GEC Reliance Ltd* [1988] 1 All ER 626, HL; *Finnegan* v *Clowney Youth Training Programme Ltd* [1990] 2 All ER 546, HL).

The European Communities Act 1972 does not *expressly* forbid Parliament from amending or repealing that Act itself. The effect on parliamentary sovereignty of the accession of the United Kingdom to the European Union was questioned in *Blackburn* v *Attorney-General* [1971] 2 All ER 1380, CA (see also *McWhirter* v *Attorney-General* [1972] CMLR 882, CA).

Mr Blackburn applied for a declaration that the United Kingdom government would, by signing the Treaty of Rome, surrender in part the sovereignty of Parliament and would surrender that part for ever, which, he argued, would be in breach of the law. The Court of Appeal held that Mr Blackburn's statement of claim disclosed no cause of action and should be struck out. The Court said that making treaties was not the function of Parliament but of the Crown acting through government ministers. It was a matter of the royal prerogative, the exercise of which could not be challenged in the courts. On the binding of Parliament's successors, the Court of Appeal said that in legal theory it was not possible, but what the legal position would be if Parliament did try to bind its successors was a hypothetical question and should not be decided in the present case.

The point arose again in *Macarthys Ltd* v *Smith* [1979] 3 All ER 325, CA. On this occasion, although the question of a possible conflict between United Kingdom legislation and European Community law was still hypothetical, the Court of Appeal was much more vigorous in its response. Lord Denning MR said this (at p. 329):

If on close investigation it should appear that our legislation is deficient or is inconsistent with Community law by some oversight of our draftsmen then it is our bounden duty to give priority to Community law. Such is the result of s. 2(1) and (4) of the European Communities Act 1972 . . .

Thus far I have assumed that our Parliament, whenever it passes legislation, intends to fulfil its obligations under the Treaty. If the time should come when our Parliament deliberately passes an Act with the intention of repudiating the Treaty or any provision in it or intentionally of acting inconsistently with it and says so in express terms then I should have thought that it would be the duty of our courts to follow the statute of our Parliament. I do not however envisage any such situation.

Lawton LJ (at p. 334) agreed:

Parliament's recognition of European Community law and of the jurisdiction of the European Court of Justice by one enactment can be withdrawn by another.

The question of *implied*, as opposed to *express*, repeal of the European Communities Act 1972 was raised in *Thoburn* v *Sunderland City Council* [2002] 3 WLR 247, DC, a case on the unlawful use of a weighing machine calibrated in pounds and ounces but not in metric units. It was here argued that the Weights and Measures Act 1985 had, by implication, partially repealed s. 2(2) of the European Communities Act 1972 (power to implement Community obligations by subordinate legislation). The argument was rejected, it being held that the 1972 Act is, by force of common law, a 'constitutional' statute and, unlike an 'ordinary' statute, cannot be *impliedly* repealed. A 'constitutional' statute can only be repealed *expressly*.

According to Laws LJ in the *Thoburn* case ([2002] 3 WLR 247 at [62]), a 'constitutional' statute is one which '(a) conditions the legal relationship between citizen and State in some general, overarching manner, or (b) enlarges or diminishes the scope of . . . fundamental constitutional rights'. Other examples of constitutional statutes given are the Magna Carta, the Bill of Rights, the Act of Union, the Reform Acts which distributed and enlarged the franchise, the Human Rights Act 1998, and the devolution statutes passed in 1998 affecting Scotland and Wales.

In *R* v *Secretary of State for Foreign and Commonwealth Affairs (ex parte Rees-Mogg)* [1994] 1 All ER 457, DC, an unsuccessful attempt was made to demonstrate that the United Kingdom government's ratification of the Maastricht Treaty (the Treaty on European Union signed at Maastricht on 7 February 1992) would be unlawful on the grounds, *inter alia*, that it would be in breach of what was then s. 6 of the European Parliamentary Elections Act 1978, and that it would involve transferring to Europe without statutory authority that part of the royal prerogative relating to the conduct of foreign and security policy.

These objections were dismissed by the court, which held that ratification of the Treaty would not be unlawful. Section 6 of the European Parliamentary Elections Act 1978 (now s. 12 of the European Parliamentary Elections Act 2002) provided that a treaty increasing the powers of the European Parliament could not be ratified by the United Kingdom unless it had been approved by an Act of the United Kingdom Parliament. It was held that ratification would not violate s. 6 of the 1978 Act because the Maastricht Treaty (including all its titles, protocols, and declarations) had been approved for the purposes of that section by s. 1(2) of the European Communities (Amendment) Act 1993.

The court was doubtful whether the prerogative point was justiciable in the courts. It was decided that, in any event, the point was without merit since ratification of the Maastricht Treaty would not involve a *transfer*, but an *exercise*, of prerogative power similar to that involved in membership of bodies like the United Nations Organisation or the North Atlantic Treaty Organisation. To emphasise the fact that the prerogative was not being transferred or abandoned, the court said that it would presumably be open to the government, as a last resort, to denounce the Treaty or at least to fail to honour its obligations under the provisions concerning foreign and security policy.

4.7.2 The *Factortame* litigation

Since a deliberately engineered conflict (of the type referred to by Lord Denning in *Macarthys* v *Smith* [1979] 3 All ER 325, CA, para 4.7.1 above) between English domestic law and Community law (in which, for example, a statute of the United Kingdom Parliament *expressly* provides that it is to override Community law) has not yet arisen, its legal consequences remain unresolved. In the meantime, the supremacy of directly applicable Community law over what may be described (perhaps euphemistically) as 'inadvertently' inconsistent English domestic law is well illustrated by the *Factortame* litigation concerning the Spanish fishing industry.

In the field of employment, Community law makes directly effective provision for the removal of discrimination between citizens of Member States on grounds of nationality. It also recognises the right of vessels registered in the Community to fish in any Community waters, but lays down a system of fishing quotas. Dissatisfied with the allocation of quotas to Spain, a number of Spanish fishing companies began 'quota-hopping'. They turned their attentions to British quotas by purchasing vessels already registered as British, or by re-registering their own vessels under the British flag.

The British government reacted by securing the passage through Parliament of the Merchant Shipping Act 1988. Part II of the Act, and regulations made thereunder, introduced new registration conditions so that some of the fishing boats belonging to the Spanish companies no longer qualified for British registration—with potentially disastrous consequences for the Spanish companies and their employees since the boats were not eligible to resume fishing against the Spanish quotas. The Spanish companies argued that the 1988 Act, and the regulations made under it, contravened their rights under Community law, and 95 of them applied to the High Court for judicial review of the validity of the legislation.

The High Court decided to request a preliminary ruling from the European Court of Justice under art. 234 (ex art. 177) of the EC Treaty on, *inter alia*, the interpretation of Community law provisions relating to non-discrimination on grounds of nationality and the right to establish a business in another Member State. Because it would take two years to get a ruling (by which time the Spanish companies could have suffered irreparable damage), the High Court also decided to grant an interim injunction against the Crown to prevent the application of the 1988 Act to the Spanish companies in the meantime.

The Court of Appeal allowed the Crown's appeal and set aside the interim injunction. On a further appeal, the House of Lords affirmed the decision of the Court of Appeal and held (in accordance with conventional constitutional wisdom) that English courts had no power to suspend the operation of a statute and no power to grant an interim injunction against the Crown (*Factortame Ltd v Secretary of State for Transport (No. 1)* [1989] 2 All ER 692, HL). However, the House of Lords itself referred the following question to the European Court of Justice under art. 234: 'under Community law, must a national court ignore its own national law and provide interim relief for a person with directly enforceable Community law rights who would otherwise suffer irreversible damage because of delay in having those rights determined?'

In *Factortame Ltd v Secretary of State for Transport (No. 2)* [1991] 1 All ER 70, CJEC and HL, the European Court gave an affirmative answer to the question referred. A principle found in earlier European Court case law (see, for example, *Amministrazione delle Finanze dello Stato v Simmenthal SpA* [1978] 3 CMLR 263, CJEC) was re-emphasised and applied. This principle is that, under Community law, a national law (whether legislative, judicial, or administrative in character) must be set aside by a national court if it prevents the application of Community law. The principle ensures that the operation of Community law is not frustrated by an inconsistent national law. The House of Lords implemented the European Court's decision by granting an interim injunction against the Secretary of State for Transport (in effect, the Crown) to last until such time as the European Court made a decision on the merits of the Spanish companies' case. This injunction meant that, for the first time ever, the operation of a statute of the United Kingdom Parliament had been suspended by an English court of law.

In *R v Secretary of State for Transport (ex parte Factortame Ltd)* [1991] 3 All ER 769, CJEC, the European Court, in answer to the reference made to it by the High Court in 1989, held that the registration provisions contained in part II of the Merchant Shipping Act 1988, and in regulations made under it, were contrary to Community law, and, therefore, ineffective against nationals of other Member States.

The European Court gave its decision in *Factortame Ltd v Secretary of State for Transport (No. 2)* [1991] 1 All ER 70 in June 1990, and, in response to it, the House of

Lords decided to grant the interim injunction in July 1990 but reserved its reasons for doing so until October 1990. In the intervening period, the decision of the European Court was subjected to considerable criticism.

In view of the effect of the decision on the sovereignty of the United Kingdom Parliament, it was greeted with some disbelief at Westminster, and in the media, by those who were either anti-European or less than enthusiastically pro-European.

In a famous dictum uttered in 1974, shortly after the United Kingdom became a member of the European Union, Lord Denning MR had likened the impact of Community law to an incoming tide, flowing into our estuaries and up our rivers, which could not be held back (*H.P. Bulmer Ltd* v *J. Bollinger SA* [1974] Ch 401, CA, at p. 418). By 1990, in retirement and obviously concerned as the full implications of membership of the Communities became clearer from cases like *Factortame (No. 2)*, he had rephrased his dictum as follows:

No longer is European law an incoming tide flowing up the estuaries of England. It is now like a tidal wave bringing down our sea walls and flowing inland over our fields and houses—to the dismay of all.

(Quoted in *The Independent*, 16 July 1990; see also Crossick, 'The ebb and flow of Lord Denning' (1990) 140 New LJ 1431.)

He accused the European Court (which, he said, was manned by pan-Europeans whose ideology is allowed to influence their decisions; see Crossick, op. cit.) of interfering with parliamentary sovereignty in the absence of express authority conferred by the EC Treaty. He advocated the amendment of the European Communities Act 1972 to ensure that European Court decisions and Community directives would not be binding unless approved by, respectively, the House of Lords and the relevant British government minister.

Others accused the European Court of being a political institution guilty of abusing its powers by moving away from merely interpreting law to actually creating it in order to promote the ideal of European political union (see *The Independent*, 16 July 1990).

Referring to the criticisms of the European Court's decision, Lord Bridge said in *Factortame (No. 2)* [1991] 1 All ER 70, HL, at pp. 107–8, that they were based on a 'misconception':

If the supremacy within the European Community of Community law over the national law of Member States was not always inherent in the EEC Treaty it was certainly well established in the jurisprudence of the Court of Justice long before the United Kingdom joined the Community. Thus, whatever limitation of its sovereignty Parliament accepted when it enacted the European Communities Act 1972 was entirely voluntary. Under the terms of the 1972 Act it has always been clear that it was the duty of a United Kingdom court, when delivering final judgment, to override any rule of national law found to be in conflict with any directly enforceable rule of Community law . . . Thus there is nothing in any way novel in according supremacy to rules of Community law in those areas to which they apply and to insist that, in the protection of rights under Community law, national courts must not be inhibited by rules of national law from granting interim relief in appropriate cases is no more than a logical recognition of that supremacy.

4.8 Delegated (or subordinate) legislation

4.8.1 Types of delegated legislation

Delegated legislation is law made by some person or body other than Parliament but with the authority of Parliament. It can be made by the following persons and bodies.

(a) Ministers of the Crown, in the form of regulations or orders made by statutory instrument. This is a particularly well-used method of making law, and statutory instruments are numerous. For example, in 2005 Parliament passed 24 public general Acts; in the same year, over 3,330 statutory instruments were made by United Kingdom government departments and other United Kingdom authorities (excluding the Scottish administration and the National Assembly for Wales).

One of many uses for this method of law-making enables ministers to implement Community obligations without the need for primary legislation passed by Parliament (European Communities Act 1972, s. 2(2)). In addition, in the field of human rights they can make 'remedial orders' by statutory instrument to amend legislation which has been found to be incompatible with a right enshrined in the European Convention (Human Rights Act 1998, s. 10).

If the Legislative and Regulatory Reform Bill of 2006 is enacted, they will be given even wider powers to make law. A minister would be allowed to make an order by statutory instrument amending, repealing, or replacing *any* legislation (primary as well as subordinate) for the purposes of either reforming that legislation or of implementing recommendations of the Law Commission. In the latter eventuality, the minister's order would, in addition to reforming any relevant legislation, be capable of amending, abolishing, or codifying rules of the *common law*.

The government's justification for the Bill is that it is a measure designed to accelerate, first, the process of removing burdens caused by too much regulation, and, secondly, the translation of Law Commission reports into law. Despite the safeguards contained within it, the Bill has proved highly controversial and has attracted widespread criticism.

(b) The Privy Council (in effect the government for this purpose), in the form of Orders in Council. Thus, emergency regulations having the force of law can be made by Orders in Council under the Civil Contingencies Act 2004. This method of legislation was also used to impose economic sanctions under the Southern Rhodesia Act 1965. In addition, it is used under the provisions of s. 1 of the United Nations Act 1946 to give effect to resolutions of the United Nations Security Council, such as those relating to international terrorist organisations and Iraq.

(c) Local authorities, in the form of by-laws under such enabling statutes as the Public Health Act 1936 and the Local Government Act 1972.

(d) Public corporations, in the form of by-laws.

(e) Court rule committees, in the form of rules of court governing practice and procedure. Examples of such committees are the Civil Procedure Rule Committee (established by the Civil Procedure Act 1997, as amended by the Courts Act 2003); the Criminal Procedure Rule Committee (established by the Courts Act 2003 to replace the Crown Court Rule Committee), whose jurisdiction extends to the Court of Appeal (criminal division), the Crown Court, and criminal proceedings in the magistrates' courts; and the Family Procedure Rule Committee (established by the Courts Act 2003 to replace the Family Proceedings Rule Committee).

4.8.2 Reasons for delegated legislative powers

Delegated legislative powers are conferred by Parliament for a number of reasons. For example, Parliament does not have the time to discuss all Bills in detail. Members of Parliament may not have the necessary knowledge to deal with the details of technical Bills. In an emergency a law can usually be introduced more quickly by, say, ministerial regulation than by Act of Parliament.

On occasions, however, Parliament itself can act with surprising swiftness, as in 1965 when the Southern Rhodesia Act 1965 went through all its legislative stages in one day in response to the illegal unilateral declaration of independence in what is now Zimbabwe. The Imprisonment (Temporary Provisions) Act 1980 was passed in one day to deal with the consequences of industrial action by prison officers.

4.8.3 Control of delegated legislation by the courts

One important difference between parliamentary legislation and delegated legislation is that whereas a court of law cannot question the validity of an Act of Parliament (para 4.6 above), it *can* question the validity of delegated legislation. This is because of the doctrine of *ultra vires*. If, for instance, a minister acts *ultra vires* (beyond his powers) by making a regulation which he has no power to make, then that regulation can be declared void by the court, as in *Chester* v *Bateson* [1920] 1 KB 829, DC, and *Commissioners of Customs & Excise* v *Cure & Deeley Ltd* [1962] 1 QB 340 (para 10.2.2.3.2).

It should be noted that the court has the same power of control where the minister's excess of power arises out of a document which does not necessarily have the force of law. Thus, in *Laker Airways Ltd* v *Department of Trade* [1977] QB 643, CA, a minister was held by the Court of Appeal to have acted *ultra vires* when he issued a policy directive which had the effect of contradicting the express provisions of the statute which had given him the power to issue the directive.

Similarly, it was held in *R* v *Secretary of State for the Environment (ex parte Lancashire County Council)* [1994] 4 All ER 165 that the Environment Secretary had acted *ultra vires* when he issued to the Local Government Commission policy guidance (on replacing the two-tier structure of local government with unitary authorities) which undermined the provisions of a statute. (A revised version of the policy guidance was upheld as lawful in *R* v *Secretary of State for the Environment (ex parte Lancashire County Council) (No. 2)* (1995) *The Times*, 9 December.)

Regulations can also be declared *ultra vires* if they conflict with statutory rights already conferred by previous primary legislation (*R* v *Secretary* of *State for Social Security (ex parte Joint Council for the Welfare of Immigrants)* [1996] 4 All ER 385, CA).

Statutory instruments are void if they conflict with European Community legislation (European Communities Act 1972, s. 2(4)).

In addition, local authority by-laws can be declared void by the court if they are *uncertain* or *unclear* (*Percy* v *Hall* [1996] 4 All ER 523, CA), or if they are *unreasonable* (*Kruse* v *Johnson* [1898] 2 QB 91, DC). For examples of by-laws held to be unreasonable, see *Arlidge* v *Islington Corporation* [1909] 2 KB 127, DC; *Parker* v *Bournemouth Corporation* (1902) 86 LT 449, DC; and *Strickland* v *Hayes* [1896] 1 QB 290, DC, where a by-law prohibiting the singing or reciting of any obscene song or ballad, and the use of obscene language generally, was held to be unreasonable and void because it was drawn too widely in that it was not limited to public places and it did not require the prohibited acts to be done to the annoyance of the public.

4.9 **European Community legislation: primary and secondary**

Much of English domestic law, such as land law, family law, and criminal law, is unaffected by United Kingdom membership of the European Union. European Community law is concerned mainly with competition, consumer protection, employment, agriculture, fisheries, coal, steel, nuclear energy, the environment, and the free movement of labour, capital, and services.

The *primary* European Community legislation consists, *inter alia*, of the three Treaties which established the Communities, namely, the European Coal and Steel Community Treaty, the Euratom Treaty, and the European Community Treaty, together with the Single European Act, the Treaty on European Union (the Maastricht Treaty), the Treaty of Amsterdam, and those provisions of the Treaty of Nice and the Treaty of Athens which relate to the European Communities. The provisions of these Treaties are directly applicable in the United Kingdom, without the need for further legislation by the United Kingdom Parliament, by virtue of the European Communities Act 1972, s. 2(1) (para 4.7.1 above).

The *secondary* European Community legislation consists of regulations, decisions, directives, recommendations, and opinions, made or given by the Council of Ministers or the Commission.

Regulations are binding and directly applicable without the need for further legislation by the United Kingdom Parliament (EC Treaty, art. 249; European Communities Act 1972, s. 2(1)).

Decisions are binding on those to whom they are addressed (EC Treaty, art. 249). They may be addressed to any or all Member States, to enterprises, or to individuals.

Recommendations and *opinions* are not binding as they stand (EC Treaty, art. 249), although they can be given legal effect in the United Kingdom by statute, or by delegated legislation made under the European Communities Act 1972, s. 2(2).

Directives are not directly applicable, but they are binding as to the objectives to be achieved. The United Kingdom is under an obligation to implement them within the stated time-limit by means of domestic legislation.

Directives can sometimes produce direct legal effects and create rights without having been implemented by domestic legislation if they fulfil certain criteria, notably that their terms are unconditional and sufficiently precise (see, e.g., *Van Duyn* v *Home Office* [1975] Ch 358, CJEC). When directives *are* directly effective they have 'vertical direct effect' but not 'horizontal direct effect' (*Marshall* v *Southampton and South West Hampshire Area Health Authority (Teaching)* [1986] QB 401, CJEC; *Foster* v *British Gas plc* [1990] 3 All ER 897, CJEC). This means that they have direct effect in proceedings against a Member State ('vertically'), but not in proceedings between individuals or companies ('horizontally').

For this purpose, the word 'state' has been given a wide meaning (*Foster* v *British Gas plc*, above; *Marshall* v *Southampton and South West Hampshire Area Health Authority (No. 2)* [1993] 4 All ER 586, CJEC), and in the United Kingdom would include local authorities and nationalised industries. Although an unimplemented directive may thus be enforced *against* a local authority as an emanation of the state, it cannot be enforced *by* a local authority since such a body is not an 'individual' for the purposes of Community law (*Wychavon District Council* v *Secretary of State for the Environment* (1994) *The Times*, 7 January).

A Member State which is in breach of Community law by failing to implement a directive may, in certain circumstances, be liable to compensate an individual who has suffered loss as a result of the non-implementation (*Francovich* v *Italian Republic* [1992] IRLR 84, CJEC).

The purpose of this case-law development is to prevent Member States from taking advantage of their own failure to comply with Community law. A Member State may be so liable if three conditions are fulfilled (see, for example, *Francovich*, above; *Faccini Dori* v *Recreb Srl* [1995] All ER (EC) 1, CJEC; *Dillenkofer* v *Germany* [1997] QB 259, CJEC, below). First, the purpose of the directive must be to confer rights on individuals. Secondly, the content of those rights must be identifiable from the provisions of the directive. Thirdly, there must be a direct causal link between the breach of the Member State's obligation and the damage sustained by the individual.

It is no defence to a Member State which fails to implement a directive to claim that the period allowed for implementation was too short, or that other Member States did not implement the directive on time either (*EC Commission* v *Italy* [1976] ECR 277, CJEC). In the case of *Dillenkofer*, above, it was held that failure to implement a directive within the prescribed period is a serious breach of community law, and that neither liability for the breach nor the obligation to pay compensation depends on proof of fault on the part of the guilty Member State. Accordingly, the German Federal Republic was liable to compensate some holidaymakers who suffered loss resulting from Germany's failure to implement a Council Directive of 1990 on package travel, package holidays, and package tours. The Directive was intended to, and did in a clearly identifiable way, confer on purchasers of packages a right to the reimbursement of sums already paid, and repatriation costs, in the event of the insolvency of the travel organiser and/or retailer.

The principle that a Member State which violates Community law must compensate an individual who suffers resultant loss applies equally to a breach of the Community

Treaties, whether that breach is perpetrated by the executive, the judiciary, or the legislature.

In *R v Secretary of State for Transport (ex parte Factortame Ltd)* [1991] 3 All ER 769, CJEC, it had been established by the European Court that the United Kingdom was in breach of the EC Treaty in its dealings with the Spanish fishermen (see para 4.7.2 above). In *R v Secretary of State for Transport (ex parte Factortame Ltd) (No. 4)* [1996] QB 404, CJEC, it was further held that the fishermen, who had sustained considerable financial loss, would be entitled to claim damages against the United Kingdom in the English courts—and it made no difference that it was Parliament which was responsible for the breach of the Treaty by passing the Merchant Shipping Act 1988—if the necessary requirements for state liability (to which absence of fault is no defence) were satisfied. These requirements are that the relevant provision of the EC Treaty was intended to confer rights on the fishermen; the breach was sufficiently serious; and there was a direct causal link between the breach and the damage suffered by the fishermen.

It was also laid down that exemplary damages can be awarded for breach of Community law if they could be awarded for similar claims founded on domestic law. Accordingly, in *R v Secretary of State for Transport (ex parte Factortame Ltd) (No 5)* (1997) *The Times*, 11 September, DC, the fishermen argued that they would be entitled to exemplary damages under English law if they could show that the United Kingdom had acted in an oppressive, arbitrary, or unconstitutional manner (see para 9.2.2).

It was held that they were entitled to claim *compensatory* damages because, subject to the issue of causation (which was left undecided pending a determination of the seriousness of the breach), the requirements for state liability had been satisfied. However, they were not entitled to *exemplary* damages because they would not have been awarded in a similar claim based on English domestic law. The court said that a breach of Community law was akin to a breach of statutory duty, and, in English law, exemplary damages are not available for a breach of a statutory duty unless the particular statute expressly says so, and in the present case there was no such express statutory provision.

The fishermen's claim for *exemplary* damages was not pursued when *R v Secretary of State for Transport (ex parte Factortame Ltd) (No. 5)* subsequently went to the Court of Appeal and the House of Lords. On the claim for *compensatory* damages, both the Court of Appeal ((1998) *The* Times, 28 April) and the House of Lords ([1999] 4 All ER 906) were unanimous in holding that the breach was sufficiently serious to entitle the fishermen to compensation for loss directly caused by the breach. The issue of causation was to be tried later.

Secondary European Community legislation is analogous to domestic delegated legislation, and it is subject to a similar method of judicial control. The *validity* of the *primary* European Community legislation (the Treaties) is not subject to review either in national courts or in the European Court of Justice. The power of the courts in relation to the Treaties is limited to *interpretation* (EC Treaty, art. 234; see para 2.1.3.2.1). But the *validity*, as well as the meaning or effect, of *secondary* Community legislation can be challenged both in national courts and in the European Court of Justice (EC Treaty, art. 234). A court in the United Kingdom, instead of determining the point itself, may decide to refer any question of the validity of secondary Community legislation to the European Court of Justice (ibid.).

4.10 Statutory interpretation

4.10.1 Introduction

As was mentioned in para 4.1, the function of the judges in relation to legislation is to apply it. If, however, the wording of the legislation is ambiguous, or its extent uncertain, its meaning or scope will need to be interpreted or construed first. Legislation is expressed in words, and words are an imperfect means of communication. Legislation may need interpretation because, for example, the Act is badly drafted, or because the subject-matter of the Act is so complex that errors are inevitable, or because the Act fails to provide for all possible contingencies.

In his task of statutory interpretation a judge is assisted by some so-called 'rules' and by some presumptions. In addition, there are available, within limits, some intrinsic and extrinsic aids to construction. These rules, presumptions, and aids may together be referred to as the principles of statutory interpretation.

Whatever rule, presumption, or aid the judge is minded to apply, it must always be remembered that his task is the same, a task which Lord Nicholls explained in the following way in *R* v *Secretary of State for the Environment, Transport and the Regions (ex parte Spath Holme Ltd)* [2001] 2 WLR 15, HL, at p. 37.

Statutory interpretation is an exercise which requires the court to identify the meaning borne by the words in question in the particular context. The task of the court is often said to be to ascertain the intention of Parliament expressed in the language under consideration. This is correct and may be helpful, so long as it is remembered that the 'intention of Parliament' is an objective concept, not subjective. The phrase is a shorthand reference to the intention which the court reasonably imputes to Parliament in respect of the language used. It is not the subjective intention of the minister or other persons who promoted the legislation. Nor is it the subjective intention of the draftsman, or of individual members or even of a majority of individual members of either House. These individuals will often have widely varying intentions. Their understanding of the legislation and the words used may be impressively complete or woefully inadequate. Thus, when courts say that such-and-such a meaning 'cannot be what Parliament intended', they are saying only that the words under consideration cannot reasonably be taken as used by Parliament with that meaning. As Lord Reid said in *Black-Clawson International Ltd* v *Papierwerke Waldhof-Aschaffenburg AG* [1975] AC 591, HL, at 613: 'We often say that we are looking for the intention of Parliament, but that is not quite accurate. We are seeking the meaning of the words which Parliament used'.

4.10.2 Rules

4.10.2.1 The literal rule

By the literal rule, the words used in a statute must be given their plain, ordinary, or literal meaning. The objective of the court is to discover the intention of Parliament as expressed in the words used (*Stock* v *Frank Jones (Tipton) Ltd* [1978] 1 All ER 948, HL, *per* Viscount Dilhorne at p. 951; see also the explanation of the literal rule given by Lord Diplock in *Duport Steels Ltd* v *Sirs* [1980] 1 All ER 529, HL, at p. 541).

The literal rule was applied in *Cutter* v *Eagle Star Insurance Co. Ltd* [1998] 4 All ER 417, HL, in which the House of Lords was concerned with the meaning of the word 'road' in the Road Traffic Act 1988. Their Lordships rejected a purposive approach in favour of giving the word 'road' its ordinary, literal meaning. The claimant was sitting in the front passenger seat of his friend's car parked in a multi-storey car park. Inflammable gas leaked inside the car from a can of lighter fuel, and when the driver returned to the car and lit a cigarette the gas was ignited and the claimant was injured. The claimant sued the driver for negligence and won the case, but the driver had no money with which to pay damages.

As required by law, the driver was insured under a motor vehicle policy against any liability for death or bodily injury to any person arising out of the use of his car on a 'road'. The Road Traffic Act 1988 defines the word 'road' as 'any highway and any other road to which the public has access'. The House of Lords held that the car park was not a road within the definition because a road provides for cars to move along it to a destination. A car park simply enables cars to stand and wait. The fact that a car can be driven across a car park does not make it a 'road' because that is merely incidental to the main function of parking. Similarly, the fact that a car is parked on a road does not make the road a 'car park'. Accordingly, the claimant had not been injured due to the use of the car on a 'road', and the insurance company was not liable to pay out on the driver's policy. (See further on this case, para 4.10.6 below.)

If the words used in the statute are quite clear, they must be applied even though the result is absurd, or even though one may dislike the statute (*Stock* v *Frank Jones (Tipton) Ltd* [1978] 1 All ER 948, HL, *per* Lord Edmund-Davies at p. 954), or even though the interpretation may inflict hardship on those affected by the legislation. Thus, in *Leedale* v *Lewis* [1982] 3 All ER 808, HL, the unanimous opinion of the House of Lords was that if the meaning of a taxation statute is *clear* it must be given effect to, and the court should not seek to discover some alternative (*ex hypothesi* wrong) interpretation merely because the true meaning involves hardship for the taxpayer (or, it may be added, because it might leave a loophole for frauds against the revenue: *Barnard* v *Gorman* [1941] AC 378, HL, *per* Viscount Simon LC at p. 384).

Similarly, it has been held that a person is not homeless for the purposes of the Housing (Homeless Persons) Act 1977 (now part VII of the Housing Act 1996) if he is occupying 'accommodation' within the ordinary meaning of that word, even though the place which he occupies is lacking in cooking and washing facilities and he is thereby compelled to eat out and to use a launderette for washing his clothes (*Puhlhofer* v *Hillingdon London Borough Council* [1986] 1 All ER 467, HL). It might be otherwise where the 'accommodation' is not, by reason of its size, capable of accommodating a person together with others who normally reside with him as members of his family (ibid., *per* Lord Brightman at p. 474). Such a place could not be described as 'accommodation' in any meaningful sense.

In *Smith* v *Director of Serious Fraud Office* [1992] 3 All ER 456, HL, it was held that a statute had taken away the right to silence in the circumstances specified. The case was concerned with some of the provisions of the Criminal Justice Act 1987 relating to serious or complex fraud. By s. 2(2) and (13), a person who without reasonable excuse fails to answer questions or provide information on matters relevant to an investigation by the Director of the Serious Fraud Office is guilty of an offence and liable to be imprisoned, or fined, or both. The question arose whether a defendant can rely on the privilege against self-incrimination and refuse to answer questions.

It was decided unanimously by the House of Lords that the right to silence was not to be implied into the express, clear wording of the 1987 Act, which showed that Parliament had intended to establish an inquisitorial system in cases of serious or complex fraud whereby the Director of the Serious Fraud Office can, in the questioning process, *require* a defendant to provide answers which might be self-incriminating. It was held that the defendant's right to silence has been removed in such cases, with the result that the Director can continue to question him after he has been charged and without giving a further caution. (It should be noted that a statement made by the defendant in response to a requirement imposed on him under s. 2 is only later admissible in evidence against him in a prosecution for making a false or misleading statement, or in a prosecution for some other offence where in his evidence he makes an inconsistent statement (s. 2(8)).)

Lord Mustill (with whose speech the other Law Lords agreed) pointed out that there is a strong presumption of interpretation against taking away the right to silence but that 'statutory interference with the right is almost as old as the right itself' ([1992] 3 All ER 456, HL, at pp. 471–2; see also *A.T. & T. Istel Ltd* v *Tully* [1992] 3 All ER 523, HL, another case in which the right to silence was under attack).

Sometimes the search for the ordinary, natural meaning of words can produce disagreement and surprising results at the highest level. The case of *R* v *Maginnis* [1987] 1 All ER 907, HL, was concerned with the interpretation of the Misuse of Drugs Act 1971, by s. 5(3) of which

. . . it is an offence for a person to have a controlled drug in his possession, whether lawfully or not, with intent to supply it to another . . .

The police had found a package of cannabis resin in the defendant's car. He said that the package was not his but had been left in his car by a friend for collection later. The defendant was convicted of the offence, and he appealed on the ground that his intention to return the drug to its owner did not amount to an intention to 'supply' the drug within the meaning of the statute. The Court of Appeal allowed his appeal, and the Crown appealed to the House of Lords. Their Lordships held, by a majority of four to one (Lord Goff dissenting), that the defendant was guilty of the offence because a person in unlawful possession of a controlled drug left with him by another person for safekeeping had the necessary 'intent to supply it to another' (even though the supply was not being made from the provider's own resources) if his intention was to return it to the other person and for that other person's purposes.

The majority of their Lordships purported to apply the ordinary, natural meaning of the word 'supply'. Lord Goff, however, dissented on that very point and referred to definitions of the word given in *The Shorter Oxford English Dictionary*. In his view, which would seem clearly preferable to that of the majority, the word 'supply' was not apt to describe a transaction in which A handed back to B goods which B had previously left with A. Thus, the cloakroom attendant, left luggage officer, warehouseman and shoe repairer do not, in ordinary parlance, 'supply' their customers.

Lord Goff was further of the opinion that the particular offence in question was aimed at drug 'pushers'. The defendant was not a 'pusher' and should have been charged with the lesser offence of 'unlawful possession'. If, he said, persons in the position of the defendant were to be convicted of 'possession with intent to supply', it was up to Parliament and not the courts to enlarge the definition of 'supply'. (See [1987] 1 All ER 907 at pp. 913–15.)

The application of the literal rule produced disagreement among their Lordships again in *R v Brown* [1996] 1 All ER 545, HL. Here, a majority of the House held that the offence under s. 5(2)(b) of the Data Protection Act 1984 (since repealed by the Data Protection Act 1998) of knowingly or recklessly using personal data other than for the purposes described in the relevant entry in the register of data users is not committed by a person who merely accesses information on a computer in order to read it on screen or by means of a print-out. The data is only 'used', and the offence committed, if the defendant thereafter goes on to make unauthorised use of it, such as by passing it on to someone else.

This conclusion was reached by giving the word 'use' in s. 5(2)(b) its natural and ordinary meaning of 'make use of' or 'employ for a purpose'. Since there was no evidence that the defendant, a police officer who had twice used the police national computer to check the registration numbers of vehicles owned by debtors of clients of his friend's debt collection company, had passed on the data to his friend he was not guilty of the offence.

Lords Griffiths and Jauncey dissented on the ground that the word 'use' required a broad construction in order to achieve the purpose of the 1984 Act and to prevent a serious gap in the protection it seeks to provide. The Act was passed to give effect to the United Kingdom's obligations under the Council of Europe Convention for the Protection of Individuals with regard to Automatic Processing of Personal Data.

The two dissentients adopted a purposive approach, preferring an interpretation of s. 5 which accorded with the purpose of the Convention, namely, to secure respect for individual rights and fundamental freedoms (especially the right to privacy) in relation to the automatic processing of personal data. In their view, a person 'uses' data stored in a computer if he informs himself of its content, and the section should be construed widely so as to cover not only the use of data for an unauthorised purpose but also the invasion of privacy resulting merely from the unauthorised display of data about other people (see especially *per* Lord Griffiths at pp. 554–5).

It must be appreciated in relation to the literal rule of interpretation that the ordinary meaning of a technical word is something technical. For example, the case of *Unwin v Hanson* [1891] 2 QB 115, CA, was concerned with the Highways Act 1835, which empowered the borough surveyor to prune and lop trees which excluded light from the highway. The claimant sued a borough surveyor who had cut the tops off the claimant's trees. It was shown in evidence that in forestry terminology 'prune' meant removing surplus branches to improve growth, and 'lop' meant to cut branches from the side of the tree. The defendant had done neither of these things. He had 'topped' the trees when he had no power to do so.

In *Munby v Furlong* [1977] 2 All ER 953, CA, the Court of Appeal held that the word 'plant' in a taxation statute covered books bought by a barrister for the purpose of his practice, and that, therefore, he was entitled to an allowance for capital expenditure. Lord Denning MR said (at p. 956) that the word 'plant' in a taxation statute did not mean what the ordinary English person thought it meant. It had acquired a special meaning in tax cases. It was not limited to things used physically, like machinery, but covered the 'intellectual storehouse' which a professional person has in exercising his profession.

In order to arrive at the proper meaning of the words used in a statute, a judge may have recourse to dictionaries, to the definition section (if any) in the particular statute, to the Interpretation Act 1978, and to previous cases decided on the meaning of similar words.

The use of dictionaries does not always yield helpful results, however. In *Coltman* v *Bibby Tankers Ltd* [1987] 3 All ER 1068, HL, the House of Lords decided that a ship fell within the definition of 'equipment' given in s. 1(3) of the Employer's Liability (Defective Equipment) Act 1969, where the word 'equipment' is defined as including 'any plant and machinery, vehicle, aircraft and clothing'. In so holding, their Lordships unanimously reversed the decision of the majority of the Court of Appeal and expressed disagreement with the conclusion reached by O'Connor LJ in the court below on the dictionary definition of 'equipment'.

Referring to the *Oxford English Dictionary*, O'Connor LJ had held that 'equipment' denoted only 'something ancillary to something else', and did not, therefore, include an entire ship, though it would cover machinery attached to the ship. Speaking for the House, Lord Oliver said (at p. 1071) that there was nothing in the *Oxford English Dictionary* definition which necessarily limited 'equipment' to 'parts of a larger whole'. One would, he said, refer without any misuse of language to a shipowner's fleet of ships as the 'equipment of his business'.

Coltman v *Bibby Tankers Ltd* [1987] 3 All ER 1068, HL, was considered in *Knowles* v *Liverpool City Council* [1993] 4 All ER 321, HL, where it was held that it was consistent both with the purpose of the 1969 Act, and with an ordinary interpretation of the word 'equipment', to construe that word as including the material with which an employee works in his employer's business. Accordingly, the defendant council was held liable when one of its employees who was employed to lay flagstones was injured by a defective flagstone supplied to the council by a third party. It was also said, *obiter*, that defective equipment which causes injury to an employee will be within the scope of the 1969 Act even though the employee was not required to use it, and had not, in fact, used it, since s. 1(1) of the Act does not require that the equipment has been provided for the employee's use but only that it has been provided 'for the purposes of the employer's business' (see [1993] 4 All ER 321 *per* Lord Jauncey at p. 327, the other Law Lords agreeing with him).

Where a statute defines a word in a certain way, Parliament is presumed to have intended that definition to be used in preference to the ordinary meaning of the word (as given, for example, in a dictionary) unless there is some clear indication to the contrary. Thus, in *Wyre Forest District Council* v *Secretary of State for the Environment* [1990] 1 All ER 780, HL, there being no clear indication to the contrary, a chalet structure was held by the House of Lords to come within the definition of 'caravan' given in s. 29(1) of the Caravan Sites and Control of Development

Act 1960 even though a chalet structure is not a caravan within the ordinary meaning of that word.

The Interpretation Act 1978 provides definitions of many words and expressions. These definitions are to be used in construing any Acts which contain those words or expressions (Interpretation Act 1978, s. 5 and sch. 1). The 1978 Act also states that, unless the contrary intention appears, words in a statute importing the masculine gender include the feminine, and *vice versa*, and that words in the singular include the plural, and *vice versa* (ibid., s. 6).

If these various aids to construction are of little or no assistance to the judge in a particular case then he must simply decide for himself what the words mean. The case of *Mandla* v *Dowell Lee* [1983] 1 All ER 1062, HL (the 'Sikh turban case') turned on the meaning of the word 'ethnic' in s. 3 of the Race Relations Act 1976. The word is not defined in the Act itself, and recourse was had, in both the Court of Appeal and the House of Lords, to dictionary definitions, all of which were rejected by the House of Lords which eventually held (reversing the Court of Appeal) that Sikhs are a racial, and not solely a religious, group for the purposes of the race relations legislation (see *per* Lord Fraser at p. 1066; Lord Fraser's tests were applied in *Crown Suppliers (PSA)* v *Dawkins* (1993) *The Times*, 4 February, CA, in which the Court of Appeal held that Rastafarians are solely a religious sect and not a racial group entitled to the protection of the legislation).

4.10.2.2 The golden rule

The so-called 'golden' rule of interpretation may be used either in a narrow way or a broad way. It is mostly used in a narrow way as a means of modifying the literal rule.

In its narrow application, the golden rule lays down that if the words used in the statute are ambiguous the court should adopt an interpretation which avoids an absurd result. For example, *Adler* v *George* [1964] 2 QB 7, DC, concerned a prosecution under the Official Secrets Act 1920 which made it an offence to obstruct HM Forces 'in the vicinity of' a prohibited place. The defendants had obstructed HM Forces *in* a prohibited place, and they were held to be guilty of the offence. The Divisional Court said that the words, 'in the vicinity of', were to be read as 'in or in the vicinity of', holding that it would be absurd if an offence was committed when the obstruction took place *outside*, albeit in the vicinity of, the prohibited place, but no offence was committed when the obstruction occurred *inside* the prohibited place.

In its second, broader, application the golden rule is sometimes used in preference to the literal rule where the words used can have only one literal meaning, and are not, therefore, ambiguous. This is especially so where considerations of public policy intervene to discourage the adoption of an interpretation which may produce an obnoxious result.

There is, for example, a principle grounded in public policy which precludes a murderer from benefiting under his victim's *will*. In *Re Sigsworth* [1935] Ch 89, the judge had to decide whether the same principle applied so as to preclude a murderer from claiming a benefit conferred on him *by statute* where the victim died *intestate*. By the Administration of Estates Act 1925, the residuary estate of a person dying intestate was to be divided among the 'issue'. Mrs Sigsworth died intestate leaving only a son who, a coroner's jury found, had murdered her. The question for the court was whether the son as 'issue' could succeed on the mother's intestacy. Clauson J held that he could not. He said that the principle of public policy which prevents a murderer from reaping the fruits of his crime must be applied in the construction of Acts of Parliament so as to avoid conclusions which are obnoxious to that principle. While not mentioning any so-called 'rule' of interpretation by name, Clauson J in *Re Sigsworth* in effect applied the golden rule *in preference* to the literal rule since the only possible literal interpretation of 'issue' must include a son.

The importance of considerations of public policy, exemplified in *Re Sigsworth*, was reaffirmed in two later decisions of the Divisional Court of the Queen's Bench Division. In both cases the wording of the relevant statute was clear, and yet, on grounds of public policy, an alternative meaning was found. These cases show that a statute is to be construed in the light of the principles of public policy accepted by the courts at the time when the statute was passed and which the draftsman and Parliament must have had in mind.

In the first of these cases, *R v National Insurance Commissioner (ex parte Connor)* [1981] 1 All ER 769, DC, Mrs Connor had been acquitted by a jury of the murder of her husband but found guilty of his manslaughter and placed on probation for two years. She applied for a widow's allowance under what was then the Social Security Act 1975, s. 24(1), which provided that:

A woman who has been widowed shall be entitled to a widow's allowance . . . if:

(a) she was under pensionable age at the time when her late husband died; . . .

(b) her late husband satisfied the contribution condition for a widow's allowance.

In Mrs Connor's case the necessary conditions were satisfied, but it was decided that public policy disentitled her from receiving a widow's allowance. The Divisional Court held that although the Social Security Act 1975 laid down a self-contained code, it was to be applied subject to the rules of public policy which demanded that she should not receive a widow's allowance on account of her deliberate and conscious killing of her husband, even though the punishment for that crime had been only a probation order. Lord Lane CJ said (at p. 774):

The fact that there is no specific mention in the Act of disentitlement so far as a widow is concerned if she were to commit this sort of offence and so become a widow is merely an indication . . . that the draftsman realised perfectly well that he was drawing this Act against the background of the law as it stood at the time.

R v National Insurance Commissioner (ex parte Connor), above, was applied in *R v Secretary of State for the Home Department (ex parte Puttick)* [1981] 1 All ER 776, DC. The question before the court in this case was whether Astrid Puttick (née Proll), who had contracted a valid marriage to a United Kingdom citizen by means of fraud, forgery, and perjury, was entitled to be registered as a United Kingdom citizen under the British Nationality Act 1948, s. 6(2) (since repealed), which (as amended) provided that:

. . . a woman who has been married to a citizen of the United Kingdom and Colonies shall be entitled, on making application therefor to the Secretary of State in the prescribed manner, to be registered as a citizen of the United Kingdom and Colonies.

Mrs Puttick's circumstances fell fairly and squarely within s. 6(2), but it was held that the Home Secretary was entitled, despite the mandatory wording of s. 6(2), to refuse to register Mrs Puttick as a United Kingdom citizen. She had secured her marriage, and therefore her entitlement to registration under s. 6(2), by illegal means, and public policy would not allow her to benefit from her own criminality. Donaldson LJ said (at p. 781) that:

when the British Nationality Act 1948 was enacted it was well established that public policy required the courts to refuse to assist a criminal to benefit from his crime at least in serious cases and that Parliament must be deemed to have been aware of this.

The cases of *Connor* and *Puttick*, above, were considered (and, in effect, extended) by the Court of Appeal in the highly unusual case of *R* v *Registrar General (ex parte Smith)* [1991] 2 All ER 88, CA. Section 51(1) of the Adoption Act 1976 (now the Adoption and Children Act 2002, sch. 2, para 1) provided that:

... the Registrar General shall on an application made in the prescribed manner by an adopted person a record of whose birth is kept by the Registrar General and who has attained the age of 18 years supply to that person on payment of the prescribed fee (if any) such information as is necessary to enable that person to obtain a certified copy of the record of his birth.

The applicant was detained in a secure mental hospital, having been convicted of the murder of a stranger in a public park and of the manslaughter of a cellmate in prison. The applicant had been adopted when only a few weeks old. He had later expressed extreme hatred for his adoptive parents, and had killed his cellmate while under the delusion that he was killing his adoptive mother. He now wished to obtain a copy of his birth certificate.

He satisfied the statutory conditions, but the Registrar General refused to supply him with the necessary information on grounds of public policy. The medical advice given to the Registrar General pointed to a possibility that the applicant's natural mother might be placed in danger if the applicant was ever released and he knew her identity. Counsel for the applicant argued that the statute conferred an absolute right to the information, and sought to distinguish the *Connor* and *Puttick* cases on the ground that the public policy principle did not apply to possible *future* criminal conduct.

It was held that, notwithstanding the absolute nature of the duty in s. 51(1), the Registrar General had not acted unlawfully in refusing to supply the necessary information. An absolute duty was, as a matter of statutory interpretation, subject to the principle of public policy that performance of it was not required if to do so would enable a person to benefit from a serious crime already committed by him, or which he intended to commit, or if there was a significant risk that performance of the duty would facilitate future criminal conduct resulting in danger to life.

The cases of *Connor*, *Puttick*, and *Smith*, above, were applied in *Whiston* v *Whiston* (1995) [1998] 1 All ER 423, CA, where it was held that, as a matter of public policy, a person who knowingly contracts a bigamous marriage cannot, after the marriage is declared void, claim financial relief under the Matrimonial Causes Act 1973 (despite its clear wording) because a person cannot be allowed to benefit from his own crime, and bigamy is a serious criminal offence which strikes at the root of the institution of marriage.

In the later case of *S-T* v *J* (1996) [1998] 1 All ER 431, CA, which was concerned with a marriage contracted by a transsexual, a majority of the Court of Appeal restricted the effect of *Whiston* to the crime of bigamy where it is the marriage itself which constitutes the criminal act. It was held that the commission of other serious criminal offences in relation to the marriage, such as perjury, would not, on public policy grounds, automatically bar a claim for financial relief. The court was unanimous, however, in dismissing the transsexual's claim in the exercise of the statutory discretion conferred by the Matrimonial Causes Act 1973 to take into account all the circumstances of the case.

S-T v *J* was applied, and *Whiston* distinguished, in *Rampal* v *Rampal* (*No. 2*) [2001] 3 WLR 795, CA, in which the Court of Appeal denied that *Whiston* had established a universal rule of public policy that no bigamist could ever claim financial relief.

In *R (Crown Prosecution Service)* v *Registrar General of Births, Deaths and Marriages* [2003] 2 WLR 504, CA, the Court of Appeal declined to extend *R* v *Registrar General (ex parte Smith)* [1991] 2 All ER 88, CA, above, any further. It was pointed out that the *Smith* case was decided 'on very extreme facts, where the provision of information amounted to the signing of the adoptive mother's death warrant' ([2003] 2 WLR 504, CA, *per* Waller LJ at [16]).

In *R (Crown Prosecution Service)* v *Registrar General of Births, Deaths and Marriages*, above, a prisoner awaiting trial for murder asked the prison director for permission to marry his partner, who was an important prosecution witness in the trial. If they married, she would cease to be a compellable witness. Despite the attempts of the Crown Prosecution Service to persuade them to forbid the marriage until after the trial, the prison director stated formally that he had no objection to the marriage taking place in his prison, and the Registrar General decided that there were no public policy grounds which affected the absolute duty of a superintendent registrar to issue a marriage certificate under the Marriage Act 1949 once he had received a notice of marriage. The Crown Prosecution Service applied for judicial review of the decisions of the prison director and the Registrar General.

The Court of Appeal held that both officials had acted lawfully. The prison director was not bound to take into account public policy considerations in reaching his decision; he was only obliged to consider the convenience and availability of his prison. It was further decided that entering into a lawful marriage with the consequence that a potential witness would cease to be compellable did not amount to perverting, or attempting to pervert, the course of justice. It followed that it would not be contrary to public policy for a superintendent registrar to issue a marriage certificate in accordance with his absolute statutory duty.

The case of *Puttick* [1981] 1 All ER 776, DC, above, is distinguishable where the claimant has not done anything wrong in satisfying the necessary conditions for registration as a British citizen. Thus, an attempt to apply the *Connor*, *Puttick*, and *Smith* line of authority failed in *R (Hicks)* v *Secretary of State for the Home Department* [2005] EWHC 2818 (Admin). Here, the claimant was an Australian citizen who was being held by the US authorities at the Guantanamo Bay detention centre in Cuba and against whom there were allegations of terrorist activity involving Al-Qaeda and the

Taliban. His mother was born in the United Kingdom and he claimed to be registered as a British citizen. If registered, he hoped then to gain assistance from British consular officials which would lead to him being treated in the same way as other British citizens who had been held at Guantanamo Bay. They had been released to the custody of the British authorities and returned to the United Kingdom. The Home Secretary, while accepting that the conditions for registration were satisfied, argued that he could refuse to register the claimant on the ground of public policy. Collins J held that the Home Secretary's contention was wrong since the claimant (unlike Mrs Puttick) had done nothing wrong in order to establish the necessary conditions for registration as a British citizen, there being no condition based on good character.

It was thought by some that the so-called forfeiture rule of public policy operated too harshly in cases like *Connor* [1981] 1 All ER 769, DC, above. Parliament passed the Forfeiture Act 1982 in an effort to mitigate some of its harshness in individual cases. The Act confers a judicial discretion to modify the effect of the forfeiture rule in respect of specified property rights which may otherwise have been forfeited. The Act is very limited in scope and does not seek to *abolish* the forfeiture rule.

The Act defines the forfeiture rule as 'the rule of public policy which in certain circumstances precludes a person who has unlawfully killed another from acquiring a benefit in consequence of the killing' (Forfeiture Act 1982, s. 1(1)). If this rule is applicable in any case coming before a court, the judge is given a discretion to modify its effect (ibid., s. 2). In deciding how to exercise the discretion, the test is whether the justice of the case requires the effect of the forfeiture rule to be modified after considering the conduct of the killer and the deceased, together with other material circumstances (ibid., s. 2(2)). The Act does not affect the application of the forfeiture rule to murderers (ibid., s. 5). It only applies to those persons whose unlawful killing of another falls short of murder.

In *Re K (deceased)* [1985] 2 All ER 833, CA, a woman was beaten by her husband on many occasions over a six-year period. During an argument the woman killed her husband with a shotgun. At her trial she pleaded guilty to manslaughter, and was placed on probation for two years. The question arose whether she should be allowed to inherit her deceased husband's estate under his will. It was held that, on the facts, the forfeiture rule would be modified to allow her to take under the will. She had been a loyal wife who had suffered extreme violence from the deceased, and it would be unjust for her to be deprived of the benefits conferred on her by the will.

In *Dunbar v Plant* [1997] 4 All ER 289, CA, it was held that the forfeiture rule applied to the survivor of a suicide pact who had aided and abetted the suicide of her fiancé. On the facts, however, the rule was modified under the 1982 Act to allow her to succeed to her fiancé's interest in their jointly owned house and to claim the proceeds of an insurance policy on his life.

In *Dalton v Latham* [2003] EWHC 796 (Ch), modification of the rule was refused. The claimant had been found not guilty on the ground of diminished responsibility of the murder of a person under whose will he was a beneficiary. He was instead convicted of manslaughter. On his claim for modification of the forfeiture rule, the judge found that the claimant had repaid the deceased's befriending of, and generosity towards, him with 'violence and abuse, both physical and financial'. Applying s. 2(2), above, and taking into account the conduct of the killer and the deceased, together with other material circumstances (including the position of the deceased's family), it was held that the justice of the case did not require the effect of the rule to be modified.

The Social Security Commissioners have the same power to modify the effect of the forfeiture rule in relation to social security benefits as the courts have in relation to property rights (Forfeiture Act 1982, s. 4(1A)–(1H), as inserted by the Social Security Act 1986).

It must be stressed that the golden rule is used most often in its first, narrower, sense, and that, in this application, the rule can only be applied where there is a sensible alternative interpretation. If there is only one interpretation, or the second interpretation is as absurd as the first, the literal rule may be applied even though the result might be ridiculous, unless, that is, the result would be so undesirable that the court can be persuaded to adopt the golden rule in its broader sense.

4.10.2.3 The mischief rule

The mischief rule, sometimes referred to as the rule in *Heydon's case* (1584) 3 Co Rep 7a, allows the court to look at the state of the former law in order to discover the *mischief* in it which the present statute was designed to remedy. The mischief was, in former times, usually referred to in the preamble to the statute (*Black-Clawson International Ltd v Papierwerke etc. AG* [1975] AC 591, HL, *per* Lord Diplock at p. 638; *Sussex Peerage Case* (1844) 11 Cl & F 85, HL, *per* Tindal CJ, at p. 147; on the preamble, see further, para 4.10.4.3 below).

In *Heydon's case* the position was explained thus by the court:

[F]our things are to be discerned and considered:

1st What was the common law before the making of the Act.

2nd What was the mischief and defect for which the common law did not provide.

3rd What remedy the Parliament hath resolved and appointed to cure the disease of the commonwealth. And

4th The true reason of the remedy; and then the office of all the Judges is always to make such construction as shall suppress the mischief, and advance the remedy.

In *Corkery* v *Carpenter* [1951] 1 KB 102, DC, a man was arrested without a warrant for being drunk in charge of a bicycle on the highway. By the Licensing Act 1872, s. 12, a person found drunk in charge of a 'carriage' on the highway could be arrested without a warrant. The Divisional Court held that a bicycle was a 'carriage' for the purposes of the 1872 Act and so the defendant had been properly arrested. The mischief aimed at by the Act was drunken persons on the highway in charge of some form of transport. If the court had applied the literal rule instead of the mischief rule the result might well have been different since it is arguable that a bicycle is not a 'carriage' within the ordinary meaning of that word.

The mischief rule was applied, and the presumption in favour of *mens rea* for criminal liability was rebutted, in *Maidstone Borough Council* v *Mortimer* [1980] 3 All ER 552, DC (for *mens rea* and the presumption in favour of it, see para 4.10.3.5 below). The defendant cut down an oak tree which, unknown to him, was protected by a tree preservation order under what was then the Town and Country Planning Act 1971. The local council prosecuted him for an offence in the magistrates' court. The justices dismissed the case on the ground that knowledge of the preservation order was an essential element of the offence. On appeal by the council, the Divisional Court allowed the appeal and held that the offence was committed whether or not the accused had knowledge of the preservation order. The mischief which the statute, read as a whole, was designed to prevent was the cutting down of protected trees without the consent of the local planning authority.

The House of Lords applied the mischief rule in *Royal College of Nursing of the United Kingdom* v *DHSS* [1981] 1 All ER 545, HL, a case which provoked considerable judicial disagreement.

By the Abortion Act 1967, s. 1(1), it is provided that no criminal offence is committed 'when a pregnancy is terminated by a registered medical practitioner' in specified circumstances. In 1972, surgical abortions in hospitals were replaced by medically induced abortions, which were not within the contemplation of Parliament in 1967. They involved pumping a chemical fluid into the mother's womb to induce premature labour. The medical induction was in two stages. First, the insertion, *by a doctor*, of a catheter into the womb. Secondly, the administration of the fluid into the womb via the catheter by means of a pump or drip. This second stage was what actually caused premature labour and termination of the pregnancy. However, the second stage was not carried out by a doctor but *by nurses* under the doctor's instructions while the doctor was absent but on call.

The question for the court was whether, in a medically induced abortion, the pregnancy was terminated 'by a registered medical practitioner'. At first instance, Woolf J granted a declaration that it was, but the Court of Appeal unanimously reversed his decision. On a further appeal, the House of Lords, by a majority of three to two, held that the procedure was lawful. The decision of the Court of Appeal was reversed and that of the trial judge restored. On a count of heads, five out of the nine judges who sat in the three courts thought the medical induction procedure was unlawful while only four considered it to be lawful.

The decision of the majority of the House of Lords was based on two grounds. The first ground was the mischief which the Abortion Act 1967 was intended to remedy (see especially [1981] 1 All ER 545 *per* Lord Diplock at p. 567). The mischief was the unsatisfactory and uncertain state of the law prior to 1967. The second ground was the policy of the 1967 Act (see especially, ibid., *per* Lords Diplock and Keith at pp. 568–9 and 575, respectively). This was to broaden the basis on which abortions might lawfully be obtained, and to ensure that abortions were performed with proper skill in hygienic hospital conditions.

These two grounds led the majority of their Lordships to the conclusion that s. 1(1) of the 1967 Act was satisfied so long as a doctor prescribed the treatment, remained in charge and accepted responsibility throughout, and the treatment was carried out in accordance with his directions. If these conditions were fulfilled, it did not matter that the doctor was not actually present when the pregnancy was terminated.

The case of *Royal College of Nursing of the United Kingdom* v *DHSS* [1981] 1 All ER 545, HL, was one of considerable public importance. It does seem that the statutory language was strained so as to avoid the conclusion that since 1972 large numbers of doctors and nurses had been unwittingly performing illegal abortions. Lord Wilberforce and Lord Edmund-Davies dissented on the ground that it could not be said that a pregnancy was terminated 'by a registered medical practitioner' when it was plainly not terminated by a registered medical practitioner. They felt that the majority in the House were not engaged in interpretation but in rewriting, or, as Lord Edmund-Davies put it, in 'redrafting with a vengeance' (p. 573).

The application of the mischief rule was preferred to a literal approach in *Director of Public Prosecutions* v *Johnson* [1995] 4 All ER 53, DC. Here it was decided that in the context of s. 5(1) of the Road Traffic Act 1988 (part of the breathalyser law by which it is an absolute offence to drive on a road while over the prescribed alcohol limit), the expression 'consuming alcohol' is wide enough to cover the ingestion of alcohol by any method notwithstanding the fact that the primary meaning of that expression is consuming it by mouth. The mischief at which s. 5 is aimed (identified with the help of the marginal note to the section) is the danger caused by drunken driving, and the court was of the view that that mischief is better remedied by an interpretation of s. 5 which renders irrelevant the route by which alcohol reaches a driver's body, whether it be by drinking, eating, injecting, sniffing, rubbing, or in the form of a suppository.

In *Manchester City Council* v *McCann* [1999] 2 WLR 590, CA, the Court of Appeal applied the mischief rule when it decided that a threat is an 'insult' within the meaning of s. 118(1)(a) of the County Courts Act 1984 (first enacted in the County Courts Act 1846), which enables the county courts to deal with any person who 'wilfully insults the judge . . . or any juror or witness, or any officer of the court . . .'. The mischief was identified as the need to protect the persons mentioned as participants in the court process. It was said that if the court could deal with 'insults' but not 'threats', it would not be able to give immediate protection from interference to those who need it most (see [1999] 2 WLR 590 *per* Lord Woolf MR at p. 598). Although, literally, an 'insult' is not necessarily a 'threat', Parliament was taken to have intended that the protection should extend to both.

By ss. 37(5) and 40(1) of the Police and Criminal Evidence Act 1984, a person who has been arrested but not yet charged must have his detention at a police station reviewed periodically. The review officer is required to make a written record of the grounds for detention 'in the presence of the person arrested', and, at that time, to inform the detainee of those grounds.

In *R* v *Chief Constable of Kent Constabulary (ex parte Kent Police Federation Joint Branch Board)* (1999) *The Times*, 1 December, DC, the Chief Constable had proposed that such reviews should be conducted by video link in the majority of cases. It was held that this would be unlawful since video link reviews would not take place 'in the presence of' the detainee. In so holding, the court identified the mischief as the prolonged detention of suspects without periodic reviews of their detention, and emphasised the need to give a meaning to the statutory language which best effected the intention of Parliament. Although the statute did not refer to 'physical presence', 'presence' in ordinary parlance meant 'physical presence', and the court was not prepared to hold, even taking a modern and progressive approach, that a record was made by a review officer 'in the presence of' the detainee if they were not in the same place and in each other's company at the time. Parliament, it was said, had provided for a face-to-face confrontation, and if it was felt desirable to make use of modern technology it was for Parliament to make the change and not for the courts.

4.10.2.4 The statute must be read as a whole

The words used in a statute must not be interpreted out of their context. This rule is based on the maxim, *noscitur a sociis*, a word is known by the company it keeps. Each section in a statute must be read subject to every other section, which may explain or modify it. If there is an irreconcilable conflict between two sections in the same statute, or between two subsections within the same section, the correct test to be applied is to determine which is the leading provision and which the subordinate provision. The statute will then be interpreted so as to give effect to the leading provision. The principle of interpretation (if it ever existed) that the provision occurring later in the statute prevails over that occurring earlier is long since obsolete (*Re Marr and another (bankrupts)* [1990] 2 All ER 880, CA, *per* Nicholls LJ at p. 886, applying a dictum of Lord Herschell LC in *Institute of Patent Agents* v *Lockwood* [1894] AC 347, HL, at p. 360 and disapproving a dictum in *Wood* v *Riley* (1867) LR 3 CP 26 at p. 27).

Where the statute to be construed is ambiguous and forms part of a statutory code intended by Parliament to achieve a distinct purpose, the other statute(s) in the code should be looked at as well. In *Oliver Ashworth (Holdings) Ltd* v *Ballard (Kent) Ltd* [1999] 2 All ER 791, CA, it was found that s. 18 of the Distress for Rent Act 1737 was ambiguous about the extent of a landlord's right to double rent, and that, consequently, the 1737 Act and the Landlord and Tenant Act 1730, which together constituted a single code, should be read collectively as a whole. (See further on this case, para 4.10.6 below.)

The rule of interpretation that the statute must be read as a whole gives rise to a presumption that, so far as possible, every word used in the statute is to be given some effective meaning. If, however, there is a word or phrase for which no sensible meaning can be found, the court may disregard it if to do otherwise would render the statute absurd or incapable of achieving its evident purpose. (For a recent example where the House of Lords dismissed statutory words as mere surplusage, see *R* v *R (rape: marital exemption)* [1991] 4 All ER 481, HL, para 5.3.2.7.)

One particular application of the general rule that the statute must be read as a whole is *expressio unius est exclusio alterius*—the express mention of a person or thing excludes by implication other persons or things not mentioned. In *Tempest* v *Kilner* (1846) 3 CB 249, it was held that the Statute of Frauds 1677, s. 17 (now repealed), which required a contract for the sale of 'goods, wares and merchandise' for £10 or more to be evidenced in writing, did not apply to a contract for the sale of stocks and shares. The latter were not 'goods, wares and merchandise', and were excluded by implication because they received no express mention.

Another particular, and more common, application of the general rule is the so-called *ejusdem generis* ('of the same kind') rule. This provides that general words which follow particular words must be limited to meanings similar to those of the particular words, as in *Powell* v *Kempton Park Racecourse Co.* [1899] AC 143, HL. This case concerned the Betting Act 1853, which had prohibited the keeping of a 'house, office, room or other place' for the purpose of betting with people who called there. The question for the House of Lords was whether Tattersall's ring, an outdoor place at a racecourse, fell within the words 'other place'. It was held that it did not because the specific places mentioned were indoor places and the words 'other place' must be construed *ejusdem generis*.

In *DPP* v *Jordan* [1977] AC 699, HL, the appellant was charged with possessing obscene films, books, and magazines for publication for gain contrary to the Obscene Publications Act 1959. The appellant wished to rely on the defence of 'public good' and sought to adduce expert evidence that the articles, though obscene, had some psychotherapeutic value for sexual deviants. The House of Lords held that such evidence had been rightly excluded at the appellant's trial. The defence of 'public good' is contained in s. 4(1) of the 1959 Act, which says that there should be no conviction if it is proved that publication of an obscene article

is justified as being for the public good on the ground that it is in the interests of science, literature, art or learning, *or of other objects of general concern* (italics supplied).

It was not suggested that the obscene articles in question had any scientific, literary or artistic merit, and the House of Lords decided that they did not fall within the italicised words, which must be construed as falling within the same field or dimension as 'science, literature, art or learning'.

In *Wood* v *Commissioner of Police of the Metropolis* [1986] 2 All ER 570, DC, it was held that a piece of glass accidentally broken was not an offensive weapon within the meaning of the words 'any gun, pistol, hanger, cutlass, bludgeon, or other offensive weapon' in s. 4 of the Vagrancy Act 1824. The general words 'other offensive weapon' had to be construed *ejusdem generis* with the preceding words, which comprised a list of articles made or adapted for the purpose of causing injury.

By s. 67(1) of the Copyright, Designs and Patents Act 1988, it is not an infringement of copyright to play a sound recording 'as part of the activities of, or for the benefit of, a club, society or other organisation' provided that the conditions in s. 67(2) are satisfied. One of those conditions is that the organisation is non-profit-making and its main objects are charitable or otherwise concerned with the advancement of religion, education, or social welfare. In *Phonographic Performance Ltd* v *South Tyneside Metropolitan Borough Council* [2001] 1 WLR 400, a preliminary issue arose as to whether a local authority which plays sound recordings at keep-fit and aerobics classes can claim the statutory copyright exemption contained in s. 67.

Reading subss. (1) and (2) of s. 67 together, it was held that it cannot. First, the functions of a local authority are administrative and governmental rather than charitable within s. 67(2). Secondly, although it is literally an 'organisation', a local authority is not an 'other organisation' within s. 67(1) since those words must be construed *ejusdem generis* with the words 'club' and 'society'. A local authority cannot, in ordinary language, be described as a 'club' or 'society', and it seems reasonable to assume that Parliament did not intend, by use of the words 'other organisation', to extend the statutory copyright exemption to local authorities, central government, quangos, and the like.

There must be at least two particular words to constitute a category or *genus*. Thus, in *Allen* v *Emmerson* [1944] KB 362, DC, it was held that a funfair was a 'place of amusement', within the statutory expression 'theatres and other places of amusement', even though not *ejusdem generis* with the word 'theatres'.

It should be noted that the *ejusdem generis* rule cannot be applied if the particular words do not belong to any category (*R v Payne* (1866) 35 LJMC 170; *Re C (a minor) (interim care order: residential assessment)* [1996] 4 All ER 871, HL), or if the general words are unambiguous.

Nor is it easy to apply the rule if the general words are followed by a word like 'whatsoever' because such a word indicates that the general words are intended to be completely general and open-ended.

4.10.3 Presumptions

4.10.3.1 Against an alteration of the common law

Parliament is supreme and can, therefore, alter the existing common law. However, an intention to change the common law will not be *implied*. If a statute is capable of two interpretations, one involving an alteration of the common law and the other one not, the second interpretation will be preferred.

In *Beswick v Beswick* [1968] AC 58, HL, the Court of Appeal had held, in effect, that s. 56(1) of the Law of Property Act 1925 had abolished the doctrine of privity of contract, a judge-made rule that a third person cannot sue, or be sued, on a contract to which he is not a party. The House of Lords disagreed, holding that in the absence of express, clear words of abolition in s. 56(1) it was not possible to *imply* an intention on the part of Parliament to revolutionise the law of contract. (The doctrine was eventually *expressly modified* by the Contracts (Rights of Third Parties) Act 1999.)

4.10.3.2 Against deprivation of liberty

It is presumed that Parliament does not intend to deprive a person of his liberty. If it does so wish, then clear, express words must be used. Thus, in defence of individual freedom, the House of Lords, in *R v Secretary of State for the Home Department (ex parte Khawaja)* [1983] 2 WLR 321, HL, was prepared to depart from its own earlier decision in *R v Secretary of State for the Home Department (ex parte Zamir)* [1980] AC 930, HL (see para 5.3.2.2.2). In *Khawaja*, Lord Scarman said (at p. 344):

If Parliament intends to exclude effective judicial review of the exercise of a power in restraint of liberty, it must make its meaning crystal clear.

Their Lordships were of the opinion that the immigration statute with which they were concerned did not have the effect of placing the burden of proof on an immigrant to show that the decision of the Home Office to detain him was unjustified.

Even in cases where Parliament has expressly provided for deprivation of liberty, as in the Mental Health Act 1983, the words used will be construed in a way which permits the least possible interference with the liberty of the subject (*R v Hallstrom (ex parte W) (No. 2)* [1986] 2 All ER 306; *R(H) v London North and East Region Mental Health Review Tribunal* [2001] 3 WLR 512, CA, para 4.10.8.2 below— incompatibility of parts of the 1983 Act with the European Convention on Human Rights; see also *R v Governor of Durham Prison (ex parte Singh)* [1984] 1 All ER 983 — Home Secretary's power to detain a person under the Immigration Act 1971 pending deportation is subject to implied limitations).

4.10.3.3 Against deprivation of property or interference with private rights

Statutes which encroach on the rights of subjects should be interpreted wherever possible so as to respect such rights (*Glassbrook Bros* v *Leyson* [1933] 2 KB 91, CA). To this end, there is a presumption of interpretation that Parliament does not intend to deprive a person of his property or interfere with his private rights—at least, not without compensation.

In *Managers of the Metropolitan Asylum District* v *Hill* (1881) 6 App Cas 193, HL, the appellants had been given statutory authority under the Metropolitan Poor Act 1867 (since repealed) to provide asylums for the reception and relief of the sick, insane, or infirm. Acting under this authority, they built a smallpox hospital in Hampstead where it caused a nuisance to the respondent and other residents of the neighbourhood. The statute contained no provision for compensation. The House of Lords held that the appellants would be restrained by an injunction from using the hospital in such a manner as to create a nuisance. The appellants' statutory authority was permissive, not mandatory, and must be exercised so as not to interfere with private rights.

In *British Airports Authority* v *Ashton* [1983] 3 All ER 6, DC, it was held that a section in the Trade Union and Labour Relations Act 1974 (since repealed) which declared when picketing was lawful did not authorise picketing on land against the owner's wishes. Such action was not *expressly* authorised by the section, and the court said it would be 'astonishing if Parliament intended that such a right should be *implied*' (ibid., *per* Mann J at p. 13).

Cases like *Managers of the Metropolitan Asylum District* v *Hill*, above, will be rare in contemporary society because a statute will not easily achieve its object unless it confers the right to interfere with private interests. The clear wording of a statute can deprive the citizen of the full and proper enjoyment of his land by authorising the commission of a nuisance without any compensation. Where the statute is sufficiently clear, the citizen is expected simply to put up with the infringement of his rights without any financial recompense. (See, e.g., *Allen* v *Gulf Oil Refining Ltd* [1981] 1 All ER 353, HL; but cf *Tate & Lyle Industries Ltd* v *Greater London Council* [1983] 1 All ER 1159, HL.)

In a modern, controversial instance—the War Damage Act 1965 (para 4.10.3.6 below)—the words used in the statute were clear enough to abolish (and retrospectively at that) any common law right the citizen may have had to claim compensation from the Crown in respect of lawful destruction of, or damage to, his property caused on behalf of the Crown during, or in contemplation of, a war.

Subjects can only be deprived of their property by the Crown in the form of taxation under the authority of a statute expressed in clear words (*Bowles* v *Bank of England* [1913] 1 Ch 57; *Attorney-General* v *Wilts United Dairies* (1922) 91 LJKB 897, HL). If the meaning of the statute is clear, the tax must be paid, even though this may result in hardship to the individual taxpayer, for Parliament must be taken to have been aware that some hardship might be involved (*Leedale* v *Lewis* [1982] 3 All ER 808, HL, para 4.10.2.1 above). If, however, the meaning of the statute is *not* clear, it may possibly be legitimate for the court to choose an interpretation favourable to the taxpayer on the ground that he can only be taxed by *clear* words (ibid., *per* Lord Wilberforce at p. 816).

4.10.3.4 That the Crown is not bound unless the Act says so expressly or by necessary implication

This presumption applies whether or not the statute in question is one which was passed for the public benefit (*Lord Advocate* v *Dumbarton District Council* [1990] 1 All ER 1, HL, where it was held that the Crown was not bound, either expressly or by necessary implication, by the relevant Scottish planning and roads legislation to obtain local authority consent before closing a stretch of road, and storing building materials thereon, during the construction of an improved security fence at the Faslane submarine base).

It has been said to be desirable for the sake of clarity that statutes should state expressly (rather than leaving it to necessary implication) whether the Crown is to be bound by any, and if so which, of their provisions (*Lord Advocate* v *Dumbarton District Council*, above, *per* Lord Keith at p. 18).

Clarity in this matter is particularly important in those areas of law where the potential legal liability of the Crown is very wide, as in occupier's liability and employer's liability. The Crown is the occupier of a great deal of land and the employer of a great number of people. It is expressly bound by the Occupiers' Liability Acts of 1957 and 1984, by the Equal Pay Act 1970, by important parts of the Health and Safety at Work etc. Act 1974, by the Sex Discrimination Act 1975, and by some important parts of the Employment Rights Act 1996 (including the provisions on unfair dismissal, but excluding the provisions on redundancy payments).

4.10.3.5 That *mens rea* is required in criminal offences

All common law offences require proof of *mens rea* (guilty mind or intention) before a person can be convicted. In the case of statutory offences, it is *presumed* that Parliament intended no liability without proof of *mens rea*. For the vast majority of offences a person cannot be properly convicted unless he committed the *actus reus* (forbidden act) with *mens rea*.

When creating new statutory offences, Parliament regularly provides definitions which are confined to the *actus reus* and are silent about *mens rea*. Here, '[t]he common law presumes that, unless Parliament indicated otherwise, the appropriate mental element is an unexpressed ingredient of every statutory offence' (*B (a minor)* v *Director of Public Prosecutions* [2000] 1 All ER 833, HL, *per* Lord Nicholls at p. 836). The weight to be given to the presumption increases with the seriousness of the offence since the punishment is more severe and the stigma of a conviction is graver (ibid., *per* Lord Nicholls at p. 839). Parliament has been criticised for failing to state in clear terms in some statutes whether or not *mens rea* is required when creating or restating offences (ibid., *per* Lord Hutton at p. 856).

In *Harding* v *Price* [1948] 1 KB 695, DC, a motorist was prosecuted for failing to report an accident contrary to the Road Traffic Act 1930, s. 22 (see now Road Traffic Act 1988, s. 170, as amended by Road Traffic Act 1991, sch. 4). He did not know about the accident and the Divisional Court held he was not guilty of the offence because he lacked *mens rea*.

In *Sweet* v *Parsley* [1970] AC 132, HL, a schoolteacher was convicted of being concerned in the management of premises used for the smoking of cannabis contrary to s. 5 of the Dangerous Drugs Act 1965 (see now Misuse of Drugs Act 1971, s. 8). The cannabis was used by her tenants and she knew nothing about it. The House of Lords held she was not guilty of the offence, and her conviction was quashed, because she did not *know* the premises were being used for the unlawful purpose and did not *intend* them to be so used.

The presumption that proof of *mens rea* is required in statutory offences can be rebutted either by express words in a statute or by implication. Offences which do not require proof of *mens rea* are called crimes of *strict liability* and are usually created in the public interest. The question whether *mens rea* is required or not must be determined by interpretation of the language of the particular statutory provision taken in conjunction with the subject matter and the structure of the statute as a whole (*Wings Ltd* v *Ellis* [1984] 3 All ER 577, HL, *per* Lord Scarman at p. 589).

The presumption that *mens rea* is required is particularly strong where the offence is truly criminal in character, as in *Sweet* v *Parsley*, above. Indeed, it seems that the only situation in which the presumption can be rebutted is where (a) the particular statute deals with an issue of social concern, such as public safety, health, or morals, *and* (b) it can be shown that the creation of strict liability will be effective to promote the objects of the statute by encouraging greater vigilance to prevent the commission of the forbidden act (*Gammon (Hong Kong) Ltd* v *Attorney-General of Hong Kong* [1984] 2 All ER 503, PC, *per* Lord Scarman at p. 508).

In *Customs and Excise Commissioners* v *Air Canada* [1991] 1 All ER 570, CA, cargo unloaded at Heathrow Airport from an aeroplane making a regular scheduled flight was found to contain a large quantity of cannabis resin with an estimated street value of £800,000. Under s. 141(1)(a) of the Customs and Excise Management Act 1979, using an aircraft for the carriage of a thing liable to forfeiture (in this case the cannabis resin) renders the aircraft itself liable to forfeiture. The preliminary issue arose as to whether the defendants' lack of knowledge of the presence of prohibited drugs on their aeroplane would be a defence to a claim for forfeiture of the aeroplane. Tucker J held that it would.

In unanimously reversing this decision, the Court of Appeal was of the firm opinion that the wording of s. 141(1)(a) clearly and unambiguously allows for forfeiture of the aeroplane without proof of *mens rea* on the part of the defendants. The dictum of Lord Scarman in *Gammon (Hong Kong) Ltd* v *Attorney-General of Hong Kong* [1984] 2 All ER 503, PC, at p. 508 (see above) was distinguished on the ground that breach of s. 141(1)(a) of the 1979 Act does not in itself constitute a criminal offence, but gives rise instead to *civil* proceedings. Reference, therefore, to cases on the presumption of *mens rea* in relation to criminal offences is misleading (see [1991] 1 All ER 570, CA, *per* Purchas and Balcombe LJJ at pp. 586 and 587–8, respectively).

When Air Canada later took its case to the European Court of Human Rights it was held that the seizure of the aeroplane, and its subsequent release on payment of £50,000, did not amount to an unjustified interference with the peaceful enjoyment of possessions as guaranteed in art. 1 of the First Protocol of the European Convention on Human Rights. It was further held that there had been no violation of Air Canada's civil rights, guaranteed under art. 6 of the Convention, because the airline had not been denied access to a fair hearing before an independent and impartial tribunal (*Air Canada* v *United Kingdom* (1995) *The Times*, 13 May, ECHR).

Parliament has created a number of offences which do not require proof of *mens rea*; for example, speeding and parking offences and the offence of selling liquor to a person who is already drunk.

In *R v Hussain* [1981] 2 All ER 287, CA, it was decided that possessing a firearm without a certificate is, under s. 1(1)(a) of the Firearms Act 1968, an offence of strict liability and that the prosecution does not have to prove the accused knew that the article was a firearm.

It was held in *R v Bradish* [1990] 1 All ER 460, CA, that the offence of being in possession of a prohibited weapon (a spray canister containing CS gas) contrary to s. 5(1) of the Firearms Act 1968 is one of strict liability. It was, therefore, no defence for the defendant to attempt to show that, by reason of the fact that the gas was concealed within the canister, he did not know, and could not reasonably have been expected to know, that the article in his possession was a prohibited weapon. It was, in any event, further held that the weapon consisted of the canister and gas *in combination*, and not merely the noxious gas by itself. The imposition of strict liability in these circumstances furthers the purpose of the firearms legislation, namely the tight control of dangerous weapons in the public interest, and is in line with the interpretation adopted by the Court of Appeal in *R v Hussain*, above, in relation to s. 1 of the same Act.

In *Wings Ltd v Ellis* [1984] 3 All ER 577, HL, it was decided that since the purpose of the Trade Descriptions Act 1968 is the protection of the public through the maintenance of trading standards it followed that the offence, created by s. 14(1)(a) of that Act, of making a statement knowing it to be false does not require proof of *mens rea*.

In *Pharmaceutical Society of Great Britain v Storkwain Ltd* [1986] 2 All ER 635, HL, s. 58(2)(a) of the Medicines Act 1968, which prohibits the retail sale or supply of certain medicinal products except in accordance with a prescription given by an appropriate practitioner, was held to have created an offence of strict liability. It follows that a pharmacist who supplies prohibited drugs on a forged prescription, and without any fault on his part, nevertheless commits an offence.

In *Kirkland v Robinson* [1987] Crim LR 643, DC, it was decided that possession of a live wild bird, contrary to s. 1(2)(a) of the Wildlife and Countryside Act 1981, is an offence of strict liability which does not require proof of knowledge that the bird was a wild bird protected by the Act.

In *R v Brockley* (1994) 99 Cr App R 385, CA, it was held that the offence, created by s. 11 of the Company Directors (Disqualification) Act 1986, of acting as director of a company while an undischarged bankrupt except with leave of the court is one of strict liability. It followed that a mistaken but genuine belief that the bankruptcy had been discharged was no defence to the crime. Applying Lord Scarman's dictum in *Gammon (Hong Kong) Ltd v Attorney-General of Hong Kong* [1984] 2 All ER 503, PC, at p. 508 (see above), it was said that the mischief aimed at by s. 11 was clearly one of social concern, and that the creation of strict liability would promote the object of the section by obliging bankrupts to ensure that their bankruptcy was in fact discharged before they acted as company directors. (See also *R v Doring* (2003) 1 Cr App R 143, CA, in which *R v Brockley* was followed.)

Lord Scarman's dictum was applied again in *R v Bezzina* [1994] 3 All ER 964, CA, where it was held that s. 3(1) of the Dangerous Dogs Act 1991 is aimed at securing public safety, and creates an offence of strict liability committed by the owner or handler of a dog which is 'dangerously out of control in a public place'. It was also applied in *R v Blake* [1997] 1 All ER 963, CA, in which it was decided that the offence under s. 1(1) of the Wireless Telegraphy Act 1949, as amended, of using a wireless telegraphy station or apparatus without a licence was created in the interests of public safety and is, therefore, one of strict liability. It follows that a person may be convicted of the offence where he knew he was using the station or apparatus but had no intention of, and was unaware that he was, broadcasting anything.

The offence of selling tobacco 'to a person apparently under the age of 16 years', contrary to s. 7(1) of the Children and Young Persons Act 1933, was held to be one of strict liability in *St Helen's Metropolitan Borough Council v Hill* (1992) 156 JP 602, DC, where a boy aged 12 had been sold a packet of cigarettes by a shop assistant while the shopowner was in the stockroom and unaware of the sale. The shopowner was charged with an offence. As a general rule an employer is not liable for his employee's crimes committed in the course of his employment, but there is an exception in the case of crimes of strict liability and the shopowner was held to have committed the offence. (Note that s. 7 of the Children and Young Persons Act 1933 has been amended by the Children and Young Persons (Protection from Tobacco) Act 1991. The word 'apparently' is removed from s. 7(1), and a new subsection 7(1A) is added, creating a defence for a person charged under s. 7(1) if he can prove that he 'took all reasonable precautions and exercised all reasonable diligence to avoid the commission of the offence'. At the same time, the 1991 Act increased the penalties for selling tobacco to persons under the age of 16.)

In *Harrow London Borough Council v Shah* [1999] 3 All ER 302, DC, a 13 and a half year-old boy was sold a National Lottery ticket by an employee of a shopowner, who, at the time, was not in the shop but was working in a back room. The employee reasonably, but mistakenly, believed that the boy was at least 16 years old. The shopowner was charged with the offence, created by s. 13 of the National Lottery etc Act 1993 and reg. 3 of the National Lottery Regulations 1994 (SI 1994, No. 189), of selling a National Lottery ticket to a person under the age of 16.

In reaching the conclusion that the offence is one of strict liability, the court applied the dictum of Lord Scarman in *Gammon (Hong Kong) Ltd v Attorney-General of Hong Kong* [1984] 2 All ER 503, PC, at p. 508 (see above), and decided that the offence is not truly criminal in character, the statutory provisions deal with gambling, an issue of social concern, and the imposition of strict liability will encourage greater vigilance in preventing the commission of the *actus reus* (see [1999] 3 All ER 302, DC, *per* Mitchell J at pp. 306–7). Since the offence is one of strict liability, it follows that the prosecution need only prove a sale to a particular person who was under 16 at the time; it does not have to prove that the defendant or his agent was aware of the buyer's age or was reckless as to his age. Moreover, as in *St Helen's Metropolitan Borough Council v Hill* (1992) 156 JP 602, DC, above, it is no defence to the shopowner that the ticket was not sold by him but by his employee.

There is little or no moral stigma attached to offences of strict liability and, although ignorance of the facts will not prevent conviction, it can be reflected in the sentence imposed. In *Maidstone Borough Council v Mortimer* [1980] 3 All ER 552, DC, in which the defendant was held guilty of the offence of cutting down a protected tree even though he had no knowledge of the tree preservation order, the court emphasised the public interest in preserving protected trees and that the defendant's moral innocence could be reflected in the penalty imposed on him by the magistrates, to whom the case was sent back with a direction to convict.

Remitting the case to the magistrates is discretionary. In *St Helen's Metropolitan Borough Council* v *Hill* (1992) 156 JP 602, DC, above, the Divisional Court did not consider the case a suitable one to be sent back with a direction to convict, even though it was held that the magistrates were wrong to have acquitted the defendant in the first place. In the court's view, the penalty would have been a modest one as it was nearly two years since the offence was committed, and the local authority's appeal had been brought solely to establish the principle that the offence in question was one of strict liability.

In *Harrow London Borough Council* v *Shah* [1999] 3 All ER 302, DC, above, the Divisional Court sent the case back to the magistrates' court with a suggestion that leniency in sentencing would be appropriate, given the employee's state of mind and the obvious care with which the shopowner conducted the sale of National Lottery tickets.

4.10.3.6 Against retrospective effect

There is a presumption of interpretation that statutes do not operate retrospectively. This presumption is particularly important in Acts creating crimes, since it would be oppressive and abhorrent for a statute to criminalise retroactively an act which was quite lawful at the time it was done.

If a statute is repealed, it is not revived if the repealing statute is itself later repealed unless words are inserted reviving it (Interpretation Act 1978, s. 15). At the same time, if a statute abolishes a crime by repealing an existing statutory provision, the repeal does not affect the institution or continuation of legal proceedings in respect of offences alleged to have been committed before the date of the repeal unless a contrary intention appears in the repealing Act (ibid., s. 16(1)). In other words, the repeal does not automatically legalise conduct which was unlawful before the repeal took effect (see, e.g., *Commissioner of Police of the Metropolis* v *Simeon* [1982] 2 All ER 813, HL, where this rule was relied on by the House of Lords in a case which arose out of the repeal of the 'sus' law in 1981).

Since Parliament is supreme, it is within its powers to pass retrospective legislation, but the courts insist that clear, express words must be used. Retrospective tax legislation is perhaps the commonest example. Another notable example is the War Damage Act 1965, which was passed specifically to overrule retroactively the decision of the House of Lords in *Burmah Oil Co. Ltd* v *Lord Advocate* [1965] AC 75, HL, and to deprive Burmah Oil of the victory gained therein. The company's oil installations in Burma (then a British colony) had been destroyed by British forces in 1942 to prevent them being captured by invading Japanese forces. The company, which was registered in Scotland, now sued the Crown for compensation. The Crown argued that no compensation was payable when property was destroyed under the royal prerogative. The House of Lords held that compensation was payable. An inquiry as to damages was ordered, but, in the end, the taxpayer was relieved by Act of Parliament of the burden of having to pay compensation.

The War Crimes Act 1991 is expressly made retrospective in operation. This statute allows the Attorney-General to authorise proceedings for homicide committed in Germany, or German occupied territory, during the Second World War to be brought against a person in the United Kingdom regardless of his nationality at the time of the alleged offence.

There was formerly no presumption against retroactivity as regards *procedural* matters. Indeed, there was authority for the view that since statutory changes in procedure are beneficial they are always to operate retrospectively unless there is a good reason why they should not. A good reason might be that retroactivity would cause unfairness or injustice.

Doubt was cast on this authority in *L'Office Cherifien des Phosphates* v *Yamashita-Shinnihon Steamship Co. Ltd* [1994] 1 All ER 20, HL, where the House of Lords was critical of an approach for ascertaining the retrospective effect of statutes which depends on a distinction between accrued substantive rights on the one hand (where the presumption against retroactivity applies) and procedural rights on the other (where the presumption does not apply). As Lord Mustill pointed out (at p. 32, the rest of their Lordships agreeing with him), this distinction is both misleading (since some procedural rights are more valuable than some substantive rights) and difficult to apply in practice (since it proceeds on the assumption that every right is capable of classification as either substantive or procedural).

The *L'Office Cherifien* decision, above, was applied by the Court of Appeal in *Antonelli* v *Secretary of State for Trade and Industry* [1998] 1 All ER 997, CA, which was concerned with the interpretation of s. 3 of the Estate Agents Act 1979, under which (as subsequently amended by the Enterprise Act 2002) the Office of Fair Trading has power to make an order disqualifying from acting as an estate agent a person who, *inter alia*, 'has been convicted of . . . an offence involving fraud or other dishonesty or violence', and is, in the OFT's opinion, 'unfit to carry on estate agency work'.

The Act came into force in May 1982. Some ten years earlier, Mr Antonelli had been convicted in the United States of the crime of 'burning real estate other than a dwelling house'. It was held that s. 3 applies to conviction for an offence committed abroad before the Act came into force and that the OFT's disqualification order had been lawfully made. It was said to be 'quixotic' to suppose that Parliament intended to protect the public from an estate agent convicted a week *after* the Act came into force but not from one convicted a week *before* (*per* Beldam LJ at p. 1006).

In *R* v *Field* [2003] 1 WLR 882, CA, a case similar to *Antonelli* but in which *Antonelli* was not cited, the question for the Court of Appeal was whether s. 28 of the Criminal Justice and Court Services Act 2000 (which allows the court to make an order disqualifying a person from working with children if he has been convicted of certain offences against a child) applies in respect of offences committed before the Act came into force. It was held that it does. The court said that a statute does not violate the presumption against retrospective effect simply because it depends for its future application on events which took place before it came into force. Moreover, the purpose of s. 28—the protection of children—would be seriously undermined if a disqualification order could only be made in relation to offences committed after it came into operation.

4.10.4 Intrinsic aids to construction

4.10.4.1 Short title

The short title of an Act is not of much use for interpretation purposes. It does not enact anything and is merely descriptive.

4.10.4.2 Long title

The long title of an Act is part of the Act and may be looked at *in cases of ambiguity* (see, e.g., *Manuel* v *Attorney-General* [1982] 3 All ER 822, CA, *per* Slade LJ at p. 831). If, however, there is no ambiguity in the body of the Act because the words employed are clear, the long title cannot be used to restrict the scope of those words merely because they seem to be unduly wide (*R* v *Galvin* [1987] 2 All ER 851, CA, *per* Lord Lane CJ at p. 855). Whilst the long title may provide a useful guide to the general objectives of the Act, it is less helpful on specific provisions.

In *Black-Clawson International Ltd* v *Papierwerke etc. AG* [1975] AC 591, HL, Lord Simon said (at p. 647) he could see no reason for resorting to the long title only in cases of ambiguity when the long title provided the 'plainest of all guides to the general objectives of a statute'.

In *Royal College of Nursing of the United Kingdom* v *DHSS* [1981] 1 All ER 545, HL (para 4.10.2.3 above), the long title of the Abortion Act 1967 was specifically referred to by four of the five Law Lords who decided the appeal, and the long title of the Human Fertilisation and Embryology Act 1990 was quoted by the three Law Lords who delivered full judgments in *R (Quintavalle)* v *Secretary of State for Health* [2003] 2 WLR 692, HL (para 4.10.6 below).

4.10.4.3 Preamble

A preamble, which sets forth (sometimes at great length) the need for, and the intended effect of, the legislation, is not commonly found in modern public statutes. If there is a preamble it may be looked at for guidance in cases of ambiguity, subject to the caveat that it may be found to be narrower or wider than the operative words of the Act.

4.10.4.4 Punctuation

Punctuation was not used in old statutes, but modern statutes use normal punctuation. Formerly, punctuation was not regarded as part of the statute, but, more recently, there have been signs that this approach has been relaxed and that punctuation may be used as an aid to interpretation where a statutory provision is ambiguous (see *DPP* v *Schildkamp* [1971] AC 1, HL, *per* Lord Reid at p. 10).

4.10.4.5 Headings and side-notes

It was confirmed by the House of Lords in *R* v *Montila* [2004] 1 WLR 3141, HL, that headings and side-notes are part of a statute and can be used as aids to construction.

However, since they are unamendable on the passage of a Bill through Parliament, and are included for ease of reference rather than for debate, they may be given less weight as aids than those parts of the statute which were open for debate in Parliament. In their Lordships' view, headings and side-notes are as much part of the 'contextual scene' as explanatory notes, which, as mentioned in para 4.10.5.6 below, can be used as aids to construction even though they are not part of the statute to which they refer. It would be illogical to treat headings and side-notes any differently.

4.10.4.6 Schedules

Schedules to an Act are undoubtedly part of the statute, and may be looked at to cast light on any uncertainty thrown up in the main body of the legislation. Schedules are placed at the end of an Act and their principal purpose is to prevent the main body from becoming too cluttered and complex. They usually consist of transitional provisions, minor and consequential amendments, forms, and a list of enactments repealed, although, occasionally, there is evidence of operative legislative words occurring in a schedule. Examples are the annual Finance Act and several schedules in the Access to Justice Act 1999.

4.10.4.7 Examples

Occasionally a statute gives examples to illustrate the working of the Act or the use of the new terminology created by it. These examples, whether contained in sections or in schedules, are part of the statute.

The Law of Property Act 1925 (schs. 3, 4, and 5) and the Settled Land Act 1925 (sch. 1) give many examples of specimen forms and instruments which may be used to satisfy the terms of the Acts. The examples are part of the legislation and may be used as they stand or with appropriate variations or additions.

The Consumer Credit Act 1974 created a new terminology for money-lending, pawning, and hire-purchase transactions, and the statute provides 24 illustrations of its use (sch. 2). Although the examples are part of the legislation, the statute makes it clear that they are not exhaustive and that they cannot prevail in any conflict with any other provision of the Act (s. 188(2)–(3)). Nevertheless they are of substantial persuasive authority in the construction of the new terminology.

Section 44(6) of the Criminal Justice Act 2003 gives three examples of circumstances which may constitute 'evidence of a real and present danger that jury tampering would take place' such as to compel a judge to order that a trial on indictment is to be held without a jury (see para 6.5.9).

4.10.5 Extrinsic aids to construction

4.10.5.1 Introduction

Extrinsic aids are those which are not contained in the Queen's Printer's copy of the statute. The use of extrinsic material, including *travaux préparatoires* (preparatory works leading to legislation), as an aid to statutory interpretation was historically restricted.

The strict practice was that, since the judges were not allowed to inquire into the social or political history of an Act, they would not look at the reports of parliamentary debates contained in *Hansard*, ministerial pronouncements, explanatory memoranda issued with the Act, reports of commissions and official committees, or international conventions. For example, in *Ellerman Lines Ltd* v *Murray* [1931] AC

126, HL, the House of Lords refused to look at an international convention (set out in a schedule to the statute) when interpreting the Merchant Shipping (International Labour Conventions) Act 1925 which had been passed to give effect to the convention. (For the more modern approach to the use of international conventions, see *Fothergill v Monarch Airlines Ltd* [1980] 2 All ER 696, HL, para 4.10.5.2 below.)

4.10.5.2 Relaxation of the strict approach

The strict practice outlined above was gradually relaxed, and extrinsic material is now regularly consulted, especially in order to discover the mischief which the Act was intended to remedy. Thus, reports of the Law Commission, royal commissions, the Law Reform Committee and other official committees, as well as government white papers and green papers, have all been looked at by the courts (see *Black-Clawson International Ltd* v *Papierwerke etc. AG* [1975] AC 591, HL, *per* Viscount Dilhorne at p. 621).

In the *Black-Clawson* case, the House of Lords said unanimously that where a statute is ambiguous the court is entitled to look at the report of an official committee, presented to Parliament and containing proposals for legislation, for the purpose of discovering the *mischief* which the resulting statute was designed to remedy. But there was disagreement among their Lordships as to whether the report could be considered for any other purpose. Lord Reid (at p. 614) and Lord Wilberforce (at p. 629) said it was not permissible to look at the report as a guide to the *meaning* of the legislation, even though the report contains a draft Bill which is later enacted without amendment. Viscount Dilhorne and Lord Simon disagreed, the latter saying (at p. 646) that:

Where Parliament is legislating in the light of a public report I can see no reason why a court of construction should deny itself any part of that light and insist on groping for a meaning in darkness or half-light.

The view of Lords Reid and Wilberforce has prevailed over that of Viscount Dilhorne and Lord Simon (see *Attorney-General's Reference (No. 1 of 1988)* [1989] 2 All ER 1, HL, *per* Lord Lowry at p. 6). In *R v Ayres* [1984] 1 All ER 619, HL, their Lordships looked at a report of the Law Commission, and in *R v Allen* [1985] 2 All ER 641, HL, they considered a report of the Criminal Law Revision Committee. In each case this was done as an aid to discovering the mischief in the criminal law which the subsequent statute was designed to remedy. In the latter case, Lord Hailsham LC made it clear (at p. 644) that the report had been looked at solely for that purpose, and that it had not been used as a guide to the meaning of the statute.

In *Duke* v *GEC Reliance Ltd* [1988] 1 All ER 626, HL, Lord Templeman (at pp. 629–30 and 634) referred to the White Paper, *Equality for Women* (Cmnd 5724, 1974), as a guide to the intention of Parliament in enacting the Sex Discrimination Act 1975.

In *R v Burke* [1990] 2 All ER 385, HL, Lord Griffiths (at p. 389), in determining the 'social evil' which had led Parliament to create the criminal offence of unlawful harassment of a residential occupier (formerly s. 30(2) of the Rent Act 1965, now s. 1(3) of the consolidating Protection from Eviction Act 1977), and in concluding that the offence is committed notwithstanding that the harassment does not constitute a civil wrong known to the law, derived assistance from the *Report of the Committee on Housing in Greater London* (the Milner Holland Report, Cmnd 2605, 1965).

The *Report of the Committee on Homosexual Offences and Prostitution* (the Wolfenden Report, Cmnd 247, 1957), was relied on in a case where it was decided that the term 'common prostitute' in s. 1(1) of the Street Offences Act 1959 does not include a male prostitute. This *Report* had led to the passing of the Act, and had identified the mischief associated with loitering or soliciting for the purpose of prostitution as one caused by women (*Director of Public Prosecutions* v *Bull* [1994] 4 All ER 411, DC; s. 1(1) of the 1959 Act was later amended by the Sexual Offences Act 2003 so as to make it applicable to *male* as well as female prostitutes).

The danger of ignoring an extrinsic aid is illustrated by *Anderton* v *Ryan* [1985] AC 560, HL, in which the House of Lords virtually ignored the report of the Law Commission which preceded the Criminal Attempts Act 1981 (*Criminal Law: Attempt and Impossibility in Relation to Attempt, Conspiracy and Incitement*, Law Com. No. 102, 1980). This error led to a bad decision which the House found it necessary to overrule at the earliest opportunity in the case of *R* v *Shivpuri* [1986] 2 All ER 334 (see further, para 5.3.2.2.2).

Lord Denning was instrumental in persuading the courts to change the practice adopted in *Ellerman Lines Ltd* v *Murray* [1931] AC 126, HL (para 4.10.5.1 above), and to use *travaux préparatoires* as aids to the construction of an *international convention*. In *Fothergill* v *Monarch Airlines Ltd* [1979] 3 All ER 445, CA, one question before the Court of Appeal was whether the court could consider *travaux* in construing the Carriage by Air Act 1961, which contained in a schedule the Warsaw Convention of 1929 as amended in 1955. The *travaux* included the minutes of 34 meetings held at an international conference at the Hague in 1955. Lord Denning said (at p. 451) that:

These *travaux préparatoires* can be used not only to see what was the mischief needing to be remedied, not only to see what was the purpose or object of the draftsmen, but also to find out what they really meant to convey by the words they used.

Browne and Geoffrey Lane LJJ disagreed. Browne LJ said (at p. 456) that the court's power was limited to discovering the mischief of the old law. The court cannot take into account direct statements of what the Act was intended to mean.

Lord Denning's approach was completely vindicated, for when *Fothergill* v *Monarch Airlines Ltd* went to the House of Lords ([1980] 2 All ER 696) a majority of four to one of their Lordships were of the opinion that, in interpreting international conventions, the court could look at *travaux* in a general way and not merely for the specific purpose of discovering the mischief of the former law. Their Lordships were convinced that a more liberal rule of interpretation was justified in the interests of universal uniformity of application.

However, *travaux préparatoires* should be used with caution. In particular, they should be considered only where, first, they are publicly accessible, *and*, secondly, they clearly point to a definite legislative intention (*Fothergill*, above, *per* Lord Wilberforce at p. 703). *Travaux* are *aids* to the interpretation of, and not *substitutes* for, the terms of a convention.

The use of *travaux* is not mandatory. The court has a *discretion* whether to look at them and what weight to attach to them (ibid., *per* Lord Scarman at p. 715). Thus, Bingham J in *Data Card Corp.* v *Air Express International Corp.* [1983] 2 All ER 639, Hirst J in *Cia Portorafti Commerciale SA* v *Ultramar Panama Inc.* [1989] 2 All ER 54,

Lord Oliver in *Hiscox* v *Outhwaite (No. 1)* [1991] 3 All ER 641, HL, and Lord Hope in *Sidhu* v *British Airways plc* [1997] 1 All ER 193, HL, looked at the relevant *travaux préparatoires* but found them of no assistance whatever, while in *Gatoil International Inc.* v *Arkwright-Boston Manufacturers Mutual Insurance Co.* [1985] 1 All ER 129, HL, a majority of the House of Lords found the relevant *travaux* helpful as an external aid to construction.

4.10.5.3 *Hansard*

In the construction of purely *domestic* legislation, *Hansard* remained a closed book for much longer than other forms of extrinsic material. Reference to it for this purpose was not permitted until the early 1990s.

The use of *Hansard* was advocated in 1974 by Lord Simon in *Race Relations Board* v *Dockers' Labour Club & Institute* [1976] AC 285, HL, at p. 299, albeit only for the purpose of discovering the *mischief* of the former law and not the *meaning* of the present statutory language.

Lord Denning MR went the whole way in *Davis* v *Johnson* [1979] AC 264, CA and HL. In construing s. 1 of the Domestic Violence and Matrimonial Proceedings Act 1976, he looked at both the *Report of the House of Commons Select Committee on Violence in Marriage (H. of C. Papers (1974–75) 553-i)* and the reports of parliamentary debates in *Hansard (H. of C. Official Report (1975–76) Standing Committee F, Domestic Violence Bill)*. On the latter, his Lordship said (at pp. 276–7):

Some may say—and indeed have said—that judges should not pay any attention to what is said in Parliament. They should grope about in the dark for the meaning of an Act without switching on the light. I do not accede to this view . . . And it is obvious that there is nothing to prevent a judge looking at these debates himself privately and getting some guidance from them. Although it may shock the purists, I may as well confess that I have sometimes done it. I have done it in this very case. It has thrown a flood of light on the position.

Goff and Cumming-Bruce LJJ expressly dissociated themselves from Lord Denning's 'heresy'. Cumming-Bruce LJ said (at p. 316) that he was 'not alarmed by the criticism that I am a purist who prefers to shut his eyes to the guiding light shining in the reports of Parliamentary debates in *Hansard*'.

When *Davis* v *Johnson* went to the House of Lords, all five Law Lords disagreed with Lord Denning's view. Lord Scarman's opinion is particularly interesting. He said ([1979] AC 264, 349–50):

There are two good reasons why the courts should refuse to have regard to what is said in Parliament or by ministers as aids to the interpretation of a statute. First, such material is an unreliable guide to the meaning of what is enacted. It promotes confusion, not clarity. The cut and thrust of debate and the pressures of executive responsibility . . . are not always conducive to a clear and unbiased explanation of the meaning of statutory language. And the volume of Parliamentary and ministerial utterances can confuse by its very size. Secondly, counsel are not permitted to refer to *Hansard* in argument. So long as this rule is maintained by Parliament (it is not the creation of the judges), it must be wrong for the judge to make any judicial use of proceedings in Parliament for the purpose of interpreting statutes.

After the decision of the House of Lords in *Davis* v *Johnson*, given in March 1978, the matter was thought to be closed; but Lord Denning reopened it. He found a way round the prohibition on the use of *Hansard* as an aid to construction. *Direct* reference to *Hansard* was forbidden, but *indirect* reference was not.

Thus, if *Hansard* was quoted in a textbook or published speech, the textbook or speech could, according to Lord Denning, legitimately be looked at by the judge. In *R v Local Commissioner for Administration (ex parte Bradford Metropolitan City Council)* [1979] 2 All ER 881, CA, heard in July 1978, he consulted a textbook quotation of a public address, which itself contained quotations from *Hansard*, in order to clarify the meaning of the word 'maladministration' in the Local Government Act 1974, s. 26(1).

The parliamentary rule which prevented counsel from citing *Hansard* in court (referred to in Lord Scarman's speech in *Davis* v *Johnson* [1979] AC 264, HL, above) was abolished in 1980. However, this fact made no immediate difference to the use of *Hansard* as an extrinsic aid to the construction of United Kingdom domestic legislation. Lord Scarman's first objection to its use remained valid (see *Davis* v *Johnson*, above).

It was argued by some that prohibiting citation from *Hansard* conflicted with the duty of our courts to discover and give effect to the intention of Parliament when interpreting statutes, although it will be appreciated that extensive reference to *Hansard* could add considerably to the length (and, therefore, the cost) of litigation without any guarantee that it would assist in resolving the problem before the court.

The immediate post-1980 position was illustrated in the opinions of the Law Lords in *Hadmor Productions Ltd* v *Hamilton* [1982] 1 All ER 1042, HL. When that case was before the Court of Appeal (see [1981] 2 All ER 724, CA), Lord Denning, without any reference to the criticism of the practice made by the House of Lords in *Davis* v *Johnson*, proceeded to refer once again to *Hansard* (at pp. 731 and 733). He was criticised for his approach when the *Hadmor* case reached the House of Lords.

The process of legitimising reference to *Hansard*, which culminated in *Pepper* v *Hart* [1993] 1 All ER 42, HL (para 4.10.5.4 below), began when the House of Lords took the lead in adopting a more relaxed approach towards its use in cases involving the construction of *delegated legislation* approved by the United Kingdom Parliament and designed to give effect to our European Union obligations. The justification for this development was that proposed delegated legislation, in the form of draft regulations, is not (unlike a Bill) subject to the parliamentary process of consideration and amendment in committee.

For example, in *Pickstone* v *Freemans plc* [1988] 2 All ER 803, HL, Lord Templeman (at pp. 814–5) quoted from the *Hansard* report of the House of Commons' debate on some draft regulations, made under the authority of s. 2(2) of the European Communities Act 1972 and approved by Parliament in order to comply with a decision of the European Court that English law did not accord with a Directive on equal pay for equal work. He did so as an aid to discovering, from 'the explanations of the government and the criticisms voiced by members of Parliament', what was the true intention of Parliament in approving the regulations. Lord Keith adopted a similar stance (at p. 807).

4.10.5.4 *Pepper* v *Hart*

It was finally decided that reference to *Hansard* was to be allowed as an aid to the interpretation of United Kingdom domestic primary legislation in *Pepper v Hart* [1993] 1 All ER 42, HL. Here, the House of Lords sitting seven-strong declined to follow, by a majority of six to one (Lord Mackay LC dissenting), three of its earlier decisions on the use of *Hansard*—*Beswick v Beswick* [1968] AC 58, HL, *Black-Clawson International Ltd v Papierwerke Waldhof-Aschaffenburg AG* [1975] AC 591, HL, and *Davis v Johnson* [1979] AC 264, HL.

Their Lordships held in *Pepper v Hart*, above, that the exclusionary rule against the use of *Hansard* would be relaxed so as to permit reference to parliamentary materials subject to the following conditions:

(a) the legislation is ambiguous, or obscure, or its literal meaning leads to an absurdity;

(b) the material relied on consists of statements by a minister or other promoter of the Bill; and

(c) the statements relied on are clear.

Submissions were made in *Pepper v Hart* [1993] 1 All ER 42, HL, on the appropriateness of modifying the prohibition on the use of parliamentary materials, and on whether the use of *Hansard* in the courts would amount to a violation of the Bill of Rights 1688, or a breach of parliamentary privilege.

The Attorney-General, appearing on behalf of the Crown, had argued that using *Hansard* would infringe s. 1, art. 9, of the Bill of Rights 1688 ('That the freedome of speech and debates or proceedings in Parlyament ought not to be impeached or questioned in any court or place out of Parlyament'). But the House of Lords unanimously disagreed, holding that the courts, by making use of *Hansard*, would not be 'questioning' or criticising what was said in Parliament but trying to implement what was said there (see [1993] 1 All ER 42, especially *per* Lord Mackay LC, Lord Griffiths, Lord Oliver, and Lord Browne-Wilkinson at pp. 47, 50, 53, and 67–9, respectively).

The House further held that it was not inhibited from deciding the case before it by any general considerations of parliamentary privilege because no such privilege extending beyond the Bill of Rights had been identified (see especially *per* Lord Browne-Wilkinson at pp. 73–4).

The dissent of Lord Mackay LC on the use of *Hansard* is perhaps not surprising in view of his role as the government minister responsible for public expenditure on legal aid. In his opinion (expressed at p. 48), the new approach introduced 'the possibility at least of an immense increase in the cost of litigation in which statutory construction is involved'. The rest of their Lordships, however, regarded this and other objections to the use of *Hansard* as exaggerated and were prepared to sweep them aside. Lord Bridge said (at p. 49) that he found it

difficult to suppose that the additional cost of litigation or any other ground of objection can justify the court continuing to wear blinkers which, in such a case as this, conceal the vital clue to the intended meaning of an enactment.

While recognising that practitioners would in some cases incur fruitless costs in search of the 'vital clue' where none existed, he implied that there could be many more cases where the cost of litigation would be avoided because *Hansard* provided the answer.

Lord Browne-Wilkinson (at p. 63) summarised the reasons put forward for the prohibition on the use of *Hansard*. First, it preserves the 'constitutional proprieties' whereby Parliament legislates in words and the courts interpret the meaning of the words. Secondly, it avoids the practical difficulty of an expensive researching of parliamentary materials. Thirdly, *Hansard* does not provide the citizen with an accessible and defined text regulating his legal rights. Fourthly, it is in any event improbable that any helpful guidance will be found in *Hansard*.

These objections notwithstanding, Lord Browne-Wilkinson was convinced that, as a matter of law, there were now sound reasons for modifying the prohibition in a limited way. Among those reasons (explained at pp. 64–5) was that reference to *Hansard* could allow the courts more regularly to get at the underlying parliamentary purpose of specific legislation, his acceptance of the view that 'clear and unambiguous statements made by ministers in Parliament are as much the background to the enactment of legislation as white papers and parliamentary reports', and his opinion that looking at ministerial statements made in introducing regulations which cannot be amended by Parliament (authorised by *Pickstone* v *Freemans plc* [1988] 2 All ER 803, HL, above) is 'logically indistinguishable' from looking at ministerial statements made in introducing *primary* legislation which, though capable of amendment, was not amended in fact.

4.10.5.5 The use of *Hansard* after *Pepper* v *Hart*

In *Warwickshire County Council* v *Johnson* [1993] 1 All ER 299, HL, the House of Lords followed the new practice in relation to *Hansard* approved only two weeks earlier in *Pepper* v *Hart*.

The *Warwickshire County Council* case was concerned with the interpretation of s. 20(1) of the Consumer Protection Act 1987, which makes it an offence for a person to give 'in the course of any business of his' a misleading indication about the price of goods. The words 'any business of his' were ambiguous. The question was whether they meant that a branch manager of Dixons (the electrical goods retailers) who was responsible for a misleading price advertisement could be guilty of the offence as well as the company. A literal approach, simplicity, and common sense suggested that he could.

The leading opinion was delivered by Lord Roskill. He looked at the internal evidence and the legislative history of the provision, and concluded that the words 'any business of his' were meant to refer to the business of the employer. Since the employee did not have a business he could not be guilty of the offence. This view was confirmed by reference to *Hansard*, where there were clear statements by the minister at the report stage of the Bill in the House of Lords that the intention of the proposed legislation was to catch employers and not employees.

Subsequent cases in which the House of Lords has applied the new practice include *Stubbings* v *Webb* [1993] 1 All ER 322, HL (where Lord Griffiths, at pp. 328–9, referred to debates in *Hansard* on s. 2(1) of the Law Reform (Limitation of Actions etc) Act 1954 as an aid to the construction of s. 11 of the Limitation Act 1980, which contained without alteration (via the Limitation Act 1975) the wording of the 1954 Act); *Chief Adjudication Officer* v *Foster* [1993] 1 All ER 705, HL (in which Lord Bridge, at pp. 715–17, resorted to *Hansard* in confirmation of his interpretation of s. 22 of the

Social Security Act 1986); *Steele Ford & Newton* v *Crown Prosecution Service* [1993] 2 All ER 769, HL (where Lord Bridge, at p. 777, consulted *Hansard* when construing certain statutory provisions relating to the payment of costs); and *Scher* v *Policy-holders Protection Board (No. 2)* [1993] 4 All ER 840, HL (in which Lord Mustill, at p. 852, referred to debates at the committee and report stages of the passage of the Policyholders Protection Act 1975).

In *Secretary of State for Social Security* v *Remilien* [1998] 1 All ER 129, HL, the House of Lords did not find 'clear', within the meaning of *Pepper* v *Hart* [1993] 1 All ER 42, HL, and derived no assistance from, a statement about the Income Support (General) Regulations 1987 made to the House of Commons Second Standing Committee on Statutory Instruments by the Parliamentary Under-Secretary of State for Social Security.

In *R* v *Secretary of State for the Environment, Transport and the Regions (ex parte Spath Holme Ltd)* [2001] 2 WLR 15, HL, a majority of four to one in the House of Lords held that it was not appropriate to have recourse to *Hansard* as an aid to the construction of s. 31 of the Landlord and Tenant Act 1985 because the conditions laid down in *Pepper* v *Hart* were not satisfied. All their Lordships (including the dissentient Lord Cooke, who would have allowed *Hansard* to be consulted) were agreed that the *Pepper* v *Hart* conditions should be strictly insisted on, so that resort to *Hansard* as an aid to interpretation is the exception rather than the rule. (See [2001] 2 WLR 15, HL, *per* Lords Bingham, Nicholls, Cooke, and Hutton at pp. 32, 39, 44, 48, and 54, respectively.)

In *R* v *Deegan* [1998] 2 Cr App R 121, CA, the Court of Appeal, when called on to interpret the statutory expression 'folding pocketknife', found ministerial statements (made during debates on the Bill which became the Criminal Justice Act 1988) to be not clear in the sense required by *Pepper* v *Hart*, and declined to take them into account. On the other hand, in *A.E. Beckett & Sons (Lyndons) Ltd* v *Midland Electricity plc* [2001] 1 WLR 281, CA, the Court of Appeal, when construing s. 21 of the Electricity Act 1989 on the liability of a public electricity supplier for economic loss resulting from negligence, discovered that reference to *Hansard* 'immediately made clear what had previously been obscure' (*per* Lord Phillips MR, delivering the judgment of the court, at [34]).

The relaxed rule sanctioned in *Pepper* v *Hart* [1993] 1 All ER 42, HL, allows reference to be made to parliamentary materials which provide evidence of the purpose behind legislation then going through Parliament. It does *not* permit the citation of parliamentary materials which merely evidence a minister's understanding of the current state of the law (*Hillsdown Holdings plc* v *Pensions Ombudsman* [1997] 1 All ER 862, in which Knox J declined to take into account a ministerial statement made two years after the relevant statutory provision came into force).

Nor is it permissible to refer to ministerial or other statements recorded in *Hansard* which are not directly concerned with the specific statutory provision under consideration or with the problem raised by the litigation. Such statements do not assist the court, and attempts to widen the category of materials to which reference may be made can simply cause extra delay and expense (*Melluish (Inspector of Taxes)* v *BMI (No. 3) Ltd* [1995] 4 All ER 453, HL, where the House of Lords refused to allow the Crown to introduce ministerial statements which were not directly in point).

The House of Lords has held that it is not permissible to look at reports of parliamentary debates in *Hansard* with a view to deciding under s. 4 of the Human Rights Act 1998 whether a statutory provision is compatible with the European Convention on Human Rights. To do so would involve an evaluation of the policy considerations advanced by those taking part in the debates. This would amount to 'questioning' what is said in Parliament contrary to s. 1, art. 9, of the Bill of Rights 1688 (*Wilson* v *First County Trust Ltd (No. 2)* [2003] 3 WLR 568, HL).

If it is intended to refer to *Hansard* in a case before the Court of Appeal, the High Court, the Crown Court, or the county courts, copies of the relevant extract and a brief summary of the argument to be based on the extract must, unless the judge directs otherwise, be served on all other parties and on the court (*Practice Direction (Criminal Proceedings: Consolidation), Part II* [2002] 1 WLR 2870). A failure to do this may be penalised in costs (ibid.).

There is some evidence that the cost of litigation has increased after *Pepper* v *Hart* [1993] 1 All ER 42, HL, as lawyers increasingly feel obliged to spend expensive time combing through *Hansard* in appropriate cases in order to avoid the risk of being sued in negligence by their clients (see *The Times*, 6 July 1993). According to Lord Hobhouse in *Wilson* v *First County Trust Ltd (No. 2)*, above, at [140]:

the attempt by advocates to use parliamentary material from *Hansard* as an aid to statutory construction has not proved helpful and the fears of those pessimists who saw it as simply a cause of additional expense in the conduct of litigation have been proved correct.

(See also *R (Quintavalle)* v *Human Fertilisation and Embryology Authority* [2005] 2 WLR 1061, HL, *per* Lord Hoffmann at [34].)

4.10.5.6 Explanatory notes

Explanatory notes have accompanied the majority of public Bills brought before Parliament since 1999. They are prepared by the government department responsible for the Bill, but, as is made clear in the notes themselves, they are not part of it. They provide background information on the proposed legislation, and explain the effect of clauses in the Bill; however, they are politically neutral and do not seek to justify what is in the Bill.

The status of explanatory notes as extrinsic aids to construction was clarified by Lord Steyn in *R (Westminster City Council)* v *National Asylum Support Service* [2002] 1 WLR 2956, HL, at [2]–[6]. Their use is always permissible for the purpose of illuminating the 'contextual scene' of the statute and the mischief, if any, which it seeks to put right. In Lord Steyn's view, they are in this respect likely to be 'more informative and valuable' than other extrinsic aids because of their close connection with the statute in question.

Since, however, the court's task in interpreting a statute is to ascertain the intention of Parliament *as expressed in the words Parliament has chosen to use in the statute*, explanatory notes prepared by the government cannot be used to ascribe to Parliament an intention which the government may have but which Parliament does not have.

4.10.6 Reform of the principles of statutory interpretation: the 'literal' approach v the 'purposive' approach

There has been much criticism of the English principles of interpretation. In the first place, it is complained in particular that too much use has been made in the past of the literal rule at the expense of a more purposive approach to statutory construction.

Secondly, the three so-called 'rules'—literal, golden, and mischief—are not rules at all since there is no compulsion to apply them. Moreover, there is no consistency of application when they *are* applied. There is no set order of priority. In one case the court will apply the literal rule, while in another it will apply the mischief rule. Sometimes the court does not identify which rule it is applying.

Thirdly, the presumptions are similarly criticised. For instance, there is no rule about which presumption should be applied where two of them conflict.

Fourthly, some of the historic limitations on the use of extraneous material have been felt to be unduly restrictive.

As long ago as 1969, the Law Commission published its report on *The Interpretation of Statutes* (Law Com. No. 21). Its proposals have not been implemented by legislation. The Law Commission was most impressed by the mischief rule, but said that it needed to be adapted to modern conditions. It was in favour of a principle of interpretation giving preference to a construction which promoted the general legislative purpose over one which did not.

Attempts were made by Lord Scarman, a former Lord of Appeal in Ordinary, to secure legislation which would give effect to these proposals of the Law Commission, of which, at the time of its report in 1969, he was the first chairman. Lord Scarman's attempts failed. In 1980, his Interpretation of Legislation Bill was withdrawn after opposition from his judicial colleagues in the House of Lords. Only the Lord Chancellor, Lord Hailsham LC, gave any encouragement to the Bill. In 1981, Lord Scarman introduced a more comprehensive Bill which passed through the House of Lords but was rejected by the House of Commons.

Some judicial efforts, notably by Lord Denning, were made to improve interpretative technique. Writing extra-judicially, Lord Denning agreed that the object of statutory interpretation is to discover the intention of Parliament. But he argued that the actual words used in the statute are only the starting point and not the finishing point (*The Discipline of Law*, 1979, p. 9; see also *The Closing Chapter*, 1983, pp. 94–107 and 110–14). He preferred the 'purposive' approach to the literal approach. He was an 'intention seeker' rather than a 'strict literal constructionist'. The purposive approach is the European approach to statutory interpretation, and he recommended its extension to the construction of Acts of the United Kingdom Parliament.

Lord Denning also aired his view from the bench in *Nothman* v *London Borough of Barnet* [1978] 1 All ER 1243, CA, at p. 1246:

The literal method is now completely out of date. . . . In all cases now in the interpretation of statutes we adopt such a construction as will 'promote the general legislative purpose underlying the provision' [quoted from the Law Commission report No. 21 cited above]. It is no longer necessary for the judges to wring their hands and say: 'There is nothing we can do about it.' Whenever the strict interpretation of a statute gives rise to an absurd and unjust situation, the judges can and should use their good sense to remedy it—by reading words in, if necessary—so as to do what Parliament would have done had they had the situation in mind.

When the *Nothman* case reached the House of Lords ([1979] 1 All ER 142), Lord Denning's approach was criticised. In particular, his attempt, single-handed and without legislation, to implement the Law Commission's recommendations of 1969 was too much for Lord Russell, who (at p. 151) expressly disclaimed Lord Denning's 'sweeping comments'. Lord Denning's purposive approach to the interpretation of *domestic legislation* received little judicial support at the time.

A unanimous House of Lords had no hesitation in preferring a purposive approach to a literal interpretation in *R v Pigg* [1983] 1 All ER 56, HL (para 6.5.4). It was held there that, despite the clear mandatory wording of s. 17(3) of the Juries Act 1974, the foreman of a jury when announcing a majority verdict need not state expressly how many jurors *dissented from* the verdict, provided that that was clear from his statement of how many *agreed with* it. Their Lordships said it was the *substance* of the requirement that had to be complied with rather than the precise form of words by which such compliance was to be achieved.

R v Pigg was clearly an exceptional case, however. The House of Lords later said that a judge may adopt a purposive interpretation only if he can find in the statute, or in permitted extrinsic material, an expression of Parliament's purpose or policy (*Shah* v *Barnet London Borough Council* [1983] 1 All ER 226, HL). The judge is not permitted to interpret legislation in the light of his own views about policy. In essence, this represented a reaffirmation of the literal approach to statutory interpretation by which judges are positively discouraged from seeking to discover the policy underlying a piece of legislation. The only concession granted to judicial creativity by the *Shah* case was that judges may adopt a purposive approach to interpretation if the purpose or policy of Parliament is discernible from the statute itself or from material to which they are permitted by law to refer as an aid to the construction of the statute.

The House of Lords moved towards a purposive approach in the interpretation of *international conventions and treaties* (para 4.10.5.2 above) in *Fothergill* v *Monarch Airlines Ltd* [1980] 2 All ER 696, HL, where Lords Wilberforce, Diplock, and Scarman gave overt support to such a move (at pp. 700, 704, and 713, respectively).

Their Lordships later demonstrated a willingness to adopt a purposive, as opposed to a literal, approach when construing legislation of the United Kingdom Parliament passed to give effect to our obligations under European Community law (para 4.10.5.3 above).

Thus, in *Pickstone* v *Freemans plc* [1988] 2 All ER 803, HL, regulations approved by Parliament and designed to fill a gap (identified by a decision of the European Court) in the English equal pay legislation were found, on a literal reading, to be inadequate for filling that gap. A *literal* approach would have left the United Kingdom in breach of its treaty obligations to give effect to European Community directives. The House of Lords held that, in the circumstances, it was permissible to give a *purposive* construction to the regulations so as to make them comply with European Community law.

In order to achieve the manifest purpose of the regulations, and to give effect to the clear, but inadequately expressed, intention of Parliament, it was decided that certain words must be read into the regulations by necessary implication—even though that entailed departing from the strict, literal application of the words which Parliament had chosen to use.

Pickstone v *Freemans plc* was applied in *Litster* v *Forth Dry Dock and Engineering Co. Ltd* [1989] 1 All ER 1134, HL. This case was concerned with the construction of statutory regulations approved by Parliament in order to give effect to a Directive. The regulations provided that the transfer of a business did not terminate the contracts of employment of employees who were employed in the business 'immediately before the transfer', and that those contracts were to be treated after the transfer as if they had originally been made between the employees and the transferee of the business. Some employees who had been dismissed one hour before the transfer of an insolvent business to new owners claimed compensation for unfair dismissal. The transferees argued that the employees' claim must fail since, having already been dismissed, they had not been employed in the business 'immediately before' its transfer.

If this argument, based on a literal interpretation of the regulations, had prevailed, the employees in question would have been left without any effective remedy for unfair dismissal. The House of Lords was unanimous in preferring a purposive to a literal construction. It was held that the regulations applied not only to employees who were employed in the business immediately before the transfer in point of time but also to employees who would have been so employed had they not been unfairly dismissed before the transfer for a reason connected with it. The House of Lords was prepared to achieve this result by reading words into the regulations by necessary implication (see [1989] 1 All ER 1134, HL, *per* Lords Keith, Templeman, and Oliver at pp. 1136, 1139, and 1153, respectively).

The *Litster* case [1989] 1 All ER 1134, HL, was applied in *Morris Angel & Son Ltd* v *Hollande* [1993] 3 All ER 569, CA, in which the Court of Appeal adopted a purposive approach to the construction of the same regulations in relation to their application to restraint of trade clauses in contracts of employment. It was also applied by the House of Lords in *C.R. Smith Glaziers (Dunfermline) Ltd* v *Customs and Excise Commissioners* [2003] 1 WLR 656, HL.

At first, the new approach exemplified in *Litster*, above, was only applicable to the construction of primary or delegated legislation of the United Kingdom Parliament which was designed to give effect to European Community law. Accordingly, it was held that where the parliamentary legislation in question was not passed for the purpose of implementing Community law the English courts would continue to prefer a literal to a purposive approach (*Duke* v *GEC Reliance Ltd* [1988] 1 All ER 626, HL).

The new approach was extended to *all* legislation of the United Kingdom Parliament by the decision of the House of Lords in *Pepper* v *Hart* [1993] 1 All ER 42, HL (para 4.10.5.4 above). The resolution of the entire case proceeded on the basis of the purposive approach to statutory interpretation which, according to a majority of their Lordships, the courts had now adopted in order to give effect to the true intention of Parliament. Thus, in construing any domestic legislation the courts may now (subject to the limitations noted in para 4.10.5.4 above) refer to *Hansard* and other parliamentary materials. Moreover, they may do so not only for the limited purpose of discovering the mischief which the legislation was designed to cure, but also for the purpose of discovering the general legislative intention. Their Lordships' support in *Pepper* v *Hart* for the purposive approach was somewhat unexpected and surprising given the previous history of the subject in the House of Lords (see above and para 4.10.2.1).

Both Lord Griffiths and Lord Browne-Wilkinson in their speeches stressed the merits of the purposive, as opposed to the literal, approach to statutory interpretation. According to Lord Griffiths ([1993] 1 All ER 42 at p. 50),

[t]he days have long passed when the courts adopted a strict constructionist view of interpretation which required them to adopt the literal meaning of the language. The courts now adopt a purposive approach which seeks to give effect to the true purpose of legislation and are prepared to look at much extraneous material that bears on the background against which the legislation was enacted.

Lord Browne-Wilkinson (with whose opinion Lords Keith, Bridge, Ackner, and Oliver expressly agreed) said (ibid., at p. 65):

[g]iven the purposive approach to construction now adopted by the courts in order to give effect to the true intentions of the legislature, the fine distinctions between looking for the mischief and looking for the intention in using words to provide the remedy are technical and inappropriate.

According to Laws LJ in *Oliver Ashworth (Holdings) Ltd* v *Ballard (Kent) Ltd* [1999] 2 All ER 791, CA, at p. 805 (Robert Walker LJ expressly agreeing with him at p. 803), it is misleading to make any rigid distinction between a literal and a purposive approach to statutory interpretation:

The difference ... is in truth one of degree only. On received doctrine we spend our professional lives construing legislation purposively, inasmuch as we are enjoined at every turn to ascertain the intention of Parliament. The real distinction lies in the balance to be struck, in the particular case, between the literal meaning of the words on the one hand and the context and purpose of the measure in which they appear on the other. Frequently there will be no opposition between the two, and then no difficulty arises. Where there is a potential clash, the conventional English approach has been to give at least very great and often decisive weight to the literal meaning of the enacting words. This is a tradition which I think is weakening, in face of the more purposive approach enjoined for the interpretation of legislative measures of the European Union and in light of the House of Lords' decision in *Pepper* v *Hart* . . .

(See further on this case, para 4.10.2.4 above.)

Since *Pepper* v *Hart* [1993] 1 All ER 42, HL, not every opportunity has been taken to further the purposive approach to interpretation. In *R* v *Brown* [1996] 1 All ER 545, HL, for example, a purposive approach to the construction of s. 5 of the Data Protection Act 1984 (since repealed by the Data Protection Act 1998) was preferred only by a minority of their Lordships. The majority decided the appeal on a literal interpretation of the section (see further on this case, para 4.10.2.1 above).

The Court of Appeal in *Jones* v *Tower Boot Co. Ltd* [1997] 2 All ER 406, CA, noting that the purpose of the Race Relations Act 1976 was the eradication of racial discrimination, interpreted s. 32 of the Act in a liberal way so as to advance that purpose. The complainant was of mixed ethnic parentage, and had been subjected to severe verbal and physical racial abuse at work by two fellow employees. After four weeks in the job, he resigned and made a complaint to an employment tribunal against his former employers, alleging that his fellow employees' conduct amounted to racial discrimination for which the employers were liable under the Race Relations Act 1976, s. 32(1), which provides that:

Anything done by a person in the course of his employment shall be treated for the purposes of this Act (except as regards offences thereunder) as done by his employer as well as by him, whether or not it was done with the employer's knowledge or approval . . .

The employers argued that they were not liable because the employees' acts of discrimination were not committed 'in the course of their employment' since, according to the common law principles for determining an employer's vicarious liability in the law of tort, the employees were not specifically employed to engage in the racial abuse of their colleagues, and their behaviour was not merely an improper mode of performing the duties of their employment but was completely outside the scope of those duties.

It was held that the words 'in the course of his employment' are to be given their natural everyday meaning and are not to be construed restrictively in accordance with the common law principles of vicarious liability. Waite LJ ([1997] 2 All ER 406, CA, at p. 415) referred to the result of upholding the employers' contention, namely 'the more heinous the act of discrimination, the less likely it will be that the employer would be liable', and concluded that such a result ran counter to the whole legislative scheme and underlying policy of s. 32 of the 1976 Act. (See further on *Jones* v *Tower Boot Co. Ltd*, para 5.3.2.3.1.)

When later construing s. 7(1) of the same statute in *Harrods Ltd* v *Remick* [1998] 1 All ER 52, CA, the Court of Appeal applied the purposive approach adopted in *Jones* v *Tower Boot Co. Ltd*, above, so as to achieve the perceived statutory objective of providing a remedy to victims of discrimination who would otherwise be without one.

Anyanwu v *South Bank Student Union* [2001] 1 WLR 638, HL, was concerned with s. 33(1) of the 1976 Act, by which:

A person who knowingly aids another person to do an act made unlawful by this Act shall be treated for the purposes of this Act as himself doing an unlawful act of like description.

It was alleged that a university had 'knowingly aided' a student union to dismiss the applicants from their salaried posts by means of racial discrimination. The Court of Appeal held that the university could not be liable under s. 33(1) because it was the prime mover in the dismissals, and had not, therefore, 'knowingly aided' the union.

This decision was reversed by the House of Lords, which held that s. 33(1) is to be construed purposively and that the word 'aids' is a familiar word in everyday use with no special or technical meaning in the context of s. 33(1). It simply refers to the help given by one person to another, and it does not matter that the help is not substantial or productive as long as it is not negligible. Lord Bingham (at [19]) said that it is an unhelpful distraction to import expressions like 'prime mover' or 'free agent' into the statutory language, which is intended to provide a simple test of liability.

Anyanwu, above, was applied by the House of Lords in another case on s. 33(1) of the 1976 Act decided on the same day, *Hallam* v *Avery* [2001] 1 WLR 655, HL.

In *Re Ismail* [1998] 3 All ER 1007, HL, which was concerned with an application for a writ of habeas corpus, the House of Lords noted that the purpose of the Extradition

Act 1989 (since repealed and replaced by the Extradition Act 2003) is to facilitate the bringing to justice of persons accused of serious crimes, and proceeded to adopt a 'broad and purposive approach' in holding that the appellant was an 'accused' person within s. 1(1) of the 1989 Act, and, therefore, subject to extradition proceedings.

In *Cutter* v *Eagle Star Insurance Co. Ltd* [1998] 4 All ER 417, HL, the House had to construe the word 'road' in the Road Traffic Act 1988 in order to determine whether a car park is a road. Some doubt was expressed about the true purpose of this statute, although their Lordships were prepared to accept that it was designed to protect the public from dangers arising from the use of motor vehicles. In any event, they eschewed a purposive approach in favour of giving the word 'road' its ordinary, literal meaning (see para 4.10.2.1 above).

Lord Clyde, with whom the rest of their Lordships agreed, said (at [1998] 4 All ER 417, 425):

It may be perfectly proper to adopt even a strained construction to enable the object and purpose of legislation to be fulfilled. But it cannot be taken to the length of applying unnatural meanings to familiar words . . . This must particularly be so where the language has no evident ambiguity or uncertainty about it . . . [I]t has to be remembered that in many instances the purpose of the legislation is achieved by the creation of an offence. Against the employment of a broad approach to express the purpose of the Act must be put the undesirability of adopting anything beyond a strict construction of provisions which have penal consequences.

The House of Lords adopted a purposive approach to the Human Fertilisation and Embryology Act 1990 in *R (Quintavalle)* v *Secretary of State for Health* [2003] 2 WLR 692, HL, when it decided that the creation of a human embryo by cell nuclear replacement is not prohibited by the Act, but is allowed—subject to the same licensing and regulatory process as one produced by fertilisation. 'Embryo' is defined in the Act by reference to fertilisation; 'cell nuclear replacement' is a procedure which does not involve fertilisation and which was unknown in 1990. Embryos created by the two methods are similar organisms, and cell nuclear replacement does not involve any procedure for which a licence is forbidden under the Act.

Their Lordships concluded that since the purpose of the Act was to protect live human embryos created outside the body, it must have been the intention of Parliament to permit, but regulate, any practice which was not significantly different from one already regulated by the Act.

According to Lord Steyn, the 'pendulum has swung towards purposive methods of construction' ([2003] 2 WLR 692, HL, at [21]). Lord Bingham said that the basic task of the court when interpreting a statute is to 'ascertain and give effect to the true meaning of what Parliament has said'. The performance of this task should not, however, be confined to a literal approach since 'undue concentration on the minutiae of the enactment may lead the court to neglect the purpose which Parliament intended to achieve'. He went on (ibid., at [8]):

The court's task, within the permissible bounds of interpretation, is to give effect to Parliament's purpose. So the controversial provisions should be read in the context of the statute as a whole, and the statute as a whole should be read in the historical context of the situation which led to its enactment.

These words of Lord Bingham were applied in *R* v *Z* (*Attorney-General for Northern Ireland's Reference*) [2005] 2 WLR 1286, HL, in which the House of Lords held that the term 'Irish Republican Army' used in anti-terrorist legislation includes the Real Irish Republican Army. This latter body is, therefore, a proscribed organisation and membership of it is unlawful.

The fact that it is not possible to adopt a purposive approach to the interpretation of a statutory provision where the language of the statute is clear, as exemplified in *Cutter* v *Eagle Star Insurance Co. Ltd* [1998] 4 All ER 417, HL, above, was re-emphasised by the House of Lords in *R* v *Bentham* [2005] 1 WLR 1057, HL. In this case, it was decided that the defendant was not guilty of the offence of being in possession of an imitation firearm during the course of a robbery contrary to s. 17(2) of the Firearms Act 1968. The defendant had given the impression to the victim that he had a gun by concealing his hand inside his jacket and pointing his fingers through the material of the jacket. 'Imitation firearm' is defined in s. 57(4) of the 1968 Act as 'any thing which has the appearance of being a firearm . . . whether or not it is capable of discharging any shot, bullet or other missile'. It was held that a person's hand or fingers are not separate and distinct from, but part of, him, and, accordingly, he does not 'possess' them. Furthermore, his hand and fingers are not a 'thing' within the definition in s. 57(4).

Lord Bingham ([2005] 1 WLR 1057 at [9]) said that the defendant's conduct in pretending to have a gun was highly reprehensible but not criminal since falsely pretending to have a firearm is not an offence. He continued (at [10]):

Rules of statutory construction have a valuable role when the meaning of a statutory provision is doubtful, but none where, as here, the meaning is plain. Purposive construction cannot be relied on to create an offence which Parliament has not created.

4.10.7 Interpretation of European Community legislation

The traditional English principles and technique of interpretation are not suitable for the construction of European Community legislation. The latter is drafted in Continental form, which is to state broad principles and leave the details to be filled in later by the judges, who are expected to promote the general legislative purpose.

Lord Denning MR compared the English style and the style of the EC Treaty in *H.P. Bulmer Ltd* v *J. Bollinger SA* [1974] Ch 401, CA, and in *Macarthys Ltd* v *Smith* [1979] 3 All ER 325, CA, at p. 329. In the *Bulmer* case, he said (at p. 425):

The draftsmen of our statutes have striven to express themselves with the utmost exactness. They have tried to foresee all possible circumstances that may arise and to provide for them. They have sacrificed style and simplicity. They have forgone brevity. They have become long and involved. In consequence, the judges have followed suit. They interpret a statute as applying only to the circumstances covered by the very words. They give them a literal interpretation. If the words of the statute do not cover a new situation—which was not foreseen—the judges hold that they have no power to fill the gap . . .

How different is this treaty! It lays down general principles. It expresses its aims and purposes . . . It uses words and phrases without defining what they mean. An English lawyer would look for an interpretation clause, but he would look in vain. There is none. All the way through the treaty there are gaps and lacunae. These have to be filled in by the judges, or by regulations or directives. It is the European way.

Legislation in European form cannot be interpreted literally. The mischief rule is more helpful, but Continental judges go further and adopt the *purposive* approach. The aim is to apply the 'spirit' rather than the 'letter' of the law. The European Court has made it clear in its judgments that Community legislation should be interpreted in this fashion (*Da Costa en Schaake NV* v *Nederlandse Balastingadministratie* [1963] CMLR 224, CJEC; *Van Duyn* v *Home Office* [1975] Ch 358, CJEC). In the *Bulmer* case [1974] Ch 401, CA, the Court of Appeal said, through Lord Denning MR (at pp. 425–6), that English judges must do likewise:

Beyond doubt the English courts must follow the same principles as the European court. Otherwise there would be differences between the countries of the [Member States]. That would never do. All the courts of all [Member States] . . . should interpret the [EC Treaty] in the same way.

. . .

Likewise the regulations and directives . . .

[The English courts] must follow the European pattern. No longer must they examine the words in meticulous detail. No longer must they argue about the precise grammatical sense. They must look to the purpose or intent . . . They must divine the spirit of the treaty and gain inspiration from it. If they find a gap, they must fill it as best they can. They must do what the framers of the instrument would have done if they had thought about it.

Lord Denning's approach received tacit approval from Lord Scarman in *Fothergill* v *Monarch Airlines Ltd* [1980] 2 All ER 696, HL, at p. 712. Later, in *Macarthys Ltd* v *Smith* [1979] 3 All ER 325, CA, Lord Denning (at p. 329) went so far as to state that in interpreting a statute of the United Kingdom Parliament:

[W]e are entitled to look to the [EC Treaty] as an aid to its construction; but not only as an aid but as an overriding force.

However, the other two judges in the appeal apparently disagreed with this statement. Lawton and Cumming-Bruce LJJ said (at pp. 332 and 335, respectively) that it was not permissible to look outside the Act at the terms of the treaty. When *Macarthys Ltd* v *Smith* came back to the Court of Appeal from the European Court, Cumming-Bruce LJ took the opportunity to explain that what he had said earlier was on the assumption that there was no ambiguity about the United Kingdom statute. If there is such ambiguity, then it *is* appropriate to look at the treaty in order to resolve it ([1981] 1 All ER 111 at p. 121).

In *Shanning International Ltd* v *Lloyds TSB Bank plc* [2001] 1 WLR 1462, HL, the House of Lords was concerned with the interpretation of art. 2 of Council Regulation (EEC) No. 3541/92. A trade embargo had been imposed on Iraq by United Nations Resolutions following the invasion of Kuwait by Iraq in 1990. Article 2 of the Council Regulation prohibited the satisfying of Iraqi claims on commercial transactions whose performance was affected by the trade embargo, but it was silent as to whether the prohibition was or was not intended to be permanent. Their Lordships adopted a purposive approach to the construction of art. 2, and held that the prohibition was intended to be permanent. From an examination of the circumstances leading to the adoption of the Regulation and of the preparatory documents, they concluded that the purpose of art. 2 was to provide permanent protection from the risk of future claims against them to non-Iraqi parties who had been unable to perform their contractual obligations due to the trade embargo.

When an English judge is interpreting Community legislation he must follow decisions and opinions of the Court of Justice of the European Communities. This is made clear by s. 3 of the European Communities Act 1972 (as amended by the European Communities (Amendment) Act 1986):

(1) For the purposes of all legal proceedings any question as to the meaning or effect of any of the Treaties, or as to the validity, meaning or effect of any Community instrument, shall be treated as a question of law (and, if not referred to the European Court, be for determination as such in accordance with the principles laid down by and any relevant decision of the European Court or any court attached thereto).

(2) Judicial notice shall be taken of the Treaties, of the Official Journal of the Communities and of any decision of, or expression of opinion by, the European Court or any court attached thereto, on any such question as aforesaid.

Section 3 of the 1972 Act itself must be read in conjunction with art. 234 of the EC Treaty (para 2.1.3.2.1). Under art. 234, a judge at first instance and the Court of Appeal (unless its decision would be final in the case) have a *discretion* to refer any question of the interpretation of European Community legislation to the European Court for a ruling. The discretion arises if a decision on the question is necessary to enable the domestic court to give judgment in the case (para 2.1.3.2.3). On the same conditions, the House of Lords *must* refer such a question to the European Court.

When the European Court gives a ruling on the reference, its decision is then binding in that particular case. But it is not binding in future cases because the European Court is not bound by its own decisions (see, e.g., *Da Costa en Schaake NV v Nederlandse Balastingadministratie* [1963] CMLR 224, CJEC). This freedom from the rigidity of any strict doctrine of precedent is necessary to enable the European Court to take into account future policy considerations (*H.P. Bulmer Ltd v J. Bollinger SA* [1974] Ch 401, CA, *per* Lord Denning MR at p. 420). It also means that if the House of Lords considers that an earlier decision of the European Court is wrong or unsatisfactory in some way, it can refer the point again to the European Court for reconsideration in a later case.

It is also made clear by s. 3(1) of the European Communities Act 1972 that any question arising in an English court concerning the meaning or effect of any of the Community Treaties is to be treated as a question of *law*. Accordingly, such a question arising in a criminal trial should be decided by the *judge alone* and may properly be determined by him in the absence of the jury (*R v Goldstein* [1983] 1 All ER 434, HL).

4.10.8 Interpretation of United Kingdom legislation after the Human Rights Act 1998

4.10.8.1 Introduction: background to, and main provisions of, the Human Rights Act 1998

The European Convention on Human Rights (*Convention for the Protection of Human Rights and Fundamental Freedoms, agreed by the Council of Europe at Rome on 4th November 1950)* guarantees certain fundamental rights and freedoms, including the right to life, the right to liberty and security, the right to a fair trial, the right to respect for private and family life, freedom of thought, conscience and religion, freedom of expression, and freedom of assembly and association.

The United Kingdom has since 1951 been a signatory to the Convention and subject to the jurisdiction of the European Court of Human Rights at Strasbourg. The Convention, however, while binding on the United Kingdom under *international* law, was not part of English *domestic* law since it was an international treaty which had not been incorporated into our law. It was, therefore, not *directly* enforceable in our courts.

When the Human Rights Act 1998 came fully into force in England and Wales on 2 October 2000, some of the Convention rights and freedoms became part of English domestic law and, as such, directly enforceable in our courts. However, these rights are domestic rather than international rights, and their source is the 1998 Act rather than the Convention (see, e.g., *In Re McKerr* [2004] 1 WLR 807, HL, *per* Lords Nicholls, Hoffmann, and Rodger at [25], [63], and [74]–[75], respectively). The Convention itself remains unenforceable directly in our courts. According to Lord Hoffmann in *In Re McKerr*, above at [63], it is a 'misleading metaphor' to say that the Convention has been incorporated into domestic law.

Under the Human Rights Act 1998 (which was based on the Labour Government's White Paper, *Rights Brought Home: The Human Rights Bill* (Cm 3782, 1997)):

(a) some of the rights and freedoms (referred to in the Act as 'Convention rights') have been incorporated into English law (s. 1 and sch. 1);

(b) primary and delegated legislation, whenever made, will as far as possible be interpreted and applied in a way which is compatible with the Convention rights, although the validity, continuing operation, or enforcement of any incompatible legislation will, in general, be unaffected (s. 3);

(c) if a primary or delegated legislative provision is found by the House of Lords, the Judicial Committee of the Privy Council, the Courts-Martial Appeal Court, the Court of Appeal, or the High Court (but *not* the Employment Appeal Tribunal (see *Whittaker v Watson* (2002) *The Times*, 26 March, EAT), the Crown Court, a county court, or a magistrates' court) to be incompatible with one or more of the Convention rights, that court is able to make a declaration of incompatibility (s. 4(1)–(5));

(d) a declaration of incompatibility does not affect the validity, continuing operation, or enforcement of the incompatible provision, and is not binding on the parties to the proceedings in which it is made (s. 4(6));

(e) where a court is considering whether to make a declaration of incompatibility, a Minister of the Crown, or a person nominated by him, is entitled to be joined as a party to the proceedings (s. 5);

(f) if a Minister of the Crown considers that there are compelling reasons for doing so, he is able to remove the incompatibility by making a remedial order which amends (or repeals) the offending legislation and any other affected legislation (s. 10);

(g) a minister's power to make a remedial order is also exercisable where the incompatibility has been exposed in a finding of the European Court of Human Rights made after 2 October 2000 in proceedings taken against the United Kingdom under the Convention (s. 10);

(h) although not expressly stated, it is clearly the expectation that Parliament will not pass legislation which interferes with the Convention rights (but it can still do so if it wishes);

(i) a Minister of the Crown in charge of a Bill in either House of Parliament is required to make in writing, before Second Reading of the Bill, either a state-ment of compatibility to the effect that in his view the provisions of the Bill are compatible with the Convention rights, or a statement that, although he is unable to make a statement of compatibility, the government nevertheless wishes the House to proceed with the Bill (s. 19);

(j) it is, in general, unlawful for a public authority (which includes a court or tribunal and any person with some functions of a public nature, but does not include Parliament) to act in a way which is incompatible with the Convention rights (s. 6);

(k) a court or tribunal deciding a question in connection with a Convention right must 'take into account' any relevant judgments, decisions, declarations, and opinions of the European Court of Human Rights, the European Commission, and the Committee of Ministers of the Council of Europe, whenever made or given (s. 2).

The magnitude of the changes brought about by the implementation of the Human Rights Act 1998 was referred to by Lord Hope in a case decided before the Act came into force, *R* v *Director of Public Prosecutions (ex parte Kebilene)* [2000] 2 AC 326, HL, at p. 338:

It is now plain that the incorporation [sic] of the European Convention on Human Rights into our domestic law will subject the entire legal system to a fundamental process of review and, where necessary, reform by the judiciary.

And in *R* v *Lambert* [2001] 3 WLR 206, HL, Lord Slynn said (at p. 210):

It is clear that the 1998 Act must be given its full import and that long or well entrenched ideas may have to be put aside, sacred cows culled.

The requirement that legislation is to be interpreted as far as possible so as to achieve compatibility with the Convention goes further than the principle of inter-pretation applicable before 2 October 2000, namely that the courts could, *as a matter of discretion*, take the Convention into account in resolving any legislative ambiguity.

The Human Rights Act 1998 does not give the courts the power to set aside Acts of Parliament which are inconsistent with the Convention rights, but merely allows certain courts to make declarations of incompatibility. As Lord Steyn pointed out in *R* v *Director of Public Prosecutions (ex parte Kebilene)* [2000] 2 AC 326, HL, at p. 333:

It is crystal clear that the carefully and subtly drafted Human Rights Act 1998 preserves the principle of Parliamentary sovereignty. In a case of incompatibility which cannot be avoided by interpretation under section 3(1), the courts may not disapply the legislation. The court may merely issue a declaration of incompatibility which then gives rise to a power to take remedial action.

This position resulted from the government's conclusion in the White Paper, above, that the doctrine of parliamentary sovereignty was too important for English courts to be allowed to nullify statutes. In Canada, the courts can strike down any legislation which is inconsistent with the Charter of Rights and Freedoms 1982, unless the legislation expressly states that it is to apply 'notwithstanding' the provisions of the Charter. In New Zealand, the Bill of Rights Act 1990, while requiring the courts as far as possible to interpret legislation consistently with the rights contained in the Act, provides that the legislation stands if that is impossible. The Human Rights Act 1998 prefers the New Zealand model to the Canadian model. The European Communities Act 1972 provides for that part of Community law which has 'direct effect' to take priority over inconsistent English domestic law. In the White Paper, however, the government dismissed this precedent on the 'essential difference' that giving such priority is a requirement of membership of the European Union whereas there is no such requirement in the European Convention.

In addition, it is to be noted that the Human Rights Act 1998 does not entrench its own provisions to protect them from subsequent amendment or repeal. The model provided by the Constitution of the United States of guaranteeing certain rights, which can be amended or repealed only by qualified majorities in the legislature and in the states themselves, was rejected by the United Kingdom government as irreconcilable with constitutional convention, which allows any statute to be amended or repealed by a subsequent statute.

As anticipated, the Human Rights Act 1998 has proved to be a fertile source of litigation, and will, no doubt, continue to be so. Within a year of its implementation, the Court of Appeal referred somewhat despairingly to the daily increase in the volume of reported human rights case law and criticised its over-citation (*R (Al-Hasan)* v *Secretary of State for the Home Department* [2002] 1 WLR 545, CA, at [73]). In *International Transport Roth GmbH* v *Secretary of State for the Home Department* [2002] 3 WLR 344, CA, below, some 227 authorities were put before the court. In both civil and criminal cases, the 1998 Act has spawned all manner of claims (many of them spurious) that Convention rights have been violated.

Guidance on the correct application of the 1998 Act is as yet somewhat limited. This is to be expected, given that the Act is still in its infancy and that the higher courts must of necessity feel their way carefully in what is an unfamiliar area.

One matter, however, is now established beyond doubt. On several occasions, the

House of Lords has held that (with the exception of proceedings 'brought by or at the instigation of a public authority' under s. 22(4)) the 1998 Act is not retrospective (see, e.g., *R v Lambert* [2001] 3 WLR 206, HL, para 7.5.7; *R v Kansal (No. 2)* [2001] 3 WLR 1562, HL, para 7.5.7; *Wilson v First County Trust Ltd (No. 2)* [2003] 3 WLR 568, HL; and *In Re McKerr* [2004] 1 WLR 807, HL).

4.10.8.2 Interpretation and incompatibility: Human Rights Act 1998, ss. 3 and 4

The interpretation and incompatibility provisions of the 1998 Act are of the utmost importance. The following guidance on their application was offered, *obiter*, by Lord Woolf CJ when delivering the judgment of the Court of Appeal in *Poplar Housing and Regeneration Community Association Ltd v Donoghue* [2001] 3 WLR 183, CA, at p. 204:

(a) if the court uses s. 3 to interpret a legislative provision in such a way as to make it compatible with the Convention, it should bear in mind that its task is still *interpretation* and not *legislation*;

(b) the extent of any modified meaning given under s. 3 should be limited to what is necessary to achieve compatibility;

(c) if the court can only make a legislative provision compatible under s. 3 by radically altering its effect, this is an indication that it is engaging in more than interpretation;

(d) if it is not possible under s. 3 to achieve an interpretation which is compatible with the Convention, the court has a discretion, but not a duty, to make a declaration of incompatibility under s. 4.

Lord Woolf CJ's comments received support from the House of Lords in *R v A (No. 2)* [2001] 2 WLR 1546, HL, in which Lord Steyn in particular dealt with s. 3 at some length. He pointed out (at p. 1562) that the interpretative obligation imposed by s. 3 is a strong one and applies (unlike traditional methods of interpretation) even if there is no ambiguity in the statutory language. He continued (at p. 1563):

Section 3 places a duty on the court to strive to find a possible interpretation compatible with Convention rights. Under ordinary methods of interpretation a court may depart from the language of the statute to avoid absurd consequences: s. 3 goes much further. Undoubtedly, a court must always look for a contextual and purposive interpretation: s. 3 is more radical in its effect. It is a general principle of the interpretation of legal instruments that the text is the primary source of interpretation: other sources are subordinate to it. . . . Section 3 qualifies this general principle because it requires a court to find an interpretation compatible with Convention rights if it is possible to do so. . . . In accordance with the will of Parliament as reflected in s. 3 it will sometimes be necessary to adopt an interpretation which linguistically may appear strained. The techniques to be used will not only involve the reading down of express language in a statute but also the implication of provisions. A declaration of incompatibility is a measure of last resort. It must be avoided unless it is plainly impossible to do so. If a *clear* limitation on Convention rights is stated *in terms*, such an impossibility will arise. . . .

(See also *per* Lords Hope and Hutton at pp. 1582 and 1601, respectively.)

In the later case of *R v Lambert* [2001] 3 WLR 206, HL, Lord Hope (at p. 234) said that the obligation imposed by s. 3

is not to be performed without regard to its limitations. Resort to it will not be possible if the legislation contains provisions ... which expressly contradict the meaning which the enactment would have to be given to make it compatible. The same consequence will follow if legislation contains provisions which have this effect by necessary implication ... As the sidenote [to s. 3] indicates, the obligation is one which applies to the interpretation of legislation. This function belongs, as it has always done, to the judges. But it is not for them to legislate. Section 3(1) preserves the sovereignty of Parliament. It does not give power to the judges to overrule decisions which the language of the statute shows have been taken on the very point at issue by the legislator.

Lord Hope gave a warning that the technique of substituting words under s. 3 should not be used if it would only make the incompatible legislation unintelligible or unworkable. In that event, the court should instead consider making a declaration of incompatibility under s. 4, thus leaving the legislation to be amended in other ways.

Section 3 was applied by the House of Lords in *R v Lambert* [2001] 3 WLR 206, HL, so as to construe a section in the Misuse of Drugs Act 1971 as if it imposed an evidential, rather than a persuasive, burden of proof on the accused. In this way, the section was made compatible with the presumption of innocence in art. 6 of the Convention. (See, however, the exceptional case of *Wilson v First County Trust Ltd (No. 2)* [2003] 3 WLR 568, HL, in which the House of Lords held that neither s. 3 nor s. 4 was available.)

Declarations of incompatibility under s. 4 were made by the Court of Appeal in the two important cases of *R (H) v London North and East Region Mental Health Review Tribunal* [2001] 3 WLR 512, CA, and *International Transport Roth GmbH v Secretary of State for the Home Department* [2002] 3 WLR 344, CA.

In the former case, the court found itself unable to use s. 3 to interpret the relevant statutory provisions in a way which was compatible with the Convention. Instead, it was declared that ss. 72(1) and 73(1) of the Mental Health Act 1983 were incompatible with art. 5(1) and (4) of the Convention (right to liberty and security) in that they imposed a 'reversed burden of proof'. That is, before a mental health review tribunal was obliged to order a patient's discharge, the patient had to disprove the lawfulness of his own detention by showing that at least one of the criteria justifying his detention in hospital for treatment was no longer satisfied. The incompatibility was later removed when new ss. 72(1) and 73(1) were substituted by the Mental Health Act 1983 (Remedial) Order 2001 (SI 2001, No. 3712).

In the second case, *International Transport Roth GmbH v Secretary of State for the Home Department*, above, the Court of Appeal concluded that it was impossible to make use of s. 3 without trespassing beyond the boundary of interpretation into the realm of legislation. Instead, the court upheld a declaration made at first instance that the provisions of part II of the Immigration and Asylum Act 1999, which created a scheme whereby vehicles could be detained and a fixed penalty of £2,000 per entrant could be imposed by the Home Secretary on lorry drivers and haulage companies for bringing clandestine entrants into the United Kingdom, were incompatible with the Convention in that they violated art. 6 (fair trial) and art. 1 of the First Protocol (protection of property). At the same time, it was held that the provisions were *not* incompatible with European Community law since they did not contravene art. 28 (quantitative restrictions on imports) or art. 49 (restrictions on freedom to provide services) of the EC Treaty. The incompatibility with the Convention was eliminated by the Nationality, Immigration and Asylum Act 2002, which amended part II of the 1999 Act.

When in a later case the Court of Appeal used s. 3 so as to read into the Children Act 1989 a power in the court to require from a local authority a progress report on a child who is the subject of a care order, it was criticised by the House of Lords for overstepping the mark (*In Re S (minors) (care order: implementation of care plan)* [2002] 2 WLR 720, HL). The House of Lords held that this was not a legitimate use of s. 3. Citing Lord Woolf CJ in the *Poplar Housing* case [2001] 3 WLR 183, CA, above, and Lord Hope in *Lambert* [2001] 3 WLR 206, above, Lord Nicholls in *In Re S* (at [37]–[38]) described s. 3 as 'a powerful tool whose use is obligatory' but whose reach is 'not unlimited'. He said (at [38]–[43]):

Section 4 (power to make a declaration of incompatibility) and, indeed, s. 3(2)(b), presuppose that not all provisions in primary legislation can be rendered convention compliant by the application of section 3(1). . . . In applying s. 3 courts must be ever mindful of this outer limit. The Human Rights Act reserves the amendment of primary legislation to Parliament. By this means the Act seeks to preserve parliamentary sovereignty. . . . Interpretation of statutes is a matter for the courts; the enactment of statutes, and the amendment of statutes, are matters for Parliament . . . The area of real difficulty lies in identifying the limits of interpretation in a particular case . . . For present purposes it is sufficient to say that a meaning which departs substantially from a fundamental feature of an Act of Parliament is likely to have crossed the boundary between interpretation and amendment. This is especially so where the departure has important practical repercussions which the court is not equipped to evaluate. . . . I consider this judicial innovation passes well beyond the boundary of interpretation. I can see no provision in the Children Act which lends itself to the interpretation that Parliament was thereby conferring this supervisory function on the court. No such provision was identified by the Court of Appeal.

In *R (Anderson)* v *Secretary of State for the Home Department* [2002] 3 WLR 1800, HL, the House of Lords held that it was not possible to use s. 3 of the Human Rights Act 1998 so as to make s. 29 of the Crime (Sentences) Act 1997 compatible with the European Convention on Human Rights. However, a declaration was made under s. 4 of the Human Rights Act 1998 that s. 29 of the 1997 Act is incompatible with art. 6 of the Convention (right to have sentence imposed by independent and impartial tribunal) because the Home Secretary is not acting independently and impartially when he decides on the minimum period to be served by a mandatory life sentence prisoner before he is considered for release on life licence. The incompatibility was removed when s. 29 of the 1997 Act was repealed by the Criminal Justice Act 2003 and replaced by new provisions about release on licence contained therein.

In *Bellinger* v *Bellinger* [2003] 2 WLR 1174, HL, their Lordships granted a declaration under s. 4 of the Human Rights Act 1998 that s. 11(c) of the Matrimonial Causes Act 1973 (which requires the parties to a marriage to be respectively male and female) is incompatible with the Convention, arts. 8 (right to respect for private and family life) and 12 (right to marry). Section 3 of the Human Rights Act 1998 could not be used since it would not be permissible to read into s. 11(c) of the 1973 Act (which is unambiguous) additional words so as to include a transsexual person. So to do would amount to judicial legislation, not mere interpretation (see [2003] 2 WLR 1174, HL, *per* Lords Hope and Hobhouse at [67] and [78], respectively). Mrs Bellinger (a male-to-female transsexual person) thus failed in her attempt to secure legal recognition of her marriage to a man.

The decision on incompatibility in *Bellinger* was almost inevitable given the earlier conclusion of the European Court of Human Rights in *Goodwin* v *United Kingdom* (2002) 35 EHRR 18, ECHR (not a marriage case) that English law discriminated against transsexual people in other ways and breached their Convention rights under arts. 8 and 12.

In order to deal with the incompatibility identified in *Bellinger*, it was neither necessary nor appropriate to amend s. 11(c) of the 1973 Act. Instead, the Gender Recognition Act 2004 made it possible for transsexual persons to enter into valid marriages in their acquired gender.

In *Ghaidan* v *Godin-Mendoza* [2004] 3 WLR 113, HL, the House of Lords had to reconsider, in the light of the Human Rights Act 1998, its decision in *Fitzpatrick* v *Sterling Housing Association Ltd* [1999] 4 All ER 705, HL. The House had held in *Fitzpatrick* that a same-sex partner cannot succeed to a statutory tenancy on the death of the original tenant, under para 2 of sch. 1 to the Rent Act 1977, because the words, 'living with the original tenant as his or her wife or husband', are gender-specific and apply to a heterosexual relationship only.

In *Ghaidan*, above, it was decided that this provision infringes art. 14 of the Convention (enjoyment of rights and freedoms without discrimination). The House of Lords, applying s. 3 of the Human Rights Act 1998, held that the words in the 1977 Act, 'as his or her wife or husband', should be read to mean, 'as *if they were* his or her wife or husband', and that the defendant had accordingly succeeded to the statutory tenancy on the original tenant's death.

The decision in *Ghaidan* was later given statutory effect in the Civil Partnership Act 2004, which was passed to enable same-sex couples to have their relationship legally recognised as a 'civil partnership' and to provide for the legal consequences thereof. Paragraph 2 of sch. 1 to the Rent Act 1977 was amended by the 2004 Act so that a person who was living with the original tenant as if they were civil partners is treated as the civil partner of the tenant and entitled to succeed to the tenancy on the tenant's death.

The government's anti-terrorism policy was thrown into temporary disarray following the decision in *A* v *Secretary of State for the Home Department* [2005] 2 WLR 87, HL, in which the House of Lords, sitting nine-strong, decided by a majority of eight to one to declare that s. 23 of the Anti-terrorism, Crime and Security Act 2001 was incompatible with the Convention, art. 5 (right to liberty and security) and art. 14 (enjoyment of rights and freedoms without discrimination).

Section 23 of the 2001 Act was part of the United Kingdom's response to the terrorist atrocities perpetrated in the United States on 11 September 2001. It permitted the detention without trial of suspected *international* terrorists. It was held to be incompatible with the Convention because it was disproportionate and discriminatory since the power of detention without trial conferred by it was limited to foreigners and did not apply to suspected *British* terrorists. The incompatibility was removed by the Prevention of Terrorism Act 2005, which repealed s. 23 of the 2001 Act and replaced it with new powers to make control orders against suspected terrorists with the aim of protecting members of the public from risks of terrorism.

Judgments in *A v Secretary of State for the Home Department* [2005] 2 WLR 87, HL, were delivered in December 2004. Lord Hoffmann's words (ibid. at [96]–[97]) that 'terrorist violence . . . does not threaten our institutions of government or our existence as a civil community' and that the 'real threat to the life of the nation . . . comes not from terrorism but from laws such as these' came back to haunt him six months later when they attracted strong criticism from politicians and others following the London terrorist bombings of July 2005.

5

Judicial precedent

5.1 The meaning of judicial precedent

At the beginning of Chapter 4 the superiority of parliamentary legislation over judge-made law was noted (para 4.2). Despite its subservient position, however, there is still considerable scope for judicial law-making in the English legal system. When a judge applies or extends an established rule to new facts, or decides that the rule does not apply in a certain situation, he is making or changing the law.

The judge does not, of course, have the same freedom to make law as the legislature, and must not be seen to be usurping its powers. All judge-made law is inferior to, and can be overruled by, parliamentary or delegated legislation. But, unless and until they are overruled, judicial decisions on the interpretation of legislation are precedents just as much as decisions on non-statutory points of law, and, as such, are subject to the same rule of *stare decisis*. Most of English law derives from statute and the common law. The function of the judges is to interpret the one and evolve the other.

The term 'judicial precedent' has at least two meanings. First, it may mean the process whereby judges follow previously decided cases. Secondly, it may refer to the decided case itself—a 'precedent' which may be relied on in the future. The term is mainly used in the first sense and is to be so understood for most of this chapter.

The doctrine of judicial precedent is one which in English law involves an application of the principle of *stare decisis*; i.e., to stand by cases already decided. In practice, this element of compulsion means that the Court of Appeal is generally bound to follow its own previous decisions, and that each court is bound to follow the decisions of a court above it in the hierarchy. (For the position of the House of Lords, see para 5.3.2.2 below.)

It is important to appreciate that the limitations placed on the courts by the principle of *stare decisis* are, for the most part, self-imposed. With two notable exceptions, there are no *legislative* provisions on the application of judicial precedent.

The first exception is the requirement that our courts must follow decisions on Community Law given by the European Court (European Communities Act 1972, s. 3(1), as amended; see para 4.10.7).

The second exception is that, by virtue of s. 2(1) of the Human Rights Act 1998, when our courts and tribunals decide questions in connection with rights guaranteed under the European Convention on Human Rights, and given effect to in the United Kingdom by the 1998 Act, they must take into account any relevant judgments, decisions, declarations, and opinions of the European Court of Human Rights, the European Commission, and the Committee of Ministers of the Council of Europe, whenever made or given (see further, para 4.10.8). However, the fact that our courts must, under s. 2(1) of the 1998 Act, 'take into account' judgments of the European Court of Human Rights does not mean that they are bound by those decisions. Thus, on appeal in *Leeds City Council* v *Price* [2006] UKHL 10, HL, the House of Lords approved the approach of the Court of Appeal ([2005] 1 WLR 1825) that where a decision of the House of Lords conflicts with a subsequent decision of the European Court of Human Rights, courts below the House of Lords are bound by the principle of *stare decisis* to follow the decision of the House. In appropriate cases, permission to appeal can be given so that the conflict may ultimately be resolved by the House.

5.2 *Ratio decidendi* and *obiter dictum*

The decision or judgment of a judge may fall into two parts: the *ratio decidendi* and *obiter dictum*. When a judge delivers judgment in a case he outlines the facts which he finds have been proved on the evidence. Then he applies the law to those facts and arrives at a decision, for which he gives the reason (*ratio decidendi*). More precisely, the *ratio decidendi* of a case is the principle of law on which the decision is based. The judge may go on to speculate about what his decision would, or might, have been if the facts of the case had been different. This is an *obiter dictum* ('something said by the way').

The binding part (if any) of a judicial decision is the *ratio decidendi*. An *obiter dictum* is not binding in later cases because it was not strictly relevant to the matter in issue in the original case. However, an *obiter dictum* may be of *persuasive* (as opposed to *binding*) authority in later cases. This matter is examined later (para 5.4.2), as is the difficulty involved in discovering the *ratio decidendi* of a case (para 5.5).

Sometimes opinions delivered in the House of Lords contain statements which look at first sight as though they are *obiter* because of the way in which the case has been decided. But on a closer examination these statements are seen not as mere *obiter dicta* but as forming part of a *ratio decidendi*. Important examples of such cases are *Rondel* v *Worsley* [1969] 1 AC 191, HL; *Hedley Byrne & Co. Ltd* v *Heller & Partners Ltd* [1964] AC 465, HL; and *National Carriers Ltd* v *Panalpina (Northern) Ltd* [1981] 1 All ER 161, HL. Such instances are rare and occur usually where the House of Lords does not wish to lose a present opportunity for clarifying the law or providing guidance for lower courts in the future. Moreover, the House will only act in this way after full legal argument and mature consideration of the points at issue. This matter also is dealt with in more detail later (para 5.4.2).

Since the only binding part (if any) of a decided case is the principle of law contained within its *ratio decidendi*, it follows that cases are not binding on questions of *fact*. Nor are they binding on questions of law which were merely assumed to be correct without argument by counsel and without full consideration by the court. This is so regardless of the status of the court which decided the case. Thus, although normally bound by decisions of the House of Lords, a High Court judge need not follow one of their Lordships' decisions which is either essentially one of fact or is based on an unargued assumption (see, e.g., *Re Hetherington (deceased)* [1989] 2 All

ER 129, para 5.3.2.6 below). And, in *R* v *Secretary of State for the Home Department (ex parte Ku)* [1995] 2 All ER 891, CA, the Court of Appeal, which (as will be seen in para 5.3.2.3.1 below) is normally bound by its own previous decisions, declined to follow *R* v *Immigration Officer (ex parte Chan)* [1992] 2 All ER 738, CA, on the ground that it was decided on both a question of fact and on an assumption.

Decisions of the Court of Appeal given on applications for permission to appeal are not binding, and, although they are sometimes reported in specialist law reports, reference to them is not encouraged (*Clark* v *University of Lincolnshire and Humberside* [2000] 3 All ER 752, CA, *per* Lord Woolf MR at [40]–[43]).

5.3 Judicial precedent in practice

5.3.1 Reversing, overruling, and distinguishing

Reversing occurs where a court higher up in the hierarchy overturns the decision of a lower court on appeal in the same case. *Overruling* is where a principle laid down by a lower court is overturned by a higher court in a different, later case. Thus, in 1963 the House of Lords held unanimously in *Hedley Byrne & Co. Ltd* v *Heller & Partners Ltd* [1964] AC 465, HL, that there could be liability in English law for negligent misstatements. In so holding, the House of Lords overruled *Candler* v *Crane, Christmas & Co.* [1951] 2 KB 164, CA, which had represented the Court of Appeal's view of the law since 1951. At the same time, the dissenting judgment of Denning LJ in *Candler* was vindicated in *Hedley Byrne*. In domestic matters, while the House of Lords can now *overrule* itself it is not possible, strictly speaking, to *reverse* itself.

English courts have always followed the practice of retrospective overruling. This means that the overruled case is regarded as never having been law, and will not be applied either in later cases or in the instant case. Retrospective overruling accords with the declaratory theory of the common law that the judges do not make or change the law but merely declare it. As Brett MR once put it (*Munster* v *Lamb* (1883) 11 QBD 588, CA, at pp. 599–600):

The judges cannot make new law by new decisions; they do not assume a power of that kind: they only endeavour to declare what the common law is and has been from the time when it first existed. But inasmuch as new circumstances, and new complications of fact, and even new facts, are constantly arising, the judges are obliged to apply to them what they consider to have been the common law during the whole course of its existence, and therefore they seem to be laying down a new law, whereas they are merely applying old principles to a new state of facts.

In more modern times, the declaratory theory has been described as a 'fiction' by Lord Simon (*Jones* v *Secretary of State for Social Services* [1972] AC 944, HL, at p. 1026), as a 'fairy tale' by Lord Reid ((1972–73) JSPTL 22), and as 'at odds with reality' by Lord Nicholls (*In re Spectrum Plus Ltd* [2005] 3 WLR 58, HL, at [34]). (See also *R* v *Governor of Brockhill Prison (ex parte Evans) (No. 2)* [1998] 4 All ER 993, CA, *per* Lord Woolf MR at p. 1002; *Kleinwort Benson Ltd* v *Lincoln City Council* [1998] 4 All ER 513, HL, *per* Lords Browne-Wilkinson, Goff, Lloyd, and Hope at pp. 518, 535, 548, and 563, respectively.)

Lord Browne-Wilkinson has explained the modern judicial function thus (*Kleinwort Benson Ltd* v *Lincoln City Council*, above, at p. 518):

The theoretical position has been that judges do not make or change law: they discover and declare the law which is throughout the same. According to this theory, when an earlier decision is overruled the law is not changed: its true nature is disclosed, having existed in that form all along. This theoretical position is, as Lord Reid said, a fairy tale in which no one any longer believes. In truth, judges make and change the law. The whole of the common law is judge-made and only by judicial change in the law is the common law kept relevant in a changing world.

Now that the declaratory theory is generally regarded as old fashioned, and it is generally accepted that judges do make new law, the injustice arising from the practice of retrospective overruling is compounded. This injustice is that the parties in the instant case will have relied on what they understood the law to be, only to be told now that it is not, and never has been, the law.

Thus, in *R* v *Governor of Brockhill Prison (ex parte Evans)* [1997] 1 All ER 439, DC, where a prison governor had calculated a prisoner's release date in accordance with an approach sanctioned by three decisions of the Queen's Bench Divisional Court, and the court later held that its three previous decisions were wrong and should not be followed, it transpired that the prisoner, who should have been released earlier under the new approach sanctioned by the court, had been detained unlawfully for 59 days.

In *R* v *Governor of Brockhill Prison (ex parte Evans) (No. 2)* [1998] 4 All ER 993, CA, the prisoner claimed damages for false imprisonment. It was argued on behalf of the prison governor that he had had lawful justification for the detention because, at the time he relied on them, the earlier decisions had not been held to be wrong. The Court of Appeal decided that when an earlier decision is effectively overruled by a later decision, the earlier decision cannot be relied on as a correct statement of the law because the later decision operates retrospectively. It followed that the governor had had no lawful justification for the detention since, from his point of view, it was as if the earlier decisions had never existed. This decision was affirmed by the House of Lords at [2000] 4 All ER 15.

In *Kleinwort Benson Ltd* v *Lincoln City Council* [1998] 4 All ER 513, HL, the House of Lords abolished the common law rule that money paid under a mistake of law is not recoverable. It was appreciated that the retrospective effect of the decision would allow claims to be made for the recovery of money paid under a mistake of law in earlier transactions (see *per* Lords Goff, Hoffmann, and Hope at pp. 535, 554, and 563, respectively). Lord Lloyd, dissenting, was alarmed by this consequence (at p. 552), and Lord Browne-Wilkinson was similarly critical in his dissenting opinion (see especially at pp. 519 and 523). (See further on this case, para 5.3.2.2.3 below.)

In the interests of justice, the practice of *prospective* overruling has been developed in the courts of some countries, notably in the Supreme Court of India in *constitutional* cases and in the US Supreme Court (see, e.g., *Linkletter* v *Walker* (1965) 381 US 618, USA), although the latter Court no longer applies prospective overruling in *criminal* cases, and the future of the practice in *civil* cases is in some doubt (see *In re Spectrum Plus Ltd* [2005] 3 WLR 58, HL, *per* Lord Hoffmann at [18]–[20]).

Prospective overruling means that the overruled case is applied in the instant case and is inapplicable in future cases only. An injustice arising from prospective overruling is that people are treated differently. Under this system of overruling there is also a disincentive to litigate. Litigants will be reluctant to be guinea pigs—that is, to take the risk of finding out whether an overruled decision will or will not apply in their case.

In *Jones* v *Secretary of State for Social Services* [1972] AC 944, HL, Lord Simon suggested (at p. 1026) that the English courts should be given, by statute, the power of prospective overruling. (Lord Simon made the same suggestion in an afterword to his opinion in *Miliangos* v *George Frank (Textiles) Ltd* [1976] AC 443, HL, at p. 490.) Lord Diplock agreed in principle ([1972] AC 944 at p. 1014). Lord Devlin, however, expressed opposition to such a power on the ground that it 'turns judges into undisguised legislators' ('Judges and lawmakers', (1976) 39 MLR 1 at p. 11).

The practice of prospective overruling appeared to commend itself to Lord Woolf MR in *R* v *Governor of Brockhill Prison (ex parte Evans) (No. 2)* [1998] 4 All ER 993, CA, at p. 1003, but in *Kleinwort Benson Ltd* v *Lincoln City Council* [1998] 4 All ER 513, HL, Lord Goff (at p. 536) said that its adoption elsewhere had had 'somewhat controversial results' and that it 'has no place in our legal system'. The question of prospective overruling was raised, but left open, when *R* v *Governor of Brockhill Prison (ex parte Evans) (No. 2)* went to the House of Lords ([2000] 4 All ER 15). Some support for prospective overruling was expressed by Lord Slynn (at p. 19), but all their Lordships were agreed that the question was best left for another day and a more suitable case.

A more suitable occasion presented itself in *In re Spectrum Plus Ltd* [2005] 3 WLR 58, HL, in which the House of Lords made some significant comments about prospective overruling in a case where the topic was considered at length in an appeal heard by seven Law Lords. Their Lordships were of the view that, despite the objections in principle and the practical difficulties involved, prospective overruling in an 'exceptional' case may be a proper exercise of judicial power in the House of Lords. There was disagreement over whether the power could ever be exercised in a case involving a point of *statutory interpretation*. Lords Steyn and Scott (at [45] and [125], respectively) doubted that it could, but Lords Nicholls and Hope (at [39] and [74], respectively), along with Baroness Hale (at [162]), thought that it could.

In giving support in principle to prospective overruling, their Lordships were influenced by developments in the Court of Justice of the European Communities and in the European Court of Human Rights, both of which in exceptional circumstances have given their decisions prospective effect only. Furthermore, the Judicial Committee of the Privy Council has a limited *statutory* power of prospective decision-making. If, under the devolution legislation affecting Scotland, Wales, and Northern Ireland, the Judicial Committee decides that a legislative provision is not within the competence of the relevant legislative body, the Committee can make an order removing or limiting any retrospective effect of its decision or suspending the effect of the decision to enable the defect to be put right.

In *In re Spectrum Plus Ltd*, above, the House of Lords did not in fact adopt prospective overruling. Although it overruled a long-standing decision, it did so in the usual retrospective manner since it took the view that the case before it was not sufficiently 'exceptional' to justify a prospective overruling. What is meant by an 'exceptional' case remains unclear.

Distinguishing a case on its facts, or on the point of law involved, is a device resorted to by judges usually in order to avoid the consequences of an inconvenient decision which is, in strict practice, binding on them. Distinguishing is discussed again in para 5.6 below.

5.3.2 Judicial precedent in individual courts

5.3.2.1 Court of Justice of the European Communities

As was seen in para 4.10.7, decisions of the European Court are binding on all English courts, including the House of Lords, on the interpretation of the Community treaties and on the validity and interpretation of secondary Community legislation. The European Court is not bound by its own decisions.

5.3.2.2 House of Lords

5.3.2.2.1 Introduction: the 1966 Practice Statement Much of what is said here about precedent in the House of Lords will apply to the Supreme Court of the United Kingdom when it replaces the House as the highest court in the land. It is assumed that the new Court will follow the same approach to precedent as the House presently does. Furthermore, it is assumed that it will normally treat itself as bound by previous decisions of the House.

Decisions of the House of Lords are binding on all courts in the country except the House itself. Until 1966, the House of Lords was bound by its own decisions. This practice was established in the mid nineteenth century and reaffirmed in 1898 in *London Tramways Co. Ltd* v *London County Council* [1898] AC 375, HL. The reason was that it was felt that decisions of the highest appeal court should be final in the public interest so that there would be certainty in the law and an end to litigation. (See especially the speech of the Earl of Halsbury LC in the *London Tramways* case.)

There was increasing judicial criticism of the practice from the 1930s. In particular, it was said that the rule did not produce the desired certainty in the law and it had become too rigid (see, e.g., Lord Wright, 'Precedent', [1944] CLJ 118; Lord Denning, 'From precedent to precedent' (The Romanes Lecture, 1959, quoted in *The Discipline of Law*, 1979, pp. 291 et seq.); *Midland Silicones Ltd* v *Scruttons Ltd* [1962] AC 446, HL, *per* Lord Reid at p. 475). But the practice was not changed until 1966, when, in July of that year, Lord Gardiner LC made a statement on behalf of himself and the Law Lords. The *Practice Statement* [1966] 3 All ER 77 is here set out in full.

Their Lordships regard the use of precedent as an indispensable foundation upon which to decide what is the law and its application to individual cases. It provides at least some degree of certainty upon which individuals can rely in the conduct of their affairs, as well as a basis for orderly development of legal rules.

Their Lordships nevertheless recognise that too rigid adherence to precedent may lead to injustice in a particular case and also unduly restrict the proper development of the law. They propose, therefore, to modify their present practice and, while treating former decisions of this House as normally binding, to depart from a previous decision when it appears right to do so.

In this connection they will bear in mind the danger of disturbing retrospectively the basis on which contracts, settlements of property, and fiscal arrangements have been entered into and also the especial need for certainty as to the criminal law.

This announcement is not intended to affect the use of precedent elsewhere than in this House.

Lord Gardiner's statement was accompanied by a press release which emphasised the importance of, and the reasons for, the change in practice (see [1966] *Public Law* 348). It would enable the House of Lords to adapt English law to meet changing conditions and to pay more attention to decisions of superior courts in the Commonwealth, and the change would bring the House into line with the practice of superior courts in many other countries. In the USA, for instance, the US Supreme Court and state supreme courts are not bound by their own previous decisions.

The change in practice was approved by lawyers generally. But some were surprised that such an important alteration in the judicial process was brought about by a mere practice statement. It was suspected that this could only be done by legislation. Instead, the House of Lords freed itself from a self-imposed restraint by exercising its inherent jurisdiction as a court to change its own practice. It would certainly not have been appropriate to effect the change in an actual case on appeal because that case would have had no greater authority than, say, *London Tramways Co.* v *London County Council* [1898] AC 375, HL.

A party to an appeal who intends to ask the House of Lords to depart from one of its own previous decisions must draw special attention to this in the appeal documents (*Practice Direction (House of Lords: Preparation of Case)* [1971] 1 WLR 534).

5.3.2.2.2 *Use of the Practice Statement* Developments since 1966 have indicated that the freedom of the House of Lords to review its previous decisions will be used sparingly, particularly in cases involving the construction of statutes and documents (*Jones* v *Secretary of State for Social Services* [1972] AC 944, HL, *per* Lords Reid, Morris, Pearson, and Simon at pp. 966, 973, 996–7, and 1024, respectively; but see *Vestey* v *Commissioners of Inland Revenue* [1979] 3 All ER 976, HL, below).

The House will not refuse to follow its earlier decision merely because that decision was wrong. And it is nothing to the point that the earlier decision was by a narrow majority. A material change of circumstances will usually have to be shown (*Fitzleet Estates Ltd* v *Cherry* [1977] 3 All ER 996, HL; *Jones* v *Secretary of State for Social Services*, above; *R* v *Knuller* [1973] AC 435, HL; *R* v *Kansal* (*No. 2*) [2001] 3 WLR 1562, HL, below; *McDonnell* v *Congregation of Christian Brothers Trustees* [2003] 3 WLR 1627, HL, below).

Nor will the House reconsider its own previous decision merely because that decision has caused grave concern. In *Food Corp of India* v *Antclizo Shipping Corp* [1988] 2 All ER 513, HL, it was said that, in order to justify a review, the judges must feel that they are free to depart from both the reasoning and the decision in the previous case, *and* they must be satisfied that a departure from the earlier decision would help to resolve the dispute in the case presently before them. The House of Lords, although tempted to do so, refused to conduct a review of the earlier authorities on the ground that the occasion was not appropriate because of the findings of fact in the case (see [1988] 2 All ER 513 *per* Lord Goff at pp. 516 and 520–1).

Two years passed before the House of Lords first exercised the power to depart from its own previous decision. That was in *Conway* v *Rimmer* [1968] AC 910, HL, unanimously overruling *Duncan* v *Cammell, Laird & Co.* [1942] AC 624, HL, on a question of the discovery of documents.

In *Duncan*, which was decided in wartime and was probably correct on its facts, the House of Lords had held that an affidavit sworn by a government minister was sufficient to enable the Crown to claim privilege not to disclose documents in civil litigation without those documents being inspected by the court (the so-called 'public interest immunity'). In *Conway* v *Rimmer*, above, their Lordships held that the minister's affidavit was not binding on the court.

The position now is that it is up to *the court* to decide whether to order disclosure. This will involve balancing the possible prejudice to the state if disclosure is ordered against the possible injustice to the individual litigant if disclosure is withheld. The minister's affidavit is entitled to full consideration but is no longer final. If it is necessary to assist him in reaching a conclusion, the trial judge may read the documents in private. Disclosure should be ordered only if the party seeking access to the documents satisfies the court that they are likely to support his case. It is not enough merely to show that the documents are likely to affect the outcome of the case one way or the other (*Air Canada and others* v *Secretary of State for Trade* [1983] 2 WLR 494, HL; *Evans* v *Chief Constable of Surrey Constabulary (Attorney-General intervening)* [1989] 2 All ER 594).

In *British Railways Board* v *Herrington* [1972] AC 877, HL, the House of Lords overruled (or, at least, modified) *Addie & Sons* v *Dumbreck* [1929] AC 358, HL. In *Addie*, the House of Lords had held that an occupier of premises was only liable to a trespassing child who was injured by the occupier intentionally or recklessly. In *Herrington*, their Lordships held that a different approach was appropriate in the changed social and physical conditions since 1929. They propounded the test of 'common humanity', which involves an investigation of whether the occupier has done all that a humane person would have done to protect the safety of the trespasser.

In *Miliangos* v *George Frank (Textiles) Ltd* [1976] AC 443, HL, the House of Lords overruled *Re United Railways of the Havana & Regla Warehouses Ltd* [1961] AC 1007, HL. In the *United Railways* case, it had been held that damages in an English civil case could only be awarded in sterling. In *Miliangos*, the House of Lords held that damages can be awarded in the currency of any foreign country specified in the contract. A new rule was needed because of changes in foreign exchange conditions, and especially the instability of sterling, since 1961.

In *Vestey* v *Commissioners of Inland Revenue* [1979] 3 All ER 976, HL, an income tax case, the House of Lords overruled its own decision in *Congreve* v *Commissioners of Inland Revenue* [1948] 1 All ER 948, HL, which had stood for some 30 years. This action was all the more surprising because *Vestey* was a case involving the interpretation of a statute.

In the earlier case of *Jones* v *Secretary of State for Social Services* [1972] AC 944, HL (decided on the construction of the National Insurance (Industrial Injuries) Act 1946), the House of Lords, by a majority of four to three, held that one of its own decisions of only five years' standing was wrong. Nevertheless, that decision was not overruled. All three dissentients, and one of the majority, took the view that they should not depart from it and so outvoted the remaining three members of the majority on this point. Lord Reid said (at p. 966) that it should only be in rare cases involving some broad issue of justice or public policy that the House should reconsider one of its own decisions involving the interpretation of a statute.

In the *Vestey* case, the decision to overrule *Congreve* v *Commissioners of Inland Revenue* was unanimous. Their Lordships felt not only that the *Congreve* case was wrong but also that it would produce 'startling and unacceptable consequences' when applied to circumstances never contemplated when it was decided ([1979] 3 All ER 976 *per* Lord Edmund-Davies at p. 1003). This consideration brought the situation within Lord Reid's dictum about broad issues of justice or public policy (*per* Lord Keith at p. 1005).

In *Jobling* v *Associated Dairies Ltd* [1981] 2 All ER 752, HL, a case concerning damages for personal injuries, the House of Lords doubted and did not follow its own decision given ten years earlier in *Baker* v *Willoughby* [1970] AC 467, HL.

In *M.V. Yorke Motors* v *Edwards* [1982] 1 All ER 1024, HL, the House did not follow (but did not expressly overrule) its own decision in *Jacobs* v *Booth's Distillery Co.* (1901) 85 LT 262, HL.

In *Paal Wilson & Co. A/S* v *Paartenrederei Hannah Blumenthal* [1983] 1 All ER 34, HL, a case on frustration of contract, the House declined an invitation to depart from its previous decision given only two years earlier in *Bremer Vulkan* v *South India Shipping Corporation* [1981] AC 909, HL. Their Lordships were resolved that the opposition of commercial people, and of the majority of the Court of Appeal, to the *Bremer Vulkan* decision did not constitute a material change of circumstances such as to justify the House in overruling it. The opposition amounted merely to what Lord Wilberforce, in *Fitzleet Estates Ltd* v *Cherry* [1977] 3 All ER 996, described (at p. 999) as 'doubts as to the correctness' of the *Bremer Vulkan* decision.

In the interests of certainty in the field of commercial law, the *Bremer Vulkan* decision was allowed to stand (see especially *per* Lords Brandon and Roskill [1983] 1 All ER 34 at pp. 47 and 54, respectively). The efforts of the majority of the Court of Appeal in the *Paal Wilson* case [1982] 3 All ER 394 to circumvent the *Bremer Vulkan* decision were said to reflect 'greater credit on their independence of mind than on their loyalty to the established and indispensable principle of judicial precedent' ([1983] 1 All ER 34 *per* Lord Brandon at p. 46).

A second attempt to persuade the House of Lords to depart from the *Bremer Vulkan* decision was unsuccessful in *Food Corp of India* v *Antclizo Shipping Corp* [1988] 2 All ER 513, HL (see above). The facts of the *Food Corp of India* case were such that the appeal was bound to fail. To have conducted a review of the earlier authorities in those circumstances would, therefore, have been a mere academic exercise and anything their Lordships said about *Bremer Vulkan* would have been *obiter* (see [1988] 2 All ER 513 *per* Lord Goff at p. 516). (The *Bremer Vulkan* and *Paal Wilson* decisions were ultimately modified by the Arbitration Act 1950, s. 13A, added with effect from 1 January 1992 by the Courts and Legal Services Act 1990.)

In *R* v *Secretary of State for the Home Department (ex parte Khawaja)* [1983] 2 WLR 321, HL, a case on illegal immigration, the House of Lords declined to follow its own decision given in a case two and a half years earlier, *R* v *Secretary of State for the Home Department (ex parte Zamir)* [1980] AC 930, HL. The feeling of the House in the *Khawaja* case was that the power of the courts to review the detention and summary removal of an alleged illegal immigrant had been defined too narrowly in the *Zamir* case, which, in fact, had cast the main burden of proof on the immigrant to show that his detention was not justified.

In *Khawaja*, Lord Scarman said (at p. 339) that the *Practice Statement* of 1966 indicated that the House of Lords, before departing from a precedent of its own making, must be satisfied on two counts. First, that continued adherence to the precedent would involve the risk of injustice and would obstruct the proper development of the law. Secondly, that a departure from the precedent is the safe and appropriate way of remedying the injustice and developing the law.

Lord Bridge said (at p. 356) that the case did not fall into any of the categories mentioned in the *Practice Statement* as requiring special caution before departing from precedent. There was no question of altering the criminal law or disturbing retrospectively the basis of any contract, settlement of property, or fiscal arrangement. Moreover, as he pointed out (at pp. 356–7), the case concerned both a matter of 'high constitutional principle affecting the liberty of the subject and the delineation of the respective functions of the executive and the judiciary' and a broad issue of justice and public policy, within the meaning of Lord Reid's dictum in *Jones* v *Secretary of State for Social Services* [1972] AC 944, HL, at p. 966.

It is not at all decisive one way or the other that the precedent from which the House is invited to depart is relatively recent. In *Khawaja*, Lord Bridge described it (at p. 356) as a 'neutral factor', while Lord Scarman (at p. 339) inclined to the view that it did not matter at all. In *R* v *Shivpuri* [1986] 2 All ER 334, HL, in which the House of Lords overruled within 12 months its own earlier decision in *Anderton* v *Ryan* [1985] AC 560, HL, Lord Bridge (who sat in both appeals) said (at p. 345):

> . . . I am undeterred by the consideration that the decision in *Anderton* v *Ryan* was so recent. The 1966 *Practice Statement* is an effective abandonment of our pretention to infallibility. If a serious error embodied in a decision of this House has distorted the law, the sooner it is corrected the better.

(See also the same judge's equally frank admission of fallibility in *Patel* v *Immigration Appeal Tribunal* [1988] 2 All ER 378, HL, at pp. 383–4.)

In *Anderton* v *Ryan* [1985] AC 560, HL, the appellant, on a construction of the Criminal Attempts Act 1981, was held not guilty of attempting dishonestly to handle a stolen video recorder. She thought the goods had been stolen but, in fact, there was no evidence that they had been. In *R* v *Shivpuri* [1986] 2 All ER 334, HL, another case on the Criminal Attempts Act 1981, the appellant was held to be guilty of attempting to commit a drugs offence. He had been caught with a suitcase which he thought contained prohibited drugs whereas, in fact, it contained snuff or some other harmless vegetable matter. It was, therefore, impossible for him to have committed the full offence involved but he was charged with *attempting to commit* the offence of being knowingly concerned in dealing with and harbouring prohibited drugs. His conviction was upheld by the House of Lords because he had intended to commit the full offence and had done acts which were 'more than merely preparatory to' the commission of the intended offence within the meaning of s. 1(1) of the 1981 Act. Mrs Ryan had escaped conviction in spite of the clear words of s. 1(2) of the Act that 'a person may be guilty of attempting to commit an offence to which this section applies even though the facts are such that the commission of the offence is impossible'. In *Anderton* v *Ryan*, the House of Lords had held, in effect, that these plain words did not mean what they said.

In *R* v *Howe* [1987] 2 WLR 568, HL, the House of Lords decided that the defence of duress is not available to a person charged with murder, whether as a principal in the first degree (the actual killer) or as a principal in the second degree (an aider and abettor). In so holding, the House overruled its earlier decision in *Director of Public Prosecutions for Northern Ireland* v *Lynch* [1975] AC 653, HL, to the effect that duress was available as a defence to a person who had participated in a murder as an aider and abettor.

Their Lordships' decision in *R* v *Howe*, above, was based on a desire to restore this part of the criminal law to what it was generally understood to be prior to *Lynch*, even though to do so produced the illogical result that, while duress is a *complete* defence to all crimes less serious than murder, it is not even a *partial* defence to a charge of murder itself (see [1987] 2 WLR 568 *per* Lord Hailsham LC, Lord Brandon, and Lord Griffiths at pp. 579–80, 585, and 592, respectively, but note that in *R* v *Gotts* [1992] 1 All ER 832, HL, the House of Lords extended the decision in *R* v *Howe* by holding, albeit by a bare majority of three to two, that duress is not a defence to *attempted* murder).

Elements of public and social policy are clearly discernible in the opinions in *R* v *Howe* [1987] 2 WLR 568, HL. Lord Hailsham, referring to the existing *administrative* mechanisms (such as the availability of parole and the royal prerogative of mercy) which might be used in appropriate cases to alleviate any hardship suffered by persons convicted of murder committed while acting under duress, said (at p. 581):

It may well be thought that the loss of a clear right to a defence justifying or excusing the deliberate taking of an innocent life in order to emphasise to all the sanctity of a human life is not an excessive price to pay in the light of these mechanisms . . .

. . . We live in the age of the holocaust of the Jews, of international terrorism on the scale of massacre, of the explosion of aircraft in mid air, and murder sometimes at least as obscene as anything experienced in Blackstone's day.

Lord Griffiths said (at p. 590):

We face a rising tide of violence and terrorism against which the law must stand firm recognising that its highest duty is to protect the freedom and lives of those that live under it. The sanctity of human life lies at the root of this ideal and I would do nothing to undermine it, be it ever so slight.

In *Murphy* v *Brentwood District Council* [1990] 2 All ER 908, HL, the House of Lords overruled its earlier decision in *Anns* v *Merton London Borough* [1978] AC 728, HL, on the common law liability of local authorities for the inspection of building foundations. In *Anns*, the House had held that a local authority was under a common law duty to take reasonable care to ensure that the foundations of a building complied with building regulations. This duty was owed to the owner and occupier of the building, who, if the duty was broken, could sue the local authority for negligence. Any damages awarded would include a sum to cover the cost of putting right the defect.

The decision in *Anns*, above, was not greeted with universal acclaim throughout the common law world. In particular, doubts were expressed about its consistency with established principles of the law of tort. The Australian courts refused to follow it on the ground that the duty it created was too wide, while courts in Canada and New Zealand not only followed it but considerably extended it. The Australian courts later changed their minds (see below).

In *Murphy* v *Brentwood District Council* [1990] 2 All ER 908, HL, the House of Lords was invited to reconsider *Anns* [1978] AC 728, HL, and, having done so, decided unanimously to overrule it. The principal factor which drove the House to this conclusion was the reluctance of English judges to provide a remedy in tort for pure *economic* loss (i.e., loss which is 'pecuniary', 'monetary', or 'financial') as opposed to *physical* loss (i.e., damage to persons or property). (There is no such reluctance where the economic loss results from negligent misstatements. The law laid down in *Hedley Byrne & Co. Ltd* v *Heller & Partners Ltd* [1964] AC 465, HL, para 5.4.2 below, is unaffected by *Murphy* v *Brentwood District Council*, above; see *Spring* v *Guardian Assurance plc* [1994] 3 All ER 129, HL.)

In *Anns*, above, the House of Lords appears to have proceeded on the mistaken assumption that the contemplated loss was physical. In reality, the loss was economic—namely, the cost of putting right the defect in the building—no person or property having been injured (see [1990] 2 All ER 908 *per* Lords Keith and Oliver at pp. 919 and 932, respectively).

Murphy v *Brentwood District Council*, above, has been rejected by the Supreme Court of Canada (*Canadian National Railway Co.* v *Norsk Pacific Steamship Co.* (1992) 91 DLR (4th) 289; *Winnipeg Condominium Corpn No. 36* v *Bird Construction Co.* (1995) 121 DLR (4th) 193), the High Court of Australia (*Bryan* v *Maloney* (1995) 128 ALR 163), and by the New Zealand Court of Appeal (*Invercargill City Council* v *Hamlin* [1996] 1 All ER 756, PC; see further, para 5.3.2.9 below).

In the historic case of *Pepper* v *Hart* [1993] 1 All ER 42, HL, the House of Lords declined to follow dicta in *Beswick* v *Beswick* [1968] AC 58, HL, *Black-Clawson International Ltd* v *Papierwerke Waldhof-Aschaffenburg AG* [1975] AC 591, HL, and *Davis* v *Johnson* [1979] AC 264, HL, on the use of *Hansard* as an extrinsic aid to the interpretation of statutes (see further, para 4.10.5.4).

In *Westdeutsche Landesbank Girozentrale* v *Islington London Borough Council* [1996] 2 All ER 961, HL, the House of Lords declined to follow its earlier decision in *Sinclair* v *Brougham* [1914] AC 398, HL, on the nature of a claim for money had and received under an *ultra vires* contract. According to *Sinclair* v *Brougham*, such a claim is an *equitable proprietary* claim. This view was rejected by the majority in the *Westdeutsche* case, it being held that it is a *personal* action based on a total failure of consideration. *Sinclair* v *Brougham* was variously described as 'controversial', 'bewildering', and 'wrongly decided' ([1996] 2 All ER 961 *per* Lords Goff, Browne-Wilkinson, and Lloyd at pp. 970, 996, and 1018, respectively), and two of their Lordships would have overruled it altogether (ibid., *per* Lord Browne-Wilkinson at pp. 993 and 996, and *per* Lord Lloyd at p. 1018).

The well-known case of *Rondel* v *Worsley* [1969] 1 AC 191, HL, on the immunity of barristers (and, by subsequent extension, all advocates) from liability in tort for the negligent presentation of cases in court and the preliminary work connected therewith, was not followed by the House of Lords, sitting as a court of seven, in *Arthur J.S. Hall & Co.* v *Simons* [2000] 3 All ER 673, HL. While making it clear that *Rondel* v *Worsley* was correct in 1967 (the year it was decided), their Lordships felt that developments in the legal process since that date meant that the immunity should be abolished as no longer being in the public interest. The Law Lords were unanimous in getting rid of the immunity in relation to *civil* proceedings, but in respect of *criminal* proceedings it was abolished by a bare majority of four to three.

In *R* v *Kansal (No. 2)* [2001] 3 WLR 1562, HL, the House of Lords refused to depart from its decision in *R* v *Lambert* [2001] 3 WLR 206, HL, given only four months earlier, even though a majority in *Kansal (No. 2)* thought it was wrong.

R v *Lambert* had held by a majority that the Human Rights Act 1998 is not retro-spective in relation to appeals arising from criminal trials which took place before the Act came into force. In *Kansal (No. 2)*, a majority of three to two held that *R* v *Lambert* was wrongly decided, but a majority of four to one held that there was no compelling reason not to apply it. *Kansal (No. 2)* was later applied by the House of Lords in *R* v *Benjafield* [2002] 2 WLR 235, HL.

In *R* v *G* [2003] 3 WLR 1060, HL, the House overruled its decision in *R* v *Caldwell* [1982] AC 341, HL, on the meaning of the word 'reckless' in s. 1 of the Criminal Damage Act 1971. *Caldwell* had been subjected to considerable academic and judicial criticism over the years, and their Lordships felt that there were now considerable legal reasons for departing from it (see, e.g., [2003] 3 WLR 1060, HL, *per* Lords Bingham and Steyn at [34]–[35] and [57], respectively).

On the same day as it decided *R* v *G*, above, the House of Lords in *Rees* v *Darlington Memorial Hospital NHS Trust* [2003] 3 WLR 1091, HL, refused to overrule *McFarlane* v *Tayside Health Board* [2000] 2 AC 59, HL. In *McFarlane*, it had been held that the cost of bringing up a normal healthy child, born as a result of negligent sterilisation advice given to the parents, was not recoverable in the law of tort. In *Rees*, the claimant, who suffered from severe visual impairment and was afraid that she would be unable to care for a child, underwent a negligently performed sterilisation oper-ation at the defendants' hospital. She later gave birth to a normal healthy child. She claimed as damages the cost of bringing up the child, including the extra cost attribut-able to her disability, and had sought to persuade the House to depart from *McFarlane*.

Rees provoked considerable judicial disagreement. The three judges who sat in the case when it was before the Court of Appeal were divided. All seven Law Lords who heard the further appeal were satisfied that *McFarlane* was correctly decided and that it would be wrong to overrule it (see, e.g., [2003] 3 WLR 1091 *per* Lord Steyn at [32]–[33]. Nevertheless, a minority of three would have created an exception to that case so as to allow a disabled mother, who had wanted to avoid pregnancy by having a sterilisation operation, to recover the *extra cost* of child-rearing resulting from her disability. The majority disagreed on the grounds that it is impossible to quantify the benefits of parenthood and inappropriate to regard a child solely as a financial liability.

In *McDonnell* v *Congregation of Christian Brothers Trustees* [2003] 3 WLR 1627, HL, the House declined to overrule *Arnold* v *Central Electricity Generating Board* [1988] AC 228, HL. Their Lordships had doubts about the correctness of *Arnold* but held that it was not 'plainly wrong'. It was emphasised that it was a unanimous decision which had stood for 16 years, that Parliament could have, but had not, reversed it, and that, applying *Fitzleet Estates Ltd* v *Cherry* [1977] 3 All ER 996, HL, above, the House needed more than doubts about the correctness of one of its own decisions in order to justify departing from it.

In *Lagden* v *O'Connor* [2003] 3 WLR 1571, HL, the House had little hesitation in departing from *Owners of Liesbosch Dredger* v *Owners of SS Edison (The Liesbosch)* [1933] AC 449, HL, even though it had stood for 70 years. It was held that the time was now right to reject *The Liesbosch*, which had been much-criticised and qualified over the years.

5.3.2.2.3 *Judicial law-making in the House of Lords* Being free of the shackles of binding precedent, the House of Lords is uniquely placed to develop English domestic law through its judicial decisions. Care must be taken, however, that evolution does not too often turn into the sort of reform which confuses the judicial function of a court of law with the legislative function enjoyed by Parliament.

By way of example, when in the momentous case of *R* v *R (rape: marital exemption)* [1991] 4 All ER 481, HL (para 5.3.2.7 below), the House abolished altogether a husband's 250 year-old immunity from criminal liability for raping his wife, their Lordships justified the decision on the basis that the case was not concerned with the creation of a new offence but with their duty to act in order to remove from the common law a fiction which had become unacceptable (ibid., *per* Lord Keith at pp. 489–90, quoting with approval from the judgment of Lord Lane CJ, delivered when the case was before the Court of Appeal). They saw the decision as an example of the ability of the common law to evolve 'in the light of changing social, economic and cultural developments' (ibid., *per* Lord Keith at p. 483). These explanations notwithstanding, some commentators viewed the decision in *R* v *R (rape: marital exemption)* [1991] 4 All ER 481, HL, as coming close to usurping the law-making functions of Parliament.

In a number of subsequent cases the House declined to change the law on the ground that to do so was the province of Parliament. In *R* v *Clegg* [1995] 1 All ER 334, HL, it had been argued that the House should make new law by creating a new qualified defence which would have the effect of reducing murder to manslaughter. This defence, it was urged, would be one of using excessive force in self defence, or to prevent crime, or to make or assist a lawful arrest, and would be available to a soldier or police officer acting in the course of his duty. However, Lord Lloyd (speaking on behalf of their Lordships), though not averse to judicial law-making and citing *R* v *R (rape: marital exemption)*, above, as a good recent example of it, declared that he had no doubt that they should abstain from law-making in the present case since the reduction of murder to manslaughter in a particular class of case was essentially a matter for Parliament, and not for them as a court, to decide on ([1995] 1 All ER 334 at p. 346).

In *C* v *Director of Public Prosecutions* [1995] 2 All ER 43, HL, the House referred to the anomalies and absurdities produced by the rebuttable common law presumption that a child between the ages of 10 and 14 is incapable of committing a crime. Nevertheless their Lordships refused to abolish the presumption, preferring instead to call on Parliament to review it.

Lord Lowry discerned in the case law the following guidelines for judicial law-making (see [1995] 2 All ER 43 at p. 52):

(a) judges should exercise caution before imposing a remedy where the solution to a problem is doubtful;

(b) they should be cautious about making changes if Parliament has rejected opportunities of dealing with a known problem or has legislated while leaving the problem untouched;

(c) they are more suited to dealing with purely legal problems than disputed matters of social policy;

(d) fundamental legal doctrines should not lightly be set aside;

(e) judges should not change the law unless they can achieve finality and certainty.

(Parliament later abolished the presumption by s. 34 of the Crime and Disorder Act 1998 with effect from 30 September 1998.)

In *Bellinger* v *Bellinger* [2003] 2 WLR 1174, HL, the House of Lords upheld the principle that English law does not recognise a marriage between two people who were of the same gender at birth, and declined to recognise as valid a marriage where the 'wife' had been registered at birth as male but had undergone gender reassignment surgery and treatment. Their Lordships said that any other conclusion would interfere with the traditional concept of marriage and involve complex and sensitive issues. Such a fundamental change in the law, it was held, should be made by Parliament and not by judges (see [2003] 2 WLR 1174 *per* Lord Nicholls at [37]).

At the same time, their Lordships granted a declaration under s. 4 of the Human Rights Act 1998 that s. 11(c) of the Matrimonial Causes Act 1973 (which requires the parties to a marriage to be respectively male and female) is incompatible with the European Convention on Human Rights, arts. 8 (right to respect for private and family life) and 12 (right to marry). (For developments subsequent to *Bellinger*, see para 4.10.8.2.)

In *Wainwright* v *Home Office* [2003] 3 WLR 1137, HL, the House refused to create a tort of invasion of privacy on the ground that it is a matter best left to the detailed approach associated with legislation rather than to the 'broad brush' approach of the common law (see [2003] 3 WLR 1137 *per* Lord Hoffmann at [33]).

In a number of cases, the House has been prepared to engage in judicial law-making. In *Kleinwort Benson Ltd* v *Lincoln City Council* [1998] 4 All ER 513, HL, by a majority of three to two, it abolished the 200 year-old common law rule that money paid under a mistake of law is not recoverable. The majority decided that the mistake of law rule should no longer form part of English law because it did not sit well with the modern law of restitution based on the principle of unjust enrichment.

Lord Goff, an acknowledged expert on the law of restitution, noted how the rule had been criticised over the years (at pp. 528–30), how it had already been abolished, either by judicial decision or by statute, in Australia, Canada, New Zealand, Scotland, and South Africa (at pp. 530–1), and how (at p. 531) the Law Commission had recommended its abolition in 1994 in its report, *Restitution: mistakes of law and ultra vires public authority receipts and payments* (Law Com. No. 227, 1994). He concluded that there was 'no good reason for postponing the matter for legislation', especially as it was not known 'whether or, if so, when, Parliament may legislate' (at p. 532).

The two dissentients, Lords Browne-Wilkinson and Lloyd (at pp. 517–19 and 546–8, respectively), while agreeing that the law should be changed so as to allow the recovery of money paid under a mistake of law, were of the view, unlike the majority, that when the law is changed by a later decision of the courts any money paid under the former law is *not* recoverable since, at the time of payment, the payer was not acting under a mistake of law. They were disturbed by the retrospective effect of the majority decision (which would allow old transactions to be reopened: see para 5.3.1 above). In view of that, they thought that the correct course was to leave the mistake of law rule to be changed by Parliament, as recommended by the Law Commission (at pp. 523 and 552, respectively).

Another piece of judicial law-making, at least as controversial as that in *R* v *R (rape: marital exemption)* [1991] 4 All ER 481, HL (above), occurred in *Fitzpatrick* v *Sterling Housing Association Ltd* [1999] 4 All ER 705, HL. Here, the House of Lords reversed the Court of Appeal and held—again by a bare majority—that a same-sex partner was capable of succeeding to an assured tenancy on the death of the original tenant as a 'member of the original tenant's family' under para 3 of sch. 1 to the Rent Act 1977, provided that the claimant could show that the homosexual relationship was characterised by features usually denoted by the word 'family'. These are that there should be 'a degree of mutual inter-dependence, of the sharing of lives, of caring and love, of commitment and support' (*per* Lord Slynn at p. 714). On the facts, the claimant was held entitled to succeed since he and the deceased had cohabited for many years in a stable, loving homosexual partnership.

Lord Lowry's guidelines for judicial law-making, proposed in *C* v *Director of Public Prosecutions* [1995] 2 All ER 43, HL, at p. 52 (above), were not cited in *Kleinwort Benson Ltd* v *Lincoln City Council* [1998] 4 All ER 513, HL, above, or in the *Fitzpatrick* case, above.

In relation to those guidelines, it could be argued that the majority in *Fitzpatrick* ignored the fact that Parliament did not see fit to extend the right to succession to the survivor of a same-sex relationship when the Rent Act 1977 was amended as recently as 1988. By that amendment, the right to succession was extended to a person living with the tenant as his or her wife or husband, thereby putting the decision in *Dyson Holdings Ltd* v *Fox* [1976] QB 503, CA, on a statutory footing (see *per* Lord Hobhouse, dissenting in *Fitzpatrick* [1999] 4 All ER 705, at pp. 744–5).

It could be further argued that the majority were meddling in a disputed social policy matter which Parliament is more suited to formulate than the judges. The view that the matter is 'disputed' within Lord Lowry's guideline is supported by the presence of two dissenting House of Lords' opinions in *Fitzpatrick*, together with the judgments of the majority when the case was before the Court of Appeal.

These criticisms notwithstanding, the majority in the *Fitzpatrick* case were convinced that they were right to make new law themselves rather than leave the matter to Parliament. Lord Slynn said (at [1999] 4 All ER 705, 710):

It has been suggested that for your Lordships to decide this appeal in favour of the appellant would be to usurp the function of Parliament. It is trite that that is something the courts must not do. When considering social issues in particular, judges must not substitute their own views to fill gaps . . . It is, however, for the court in the first place to interpret each phrase in its statutory context. To do so is not to usurp Parliament's function; not to do so would be to abdicate the judicial function. If Parliament takes the view that the result is not what is wanted it will change the legislation.

Lords Hutton and Hobhouse dissented on the ground that the matter should have been left to Parliament. Lord Hutton (at p. 742), while accepting the strength of the argument that the law should be changed so as to give protection to the homosexual partner of a deceased tenant, thought that such a change should only be made by Parliament. To Lord Hobhouse (at p. 743), it was 'an improper usurpation of the legislative function for a court to adopt social policies which have not yet been incorporated in the relevant legislation'. (Parliament did later intervene in this area by passing the Civil Partnership Act 2004; see para 4.10.8.2.)

In *Arthur J.S. Hall & Co.* v *Simons* [2000] 3 All ER 673, HL, it was argued that only Parliament, and not the House of Lords sitting as a court of law, could abolish an advocate's immunity in the tort of negligence as exemplified in *Rondel* v *Worsley* [1969] 1 AC 191, HL. This argument was based on s. 62 of the Courts and Legal Services Act 1990, which provides that a person who is not a barrister but who lawfully provides legal services in relation to any proceedings 'shall have the same immunity from liability for negligence . . . as he would have if he were a barrister . . .'. Their Lordships were of the view that this provision does not, either expressly or by implication, give statutory approval to a barrister's immunity, but simply assumes its existence and extends it to solicitor-advocates and other authorised advocates. Accordingly, the House of Lords felt able to abolish the immunity which it had confirmed 30 years earlier in *Rondel* v *Worsley* without trespassing on the legislative function of Parliament (see, e.g., [2000] 3 All ER 673, HL, *per* Lords Steyn and Hoffmann at pp. 684 and 704, respectively).

In *Rees* v *Darlington Memorial Hospital NHS Trust* [2003] 3 WLR 1091, HL (para 5.3.2.2.2 above), a majority of the House of Lords held that it would be unjust if the claimant was refused any compensation at all for the legal wrong done to her and which had prevented her from living in the way she had planned. Controversially, a 'conventional award' of £15,000 damages was made to compensate her for that legal wrong. Two of the minority vigorously opposed such a conventional award on the ground that it was contrary to principle, unsupported by the case law, and that it was a move best left to Parliament (ibid., *per* Lords Steyn and Hope at [45]–[46] and [73]–[77], respectively).

5.3.2.3 Court of Appeal (civil division)

5.3.2.3.1 *Circumstances in which the Court of Appeal is bound by precedent:* Y oung v Bristol Aeroplane Co. Ltd   The Court of Appeal is bound by decisions of the House of Lords, even if it considers them to be wrong, unless they can be distinguished on the facts or on the law, or unless they were given *per incuriam* (*IM Properties plc* v *Cape & Dalgleish* [1998] 3 All ER 203, CA, a somewhat controversial application of the *per incuriam* doctrine, as to which see (c) below). The distinguishing of House of Lords' decisions by the Court of Appeal is dealt with in more detail later (para 5.3.2.3.2 below).

In the exceptional circumstances of *R (H) v Secretary of State for the Home Depart-ment* [2002] 3 WLR 967, CA, the Court of Appeal declined to follow the decision of the House of Lords in *R v Oxford Regional Mental Health Review Tribunal (ex parte Secretary of State for the Home Department)* [1988] AC 120, HL, on the ground that it predated the coming into force of the Human Rights Act 1998 and was in need of review in the light of that Act. On appeal in *R (H) v Secretary of State for the Home Department* ([2003] 3 WLR 1278), the House of Lords approved the Court of Appeal's approach and departed from its own decision in the *Oxford* case.

Decisions of the Court of Appeal itself are binding on the High Court and the county courts, but they do not bind the House of Lords.

The remaining question is whether the Court of Appeal is bound by its own previous decisions. In *Young v Bristol Aeroplane Co. Ltd* [1944] KB 718, CA, a 'full' Court of Appeal of six members decided that it was normally so bound subject to the following three exceptions.

(a) Where its own previous decisions conflict, the Court of Appeal must decide which to follow and which to reject.

Thus, in *Tiverton Estates Ltd v Wearwell Ltd* [1975] Ch 146, CA, the Court of Appeal refused to follow *Law v Jones* [1974] Ch 112, CA, which had been decided only six months earlier by a differently constituted Court of Appeal on the same point of law—the nature of the written memorandum required by s. 40 of the Law of Property Act 1925. *Law v Jones* was inconsistent with earlier decisions of equal authority. *Law v Jones* will probably not be followed in the future but, in strict theory, it is still open to the Court of Appeal to choose between *Law v Jones* and *Tiverton Estates v Wearwell Ltd*. *Law v Jones* was not followed by a Queen's Bench Division judge in *Cohen v Nessdale Ltd* [1981] 3 All ER 118. (Note that s. 40 of the Law of Property Act 1925 was repealed, as from 27 September 1989, by the Law of Property (Miscellaneous Provisions) Act 1989.)

In *National Westminster Bank plc v Powney* [1990] 2 All ER 416, CA, the Court of Appeal, faced with two irreconcilable decisions of its own on the question whether an application for leave to issue execution of a judgment was an 'action' capable of becoming statute-barred, chose to follow *W.T. Lamb & Sons v Rider* [1948] 2 KB 331, CA, rather than *Lougher v Donovan* [1948] 2 All ER 11, CA.

In *Finnegan v Parkside Health Authority* [1998] 1 All ER 595, CA, the Court of Appeal preferred its decisions in *Costellow v Somerset County Council* [1993] 1 All ER 952, CA, and *Mortgage Corp Ltd v Sandoes* (1996) *The Times*, 27 December, CA, to its decision in *Savill v Southend Health Authority* [1995] 1 WLR 1254, CA, on the relevance of prejudice in an application to the court for an extension of time for complying with procedural requirements.

In *Starmark Enterprises Ltd v CPL Distribution Ltd* [2002] 2 WLR 1009, CA, the Court of Appeal concluded that its decisions in *Mecca Leisure Ltd v Renown Investments (Holdings) Ltd* (1984) 49 P & CR 12, CA, and *Henry Smith's Charity Trustees v AWADA Trading & Promotions Services Ltd* (1984) 47 P & CR 607, CA, were indistinguishable and irreconcilable. It held that the *Henry Smith* decision was to be preferred, despite being the earlier of the two, because *Mecca Leisure* had been wrongly decided since it had improperly distinguished the *Henry Smith* case and was inconsistent with the approach of the House of Lords in *United Scientific Holdings Ltd v Burnley Borough Council* [1978] AC 904, HL.

(b) The Court of Appeal must refuse to follow a decision of its own which cannot stand with a decision of the House of Lords even though its decision has not been expressly overruled by the House of Lords.

Thus, in *Family Housing Association v Jones* [1990] 1 All ER 385, CA, the Court of Appeal refused on this ground to follow three of its own recent decisions on the distinction between a tenancy and a licence. Although those decisions had not been expressly overruled by the House of Lords, one of them could not be reconciled with the subsequent decision of the House in *AG Securities Ltd v Vaughan* [1988] 3 All ER 1058, HL, while the other two could not stand with the later decisions in *Street v Mountford* [1985] AC 809, HL, and *AG Securities Ltd v Vaughan*, above.

The decision of the Court of Appeal in *Solle v Butcher* [1950] 1 KB 671, CA, had stood for 50 years when, in *Great Peace Shipping Ltd v Tsavliris Salvage (International) Ltd* [2002] 3 WLR 1617, CA, the Court of Appeal declined to follow it because it could not be reconciled with the earlier decision of the House of Lords in *Bell v Lever Bros Ltd* [1932] AC 161, HL. *Solle v Butcher* could not be attacked under the *per incuriam* doctrine (see (c) below) since *Bell v Lever Bros Ltd* had been cited to the court in *Solle v Butcher*.

All three cases were concerned with the effect of common mistake in the law of contract. In *Solle v Butcher*, above, it had been held that the doctrine of equitable rescission allowed a court to set aside a contract on equitable terms on the ground of a common mistake which was not fundamental enough to make the contract void at common law. In the *Great Peace Shipping* case, above, the Court of Appeal held that the doctrine of equitable rescission allowed no such thing, that the Law Lords who decided *Bell v Lever Bros Ltd* could not have been 'oblivious to principles of equity', that *Solle v Butcher* was an attempt to 'outflank *Bell v Lever Bros Ltd*' by, in effect, declaring it to have been wrongly decided, and that the two cases were irreconcilable ([2002] 3 WLR 1617, CA, *per* Lord Phillips MR at [118], [126], and [156]–[157]). At the same time, the Court of Appeal advocated greater flexibility in the law of mistake by the introduction of legislation conferring on the court an equitable jurisdiction to grant rescission of a contract on equitable terms where it is void at common law on the ground of a common fundamental mistake (ibid., *per* Lord Phillips MR at [161]).

Although it is beyond doubt that the Court of Appeal must refuse to follow a decision of its own which cannot stand with a decision of the *House of Lords*, it appears that it does not have the same freedom where one of its decisions has been disapproved by the *Judicial Committee of the Privy Council*. The reason, according to the Court of Appeal in *In re Spectrum Plus Ltd* [2004] 3 WLR 503, CA, is that such a course is not authorised by any of the exceptions to binding precedent laid down by the Court of Appeal itself in *Young v Bristol Aeroplane Co. Ltd* [1944] KB 718, CA, and reaffirmed by the House of Lords in *Davis v Johnson* [1979] AC 264, HL.

When *In re Spectrum Plus Ltd* reached the House of Lords ([2005] 3 WLR 58, HL), three of their Lordships mentioned the point under consideration. Lords Scott and Walker (at [93] and [153], respectively) were of the opinion that the Court of Appeal's approach was correct. However, Baroness Hale (at [163]) thought that the question was an open one and did not wish to rule out the possibility that an exception 'might exist or be developed' to the effect that the Court of Appeal could refuse to follow one of its own decisions which had been disapproved by the Judicial Committee of the Privy Council.

> (c)   The Court of Appeal need not follow a decision of its own if satisfied that it was given *per incuriam* (literally, by carelessness or mistake).

The usual ground on which a decision of the Court of Appeal is regarded as *per incuriam* is that it was given in ignorance or forgetfulness of a relevant statutory provision or binding decision of the House of Lords or Court of Appeal.

Before the Court of Appeal rejects one of its own decisions on this ground, it has to be shown not only that the decision was given *per incuriam*, but also that the Court of Appeal in the earlier case *must* have reached a different conclusion had it had the uncited statute or binding decision in mind (*Duke* v *Reliance Systems Ltd* [1987] 2 All ER 858, CA; the *per incuriam* doctrine did not need to be (and was not) considered on appeal to the House of Lords in this case, reported *sub nom. Duke* v *GEC Reliance Ltd* at [1988] AC 618).

In *Rakhit* v *Carty* [1990] 2 All ER 202, CA, the Court of Appeal declined to follow two of its earlier decisions on the Rent Act 1977 on the ground that the first of them was made in ignorance of a relevant provision of the Act, while the second was based solely (and reluctantly) on the first, which had been given *per incuriam*.

In *Wellcome Trust Ltd* v *Hammad* [1998] 1 All ER 657, CA, the Court of Appeal declined to follow its earlier decision in *Pittalis* v *Grant* [1989] 2 All ER 622, CA, on the meaning of 'dwelling house' in the Rent Act 1977 on the ground that it was arrived at in ignorance of some relevant decisions referred to in the speech of Lord Wilberforce in the House of Lords' case of *Maunsell* v *Olins* [1975] AC 373, HL, and was, therefore, given *per incuriam*.

Royal Bank of Scotland v *Etridge* [1997] 3 All ER 628, CA, a decision of a *two-judge* court at an interim stage on the principles for deciding whether legal advice provided to a wife has rebutted the presumption of undue influence which arises in her favour against her husband, was not followed by a *three-judge* Court of Appeal in *Royal Bank of Scotland* v *Etridge (No. 2)* [1998] 4 All ER 705, CA, on the ground that it was wrongly decided, having been made *per incuriam* in ignorance of two of its earlier decisions, *Midland Bank plc* v *Serter* [1995] 1 FLR 1034, CA, and *Barclays Bank plc* v *Thompson* [1997] 4 All ER 816, CA.

In *Wallcite Ltd* v *Ferrishurst Ltd* [1999] 1 All ER 977, CA, the Court of Appeal refused to follow its earlier decision in *Ashburn Anstalt* v *Arnold* [1988] 2 All ER 147, CA, on the scope of overriding interests under s. 70(1)(g) of the Land Registration Act 1925, on the ground that if the court in the *Ashburn* case had been referred to the House of Lords' decision in *Williams & Glyn's Bank Ltd* v *Boland* [1981] AC 487, HL, the *Ashburn* case would have been decided differently.

In *R* v *G* (*autrefois acquit*) [2001] 1 WLR 1727, CA, the criminal division of the Court of Appeal invoked the *per incuriam* doctrine when declining to follow its earlier decision in *R* v *Brookes* [1995] Crim LR 630, CA, on the ground that it was decided in ignorance of a relevant provision of the Magistrates' Courts Act 1980.

Young v *Bristol Areoplane Co. Ltd* [1944] KB 718, CA, envisaged that there would be other grounds on which decisions might later be held to be *per incuriam*, even though those decisions do not fall strictly within the usual definition of decisions given *per incuriam* (see above). It was pointed out, however, that such cases will be 'of the rarest occurrence' ([1944] KB 718 *per* Lord Greene MR at p. 729; or 'rare and exceptional', as Evershed MR described them in *Morelle Ltd* v *Wakeling* [1955] 2 QB 379, CA, at p. 406). The Court of Appeal has consistently refused to define what these other grounds might be (see, e.g., *Rickards* v *Rickards* [1989] 3 All ER 193, CA, *per* Lord Donaldson MR at p. 199 and Balcombe LJ at p. 201).

It is certainly no ground for arguing that a decision was *per incuriam* to show that a necessary party to the proceedings was not before the court on the earlier occasion (*Morelle Ltd* v *Wakeling*, CA, above), or that the court had not had the benefit of the best argument from counsel (ibid.), or that the court *might* have reached a different conclusion if other arguments or material had been placed before it (*Duke* v *Reliance Systems Ltd* [1987] 2 All ER 858, CA, above). It has, however, been held that a decision may be *per incuriam* through incomplete reference during argument to the reports of a particular case which is otherwise binding on the Court of Appeal (*Industrial Properties (Barton Hill) Ltd* v *Associated Electrical Industries Ltd* [1977] QB 580, CA).

Subsequent developments have shown that the *per incuriam* doctrine applies in exceptional circumstances where the Court of Appeal's earlier decision involved a 'manifest slip or error'.

Thus, in *Williams* v *Fawcett* [1985] 1 All ER 787, CA, the Court of Appeal held in a contempt of court case that there were 'exceptional circumstances' for treating as *per incuriam* (and, therefore, not following) two of its own previous decisions reported in 1984 and 1985. Those circumstances were that (a) the growth of the error in the previous cases could be clearly detected; (b) the cases concerned the liberty of the subject; and (c) the cases were of such a nature that it was unlikely that they would ever reach the House of Lords, which would not, therefore, have an opportunity to rectify the error which had crept into the law (see [1985] 1 All ER 787, CA, *per* Sir John Donaldson MR at p. 795; *Langley* v *North West Water Authority* [1991] 3 All ER 610, CA, *per* Lord Donaldson MR at pp. 621–2; see also *R* v *Parole Board (ex parte Wilson)* [1992] 2 All ER 576, CA, a case involving the liberty of the subject, where the Court of Appeal applied *Williams* v *Fawcett*, above, in refusing to follow its decision in *R* v *Secretary of State for the Home Department (ex parte Gunnell)* (1984) *The Times*, 7 November).

In *Rickards* v *Rickards* [1989] 3 All ER 193, CA, the Court of Appeal held that it could treat as *per incuriam* one of its own earlier decisions reported in 1981 in which the court had misunderstood the effect of a House of Lords' decision reported in 1891 and had consequently wrongly decided that it had no jurisdiction to do what was asked of it. This was a 'rare and exceptional' type of case because (a) a wrongful denial of jurisdiction was a serious matter amounting to a breach of statutory duty on the part of the Court of Appeal, and (b) it was most unlikely (because of the cost involved) that the House of Lords would be presented with an opportunity to correct the mistake. The same considerations apply to a case where the Court of Appeal is guilty of abuse of power by purporting to exercise a jurisdiction which it does not possess (ibid., *per* Lord Donaldson MR at p. 199).

It is arguable that the *per incuriam* doctrine was wrongly applied in *Rickards* v *Rickards*, above, since the House of Lords' decision of 1891 had been cited to the Court of Appeal in its decision of 1981. It could not, therefore, be said that the Court of Appeal had acted *in ignorance of* it (cf *Great Peace Shipping Ltd* v *Tsavliris Salvage (International) Ltd* [2002] 3 WLR 1617, CA, above at (b)).

The reasoning in *Williams* v *Fawcett* [1985] 1 All ER 787, CA, and *Rickards* v *Rickards* [1989] 3 All ER 193, CA, was applied in *Rakhit* v *Carty* [1990] 2 All ER 202, CA. Here, the Court of Appeal did not follow *Cheniston Investments Ltd* v *Waddock* [1988] 2 EGLR 136, CA, because it had been given in 'error' on the mistaken assumption that the court was bound by *Kent* v *Millmead Properties Ltd* (1982) 44 P & CR 353, CA, which itself had been given *per incuriam* in ignorance of a relevant statutory provision (see *Rakhit* v *Carty* [1990] 2 All ER 202 *per* Russell LJ at pp. 207–8, Sir Roualeyn Cumming-Bruce at p. 208, and Lord Donaldson MR at p. 208). The circumstances of *Rakhit* v *Carty* were 'exceptional' because the case concerned not only the rights of the immediate parties but also the rights of thousands of landlords and tenants throughout the country (ibid., *per* Lord Donaldson MR at p. 208).

In *R (W)* v *London Borough of Lambeth* [2002] 2 All ER 901, CA, the Court of Appeal, applying *Rickards* v *Rickards* [1989] 3 All ER 193, CA, and *Duke* v *Reliance Systems Ltd* [1987] 2 All ER 858, CA, held that it was free to depart from its decision in *R (A)* v *London Borough of Lambeth* [2001] EWCA Civ 1624, CA. That case had been decided in ignorance of relevant statutory provisions. If those provisions had been brought to the attention of the court the case would have been decided differently.

Like other courts, the Court of Appeal is only bound, if at all, by the *ratio decidendi* of a case. It is not bound by a previous decision which is not authority for the proposition for which it is cited, or in which a proposition of law was merely assumed to be correct without the court addressing the issue.

For example, in interpreting s. 32 of the Race Relations Act 1976 in *Jones* v *Tower Boot Co. Ltd* [1997] 2 All ER 406, CA, the Court of Appeal was confronted by its own earlier decision in *Irving* v *Post Office* [1987] IRLR 289, CA, in which the Post Office was held not liable for the act of a postman who wrote a racially offensive remark on the back of an envelope addressed to his neighbours, with whom he was in dispute.

In *Jones*, however, the Court of Appeal was able to ignore *Irving*, without concluding that it had been decided *per incuriam*, and without distinguishing it, on the ground that the court in *Irving* had not been referred to s. 32(1) of the 1976 Act, but appeared to have decided the case on the unchallenged assumption that the common law principles of vicarious liability applied to it. It followed, therefore, that *Irving* did not support the contention of the employers in *Jones*. The authority of *Irving* v *Post Office* was thus considerably weakened and the case may now be regarded as having been wrongly decided.

A fourth exception appears to have been recognised since the *Bristol Aeroplane* case was decided in 1944, namely, that the Court of Appeal is not bound to follow a decision in an *interlocutory* matter made by *two* judges in the Court of Appeal (*Boys* v *Chaplin* [1968] 2 QB 1, CA).

Thus, in *Welsh Development Agency* v *Redpath Dorman Long Ltd* [1994] 4 All ER 10, CA, it was held by a three-man Court of Appeal in an interlocutory matter that, applying *Boys* v *Chaplin*, above, an earlier decision of a *two-man* Court of Appeal given in an interlocutory matter (*Kennett* v *Brown* [1988] 1 WLR 582, CA) was wrong and not binding on it.

However, the authority of *Boys* v *Chaplin* was damaged following opinions expressed, *obiter*, in *Langley* v *North West Water Authority* [1991] 3 All ER 610, CA. (See also *Limb* v *Union Jack Removals Ltd* [1998] 2 All ER 513, CA, *per* Brooke LJ at pp. 522–3.) As a result, it was established that when the Court of Appeal is dealing with *final*, as opposed to interlocutory, appeals, the authority of a two-judge court is to be regarded as the same as that of a three-judge court (*Langley* v *North West Water Authority* [1991] 3 All ER 610, CA, *per* Lord Donaldson MR at pp. 621–2, explaining *Boys* v *Chaplin*, above). In consequence, the decision of a two-man Court of Appeal given in a final appeal will be binding on a later three-man Court of Appeal subject to the exceptions laid down in the *Bristol Aeroplane* case.

The authority of *Boys* v *Chaplin*, above, was further weakened when it was subjected to criticism, and was not followed, in *Cave* v *Robinson Jarvis & Rolf* [2002] 1 WLR 581, CA. Here, notwithstanding *Boys* v *Chaplin* and *Welsh Development Agency* v *Redpath Dorman Long Ltd*, above, the Court of Appeal held itself to be bound by a decision of its own given in 1999 by a two-judge court in an *interlocutory* appeal (*Brocklesby* v *Armitage & Guest* [2001] 1 All ER 172, CA, on the interpretation of s. 32(2) of the Limitation Act 1980).

Applying *dicta* of Lord Donaldson MR in *Langley* v *North West Water Authority*, above, and of Brooke LJ in *Limb* v *Union Jack Removals Ltd*, above, the Court of Appeal in *Cave* held that a departure from any previous decision of the court is 'highly undesirable' and should only be contemplated if the previous decision is 'manifestly' or 'incontestably' wrong. That, according to *Cave*, is now the important consideration and not whether the previous decision was given by a two-judge court (see [2002] 1 WLR 581, CA, *per* Potter, Sedley, and Jonathan Parker LJJ at [17]–[24], [31], and [37]–[43], respectively).

When *Cave* went on appeal to the House of Lords ([2002] 2 WLR 1107), it was held that the interpretation of s. 32(2) of the Limitation Act 1980 preferred in *Brocklesby* was wrong, and the decision in *Cave* was accordingly reversed. However, since no argument was presented on the question whether the Court of Appeal had been correct to hold itself bound by *Brocklesby* their Lordships declined to express any opinion on the matter (see ([2002] 2 WLR 1107 *per* Lord Scott at [66]).

In the light of the approach adopted by the Court of Appeal in *Cave* [2002] 1 WLR 581, CA, it is difficult to see any future for *Boys* v *Chaplin* [1968] 2 QB 1, CA. In any event, the former distinction between interlocutory and final appeals no longer exists. Whether the Court of Appeal sits as a two-judge or a three-judge court is now determined by other considerations, namely the nature and importance of the point to be argued. It was held in *Clark* v *University of Lincolnshire and Humberside* [2000] 3 All ER 752, CA, that decisions of the Court of Appeal given on applications for permission to appeal are not binding. This remains the case after *Cave*. Although these decisions are made by one-judge or two-judge courts, the real reason why they are not binding is that they are made under time constraints and after only brief argument.

It is at least arguable that a fifth exception exists (or ought to exist) consequent on United Kingdom membership of the European Union. In order to give full effect to

Community law it would seem desirable that the Court of Appeal, like the House of Lords and the High Court, should be free to depart from a previous decision of its own which is inconsistent with Community law. Furthermore, by s. 3(1) of the European Communities Act 1972, an English court faced with a question of Community law must either refer the question to the European Court or decide the question itself in the light of any relevant decisions of the European Court. In Community law cases it follows that, since the Court of Appeal is bound both by decisions of the European Court and those of its own making, there is scope for potential conflict between decisions of the two courts.

Not surprisingly, the principles laid down in 1944 in the *Bristol Aeroplane* case [1944] KB 718, CA, do not, as generally understood, provide a direct solution to the problem where the Court of Appeal is confronted with a binding decision of its own (or, for that matter, of the House of Lords) which is, at the same time, inconsistent with a decision of the European Court. The view of the European Court is that the duty of a national court to apply Community law may involve, if necessary, ignoring national laws and practices (*Amministrazione delle Finanze dello Stato v Simmenthal SpA* [1978] 3 CMLR 263, CJEC). On the other hand, in at least one English case it has been said that United Kingdom membership of the European Union has not abrogated the doctrine of *stare decisis* in the Court of Appeal (*Duke v Reliance Systems Ltd* [1987] 2 All ER 858, CA, *per* Sir John Donaldson MR at p. 860).

One answer is to extend or adapt the existing exceptions recognised in the *Bristol Aeroplane* case. It may be, however, that s. 3(1) of the European Communities Act 1972 itself provides the answer for it appears to imply that, in the event of conflict, priority should be accorded to decisions of the European Court.

It will be readily appreciated that the precedent problem is one which does not affect only the Court of Appeal since courts and tribunals sitting at first instance (to which s. 3(1) is also applicable) must administer Community law where appropriate. It seems unlikely that s. 3(1) was intended to confer on, say, the High Court *carte blanche* to ignore binding decisions of the Court of Appeal or House of Lords where individual judges on their own initiative consider that those decisions are in conflict with the case law of the European Court.

5.3.2.3.2 *Lord Denning's approach to precedent in the Court of Appeal, civil division* Lord Denning was one of the most eminent English judges of the twentieth century. He became a judge of the High Court in 1944, a Lord Justice of Appeal in 1948, and a Lord of Appeal in Ordinary in 1957. He left the judicial House of Lords and returned to the Court of Appeal in 1962 on his appointment as Master of the Rolls, a post he held until his resignation in 1982. He often sought to achieve a just result in a case at the expense of applying the accepted law, and for this reason some of his judgments were controversial. He died in 1999 at the age of 100.

From the time of the *Practice Statement* affecting *stare decisis* in the House of Lords in 1966 (para 5.3.2.2.1 above), Lord Denning MR carried on a one-man campaign to secure a change of practice in the Court of Appeal.

The attack was on two fronts. First, he asserted that the Court of Appeal was no longer bound by decisions of the House of Lords. Secondly, he claimed that the Court of Appeal was no longer bound to follow its own decisions as a general rule and not just in the exceptional circumstances laid down in *Young* v *Bristol Aeroplane Co. Ltd* [1944] KB 718, CA. His views were based on the *Practice Statement* itself, which, he alleged, had transformed the *stare decisis* principle in the Court of Appeal as well as the House of Lords.

(a) Lord Denning's first expression of his opinion that the Court of Appeal was not bound by House of Lords' decisions was in the cases of *Conway* v *Rimmer* [1967] 2 All ER 1260, CA, and *Broome* v *Cassell & Co. Ltd* [1971] 2 QB 354, CA. When these cases reached the House of Lords his approach was criticised on each occasion. In the latter case (*sub nom. Cassell & Co. Ltd* v *Broome* [1972] AC 1027), he was castigated for his disloyalty. Lord Hailsham LC said (at p. 1054):

[I]t is not open to the Court of Appeal to give gratuitous advice to judges of first instance to ignore decisions of the House of Lords in this way and, if it were open to the Court of Appeal to do so, it would be highly undesirable . . .

The fact is, and I hope it will never be necessary to say so again, that, in the hierarchical system of courts which exists in this country, it is necessary for each lower tier, including the Court of Appeal, to accept loyally the decisions of the higher tiers.

Lord Denning returned to the attack in *Schorsch Meier GmbH* v *Hennin* [1975] QB 416, CA, a case in which the Court of Appeal held that an English court had power to award damages for breach of contract in a foreign currency which was the currency of the contract. In so deciding, the Court of Appeal did not follow a decision of the House of Lords—*Re United Railways of the Havana & Regla Warehouses Ltd* [1961] AC 1007, HL—in which it had been laid down that damages could be awarded only in sterling.

The *Schorsch Meier* case did not go to the House of Lords, but another case on the same point did. This was *Miliangos* v *George Frank (Textiles) Ltd* [1976] AC 443, HL, in which the House of Lords, while overruling the *United Railways* case itself, once again put Lord Denning in his place on the issue of *stare decisis*. Lord Cross was particularly scathing in his criticism (at p. 496):

It is not for any inferior court—be it a county court or a division of the Court of Appeal presided over by Lord Denning—to review decisions of this House. Such a review can only be undertaken by this House itself under the declaration of 1966.

While it is true that the Court of Appeal is not bound by a House of Lords' decision if it is distinguishable on the facts or on the law, care must be taken to avoid using the process of distinguishing as a means of circumventing a House of Lords' decision which is felt to be wrong or, at least, inconvenient. Mindful of Lord Hailsham's rebuke in *Cassell & Co. Ltd* v *Broome* (above), a majority of the Court of Appeal (including Lord Denning) in *Paal Wilson & Co. A/S* v *Paartenrederei Hannah Blumenthal* [1982] 3 All ER 394, CA, sought to avoid the consequences of the House of Lords' decision in the *Bremer Vulkan* case [1981] AC 909, HL, by distinguishing it rather than openly declaring it to have been wrongly decided. Lord Denning MR and Kerr LJ held in *Paal Wilson* that an important passage in the speech of Lord Diplock, speaking on behalf of the other Law Lords in *Bremer Vulkan*, was *obiter dictum* and not part of the *ratio decidendi* of that case. Since the Court of Appeal is only bound by *rationes decidendi,* as opposed to *obiter dicta,* of the House of Lords, Lord Denning MR and Kerr LJ concluded that the *Bremer Vulkan* case was distinguishable. Griffiths LJ dissented. While not enthusiastic about the *Bremer Vulkan* decision, he thought that it was not legitimately distinguishable and should therefore be loyally followed.

When the *Paal Wilson* case reached the House of Lords ([1983] 1 All ER 34), the decision of the majority was reversed. The House held that *Bremer Vulkan* was not distinguishable from the present case and was binding on the Court of Appeal because the relevant passage in Lord Diplock's speech was not merely *obiter* but formed part of the *ratio decidendi* of the case. Furthermore, the House of Lords in *Paal Wilson* refused to reconsider its decision in the *Bremer Vulkan* case (para 5.3.2.2.2 above).

> (b) The second front of Lord Denning's attack on *stare decisis*—the assertion that the Court of Appeal was no longer bound rigidly to follow its own previous decisions—began in 1969 in *Gallie v Lee* [1969] 2 Ch 17, CA, and was continued in *Tiverton Estates Ltd v Wearwell Ltd* [1975] Ch 146, CA. However, he was unable to carry the other members of the court with him. In the latter case, Scarman LJ said (at pp. 172–3):

The Court of Appeal occupies a central, but . . . an intermediate position in our legal system. To a large extent, the consistency and certainty of the law depend on it. It sits almost always in divisions of three . . . If, therefore, throwing aside the restraints of *Young v Bristol Aeroplane Co. Ltd,* one division of the court should refuse to follow another because it believed the other's decision to be wrong, there would be a risk of confusion and doubt arising where there should be consistency and certainty. The appropriate forum for the correction of the Court of Appeal's errors is the House of Lords, where the decision will at least have the merit of being final and binding—subject only to the House's power to review its own decisions. The House of Lords, as the court of last resort, needs this power of review: it does not follow that an intermediate appellate court needs it.

By this time Lord Denning seemed to have learnt his lesson. He appeared to capitulate and to accept the orthodox view in *Miliangos v George Frank (Textiles) Ltd* [1975] QB 487, CA. But the whole matter was reopened in *Davis v Johnson* [1979] AC 264, CA and HL. This case concerned the scope of s. 1 of the Domestic Violence and Matrimonial Proceedings Act 1976 (since repealed and replaced by the Family Law Act 1996). The matter had come before the Court of Appeal on two occasions only a few months earlier in *B v B* [1978] Fam 26, CA, and *Cantliff v Jenkins* [1978] QB 47, CA. The Court of Appeal had held that the 1976 Act did not protect a female cohabitant where the parties were joint tenants or joint owners but only where she was the sole tenant or sole owner of the property.

These decisions would have had the effect of destroying what most people regarded as one of the main objects of the 1976 Act. *Davis v Johnson,* above, was heard by a 'full' court of five judges in the Court of Appeal. It was held by a majority of three to two that the 1976 Act did protect a female cohabitant even where she was not a tenant at all or only a joint tenant. *B v B* and *Cantliff v Jenkins,* above, were declared to be wrong and were not followed.

On the question of *stare decisis* in the Court of Appeal Lord Denning had this to say (at p. 278):

It is said that if an error has been made, this court has no option but to continue the error and leave it to be corrected by the House of Lords. The answer is this: the House of Lords may never have an opportunity to correct the error; and thus it may be perpetuated indefinitely, perhaps for ever.

On a further appeal in *Davis* v *Johnson*, the decision of the majority in the Court of Appeal was upheld ([1979] AC 264) and the House of Lords overruled *B* v *B* and *Cantliff* v *Jenkins*. But their Lordships rejected most of what had been said about *stare decisis* in the Court of Appeal. Lord Diplock, with whom the other four judges expressly agreed, was especially critical. Referring to what he described as Lord Denning's 'one-man crusade', he said (at p. 328):

In my opinion, this House should take this occasion to reaffirm expressly, unequivocally and unanimously that the rule laid down in the *Bristol Aeroplane* case as to *stare decisis* is still binding on the Court of Appeal.

After the decision of the House of Lords in *Davis* v *Johnson* [1979] AC 264, HL, it is beyond doubt that the true position with regard to *stare decisis* in the Court of Appeal is that, first, the Court of Appeal is bound by decisions of the House of Lords even if they are wrong (unless they can be legitimately distinguished), and, secondly, the Court of Appeal is bound by its own decisions subject only to the exceptions laid down in *Young* v *Bristol Aeroplane Co. Ltd* [1944] KB 718, CA.

5.3.2.4 Court of Appeal (criminal division)

In principle, there is no difference in the application of *stare decisis* as between the civil and criminal divisions of the Court of Appeal (*R* v *Spencer* [1985] 1 All ER 673, CA, *per* May LJ at p. 678, not overruled on this point by the House of Lords at [1986] 2 All ER 928). In practice, however, because a person's liberty may be at stake, precedent is not followed as rigidly in the criminal division.

The criminal division is not bound to follow its own previous decision if satisfied that the law was misapplied or misunderstood in it, even though the case is not within one of the exceptions in *Young* v *Bristol Aeroplane Co. Ltd* [1944] KB 718, CA (*R* v *Taylor* [1950] 2 KB 368, CCA; *R* v *Gould* [1969] 2 QB 65, CA; *R* v *Simpson* [2003] 3 WLR 337, CA, below). The same flexibility is applied to decisions of the old Court of Criminal Appeal, which was created by the Criminal Appeal Act 1907 and abolished by the Criminal Appeal Act 1966 (*R* v *Gould*, above).

In *R* v *Chalkley* [1998] 2 All ER 155, CA, the court decided that the earlier case of *R* v *Bloomfield* [1997] 1 Cr App R 135, CA, on the 'unsafeness' of a conviction within the meaning of the amended s. 2(1) of the Criminal Appeal Act 1968, was wrongly decided and declined to follow it for that reason.

R v *Popat* [1998] 2 Cr App R 208, CA, a case on identity parades, was criticised and not followed in *R* v *Forbes* [1999] 2 Cr App R 501, CA. *R* v *Forbes* itself was then criticised in *R* v *Khan* (19 August 1999, CA, unreported) and *R* v *Popat (No. 2)* [2000] 1 Cr App R 387, CA. In *R* v *Ryan* (1999) *The Times*, 13 October, CA, it was again held that *R* v *Popat* was correct and was to be preferred to *R* v *Forbes*. When *Forbes* went on further appeal, the House of Lords ([2001] 2 WLR 1) disapproved *Popat* and decided that the approach of the Court of Appeal in *Forbes* was the correct one.

The proper approach to precedent in the criminal division of the Court of Appeal was reviewed in *R* v *Simpson* [2003] 3 WLR 337, CA, which arose out of the necessity to consider whether the court was right to conclude in *R* v *Sekhon* [2003] 1 WLR 1655, CA, that *R* v *Palmer* (2002) *The Times*, 5 November, CA, had been wrongly decided. All three cases were concerned with the jurisdiction of the Crown Court to make a confiscation order in circumstances where there had been a procedural defect of a technical nature which caused no injustice to the defendant.

In *Palmer*, it had been held by the Court of Appeal that the defect was fatal to the Crown Court's jurisdiction. In *Sekhon*, it was held that *Palmer* was wrong and that the defect should only cause loss of jurisdiction if that was a necessary result in the interests of justice. In *Simpson*, it was held by a *full court of five* that *Sekhon* represented the correct approach to the problem and that the law had been misunderstood and misapplied in *Palmer*, which, in consequence, had been wrongly decided (see [2003] 3 WLR 337, CA, *per* Lord Woolf CJ at [24]–[27] and [32]–[38]). On an appeal to the House of Lords from the decision of the Court of Appeal in *Sekhon*, the House affirmed that decision under the name of *R* v *Knights* [2005] 3 WLR 330, HL. It follows that *Palmer* is now completely without authority.

In *R* v *R* [2004] 1 WLR 490, CA, the Court of Appeal, applying *R* v *Simpson*, above, held that its earlier decision in *R* v *T* [2003] 4 All ER 877, CA, had been given *per incuriam* because only the appellant had been represented and not all the relevant authorities had been put before the court.

5.3.2.5 Divisional Courts

A Divisional Court is bound by decisions of the House of Lords and Court of Appeal in both civil and criminal cases, unless it can distinguish them on the facts or on the law or it is convinced that a particular decision was given *per incuriam*. In *Hughes* v *Kingston upon Hull City Council* [1999] 2 All ER 49, DC, the Queen's Bench Divisional Court refused to follow the Court of Appeal case of *Thai Trading Co.* v *Taylor* [1998] QB 781, CA (on the enforceability of contingency fee agreements entered into by solicitors), on the ground that it was decided *per incuriam* in ignorance of the House of Lords' decision in *Swain* v *Law Society* [1983] 1 AC 598, HL.

A Divisional Court is also normally bound by its own previous decisions but subject to the same exceptions laid down for the Court of Appeal, civil division, in *Young* v *Bristol Aeroplane Co. Ltd* [1944] KB 718, CA. This means, for example, that, if its previous decisions conflict, a Divisional Court may choose which to follow and which to reject. It also means that it may refuse to follow a decision of its own if it is now satisfied that the earlier decision was given *per incuriam*.

The Queen's Bench Divisional Court has claimed that, like the Court of Appeal, criminal division, it can refuse to follow its own earlier decision, if convinced that the first decision is wrong, in two cases—(a) when hearing criminal appeals from magistrates' courts, and (b) when considering applications for judicial review. This power was claimed in *R* v *Greater Manchester Coroner (ex parte Tal)* [1984] 3 All ER 240, DC, not following *R* v *Surrey Coroner (ex parte Campbell)* [1982] 2 All ER 545, DC, on the ground that the earlier case was incorrectly decided.

In *R* v *Stafford Justices (ex parte Customs and Excise Commissioners)* [1991] 2 All ER 201, DC, the Divisional Court, on the authority of *R* v *Greater Manchester Coroner (ex parte Tal)*, DC, above, refused to follow *R* v *Ealing Magistrates' Court (ex parte Dixon)* [1989] 2 All ER 1050, DC, on the ground that it was wrongly decided.

R v Greater Manchester Coroner (ex parte Tal), above, was applied by the Divisional Court in *R v Hendon Justices (ex parte Director of Public Prosecutions)* [1993] 1 All ER 411, DC, in order to disown part of the decision (which it now regarded as wrong) in *R v Sutton Justices (ex parte Director of Public Prosecutions)* [1992] 2 All ER 129, DC, to the effect that an acquittal by a magistrates' court cannot be quashed by a quashing order.

In *Shaw v Director of Public Prosecutions* [1993] 1 All ER 918, DC, the Divisional Court refused to follow *Director of Public Prosecutions v Corcoran* [1993] 1 All ER 912, DC, a decision of its own given only four months earlier, on the grounds that it was decided *per incuriam* (in ignorance of *Metropolitan Police Commissioner v Curran* [1976] 1 All ER 162, HL, and *Roberts v Griffiths* [1978] RTR 362, DC, neither of which was cited to the court in *Director of Public Prosecutions v Corcoran*), *and*, applying *ex parte Tal*, that it was decided wrongly.

The matter did not rest there because in the wake of *Shaw* a number of attempts were made in lower courts to persuade magistrates and others that *Corcoran* was still good law. One such attempt which reached the Divisional Court was in *Butterworth v Director of Public Prosecutions* [1994] RTR 181, DC. Here, the Divisional Court preferred *Shaw* to *Corcoran* and decided that, in view of the confusion, it was time for the relevant law to be clarified by the House of Lords. Accordingly, *Butterworth* was certified as involving points of law of general public importance but leave to appeal was not granted, it being left to the House itself to decide whether it wanted to entertain an appeal. The Appeal Committee of the House granted leave to appeal, and in *Director of Public Prosecutions v Butterworth* [1994] 3 All ER 289, HL, the House of Lords affirmed the decision of the Divisional Court in *Butterworth v Director of Public Prosecutions*, approved *Shaw*, and overruled *Corcoran*.

In the judicial review case of *R (Snelgrove) v Crown Court at Woolwich* [2005] 1 WLR 3223, DC, the Queen's Bench Divisional Court declined to follow its earlier decision in *R v Central Criminal Court (ex parte Director of Serious Fraud Office)* [1993] 2 All ER 399, DC, on the ground that it was inconsistent with decisions of the House of Lords. (See further on the *Snelgrove* case, para 10.5.2.3.5.)

5.3.2.6 High Court of Justice

Decisions of individual High Court judges are binding on the county courts but not on other High Court judges. However, they are of strong persuasive authority in the High Court and are usually followed. If a High Court judge feels that he cannot follow a colleague's decision it is always with reluctance and he will usually state his reasons clearly and fully. He cannot 'overrule' it, but is limited to 'disapproving' or 'not following' it.

Conflict between High Court decisions produces uncertainty in the law. The Court of Appeal may resolve the conflict one way or the other, and thereby remove the uncertainty, but its ability to do this is dependent on an appeal being taken there by a litigant. In the early 1970s, long before the wearing of seat-belts in cars became compulsory, the question arose whether the damages awarded to a passenger injured in a road accident caused by the driver's negligence should be reduced for contributory negligence in the case of a passenger who was not wearing a seat-belt. Between

1973 and 1975 there were a dozen or so decisions on this point. Some judges held that failure to wear a seat-belt was not contributory negligence at all; others held that it was, but disagreed about the percentage by which the damages should be reduced.

Eventually, the difference of opinion was resolved by the Court of Appeal in 1975 in *Froom* v *Butcher* [1976] QB 286, CA, where it was held that in most cases failure to wear a seat-belt is contributory negligence if use of a belt would have avoided or lessened the injuries sustained in the accident. It was further suggested that, in general, the appropriate reduction is 25 per cent if the injuries would have been prevented altogether by the use of a seat-belt, or 15 per cent if they would nevertheless have occurred but would have been less severe.

A solution to the problem of conflicting High Court decisions was suggested by Nourse J in *Colchester Estates* v *Carlton Industries plc* [1984] 2 All ER 601, applying a dictum of Denning J (as he then was) in *Minister of Pensions* v *Higham* [1948] 2 KB 153.

Nourse J said that when a decision of a High Court judge has been fully considered, but not followed, by a second High Court judge, the second decision should normally be preferred by a third High Court judge, except in the rare case where the third judge is convinced that the second judge was wrong not to follow the first judge. The 'rare case' might be, for example, where some binding or persuasive authority was not cited in either of the first two cases. In the ordinary case, the second judge's decision will be regarded as conclusive of the point at first instance, and if a party in the third case is dissatisfied he will have to take the point on appeal to the Court of Appeal (see [1984] 2 All ER 601, at pp. 604–5).

A High Court judge is normally bound by decisions of the House of Lords, the Court of Appeal (including an *ex parte* decision of the Court of Appeal: *The Alexandros P* [1986] 1 All ER 278), and the Divisional Court of his particular Division. He is probably also bound by the decision of a Divisional Court of another Division. For instance, a decision of the Divisional Court of the Queen's Bench Division is probably binding on a judge of the Chancery Division.

A High Court judge is not bound to follow the earlier decision of a higher court if he can *distinguish* it on the facts or on the law.

Nor is he bound by such a decision on a proposition of law which was merely *assumed* to be correct without the higher court addressing its mind to the issue (*Re Hetherington (deceased)* [1989] 2 All ER 129, in which Sir Nicolas Browne-Wilkinson V-C held that he was not bound by certain observations of the House of Lords in *Bourne* v *Keane* [1919] AC 815, HL; see also *Baker* v *R* [1975] AC 774, PC, *per* Lord Diplock at p. 788; *Barrs* v *Bethell* [1982] 1 All ER 106 *per* Warner J at p. 116; *R* v *Secretary of State for the Home Department (ex parte Ku)* [1995] 2 All ER 891, CA, *per* Hobhouse LJ at p. 898).

The same principle is capable of applying to other courts. *Re Hetherington (deceased)*, above, and *Barrs* v *Bethell*, above, were approved in *R (Kadhim)* v *Brent London Borough Council Housing Benefit Review Board* [2001] 2 WLR 1674, CA, in which the Court of Appeal applied the principle to itself. In this case it was held that a previous decision of its own, although not given *per incuriam* since the relevant statutory regulations had been taken into account, was nevertheless not binding because it was based on an assumption (that the statutory regulations covered the situation before the court) which had not been the subject of argument or consideration.

It was pointed out, however, that the 'assumption without argument' principle must be applied carefully and only in the most obvious of cases; i.e., where the assumed point has not been raised before the court and there has been no argument on it. Even then, bearing in mind that a later court will always assume that an earlier court considered all the issues essential for its decision, the principle will not apply where it is clear from an examination of the earlier decision that the court's acceptance of the disputed point went beyond mere assumption ([2001] 2 WLR 1674, CA, *per* Buxton LJ, delivering the judgment of the court, at [38]–[39]).

A High Court judge confronted with *conflicting* decisions of the Court of Appeal, in the later of which the earlier decision was fully considered and rejected, is bound to follow the later decision. This is so even if he considers that the later decision is wrong and the earlier decision preferable. While he is free to reject the decisions of other High Court judges (see, e.g., *Colchester Estates* v *Carlton Industries plc*, above), he does not enjoy the same freedom in relation to Court of Appeal decisions which he considers to be wrong (*Re Smith (a bankrupt)* [1988] 3 All ER 203 *per* Warner J at p. 215).

5.3.2.7 Crown Court

Decisions made on points of law by judges sitting at the Crown Court, though they are of persuasive authority (especially when made by judges of the High Court), are not binding precedents. There is, therefore, no obligation on the part of other Crown Court judges to follow them. Since inconsistent Crown Court decisions may produce uncertainty in the criminal law, it is desirable that any conflict be resolved by an appellate court as quickly as possible.

This process may be illustrated by the difference of judicial opinion which surfaced in relation to the controversial subject of marital rape.

At the heart of the controversy was an extra-judicial statement of Sir Matthew Hale (Chief Justice of the Court of King's Bench between 1671 and 1676) contained in his *History of the Pleas of the Crown*, published posthumously in 1736, that a husband could not be guilty of raping his wife because on marriage she was deemed to have given him an irrevocable consent to sexual intercourse (1 Hale PC 629). For over 250 years Hale CJ's proposition, though doubted in 1888 by some of the judges who decided *R* v *Clarence* (1888) 22 QBD 23 (a case not directly on marital rape), was accepted as representing the common law of England and Wales. In order to mitigate the harshness of this rule, some exceptions to it were developed by the judges from 1949 onwards. While leaving the basic rule itself intact, the exceptions recognised that there were circumstances in which the wife's implied consent was terminated.

A similar marital exemption rule had operated in Scots law since 1797. The Scottish courts rejected the rule altogether in 1989 (*S* v *HM Advocate* 1989 SLT 469, High Court of Justiciary). In the following year, the English rule came under judicial attack in the Crown Court.

In July 1990, Owen J reluctantly accepted that part of Hale CJ's proposition which laid down that consent to intercourse was to be implied from the fact of marriage, but he denied that that consent was irrevocable (*R* v *R (rape: marital exemption)* [1991] 1 All ER 747, Crown Court sitting at Leicester).

In October of the same year, Simon Brown J went further and rejected Hale CJ's proposition in its entirety (*R* v *C* (*rape: marital exemption*) [1991] 1 All ER 755, Crown Court sitting at Sheffield). He declined to follow the decision of Owen J in *R* v *R* (*rape: marital exemption*) and held that a husband could be guilty of raping his wife.

In the following month, Rougier J (to whom new arguments had been addressed) decided that the inclusion of the word 'unlawful' in the statutory definition of rape as 'unlawful sexual intercourse with a woman who at the time of the intercourse does not consent to it' (Sexual Offences (Amendment) Act 1976, s. 1(1)(a)) meant that Parliament had intended to preserve the marital immunity rule, which was, moreover, subject only to the exceptions existing at the time the statute was passed (*R* v *J* (*rape: marital exemption*) [1991] 1 All ER 759, Crown Court sitting at Teesside, not following *R* v *C* (*rape: marital exemption*) and doubting the correctness of *R* v *R* (*rape: marital exemption*); *R* v *J* (*rape: marital exemption*) was followed by Swinton Thomas J in *R* v *S*, 15 January 1991, unreported, Crown Court sitting at Stafford).

The case of *R* v *R* (*rape: marital exemption*) was taken on appeal ([1991] 4 All ER 481). The House of Lords, following the Scottish case of *S* v *HM Advocate* 1989 SLT 469 and affirming the decision of a full Court of Appeal of five members, held that, although Hale CJ's proposition had become an accepted part of the common law and reflected the status of wives at the time it was put forward, the marital exemption rule no longer existed in English law since it was an anachronistic and offensive common law fiction which did not reflect the status of wives in contemporary society.

In reaching this conclusion, the House of Lords decided that the word 'unlawful' in s. 1(1)(a) of the Sexual Offences (Amendment) Act 1976 did not mean 'outside marriage'. It was held that the word was, in effect, meaningless in the context, and was to be treated as mere surplusage, since it is clearly unlawful to have sexual intercourse with any woman without her consent (*R* v *R* (*rape: marital exemption*) [1991] 4 All ER 481 *per* Lord Keith at pp. 487–9, the other Law Lords agreeing with him).

In the light of the House of Lords' decision in *R* v *R* (*rape: marital exemption*), above, and a subsequent recommendation by the Law Commission in its Report, *Rape within Marriage* (Law Com No. 205, 1992), Parliament redefined rape by removing the word 'unlawful'—thus making it clear that rape within marriage is a crime (Criminal Justice and Public Order Act 1994, repealing s. 1(1) of the Sexual Offences (Amendment) Act 1976 and substituting a new s. 1 of the Sexual Offences Act 1956; s. 1 of the 1956 Act was itself repealed by the Sexual Offences Act 2003 and replaced by the definition of rape contained therein).

It was later held that the House of Lords' abolition of the marital exemption rule had not violated the rights of Mr R (the defendant in *R* v *R* (*rape: marital exemption*)) under art. 7.1 of the European Convention on Human Rights, which provides that no one should be held guilty of a criminal offence in respect of an act which was not classified as criminal at the time it was committed. The European Court of Human Rights said that the decision of the House of Lords was simply a continuation of a perceptible line of case law dismantling a husband's immunity from prosecution for raping his wife. Furnishing a gradual clarification of the rules of criminal liability through judicial decision on a case-by-case basis was declared not to be a breach of art. 7.1 so long as the resulting development is consistent with the offence and can reasonably be foreseen (*CR* v *United Kingdom* [1996] 1 FLR 434, ECHR).

The Crown Court is bound by decisions of the House of Lords, the Court of Appeal, and the Queen's Bench Divisional Court.

5.3.2.8 County courts and magistrates' courts

The decisions of these courts are not binding. They are rarely important in law and are not usually reported in the law reports.

5.3.2.9 Judicial Committee of the Privy Council

Decisions (technically, 'advice') of the Judicial Committee are not binding on the English courts but they are of strong persuasive authority. Thus, the decision of the Judicial Committee in *Mutual Life & Citizens' Assurance Co. Ltd* v *Evatt* [1971] AC 793, PC, on the question of liability for negligent misstatements, was not followed by the Court of Appeal in *Esso Petroleum Co. Ltd* v *Mardon* [1976] 1 QB 801, CA. But the decision of the Judicial Committee in *Overseas Tankship (UK) Ltd* v *Morts Dock & Engineering Co. Ltd (The Wagon Mound)* [1961] AC 388, PC, on the question of remoteness of damage in negligence was followed by the Court of Appeal in *Doughty* v *Turner Manufacturing Co. Ltd* [1964] 1 QB 518, CA, in preference to one of its own decisions, *Re Polemis* [1921] 3 KB 560, CA, which was regarded as no longer good law.

In *R* v *Smith* [2000] 4 All ER 289, HL, the House of Lords declined to follow the majority decision of the Judicial Committee in *Luc Thiet Thuan* v *R* [1997] AC 131, PC, on the defence of provocation in the law of murder, the Court of Appeal having already refused to follow it in *R* v *Campbell* [1997] 1 Cr App R 199, CA, *R* v *Parker* [1997] Crim LR 760, CA, and in *R* v *Smith* itself ([1998] 4 All ER 387, CA). The judicial disagreement over whether the law on provocation is correctly stated in the decision of the House of Lords in *R* v *Smith*, above, or in the decision of the Judicial Committee in *Luc Thiet Thuan* v *R*, above, resurfaced in *Attorney-General for Jersey* v *Holley* [2005] 3 WLR 29, PC. Here, the Judicial Committee sat nine-strong and, by a majority of six to three, disapproved *Smith* and preferred its own decision in the *Luc Thiet Thuan* case.

Decisions of the Judicial Committee are binding in the country from which the appeal came and, possibly, in other countries subject to its jurisdiction where the law on the particular point is the same (*Fatuma Binti Mohamed Bin Salim Bakhshuwen* v *Mohamed Bin Salim Bakhshuwen* [1952] AC 1, PC). In relation to the country from which the appeal came, the Judicial Committee has the concomitant power to overrule the relevant case law of that country. Thus, in *Reyes* v *The Queen* [2002] 2 WLR 1034, PC, and *R* v *Gilbert* [2002] 2 WLR 1498, PC, the Committee took the opportunity to overrule earlier inconsistent decisions of, respectively, the Court of Appeal of Belize and the Eastern Caribbean Court of Appeal.

The Judicial Committee is not strictly bound by decisions of the House of Lords because the common law in the Commonwealth is not necessarily the same as in England. It would be wrong, for instance, for the Judicial Committee to inflict on, say, New Zealand, decisions of the House of Lords on the English common law. Accordingly, in *Invercargill City Council* v *Hamlin* [1996] 1 All ER 756, PC, the Court of Appeal of New Zealand was held to have been correct in declining to follow *Murphy* v *Brentwood District Council* [1990] 2 All ER 908, HL, a House of Lords' decision on the liability of local authorities at common law for economic loss caused by the negligent inspection of building foundations (see para 5.3.2.2.2 above).

The Judicial Committee said that the New Zealand courts are entitled to develop unsettled areas of the common law of New Zealand in accordance with conditions and policy considerations existing locally, which it would be rash for the Judicial Committee to ignore. As Lord Lloyd put it ([1996] 1 All ER 756 at pp. 764–5):

The ability of the common law to adapt itself to the differing circumstances of the countries in which it has taken root, is not a weakness, but one of its great strengths. Were it not so, the common law would not have flourished as it has, with all the common law countries learning from each other.

Similarly, in *B v Auckland District Law Society* [2003] 3 WLR 858, PC, Lord Millett said (at [55]):

The common law is no longer monolithic, and it [is] open to the New Zealand Court of Appeal to make a deliberate policy decision to depart from the English approach on the ground that it is not appropriate to conditions in New Zealand.

In *Australian Consolidated Press Ltd v Uren* [1969] 1 AC 590, PC, it was held that the High Court of Australia was right not to follow *Rookes v Barnard* [1964] AC 1129, HL, a House of Lords' decision on exemplary damages.

In *Parker v R* [1963] Crim LR 569 the High Court of Australia refused to follow the notorious decision in *DPP v Smith* [1961] AC 290, HL, on the mental element in the crime of murder.

Where, however, it is decided or accepted that English law *is* applicable in a particular case, the Judicial Committee will consider itself bound to follow the House of Lords' decision which covers the point in issue (*Tai Hing Cotton Mill Ltd v Liu Chong Hing Bank Ltd* [1985] 2 All ER 947, PC, *per* Lord Scarman at p. 958, a case in which the Judicial Committee, in an appeal from Hong Kong, followed the House of Lords' decision in *London Joint Stock Bank Ltd v Macmillan* [1918] AC 777, HL; *Invercargill City Council v Hamlin*, PC, above).

The Judicial Committee is not strictly bound by its own previous decisions, although it rarely refuses to follow them (the *Bakhshuwen* case [1952] AC 1, PC, above). When in *Pratt v Attorney-General for Jamaica* [1993] 4 All ER 769, PC, the Committee held that failure to carry out a death sentence as swiftly as practicable after sentence amounted to a violation of the Jamaican constitution in that it was 'inhuman or degrading punishment or other treatment', it declined to follow two of its own earlier decisions (*Abbott v Attorney-General of Trinidad and Tobago* [1979] 1 WLR 1342, PC, and *Riley v Attorney-General of Jamaica* [1983] 1 AC 719, PC) which had held that prolonged delay in carrying out a death sentence did not contravene a person's constitutional rights.

In *Lewis v Attorney-General of Jamaica* [2001] 2 AC 50, PC, the Committee held that the exercise of the prerogative of mercy in Jamaica is justiciable and subject to minimum standards of fairness. In doing so, it declined to follow its earlier decisions in *De Freitas v Benny* [1976] AC 239, PC, and *Reckley v Minister of Public Safety and Immigration (No. 2)* [1996] AC 527, PC.

Similarly, when quashing death sentences imposed by a court in Belize in *Reyes v*

The Queen [2002] 2 WLR 1034, PC, the Committee declined to follow its earlier decisions in *Runyowa* v *The Queen* [1967] 1 AC 26, PC, and *Ong Ah Chuan* v *Public Prosecutor* [1981] AC 648, PC.

Decisions of the English Court of Appeal have only persuasive authority in the Commonwealth (*Robins* v *National Trust Co.* [1927] AC 515, PC; *De Lasala* v *De Lasala* [1980] AC 546, PC). In *Reyes* v *The Queen*, PC, above, and *R* v *Gilbert* [2002] 2 WLR 1498, PC, the Judicial Committee preferred to follow decisions of the Court of Appeal rather than its own. However, in *Gleaner Co. Ltd* v *Abrahams* [2003] 3 WLR 1038, PC, an appeal from Jamaica, the Committee declined to follow *John* v *MGN Ltd* [1996] 2 All ER 35, CA, on the ground that it would be wrong to impose on Jamaica the English rule that juries in libel cases may be referred by the trial judge to awards in personal injury cases by way of guidance since that rule involves simply a question of policy but no question of legal principle. In *Agnew* v *Commissioner of Inland Revenue* [2001] 3 WLR 454, PC, the Committee disapproved, and refused to follow, the decision of the Court of Appeal on floating and fixed charges given in *In re New Bullas Trading Ltd* [1994] 1 BCLC 485, CA. The latter case was eventually overruled, and the *Agnew* decision approved, by the House of Lords in *In re Spectrum Plus Ltd* [2005] 3 WLR 58, HL.

A decision of the House of Lords on the *common law* also has only persuasive authority, although the authority is very strong because of the common membership of the House of Lords and the Judicial Committee (*De Lasala* v *De Lasala* [1980] AC 546, PC, *per* Lord Diplock at p. 557). In *Badry* v *Director of Public Prosecutions of Mauritius* [1982] 3 All ER 973, PC, the Committee regarded the decision of the House of Lords in *Attorney-General* v *British Broadcasting Corporation* [1981] AC 303, HL, as 'conclusive authority' on a point in the common law relating to contempt of court.

A decision of the House of Lords on the *interpretation of legislation* common to the overseas territory and England is also, in theory, persuasive only. But, realistically, it has the same practical effect as if it were absolutely binding (*De Lasala* v *De Lasala* [1980] AC 546, PC, at p. 558). In the case of common legislation, unlike judge-made law, there has been no divergent development and the overseas legislation would be interpreted according to English principles of interpretation. In *De Lasala* v *De Lasala*, above, an appeal from Hong Kong on the interpretation of a piece of legislation virtually identical to the English Matrimonial Causes Act 1973, the Judicial Committee held that the Hong Kong Court of Appeal was effectively bound by the decision of the House of Lords in *Minton* v *Minton* [1979] AC 593, HL, on the point.

5.4 Binding and persuasive precedents

5.4.1 Binding precedents

A *binding* precedent is a decided case which a court *must* follow even though it is considered to have been wrongly decided. But a previous case is only binding in a later case if the legal principle involved is the same and the facts are similar. An inconvenient precedent which would otherwise be binding can be circumvented by *distinguishing* it on the facts or on the legal principle involved.

5.4.2 **Persuasive precedents**

A *persuasive* precedent is one which is not absolutely binding on a court but which *may* be applied. The following are some examples of persuasive precedents.

(a) Decisions of English courts lower in the hierarchy.

For example, the House of Lords *may* follow a Court of Appeal decision, and the Court of Appeal *may* follow a High Court decision, although not strictly bound to do so.

(b) Decisions of the Judicial Committee of the Privy Council (para 5.3.2.9 above).

(c) Decisions of the courts in Scotland, Ireland, the Commonwealth (especially Australia, Canada, and New Zealand), and the USA.

It is permissible for counsel to cite decisions of courts outside England and Wales. This is particularly appropriate where there is a shortage or total lack of English authority on the point. In *Conway* v *Rimmer* [1968] AC 910, HL (para 5.3.2.2.2 above), Scottish, Australian, and American decisions were cited. In *Murphy* v *Brentwood District Council* [1990] 2 All ER 908, HL (para 5.3.2.2.2 above), in which decisions of courts in Australia, Canada, New Zealand, and the USA were cited, the House of Lords preferred the decision of the High Court of Australia in *Sutherland Shire Council* v *Heyman* (1985) 60 ALR 1 to its own decision in *Anns* v *Merton London Borough* [1978] AC 728, HL.

(d) *Obiter dicta* of English judges.

Dicta are of different kinds and of varying degrees of weight. Some are casual expressions of opinion on a point which has not been raised in the case. Other dicta, although not necessary for the decision of the case, are deliberate expressions of opinion given after consideration of a point clearly put and argued before the court. These dicta carry much greater weight than casual expressions of opinion (see *Slack* v *Leeds Industrial Cooperative Society Ltd* [1923] 1 Ch 431, CA, *per* Lord Sterndale MR at p. 451).

House of Lords' dicta are the most persuasive and especially so if there was general agreement on the point in question. In *Rondel* v *Worsley* [1969] 1 AC 191, HL, the House of Lords, after seven days of legal argument and the citation of 92 cases, held unanimously that a barrister is not liable in tort for the negligent presentation of a case in court and the preliminary work connected therewith, such as the drafting of pleadings. Four of their Lordships said, *obiter*, that a solicitor acting as an advocate was entitled to the same immunity ([1969] 1 AC 191 *per* Lords Reid, Morris, Pearce, and Upjohn at pp. 232, 243, 267, and 284, respectively). There were dicta to the same effect by three different Law Lords in the later case of *Saif Ali* v *Sydney Mitchell & Co.* [1980] AC 198, HL, and they were accepted as accurately representing the law (ibid., *per* Lords Wilberforce, Diplock, and Scarman at pp. 215, 224, and 227, respectively).

In *Rondel* v *Worsley*, three of their Lordships said, *obiter*, that a barrister would not be immune from an action in negligence in relation to matters unconnected with cases in court and the preliminary work connected therewith, although the point had not been fully argued by counsel ([1969] 1 AC 191 *per* Lords Reid, Morris, and Upjohn at pp. 231–2, 244, and 286, respectively).

In *Saif Ali* v *Sydney Mitchell & Co.* [1980] AC 198, HL, these dicta were followed by the majority and elevated to the status of *ratio decidendi*. Lord Wilberforce said (at pp. 212–13) that:

not all *obiter dicta* have the same weight, or lack of weight, in later cases. Of those then made in the House [in *Rondel* v *Worsley*] two things may be said. First, they were considered and deliberate observations after discussion of the same matters had taken place in the Court of Appeal and in the light of judgments in the Court of Appeal. It may be true that the counsel in the case did not present detailed arguments as to the position outside the court room— they had no interest in doing so—but I cannot agree that this invalidates or weakens judicial pronouncements. Judges are more than mere selectors between rival views—they are entitled to and do think for themselves. Secondly, it would have been impossible for their Lordships to have dealt with the extent of barristers' immunity for acts in court without relating this to their immunity for other acts. . . . These factors, in my opinion, tell in favour of giving considerably more weight to their Lordships' expressions of opinion than *obiter dicta* normally receive. We may clarify them, but we should hesitate before disregarding them.

Rondel v *Worsley* [1969] 1 AC 191, HL, was decided in 1967 and stood for three decades until the House of Lords decided in *Arthur J.S. Hall & Co.* v *Simons* [2000] 3 All ER 673, HL, that it should no longer be followed (see para 5.3.2.2.2 above). The immunity rule was thus abolished by the same body that had earlier confirmed it and continued its existence.

Some *obiter dicta* in House of Lords' cases are indistinguishable from *ratio decidendi*. In *Hedley Byrne & Co. Ltd* v *Heller & Partners Ltd* [1964] AC 465, HL, the House of Lords said unanimously that there was a legal duty of care in making statements whenever there was a special relationship between the parties and the duty had not been excluded by a disclaimer of responsibility. But, on the facts, it was held that there had been an effective disclaimer of responsibility so that, on a strict application of the distinction between *ratio decidendi* and *obiter dictum*, what their Lordships said about the duty of care was *obiter*.

Nevertheless, these dicta appear in the reserved opinions of five Law Lords delivered after listening to eight days of argument. It would be churlish to regard them as anything other than *ratio decidendi*, and they have been followed as such in subsequent cases.

The same considerations apply to the decision in *National Carriers Ltd* v *Panalpina (Northern) Ltd* [1981] 1 All ER 161, HL, concerning the applicability of the doctrine of frustration to leases of land. The House of Lords held unanimously that, on the facts, the lease had not been frustrated. Nevertheless, because the point had been fully argued, four out of five of their Lordships delivered reserved opinions holding that the doctrine of frustration is capable of applying to an executed lease of land.

5.4.3 The weight of persuasive precedents

The weight to be attached to any individual persuasive precedent will depend on several factors, such as the rank of the court in the hierarchy, the prestige of the judge(s) involved, the date of the case, whether judgment was reserved or given *ex tempore*, whether there was any dissenting opinion, whether the case was contested, and whether the point in question was argued or merely conceded by counsel.

The higher the court the greater the weight given to its pronouncements. For instance, the House of Lords would normally find decisions of the Court of Appeal more persuasive than decisions of High Court judges. The prestige of individual judges is a factor which can only rarely be taken into account because, in theory, all judges are of equal status. Lord Diplock has described as 'invidious' any distinction made between individual judges which is based on their reputations as jurists (*Saif Ali* v *Sydney Mitchell & Co.* [1980] AC 198 at p. 217).

Nevertheless, in practice the judgments of an acknowledged master of the law are likely to be treated with more respect than those of judges in general. From the nineteenth century the judgments of Lord Eldon, Lord Bowen, Lord Esher, Lord Lindley, and Sir George Jessell MR would fall into this category, and, from the last century, those of Lord Justice Scrutton, Lord Atkin, Lord Wright, Lord Reid, and Lord Radcliffe.

A good example of a highly influential dictum is the statement of Lord Atkin in *Donoghue* v *Stevenson* [1932] AC 562, HL (the case of the snail in the ginger-beer bottle), at p. 580, where he attempted to lay down a general test for determining when a notional duty of care arises in the tort of negligence. His dictum has become known as the 'neighbour test' and was expressed in these words:

You must take reasonable care to avoid acts or omissions which you can reasonably foresee would be likely to injure your neighbour. Who, then, in law is my neighbour? The answer seems to be—persons who are so closely and directly affected by my act that I ought reasonably to have them in contemplation as being so affected when I am directing my mind to the acts or omissions which are called in question.

This dictum, though clearly *obiter*, has been adopted in subsequent cases. In *Home Office* v *Dorset Yacht Co. Ltd* [1970] AC 1004, HL, Lord Reid said (at p. 1027) that it represented a statement of principle and should be applied unless there was some justification for its exclusion. Lord Pearson agreed (at p. 1054), but Lord Diplock was more cautious, describing Lord Atkin's dictum as a guide rather than a principle of universal application (at p. 1060). The dictum was applied in the case to make the Home Office liable for damage done to neighbouring property by absconding borstal boys.

The date of a case is a factor which can be used to support or to weaken it as a precedent. If the case is old, it can be argued that it has stood the test of time and now represents well settled law. On the contrary, it can be argued that it is now out of touch with changed conditions. If the case is recent, it can be hailed as the most up-to-date pronouncement on the matter. On the contrary, it can be attacked as being of insufficient antiquity.

A considered judgment delivered after being reserved will usually carry more weight than one delivered *ex tempore*, 'off the cuff', at the conclusion of counsel's argument. With a reserved judgment, the judge has had more time for reflection and to consider the authorities.

In the Court of Appeal and the High Court, judgment may be reserved or given *ex tempore* depending on the complexity or length of the case. A reserved judgment is signified in the law reports by the words *curia advisari vult* ('the court wishes to consider the matter'). The Latin words can often be seen abbreviated to *cur. adv. vult* or *C.A.V.*

Judgments in the House of Lords are always reserved, although sometimes their Lordships announce the actual decision at the end of counsel's argument without giving their reasons at that time. In *Royal College of Nursing of the United Kingdom* v *DHSS* [1981] 1 All ER 545, HL, a case concerning the legality of medically induced abortion carried out by nurses under the instructions of a doctor but in his absence, the House announced its decision immediately on the close of argument. Lord Wilberforce said the matter was urgent and it was in the interests of the National Health Service to announce the decision quickly, although he stressed that it was an unusual course and not to be taken as a precedent. Written opinions were delivered two months later.

The weight to be given to a persuasive precedent will be reduced if any dissenting opinion was given in the case and, in particular, by a judge whose views command the highest respect. Persuasiveness is also affected if the case was not contested, or if the point in issue was decided by the judge after being conceded by counsel. Such a case has not been subjected to the close scrutiny and refining process associated with skilled legal argument.

5.4.4 Other persuasive authorities

Where there is no direct authority in the form of decided cases, persuasive authority may be found in Roman law or in legal writings in textbooks and periodicals.

Roman law, and especially the Digest of the Emperor Justinian (promulgated in AD 533), has been resorted to on a number of occasions for a solution to English legal problems. The judgment of Blackburn J in *Taylor* v *Caldwell* (1863) 3 B & S 826, which began the modern doctrine of frustration of contract, was based largely on Roman doctrines. In *Kearry* v *Pattinson* [1939] 1 KB 471, CA, it was held, applying the Roman rule, that bees are animals *ferae naturae* ('wild by nature') so that, if they swarm, they belong to the first person who captures them. A Roman rule was applied again in *Tucker* v *Farm & General Investment Trust Ltd* [1966] 2 QB 421, CA, in which the Court of Appeal held that lambs born to ewes which are being acquired on hire-purchase belong to the hirer and not to the owner of the ewes.

Legal writings have not played a particularly influential part in the judicial process in England because of the attitude of the judges that English law should be found in statutes and decided cases rather than in the writings of academic lawyers. Denning J (as he was then) was probably exaggerating when in 1947 he wrote that the 'influence of the academic lawyers is greater now than it has ever been and is greater than they themselves realise' ((1947) 63 LQR 516). (For a more recent assessment of the influence of academic legal literature, see Birks, 'Adjudication and Interpretation in the Common Law', (1994) 14 Legal Studies 156.)

The English attitude contrasts with the position on the Continent of Europe where academic commentaries on the Codes carry great persuasive weight. In France, for example, the enacted law (*la loi*) is supreme. But it is often laconic and in need of clarification and interpretation by cases and criticism. Decisions of the courts (*la jurisprudence*) are only persuasive. Legal scholarship is held in high regard and the opinions of jurists (*la doctrine*) are as persuasive as decided cases. Judicial decisions are criticised by academic writers in textbooks and periodicals and in case notes appended to law reports. In the USA, too, legal writings are prestigious and academic commentaries are relied on by attorneys and judges alike.

It used to be the practice in England that only deceased authors could be cited in court—presumably because they could no longer change their minds and could not be effectively contradicted. This is no longer the practice as living authors may now be cited (Denning, loc. cit.).

The fact that the author happens to be a judge does not, in theory, give any added weight to his extra-judicial writings. In *Cordell* v *Second Clanfield Properties* [1969] 2 Ch 9, Megarry J, referring to the third edition of *Megarry and Wade's Real Property*, said this (at pp. 16–17):

It seems to me that words in a book written or subscribed to by an author who is or becomes a judge have the same value as words written by any other reputable author, neither more nor less. The process of authorship is entirely different from that of judicial decision. The author, no doubt, has the benefit of a broad and comprehensive survey of his chosen subject as a whole, together with a lengthy period of gestation, and intermittent opportunities for reconsideration. But he is exposed to the peril of yielding to preconceptions, and he lacks the advantage of that impact and sharpening of focus which the detailed facts of a particular case bring to the judge. Above all, he has to form his ideas without the aid of the purifying ordeal of skilled argument on the specific facts of a contested case. Argued law is tough law . . . Today, as of old, by good disputing shall the law be well known.

Formerly, some authors were cited in court with such frequency that their works were regarded as 'authoritative' in a special sense. These authors were listed by Blackstone, writing in the eighteenth century (1 Bl Comm 72–3), as Glanville (twelfth century), Bracton, Britton, Fleta, Hengham (thirteenth century), Littleton, Statham (fifteenth century), Brooke, Fitzherbert, Staundforde (sixteenth century), and Coke, whose *Reports* and four volumes of *Institutes* were published in the seventeenth century.

To this list may be added the names of Plowden (sixteenth century), Hale (seventeenth century), Hawkins, Foster (eighteenth century), and that of Blackstone himself. Sir William Blackstone (1723–80) became the first Vinerian Professor of English Law at Oxford in 1758. His *Commentaries* in four books were based on the lectures he gave at Oxford and were published between 1756 and 1769. He entered the House of Commons (1761), became Solicitor-General (1763), gave up his chair (1766), and was appointed a judge of the Court of King's Bench (1770) but in the same year transferred to the Court of Common Pleas.

Although these old works are still authoritative, it is doubtful whether they now carry any greater weight than the works of more modern authors. The attitude of the Court of Appeal in *R* v *Richards* [1974] QB 776, CA, where Hawkins's *Pleas of the Crown* (8th ed., 1824) was preferred to Smith and Hogan's *Criminal Law* (3rd ed., 1973) even though Hawkins did not make sense, appears to have been an isolated incident. What appears to be more important today than the antiquity of the old books is the wisdom and correctness of what they say. They are certainly still cited occasionally, although not always with approval. Thus, Blackstone was criticised in *Button* v *Director of Public Prosecutions* [1966] AC 591, HL, on the definition of the crime of affray and in *Reid* v *Commissioner of Police of the Metropolis* [1973] QB 551, CA, on the market overt rule in the sale of goods. In *Attorney-General of the Duchy of Lancaster* v *G.E. Overton (Farms) Ltd* [1982] 1 All ER 524, 526–9, CA, Lord Denning MR quoted the definitions of treasure trove given by Bracton, Staundforde, Coke, Comyn, and Blackstone, and preferred Coke's definition to that of all the others.

In modern times the works of other authors have been cited frequently in court, both by counsel and by judges in judgments. Among the most influential of these works are Pollock's books on *Tort, Contract*, and *Partnership*; Dicey's *Law of the Constitution* and *Conflict of Laws*; Kenny's *Outlines of Criminal Law*; the books of Salmond and Winfield on *Tort*; Cheshire on *Private International Law, Modern Real Property*, and (with Fifoot) the *Law of Contract*; Rayden on *Divorce*; Cross on *Evidence*; Wade's *Administrative Law*; and the book (first published in 1965) by Smith and Hogan on *Criminal Law*.

In addition to academic and practitioner textbooks, citation is allowed from legal periodicals which publish, *inter alia*, learned articles suggesting what the law should be on contentious matters not covered by any statute or decided case. Among the periodicals most commonly relied on are the *Law Quarterly Review* (LQR, first published in 1885), the *Cambridge Law Journal* (CLJ, 1921), the *Modern Law Review* (MLR, 1937), and the *Criminal Law Review* (Crim LR, 1954).

5.5 Discovering the *ratio decidendi* of a case

The *ratio decidendi* of a case is the principle of law which runs through the case and on which the decision is based. The *ratio* is not the decision itself. Only the litigating parties are bound by the actual decision in a case whereas the *ratio* of a case states the law for all persons and may be binding in later cases. The actual decision in *Ashton* v *Turner* [1980] 3 All ER 870, for example, is simply that the defendant was not liable to the claimant. The *rationes* on which that decision is based, three doctrines well known to the common law, are much more detailed in their articulation and far-reaching in their implications than the actual decision (see below).

Delivering the judgment of the Court of Appeal in *R (Kadhim)* v *Brent London Borough Council Housing Benefit Review Board* [2001] 2 WLR 1674, CA, Buxton LJ put the matter like this (at [16]):

Cases as such do not bind; their *rationes decidendi* do. While there has been much academic discussion of the proper way of determining the *ratio* of a case, we find the clearest and most persuasive guidance, at least in a case such as the present where one is dealing with a single judgment, to be that of Professor Cross in Cross and Harris, *Precedent in English Law*, 4th ed., 1991, p. 72: 'The *ratio decidendi* of a case is any rule of law expressly or impliedly treated by the judge as a necessary step in reaching his conclusion, having regard to the line of reasoning adopted by him'.

The traditional view was that the *ratio decidendi* of a case was what it was perceived to be by the judge who decided the case. But the difficulty with this approach is that the judge may have expressed the principle too widely or too narrowly for it to be useful in a later case. Moreover, in appellate courts each judge may state the principle in different language and so it is not always easy to discover the *ratio* by a quotation from any one judgment.

The modern, generally accepted view is that the *ratio decidendi* of a case is what it is determined to be by a court in a later case and not what the judge in the original case considered it to be. This objective approach towards finding the *ratio* of a case makes it possible for a judge in a later case to relegate to the status of *obiter dicta* statements which had hitherto been thought to be *ratio*. It also means that, since the facts of two cases are unlikely to be identical, the judge in the later case usually has the task of either restricting or enlarging the *ratio* of the earlier case. If he decides that the *ratio* does not apply to the facts before him he is restricting its scope. If he decides that the *ratio* does apply to the different factual situation he is enlarging its scope.

Discovering the *ratio decidendi* of a case is often difficult because it may involve the separation of the relevant and irrelevant parts of a judgment. The *ratio* will hardly ever be stated explicitly in the judgment but will be found buried among a mass of dicta. In some respects the old cases are easier to read because the reporter did not always trouble to report what he considered to be mere *obiter dicta*. The *ratio* may not be accurately encapsulated in the headnote to the law report. The reporter may have misinterpreted the decision and attempted to state the *ratio* too widely or too narrowly.

The decision in a case and, therefore, the *ratio decidendi* must always depend on the particular facts of the individual case. To discover the *ratio* of a case, all the facts found by the judge to be material must be considered. Whatever words are not necessary for the decision must be *obiter*. However, a judge may give two or more reasons for his decision in which event they are both or all *rationes decidendi* and not mere *obiter dicta*. As Lord Simonds said in *Jacobs* v *London County Council* [1950] AC 361, HL, at p. 369, the other Law Lords agreeing with him:

[T]here is in my opinion no justification for regarding as *obiter dictum* a reason given by a judge for his decision, because he has given another reason also.

Thus, in *Ashton* v *Turner* [1980] 3 All ER 870, where the claimant had been injured due to the negligent driving of a known drunken driver with whom he was engaged in criminal activity, Ewbank J found against the claimant on three separate grounds. They were *ex turpi causa non oritur actio* ('no action can be brought on an illegal or immoral cause'), *volenti non fit injuria* ('no injury is done to one who consents'), and contributory negligence to the extent of 50 per cent. Either of the first two grounds by itself would have been sufficient to dispose of the claimant's claim without compensation, while the third ground would have reduced his damages by one-half.

Lord Denning MR favoured a different approach to multiple *rationes* in the case of the Court of Appeal. His view was that where a Court of Appeal decision involves two or more *rationes*, that Court in a later case is not bound by all of them but may choose to ignore any particular *ratio* (*Dixon* v *British Broadcasting Corporation* [1979] 2 All ER 112, CA, at p. 116; *Ministry of Defence* v *Jeremiah* [1980] 1 QB 87, CA, at p. 98). This unorthodox view, which was not shared by Lord Denning's colleagues in the Court of Appeal, conflicts with the opinion of Lord Simonds in *Jacobs* v *London County Council*, quoted above, and with other dicta in the House of Lords.

All facts which the judge expressly or impliedly treats as immaterial must be regarded as immaterial. All facts which the judge expressly or impliedly treats as material must be regarded as material. There is a presumption against wide principles of law, and it must be remembered that the *ratio* widens or narrows according to how few or how many facts are considered to be material. If the judge decides that facts A, B, C, D, and E exist but that only facts A, B, and C are material, the *ratio* is wider than if he had decided that all the facts were material. For example, the judge decides that facts A, B, C, D, and E exist. He then excludes facts D and E as immaterial. On the remaining facts, A, B, and C, he reaches conclusion X. In any future case where the facts are A, B, C, D, and E, conclusion X must be reached. But, in reality, the *ratio* is that where facts A, B, and C exist the conclusion must be X, facts D and E having been discounted as immaterial.

The process of discovering the *ratio decidendi* of a case may be illustrated by reference to two well known decisions of the House of Lords: *Rylands v Fletcher* (1868) LR 3 HL 330, HL, and *Donoghue v Stevenson* [1932] AC 562, HL.

The facts of *Rylands v Fletcher* were that:

(a) the defendant (D) had a reservoir built on his land;

(b) the contractor was negligent in building the reservoir; and

(c) water escaped and flooded the mine of the claimant (C).

The conclusion on these facts was that D was liable to C for the damage. But the facts considered to be *material* by the court were only (a) D had a reservoir built on his land, and (c) water escaped and flooded C's mine. Fact (b) was considered immaterial, and, by discounting it, the House of Lords formulated a wide *ratio* based on strict liability for the escape of water to which lack of negligence or the employment of a contractor is no defence.

Nor did the House stop there. It was made clear that the *ratio* was not limited to the building of reservoirs and the escape of water, although, in strict theory, a later court could have so restricted it but only by a rigid and mechanical application of the doctrine of precedent. The following statement of the rule was laid down by Blackburn J ((1866) LR 1 Ex 265 at pp. 279–80) and was approved by the House of Lords:

[T]he person who for his own purposes brings on his land and collects and keeps there anything likely to do mischief if it escapes, must keep it in at his peril, and, if he does not do so, is prima facie answerable for all the damage which is the natural consequence of its escape.

In subsequent cases, the rule in *Rylands v Fletcher* has been extended to escapes of fire, gas, oil, electricity, explosions, noxious fumes, colliery spoil, poisonous vegetation, and a 'chair-o-plane' at a funfair. (For a modern case on the rule in *Rylands v Fletcher* in which, on the facts, the defendants were held not liable, see *Cambridge Water Co. Ltd v Eastern Counties Leather plc* [1994] 1 All ER 53, HL.)

In *Donoghue v Stevenson* [1932] AC 562, HL, a Scottish case which went to the House of Lords on a preliminary point of law, the assumed facts were as follows. The pursuer (the Scots law equivalent of claimant) and a friend visited a café where the friend bought for the pursuer a bottle of ginger beer manufactured by the defender (defendant). The bottle was of opaque glass through which it was impossible to see the contents. The café owner opened the bottle and poured some of the ginger beer into a tumbler. The pursuer drank some of the ginger beer, and the friend poured the remainder of the ginger beer into the tumbler from the bottle. With it emerged the decomposed remains of a dead snail. The pursuer suffered shock and gastric illness. Assuming these facts to be proved, the House of Lords held that the pursuer would be entitled to recover damages from the defender, even though there was no contract between them.

For the purposes of the decision, the House of Lords considered as *material* only that the pursuer had been injured through consuming ginger beer manufactured by the defender and bottled in glass through which the contents could not be seen and which contained a dead snail. It was *not material* that the ginger beer had been bought by the friend in a café or that it had been poured into the tumbler by the friend and the café owner. The House regarded as particularly *immaterial* the fact that there was no contractual relationship between the pursuer and the defender, and in this way a wider *ratio* was laid down.

It was made clear that the *ratio* was not to be limited to cases involving snails in ginger-beer bottles. The rule was laid down thus in the words of Lord Atkin ([1932] AC 562 at p. 599):

[A] manufacturer of products, which he sells in such a form as to show that he intends them to reach the ultimate consumer in the form in which they left him with no reasonable possibility of intermediate examination, and with the knowledge that the absence of reasonable care in the preparation or putting up of the products will result in an injury to the consumer's life or property, owes a duty to the consumer to take that reasonable care.

In subsequent cases, the *ratio decidendi* of *Donoghue* v *Stevenson* has been extended to motor cars, lifts, hair dye, industrial chemicals, and irritant chemicals in underpants. The category of persons potentially liable has been extended to include repairers, erectors, and assemblers.

In appeal cases in the Court of Appeal and House of Lords, three or even five separate judgments may be given. It is possible for all the judges in an appeal to find for the same party but for different reasons. In this event, the *ratio decidendi* is whatever is agreed on by the majority. If there is no majority in favour of any one *ratio* the case loses much of its value as a precedent, and may not be considered binding, even if it is a decision of the House of Lords. A lower court faced later with the same issue can only discuss the different *rationes* of the higher court and then decide the matter for itself *de novo*.

The House of Lords' decision in *Bell* v *Lever Bros Ltd* [1932] AC 161, HL, is a notorious illustration of the problems involved in seeking to discover what a particular case actually decided. In that case, after wide divergences of opinion in the High Court and the Court of Appeal, the appellant ultimately secured judgment by a majority of three to two in the House of Lords. But only two (Lords Atkin and Thankerton) of the majority of three decided on the same *ratio*, namely, that there was, on the facts, no fundamental mistake to avoid the contract. The opinion of the third member of the majority (Lord Blanesburgh) was based on a different, procedural, ground.

In recent years there has been a tendency in appeals before the House of Lords for only one speech to be delivered in which the other Law Lords simply concur. The object is to avoid the confusion which can result from different reasons contained in multiple judgments, as in *Bell* v *Lever Bros Ltd*. Nevertheless, the single-judgment approach itself may cause difficulties.

In *Saunders* v *Anglia Building Society* [1971] AC 1004, HL, Lord Reid (at p. 1015) warned that 'there are dangers in there being only one speech in this House' because 'statements in it have often tended to be treated as definitions and it is not the function of a court or of this House to frame definitions'. Lord Denning MR spoke of the difficulties in *Paal Wilson & Co.* v *Blumenthal* [1982] 3 All ER 394, CA. He described the task of distinguishing between *ratio decidendi* and *obiter dicta* as 'formidable' and said that, occasionally, it is more difficult to distinguish them in a single speech than in multiple opinions. Lower courts are often tempted to treat the words of a single speech almost as if they were contained in a statute and, in so doing without attempting to distil the *ratio decidendi*, they are likely to be led astray (at p. 400).

In an appeal case, if a judge says 'I agree' that means he agrees with the decision and proposed order but not necessarily that he agrees with the reasoning of his judicial colleagues. If a judge is silent about a statement of one of his colleagues, his silence does not imply agreement with it. Such short pronouncements, especially in unreasoned judgments, do not carry much weight.

5.6 Advantages and disadvantages of the doctrine of judicial precedent

The following of precedent is a *convenient time-saving device*. If a problem has already arisen and been solved in a certain way, it is natural in every walk of life to reach the same conclusion on the same problem without too much reconsideration. All legal systems follow precedent to a greater or lesser extent. The English legal system differs from most of the others in that we have a doctrine of *binding* precedent under which, more often than not, the previous case *must* be followed in the subsequent case.

But it must always be remembered that decided cases are illustrations of principles of law which a judge may turn to in deciding the case before him. Precedents should be used as 'stepping-stones' rather than as 'halting-places' (*Birch* v *Brown* [1931] AC 605, HL, *per* Lord Macmillan at p. 631). The convenience of following precedent should not be allowed to degenerate into a mere mechanical exercise performed without any thought.

So, too, the citation of authority in court should be kept within reasonable bounds because it can be costly in terms of time and money. Lord Diplock has warned of the 'danger of so blinding the court with case law that it has difficulty in seeing the wood of legal principle for the trees of paraphrase' (*Lambert* v *Lewis* [1981] 1 All ER 1185, HL, at pp. 1189–90; see also *Pioneer Shipping Ltd* v *B.T.P. Tioxide Ltd* [1981] 2 All ER 1030, HL, *per* Lord Roskill at p. 1046, the other Law Lords agreeing with him on the point).

The House of Lords has decided that it will not allow transcripts of *unreported* judgments of the Court of Appeal, civil division, to be cited before the House except with its leave. Permission will only be granted if counsel gives an assurance that the transcript contains a statement of some principle of law which is not to be found in any *reported* Court of Appeal decision (*Roberts Petroleum Ltd* v *Bernard Kenny Ltd* [1983] 1 All ER 564, HL). The hard line thus taken by the House of Lords was in response to the citation of four unreported Court of Appeal decisions, none of which was of assistance to their Lordships in determining the appeal before them. None of these unreported decisions laid down any principle of law which could not be found in reported cases, and the only result of referring to them had been to lengthen the hearing of the appeal unnecessarily. (See also *Stanley* v *International Harvester Co. of Great Britain Ltd* (1983) *The Times*, 7 February, CA, *per* Sir John Donaldson MR, and note that the Court of Appeal operates a similar practice: *Practice Direction (Court of Appeal: procedure)* [1999] 2 All ER 490.) The problem of over-citation of authority is raised again in para 5.7.6 below.

Greater *certainty* in the law is perhaps the most important advantage claimed for the doctrine of judicial precedent. From this advantage other benefits flow. If case X was decided in a particular way 20 years ago by the Court of Appeal or the House of Lords, the doctrine of *stare decisis* would normally demand that a court today faced with the same problem as in case X should decide it the same way. The existence of a precedent may prevent a judge making a mistake which he might have made if he had been left on his own without any guidance. It may also allow persons generally to order their affairs and come to settlements with a certain amount of confidence. But the advantage of certainty is lost where there are too many cases or they are too confusing. This can arise through the process of distinguishing cases and over-refining the principles embodied in them. Certainty in the law can only result from a large body of case law if the cases are uniform in outcome and not irreconcilable.

The doctrine of precedent may serve the *interests of justice*. If the decision in a case between A and B was X, it would be unjust to reach decision Y the next day in a similar case between C and D. This is one of the objections to the prospective overruling of cases, discussed earlier (para 5.3.1 above), but retrospective overruling is also affected by it. The overruling of an earlier case may cause injustice to those who have ordered their affairs in reliance on it. It is partly to prevent this sort of injustice that the Court of Appeal must normally adhere to its own previous decisions and the House of Lords should be circumspect in departing from its previous decisions. The problem is that the application of precedent may produce justice in the individual case but injustice in the generality of cases. It would be undesirable to treat a number of claimants unjustly simply because one binding case had laid down an unjust rule.

The interests of justice also demand impartiality from the judge. This may be assured by the existence of a binding precedent which he must follow unless it is distinguishable. If he tries to distinguish an indistinguishable case his attempt will be obvious.

Case law is *practical* in character. It is based on the experience of actual cases brought before the courts rather than on logic or theory. In this respect case law differs from statute law, which is often based on *a priori* theories. The doctrine of *stare decisis* is a limiting factor in the development of judge-made law. Practical law is founded on experience, but the scope for further experience is restricted if the first case is binding.

The making of law in decided cases offers opportunities for *growth and legal development* which could not be provided by Parliament. The courts can more quickly lay down new principles, or extend old principles, to meet novel circumstances. There has built up over the centuries a mass of cases illustrative of a vast number of the principles of English law. The cases exemplify the law in the sort of detail that could not be achieved in a long code of the Continental type. But therein lies another weakness of case law. Its very *bulk and complexity* make it increasingly difficult to find the law.

The case-law method is sometimes said to be *flexible*, but this presupposes that a judge is free to lay down whatever rule he considers desirable in order to keep the law in step with changing social and economic conditions. However, a judge is not thus free where there is a binding precedent. Unless it can be distinguished he must follow it, even though he dislikes it or considers it bad law. His discretion is thereby limited and the alleged flexibility of case law becomes *rigidity*. Judicial mistakes of the past are perpetuated unless bad decisions happen to come before the House of Lords for reconsideration. In any event, flexibility and certainty are incompatible features of judge-made law. A system that was truly flexible could not at the same time be certain because no one can predict when and how legal development will take place.

It has been seen that for almost every advantage claimed for the doctrine of precedent there is a corresponding disadvantage. This is not necessarily to say, however, that the advantages are outweighed by the disadvantages.

The following of precedent is easier in England than in many other countries because England has a centralised legal system with only a small number of courts. In the USA, where there are federal and state courts and legislatures, the number of reported cases runs into millions and this has led to a modification there of the doctrine of *stare decisis*. Decisions of the US Supreme Court are binding on courts other than itself, but the Supreme Court tends to decide only constitutional issues. In all other cases, US Federal and District Courts are not bound by precedent. Previous decisions are of persuasive authority only and may be rejected. Particularly persuasive is an established *line* or *trend* of decisions rather than a single decision.

The English doctrine of *stare decisis* is based on the binding effect of a single decision, which must be followed even when manifestly unjust or wrong. Such an approach is alien to Continental systems of law wherein no single decision, even of the highest court, is absolutely binding (Goodhart, 'Precedent in English and Continental law', (1934) 50 LQR 40). Continental law tends to be codified and the code is all-important so that cases decided on the interpretation of the code are of persuasive authority only. In practice, however, a long line or trend of decisions (what the French call *la jurisprudence constante*) is highly persuasive and will rarely be departed from.

Another reason for the relative unimportance of the single decision on the Continent is the great weight attached to *la doctrine*. The writings of jurists played a major part in the development of Roman law, which, in turn, influenced modern Continental systems of law. In England, because judges commanded greater respect than jurists, lawyers looked to judicial decisions rather than to textbooks for authority.

5.7 Law reporting

5.7.1 Introduction

The effectiveness of a doctrine of precedent based on *stare decisis* depends in large measure on the availability of full and accurate reports of decided cases. There has never been in England any official or systematic attempt at compiling law reports. The law reporting that exists today has simply evolved by private enterprise through three periods of development.

5.7.2 Year Books (about 1275 to 1535)

The Year Books are anonymous reports, compiled annually, and written by hand in law French. Some were later printed but most remained in manuscript. The Year Books are rarely cited in court now as they are of no practical use in modern times. They are, however, useful in the study of the medieval common law.

5.7.3 Private (or named) reports (1535 to 1865)

The private reports were compiled by individuals for commercial publication. Most of the private reports are referred to by the name of the reporter. They are cited by recognised abbreviations. Thus the reports compiled by Sir Edward Coke between 1572 and 1616 are known as Coke's Reports, abbreviated to Co Rep.

The private reports show considerable variation in style and accuracy. The famous outburst of Holt CJ in *Slater* v *May* (1704) 2 Ld Raym 1071 exemplifies the judicial frustration with bad private reporting: 'See the inconveniences of these scrambling reports, they will make us appear to posterity for a parcel of blockheads'.

The reports of Plowden, Coke, and Burrow are regarded as among the most reliable, while the most vilified are probably those of Espinasse, who reported *nisi prius* cases between 1793 and 1807. Espinasse's reports have been criticised as 'imperfect and misleading', and one case reported by him in 1799 was disapproved on the ground that the report was probably inaccurate (*Wessex Dairies Ltd* v *Smith* [1935] 2 KB 80, CA, disapproving *Nichol* v *Martyn* (1799) 2 Esp 732).

The better private reports give the reasons for decisions, unlike most of the Year Books. Many of the private reports have been republished. Perhaps the best collection is the *English Reports* (ER, published between 1900 and 1932) containing in 176 volumes reprints of cases reported between 1220 and 1866. Volumes 177 and 178 are index volumes and there is an index wall-chart indicating where each volume of the old reports can be found reprinted in the *English Reports*. They are annotated with references to subsequent cases and this makes them more useful than the original reports.

5.7.4 Modern reports (1865 to the present)

The private reports were often criticised. They were expensive to buy. Some of them were never printed but had to be cited in manuscript form. There was too much overlapping in that the same case might be reported in two or more series. Their usefulness to the legal profession was reduced by the inordinate length of time taken to report some important decisions. They were, for the most part, unreliable.

As a result of dissatisfaction with the private reports, a council was established comprising representatives of the four Inns of Court and the Law Society, with the Attorney-General and Solicitor-General as *ex officio* members. The council's reports, called *The Law Reports*, were first published in 1865 and eventually absorbed the private reports. In 1870, the council was incorporated as a company limited by guarantee and became known as the Incorporated Council of Law Reporting for England and Wales.

The Incorporated Council employs paid law reporters and sells its reports to subscribers, but it is a non-profit-making, charitable association whose purposes are regarded as beneficial to the community and for the advancement of education (*Incorporated Council of Law Reporting for England & Wales* v *Attorney-General* [1972] Ch 73, CA).

The Law Reports are not official or monopolistic. But they are usually more accurate and reliable than other series. Counsel's argument is summarised and the judgment is revised by the judge before publication.

If a case is reported in *The Law Reports* it should, as a matter of practice rather than law, be cited in court from that series rather than from any other series in which it may be reported. However, counsel appearing in the *House of Lords* may, if they consider it more convenient to do so, cite a case from the Inland Revenue's series, *Reports of Tax Cases*, even though the case is also reported in *The Law Reports* as long as the reference to the case in *The Law Reports* is also given (*Bray (Inspector of Taxes)* v *Best* [1989] 1 All ER 969, HL, *per* Lord Mackay LC at p. 971).

In the *Court of Appeal* and the *High Court* a similar practice exists in all cases. The general rule applies that, where there is a choice, citation from *The Law Reports* is preferred to citation from other series of reports because other series, although they provide a useful service (in, for example, reporting cases not reported elsewhere), do not contain a summary of counsel's argument and may not be readily available to the court (see *Practice Direction (Court of Appeal: procedure)* [1999] 2 All ER 490; *Practice Direction (Judgments: Form and Citation)* [2001] 1 WLR 194).

The Law Reports is a general series which reports decisions of the superior and appellate courts in England and Wales. The current series of *The Law Reports* began in 1891. The reports are issued in four parts per year:

AC   Appeal Cases in the House of Lords and the Judicial Committee of the Privy Council.

QB   Cases decided in the Queen's Bench Division of the High Court and on appeal therefrom to the Court of Appeal.

Ch   Cases decided in the Chancery Division and on appeal therefrom to the Court of Appeal.

Fam   Cases decided in the Family Division and on appeal therefrom to the Court of Appeal.

Since 1891, the year of publication of a volume of *The Law Reports* has appeared in square brackets and is part of the reference to that volume without which a case cannot be traced. Thus, the reference to the House of Lords' case of *Smith* v *Baker & Sons* is [1891] AC 325.

The date given in the reference for a case is the year in which it is reported and not the year in which it is decided. Thus, the reference to the House of Lords' case of *Bell* v *Lever Bros Ltd*, which was reported in *The Law Reports* in 1932, is [1932] AC 161, although the case was actually decided at the end of 1931.

Since 1953, the Incorporated Council of Law Reporting for England and Wales has also published *The Weekly Law Reports* (WLR). This is another general series. The reports are published more quickly than *The Law Reports*, although it has not been possible to keep to the original aim of making reports available within about three weeks of judgment. The only decisions likely to be reported within three to four weeks are those of the House of Lords. In *The Weekly Law Reports* counsel's argument is left out and the judgments are not necessarily revised for publication. *The Weekly Law Reports* consist of three volumes per year. Volumes 2 and 3 contain reports of cases which will later appear in *The Law Reports*. Volume 1 contains reports of less important cases which are not intended for publication in *The Law Reports* and cases which are likely to go to appeal.

Another popular general series of reports is the *All England Law Reports* (All ER) published since 1936 by Butterworths. There are four volumes per year, containing in excess of 500 cases. Counsel's argument is not reported. Judgments are revised by the judges before publication. Cases are usually reported within four or five months of judgment, although decisions of the House of Lords are often reported within three to four weeks. The *All England Law Reports Reprint* (All ER Rep) is a reprint of some 4,000 to 5,000 important cases from 1558 to 1935, contained in 36 volumes and an index volume.

Other general series of reports include those published in daily newspapers, such as *The Times* and *The Daily Telegraph*. In common with the other law reports mentioned, these reports are authenticated by the signature of a reporter who was present throughout while judgment was delivered. They may be cited in court in the absence of any other approved report of a case, although they occasionally mislead by reason of their brevity (see, e.g., *Summers* v *Summers* (1987) *The Times*, 19 May, CA, referring to a misleading *Times* report).

The rule that only reports of cases made by a barrister could be cited in court was abolished by s. 115 of the Courts and Legal Services Act 1990. Now reports of cases can be cited if made by a barrister, a solicitor, or by a person who has a Supreme Court qualification within the meaning of s. 71 of the 1990 Act.

There are many specialist series of law reports which concentrate on decisions of specific courts or on specific areas of law. Some of these specialist series are official. Among these are *Reports of Tax Cases* (TC or Tax Cas) published by the Inland Revenue, and *Reports of Patent, Design and Trade Mark Cases* (RPC) published by the Patent Office. The specialist series published privately include *Lloyds Law Reports* (Lloyd's Rep), *Knight's Local Government Reports* (LGR), *Knight's Industrial Reports* (KIR), *Industrial Cases Reports* (ICR, published by the Incorporated Council of Law Reporting), the *Common Market Law Reports* (CMLR), the *Criminal Appeal Reports* (Cr App R), the *Criminal Appeal Reports (Sentencing)* (Cr App R (S)), the *Family Law Reports* (FLR), and, beginning in 1988, the *Crown Office Digest* (COD, whose reports of cases in the Administrative Court include applications for judicial review). Legal periodicals such as the *Criminal Law Review* (Crim LR) and *Family Law* (Fam Law) report cases on all aspects of criminal law and family law, respectively.

Without discriminating between them, Lord Diplock has said that the various series of specialist reports are of limited value because they contain only a small minority of leading judgments in which some new principle of law is authoritatively propounded, as distinct from the application of some previously accepted principle to the facts of a particular case. In the main, their usefulness is restricted to helping lawyers to predict the likely outcome of future cases by showing how cases with analogous facts had actually been decided in the past (*Roberts Petroleum Ltd* v *Bernard Kenny Ltd* [1983] 1 All ER 564, HL, at p. 567).

Although, in general, the modern reports are far superior in quality and reliability than their predecessors, dissatisfaction with them is sometimes expressed. The main criticisms are that too much is reported; too little is reported; there is too much overlapping in what is reported in the various series; and that the reports are not made available quickly enough.

It is the duty of counsel to bring to the attention of the court all relevant authority, even if it is adverse to his case. The failure of counsel to cite to the trial judge a recent relevant decision of the Court of Appeal, reported some months earlier in the weekly parts of the *All England Reports* and *The Weekly Law Reports*, was criticised as a matter of 'very great concern' by Buxton LJ in *Copeland* v *Smith* [2000] 1 All ER 457, CA, at p. 459. Counsel were reminded by Brooke LJ (at p. 462) of their responsibility to bring and keep themselves up to date with new case law, not only in the specialist series of law reports but also in the general series, such as the *All England Reports* and *The Weekly Law Reports*. (Buxton LJ's reference (at p. 459) to *The Weekly Law Reports* as 'official' must be a slip of the tongue.)

The state of law reporting has not been officially investigated since 1940, when the *ad hoc* Law Reporting Committee, under the chairmanship of a High Court judge, Simonds J (as he then was), presented a report to the Lord Chancellor (*Report of the Committee on Law Reporting*, HMSO, Lord Chancellor's Office, 1940).

The report contained a concise account of the history of law reporting and discussed the main criticisms which have been made of the present system. Accuracy was identified as the most important attribute of a law report. Speed of publication is also a consideration, and, while it is not as essential as accuracy, the committee said that something should be done to hasten publication of *The Law Reports*.

The committee rejected the suggestion that a monopoly of law reporting should be granted to the Incorporated Council or to the publisher of any other series of reports. Such a monopoly would be difficult to maintain in practice. English law is what it is, not because it has been so *reported*, but because it has been so *decided*. To insist that only *The Law Reports* could be cited in court would be to deny the authority of cases not reported in that series.

The majority of the committee rejected the suggestion that there should be a written record, to be revised by the judge, of every single judgment. This idea was dismissed on grounds of cost and the additional burden it would impose on the judiciary.

Professor A.L. Goodhart (then Professor of Jurisprudence at Oxford and editor of the *Law Quarterly Review*) presented a strong dissent on this point. He was particularly concerned that many reports were not revised by the judges, thus leading to published inaccuracies, and that many decisions were not reported at all but were nevertheless authoritative. He recommended that official shorthand writers should take down and transcribe all judgments in the superior courts. The transcripts should then be sent to the judges for revision, and returned by them, preferably within one week, for filing in the records of the court.

Professor Goodhart's proposal was never implemented in full, although an improvement was made in relation to the availability of Court of Appeal decisions several years later. In *Gibson* v *South American Stores (Gath & Chaves) Ltd* [1950] Ch 177, CA, Sir Raymond Evershed MR had remarked (at p. 195) on the

peculiar and unfortunate characteristic of our system that, although in the great majority of cases which come before it, this court is the final court of appeal for England, no provision whatever is made for taking a note or making a record of the judgments of the court.

This omission was rectified in 1951, when the practice began of placing transcripts of all judgments of the Court of Appeal, civil division, in the Bar Library or, since 1978, in the Supreme Court Library in the Royal Courts of Justice.

In addition, all *unreported* judgments of the Court of Appeal, civil division, delivered since the beginning of 1980 are stored in *LexisNexis*, a computerised database operated by the LexisNexis Group, a division of Reed Elsevier (UK) Ltd.

The development of information communication technology has led to a considerable expansion of *LexisNexis* and the introduction of another computerised database, *Westlaw UK*, from the legal publishers, Sweet & Maxwell Ltd. *Justis*, operated by Context Ltd, offers the full archives of *The Law Reports, The Weekly Law Reports, The Industrial Cases Reports*, and *The WLR Daily* (before 2005 known as *The Daily Law Notes*) via the internet and on CD-ROM.

LexisNexis, Westlaw UK, and *Justis* are accessible by subscription. In addition, there is a plethora of websites giving *free* access to case law.

With effect from 14 November 1996, all opinions given in House of Lords' cases have appeared in full-text version on the House's website, usually on the same day as delivery.

Other freely accessible websites include the Judicial Committee of the Privy Council, the Court of Justice of the European Communities, the European Court of Human Rights, the Employment Appeal Tribunal, the Court Service (High Court and both divisions of the Court of Appeal), *The WLR Daily* (High Court, Court of Appeal, House of Lords, Privy Council, Court of Justice of the European Communities), and BAILLI (British and Irish Legal Information Institute).

5.7.5 Neutral citation of judgments

As part of a process of modernisation, and with a view to making it easier to store, distribute, search, and retrieve court judgments on the World Wide Web, a system of neutral citation of judgments in the Court of Appeal and the Administrative Court was established for England and Wales ('EW') in January 2001 by a *Practice Direction*. The system was extended to the rest of the High Court in January 2002 by another *Practice Direction*. These two *Practice Directions* are now incorporated in *Practice Direction (Criminal Proceedings: Consolidation), Part I* [2002] 1 WLR 2870.

Although the *Practice Direction* does not apply to courts other than the Court of Appeal and the High Court, the new practice was adopted by the House of Lords and the Judicial Committee of the Privy Council in January 2001, and county court judgments now also contain paragraph numbering.

Under the system of neutral citation, each judgment is prepared for delivery, or issued as an approved judgment, by the judge with paragraph numbering in the margin but no page numbering. In courts with more than one judge, the paragraph numbering continues sequentially through each judgment. Each case is assigned a unique number by the official court shorthand writer.

Examples of the numbering of judgments are:

House of Lords	[2004] UKHL 1, 2, etc
Court of Appeal (Civil Division)	[2004] EWCA Civ 1, 2
Court of Appeal (Criminal Division)	[2004] EWCA Crim 1, 2
High Court (Chancery Division)	[2004] EWHC 1, 2 (Ch)
High Court (Queen's Bench Division)	[2004] EWHC 1, 2 (QB)
High Court (Family Division)	[2004] EWHC 1, 2 (Fam)
High Court (Administrative Court)	[2004] EWHC 1, 2 (Admin)
Judicial Committee of the Privy Council	[2004] UKPC 1, 2

The neutral citation for, say, *R (Daly)* v *Secretary of State for the Home Department*, the 26th case decided by the House of Lords in the year 2001, is [2001] UKHL 26. Paragraph numbering ensures that a passage in a particular judgment can be found in the same place in every report of the case regardless of page numbering. Under the new citation practice, paragraph 30 in a case is cited as: *R (Daly)* v *Secretary of State for the Home Department* [2001] UKHL 26 at [30], or [2001] 2 AC 532 at [30].

If a judgment has been reported in one or more series of law reports, its neutral citation in a list of cases appears in front of the usual citation used by that particular series of reports. Thus, *R (Daly)* v *Secretary of State for the Home Department* might be listed as: [2001] UKHL 26; [2001] 2 AC 532; [2001] 2 WLR 1622; [2001] 3 All ER 433, and the Court of Appeal decision in *Taylor* v *Lawrence* might be listed as: [2002] EWCA Civ 90; [2003] QB 528; [2002] 3 WLR 640; [2002] 2 All ER 353.

Practice Direction (Judgments: Form and Citation) [2001] 1 WLR 194 re-emphasises for the avoidance of doubt that in the High Court and the Court of Appeal a case must be cited from *The Law Reports* published by the Incorporated Council if it has been reported there. It can only be cited from other series of reports if it has not been reported in *The Law Reports*.

5.7.6 Over-citation of judgments

The problem of over-citation of authority, together with some of the steps taken to curb it, was noted in para 5.6 above. The practice is wasteful in terms of time and money, and, in the main, is the result of the unwillingness, or inability, of undiscerning counsel and other advocates to keep 'the kitchen sink' out of their argument. The problem exists largely in relation to the citation of *unreported* decisions culled from electronic sources and court transcripts. The recent proliferation of computerised databases has exacerbated the situation, as has the growth in the number of cases being decided.

The practice of citing unreported Court of Appeal decisions before the House of Lords and the Court of Appeal was criticised by the House of Lords many years ago. Indeed, the House went so far as to prohibit their citation before itself except with leave, which will only be granted in limited circumstances (*Roberts Petroleum Ltd* v *Bernard Kenny Ltd* [1983] 1 All ER 564, HL, para 5.6 above; see also *Stanley* v *International Harvester Co. of Great Britain Ltd* (1983) *The Times*, 7 February, CA, *per* Sir John Donaldson MR, complaining about the indiscriminate citing of computer-recorded cases which contain no new law). The Court of Appeal also discourages the citation of unreported cases (*Practice Direction (Court of Appeal: procedure)* [1999] 2 All ER 490).

In a postscript to his judgment in *Michaels* v *Taylor Woodrow Developments Ltd* [2001] 2 WLR 224, Laddie J complained (at [79]–[87]) that

the common law system, which places such reliance on judicial authority, stands the risk of being swamped by a torrent of material . . . Sooner rather than later this problem must be tackled if the increasing ease with which prior decisions can be accessed is not going to choke the system.

The Human Rights Act 1998 came fully into force in October 2000, and within a year the Court of Appeal found it necessary, when giving judgment in July 2001 in *R (Al-Hasan)* v *Secretary of State for the Home Department* [2002] 1 WLR 545, CA, to complain (at [73]) about the over-citation of human rights cases.

Practice Direction (Citation of Authorities) [2001] 1 WLR 1001, which applies to all civil courts but not to criminal courts, represents another attempt to address the problem of over-citation of authority. The *Practice Direction* asserts that the continuing efforts to increase the efficiency of litigation, and thereby reduce its cost, will be under threat if courts are burdened with 'inappropriate and unnecessary authority'. It imposes for the first time a uniform practice under which the same rules are to be followed by all civil courts.

A judgment falling within certain specified categories (e.g., one given on an application for permission to appeal) cannot be cited at all unless it clearly indicates that it purports to establish a new principle of law or to extend the present law. The indication must, in the case of judgments delivered *after* the date of the *Practice Direction* (9 April 2001), be in the form of an express statement to that effect in the judgment. In the case of judgments delivered *before* that date, the indication must be present in, or clearly deducible from, the language used in the judgment.

With regard to all other categories of judgment (including decisions of the Court of Justice of the European Communities and the European Court of Human Rights), advocates are required to justify their decision to cite a previous case which simply applies decided law to new facts, or otherwise does not add to, or extend, the existing law. In addition, advocates must state what they understand to be the proposition of law demonstrated by every case they wish to cite, together with the part of the judgment which supports that proposition. If they wish to cite more than one case in support of the same proposition they must state the reason for doing so.

Reliance on case law from jurisdictions outside England and Wales which does not add to, or extend, the existing law is also discouraged by the *Practice Direction*. Accordingly, the requirements outlined in the previous paragraph apply equally to the citation of foreign cases.

6

Trial by jury

6.1 Introduction

Trial by jury is an ancient and democratic institution. It will be seen later that it is also a declining one, particularly in civil cases (para 6.6 below). The jury system provides an opportunity for the layman to participate in the administration of the legal system, to reassure the rest of us that justice is being done in individual cases, and to act as a restraining influence on the professional judiciary.

It is said that in the judicial process the members of a jury are essentially judges of *fact*. In reality, however, a jury's verdict is a decision of *mixed fact* and *law*. For example, in *criminal* proceedings in the Crown Court the jury will listen to the evidence from both sides, to the trial judge's summing-up of the evidence, and to any directions on the law given by the judge. The members of the jury then decide, in the light of their understanding of the law as explained by the judge, whether, in fact, the accused is guilty or not guilty. Although it is permissible for the judge to direct the jury, for legal reasons, to *acquit* a defendant, the House of Lords in *R* v *Wang* [2005] 1 WLR 661, HL, has reaffirmed that he has no power under any circumstances to direct a *conviction*. In those *civil* cases where a jury is still available, the jury will consider the evidence and any directions on the law given by the judge. The members of the jury then apply their understanding of the law to the facts of the case and decide whether to find for the claimant or for the defendant. If judgment is given for the claimant, the jury also decides, as a question of fact, how much the damages should be.

Juries do not have to justify or explain their verdicts to anyone. In a criminal case there is no right of appeal by the prosecution against a jury's verdict of not guilty at the Crown Court, although the Attorney-General may refer the case to the Court of Appeal under the procedure contained in s. 36 of the Criminal Justice Act 1972 (para 7.3). In civil cases a jury's verdict can only be upset by an appellate court if it was perverse; i.e., so unreasonable that no jury properly directed could reasonably have reached it on the evidence (*Mechanical & General Inventions Co. Ltd* v *Austin* [1935] AC 347, HL; *Powell* v *Streatham Manor Nursing Home* [1935] AC 243, HL, *per* Viscount Sankey LC at p. 250; *Scott* v *Musial* [1959] 2 QB 429, CA, *per* Morris LJ at p. 437).

For example, a jury's award of damages in a civil case will not be set aside merely because the amount is much more, or much less, than the appellate court itself would have awarded in the particular circumstances. Thus, in *Cassell & Co. Ltd* v *Broome* [1972] AC 1027, HL, and *Blackshaw* v *Lord* [1983] 2 All ER 311, CA, the House of

Lords and the Court of Appeal, respectively, refused to interfere with a jury's award of damages in a libel case even though the judges considered the amount awarded to be excessive.

On the other hand, in *Lewis* v *Daily Telegraph Ltd* [1964] AC 234, HL, the House of Lords ordered new trials of two libel actions in which juries had awarded damages which were so excessive in amount that no reasonable jury could reasonably have awarded them. In *Riches* v *News Group Newspapers Ltd* [1985] 2 All ER 845, CA, the Court of Appeal ordered a new trial on the issue of damages on the ground, *inter alia*, that no reasonable jury, properly directed by the judge, could have awarded such a high sum. The jury had awarded what were then record damages for libel, totalling £253,000, in an action brought by ten police detectives in respect of allegations of rape, assault, and blackmail made against them in the *News of the World*. (See also *Sutcliffe* v *Pressdram Ltd* [1990] 1 All ER 269, CA, para 6.8.2.2 below).

In *Warby* v *Cascarino* (1989) *The Times*, 27 October, CA, the Court of Appeal actually *increased* a jury's award of damages. Two sisters had successfully sued the Tesco supermarket chain, and a store detective employed by them, for libel, slander, and false imprisonment following an incident in which they had been detained and accused of dishonesty by switching price labels. Criminal charges against them were later dropped. The claimants were awarded damages of £800 each by a jury. Since this sum was less than the defendants' offer of £1,500 each of them had already rejected to settle the case out of court, they were liable to pay the defendants' costs as well as their own. The costs were estimated at £40,000. On appeal, the jury's award was set aside as being too low. Instead of ordering a new trial on the question of damages, the Court of Appeal itself, at the invitation of the parties, reassessed the damages and awarded a sum of £7,500 to each claimant, thereby incidentally relieving the sisters of the costs burden.

An appellate court also has the more fundamental power to overturn a jury's verdict on *liability* in a defamation case. The Court of Appeal did so in the case of *Grobbelaar* v *News Group Newspapers Ltd* [2001] 2 All ER 437, CA, although its decision was later reversed by the House of Lords. Bruce Grobbelaar, the former Liverpool Football Club goalkeeper, sued the defendants for libel arising out of allegations published in the *Sun* newspaper that he was guilty of accepting money for match-fixing. The jury returned a verdict in Mr Grobbelaar's favour and awarded him damages of £85,000.

On appeal by the defendants, the Court of Appeal, while emphasising that it was anxious not to usurp the jury's function and was reluctant to find a verdict perverse, made it clear that it had a duty nevertheless to intervene where the verdict was so plainly wrong that no jury, acting reasonably, could have reached it on the balance of probabilities. The defendants' appeal was successful and the jury's verdict was set aside.

On a further appeal, the House of Lords held that, in the circumstances, the Court of Appeal had been wrong to set aside the jury's verdict since the verdict was not perverse but one which it was open to the jury to find on the evidence before it ([2002] 1 WLR 3024). However, their Lordships further held that the jury's award of £85,000 in damages was unjustified as the evidence before the jury (including secretly videotaped confessions) had shown that Mr Grobbelaar's own conduct had destroyed the value of his reputation. The sum of £1 in nominal damages was substituted.

The virtual inviolability of a jury's verdict makes it possible on occasions for juries to criticise by implication unsatisfactory laws and to assist in the shaping of new laws (para 6.8.1 below). Juries have, in effect, a power (albeit rarely exercised) to nullify the law.

6.2 **Qualifications for jury service**

The law on jury service is largely consolidated in the Juries Act 1974, as amended. In particular, important amendments were made by the Criminal Justice Act 2003. These changes were based on Auld LJ's *Review of the Criminal Courts of England and Wales* (Lord Chancellor's Department, 2001) and the government's White Paper response to it, *Justice For All*, Cm 5563, 2002.

The Criminal Justice Act 2003 Act abolished the former categories of ineligibility for jury service (except in the case of mentally disordered persons) and 'excusal as of right'. Thus, certain persons who in the past were not allowed to do jury service, or could opt out of it, are now required to do it unless they can show good reason not to (see paras 6.3.3 and 6.3.4 below).

By s. 1 of the Juries Act 1974, as substituted by the Criminal Justice Act 2003, a person is qualified for jury service in the Crown Court, the High Court, and the county court if he:

(a) is aged between 18 and 70 and registered as a parliamentary or local government elector;

(b) has been ordinarily resident in the United Kingdom, the Channel Islands or the Isle of Man for at least five years since the age of 13;

(c) is not a mentally disordered person; *and*

(d) is not disqualified for jury service.

Among those ineligible for jury service until the Juries Act 1974 was amended by the Criminal Justice Act 2003 were persons involved in the administration of justice. Now, individuals such as judges, justices of the peace, solicitors, barristers, and police officers, *are* eligible.

In *R v Abdroikov* [2005] 1 WLR 3538, CA, the Court of Appeal held that the presence on a jury of a serving police officer, or a solicitor employed by the Crown Prosecution Service, would not *by itself* lead a fair-minded and informed observer to conclude that there was a real possibility that that juror was biased against the defendant. The opposite conclusion would have made a nonsense of the statutory change in the law.

However, a real possibility of bias may be made out if any such person serving on a jury fails to notify the trial judge of any special knowledge he has about individuals involved in the case or about the facts of the case (ibid.).

6.3 Mental disorder, disqualification, excusal, and discretionary deferral

6.3.1 Mental disorder

Mentally disordered persons are ineligible for jury service. This category includes a person who suffers or has suffered from mental illness, psychopathic disorder, mental handicap, or severe mental handicap, and, as a result, is either:

(a) resident in a hospital or similar institution; or

(b) regularly attending for treatment by a medical practitioner (Juries Act 1974, s. 1 and sch. 1, pt 1, as substituted by the Criminal Justice Act 2003).

The lawfulness of a conviction is not affected by reason of the presence on the jury of an ineligible person (Juries Act 1974, s. 18(1)(b)).

6.3.2 Disqualification

The following persons are *disqualified* from jury service:

(a) a person who is on bail in criminal proceedings;

(b) a person who *at any time* in the United Kingdom, the Channel Islands, or the Isle of Man has been sentenced to imprisonment for life, detention for life, or custody for life, to detention during Her Majesty's pleasure or during the pleasure of the Secretary of State, to imprisonment or detention for public protection, to an extended sentence under s. 227 or s. 228 of the Criminal Justice Act 2003 or s. 210A of the Criminal Procedure (Scotland) Act 1995, or to a term of imprisonment or a term of detention of five years or more;

(c) a person who *in the last ten years* has, in the United Kingdom, the Channel Islands, or the Isle of Man, served any part of a sentence of imprisonment or a sentence of detention, or had passed on him a suspended sentence of imprisonment, or had made in respect of him a suspended order for detention;

(d) a person who *in the last ten years* has, in England and Wales, had made in respect of him a community order under s. 177 of the Criminal Justice Act 2003, a community rehabilitation order, a community punishment order, a community punishment and rehabilitation order, a drug treatment and testing order or a drug abstinence order, or had made in respect of him any corresponding order under the law of Scotland, Northern Ireland, the Isle of Man, or any of the Channel Islands (Juries Act 1974, s. 1 and sch. 1, pt 2, as substituted by the Criminal Justice Act 2003).

The presence on the jury of a disqualified person does not affect the lawfulness of the jury's verdict (Juries Act 1974, s. 18(1)(b)). A person who serves on a jury knowing that he is disqualified is guilty of a criminal offence (ibid., s. 20, as amended by the Criminal Justice Act 2003).

6.3.3 Excusal

A person is entitled to be excused from jury service if he has (a) served on a jury (excluding a coroner's jury), or attended to serve on a jury, in the past two years, *or* (b) been excused from jury service by any court for a period which has not expired (Juries Act 1974, s. 8).

A full-time serving member of the armed forces must be excused by 'the appropriate officer' (now the Jury Central Summoning Bureau—JCSB) if his commanding officer certifies that it would be prejudicial to the efficiency of the service if he were to be required to be absent from duty (ibid., s. 9(2A), as inserted by the Criminal Justice Act 2003).

At the discretion of the JCSB, a person may be excused from jury service on a particular occasion if he shows a 'good reason' (ibid., s. 9(2)), such as illness or an arranged holiday commitment.

There is a right of appeal against the refusal of the JCSB to excuse a person (ibid., s. 9(3)). The appeal will be considered initially by the Head of the JCSB. If the original decision is upheld, the appeal can be referred to a judge at the court for a final decision. Under rules of court, the appeal must not be dismissed unless the appellant has been given an opportunity of making representations (Criminal Procedure Rules 2005 (SI 2005, No. 384), r. 39.2). The court's power when hearing such an appeal is limited to excusing the appellant or refusing to excuse him. The court cannot excuse him and then impose an obligation on him to attend on a specified future date instead (*R v Crown Court at St Alban's (ex parte Perkins)* (1981) *The Times*, 12 December, DC; but see discretionary deferral, para 6.3.4 below).

Although there is no *right* to legal representation for a person appealing to the judge against a refusal to excuse, the court has a *discretion* to allow it and should carefully and sympathetically consider exercising that discretion in cases involving conscientious objections to jury service (*R v Crown Court at Guildford (ex parte Siderfin)* [1989] 3 All ER 7, DC, *per* Watkins LJ at p. 12).

Independently of the JCSB's discretion, the court before which a person is summoned to attend for jury service has the power to excuse attendance (Juries Act 1974, s. 9(4)).

6.3.4 Discretionary deferral

A person summoned for jury service may have his attendance deferred if he shows a 'good reason' to the satisfaction of the JCSB (Juries Act 1974, s. 9A, added by the Criminal Justice Act 1988). Where deferral is granted, the days of attendance will be varied. If deferral is refused, there is a right of appeal to the court (ibid., s. 9A(3) and Criminal Procedure Rules 2005, r. 39.2).

6.3.5 Guidance in excusal and deferral cases

By s. 9AA of the Juries Act 1974, as inserted by the Criminal Justice Act 2003, the Lord Chancellor is obliged to issue guidance on how the JCSB should exercise its discretion under s. 9 (excusal) and s. 9A (deferral). The guidance, and any revised guidance, is required to be laid before each House of Parliament, and to be published. It is intended to balance the needs of those summoned for jury service with the need to have sufficient jurors available for trials.

According to the guidance (available on the Court Service website), all applications for excusal and deferral must be considered 'carefully, sympathetically and with regard to the individual circumstances of the applicant'. If a person shows good reason for not sitting on a chosen date, deferral will be considered in the first instance. Excusal is reserved for extreme circumstances. A 'good reason' would include illness, holiday commitments, difficulty in making arrangements to have young children or old people looked after, work or business commitments, interference with tuition or examinations during term time, the parliamentary duties of Members of Parliament, personal involvement in the facts of the particular case, and close connection with a party or prospective witness.

Applications for excusal on the ground of insufficient understanding of English will normally be granted. In cases of doubt, the applicant can be taken before the judge under s. 10 of the Juries Act 1974 (para 6.3.7 below). Applications for excusal from practising members of religions whose beliefs are incompatible with jury service will normally be granted. The position of those with non-religious conscientious objections to jury service is less clear under the guidance.

Guidance on how a judge should exercise his discretion to excuse or discharge a juror, or to adjourn a trial, has also been given by the Lord Chief Justice in *Practice Direction (Crown Court: Jury Service)* [2005] 1 WLR 1361.

6.3.6 Disabled persons

If there is doubt about a person's capacity to act effectively as a juror because of *physical disability* (such as deafness or blindness), the judge will decide whether or not that person should act as a juror. The judge must allow the person to act as a juror unless he is of the opinion that the person will not be capable of acting effectively by reason of his disability (Juries Act 1974, s. 9B, as inserted by the Criminal Justice and Public Order Act 1994 and amended by the Courts Act 2003).

6.3.7 Doubt about capacity

Any doubt about a person's capacity to act effectively as a juror because of *insufficient understanding of English* will be decided by the judge (Juries Act 1974, s. 10, as amended by the Courts Act 2003).

6.4 Summoning jurors

The annual electoral register is the starting-point for the provision of jurors for both criminal and civil trials. The electoral register is compiled from information supplied by occupiers of houses. The information required includes the names of persons who live in a particular house and whether any such person is over the age of 18 or over the age of 70.

From the autumn of 2000, the summoning of jurors by local court officers was replaced by a centralised system following a successful pilot scheme operated in various parts of England and Wales over a 12-month period. Under the old, localised system, the jury summoning officer at each court selected names at random from paper copies of the electoral register, laboriously produced individual summonses, and dealt with jurors' responses.

Under the new system, the Jury Central Summoning Bureau (JCSB), based at Blackfriars Crown Court Centre in London, uses a computerised system to select jurors at random from the electoral register, to issue summonses, and to deal with jurors' responses. The JCSB makes all decisions on requests to be excused from jury service or have it deferred. The system ensures that people are treated more equally, that the rules are enforced more consistently, and that the number of jurors needed by the courts is more accurately predictable so that jurors spend more time sitting on trials and less time waiting around.

From those summoned for jury service, the court officer prepares lists (called panels) of jurors. The information to be included in panels is determined by the Lord Chancellor (ibid., s. 5(1)). Reasonable facilities for inspecting a panel must be provided for a party to the proceedings and any person acting on behalf of a party. In addition, the court has a discretion to allow any other person to inspect the panel (ibid., s. 5(2)–(3)).

A particular jury is selected by ballot from the panel in open court. Each juror must be sworn separately (ibid., s. 11(1), (3)).

The power to summon and empanel jurors is an administrative matter. The power has been conferred by statute specifically on the Lord Chancellor. It follows that a trial judge has no jurisdiction to order a multiracial jury to be empanelled for a particular trial (R v Ford [1989] 3 All ER 445, CA; R v Smith [2003] 1 WLR 2229, CA). Nor, for the same reason, can the judge order that jurors be brought in from outside the normal catchment area merely because he suspects that local jurors might be intimidated (R v Tarrant (1997) The Times, 29 December, CA, in which R v Ford, above, was applied).

If at any trial the court is short of jurors it can require any qualified person who is in the vicinity to be summoned for jury service without any written notice (Juries Act 1974, s. 6).

6.5 Juries in criminal cases

The jury is used in the Crown Court. There is no jury in the magistrates' courts, which try about 98 per cent of criminal cases. The jury is thus used (theoretically) in only the 2 per cent of criminal trials which take place at the Crown Court. In practice, however, the jury is used even less than this (in about 0.8 per cent of criminal trials) because 58 per cent of defendants at the Crown Court plead guilty and do not need a jury trial (Judicial Statistics 2004, Cm 6565, 2005, p. 90).

6.5.1 Challenging jurors

Before 1989, a person being tried on indictment could make a *peremptory challenge* (i.e., a challenge without cause) to not more than three jurors. This right was abolished by the Criminal Justice Act 1988 as a result of the government's concern that the right was being abused. The accused retains the right to challenge all or any of the jurors *for cause.*

Any challenge for cause is tried as a preliminary matter by the trial judge (Juries Act 1974, s. 12(1)(a)–(b)). The challenge must be made after the juror's name has been drawn by ballot (ibid., s. 12(3)). It was held in *R v Comerford* [1998] 1 All ER 823, CA, that this provision does not make it mandatory to call out jurors' names. It follows that where jurors are called by *numbers*, their names having been withheld to prevent jury-nobbling, the ensuing trial by anonymous jury is valid provided that the defendant's right of challenge is preserved. The purpose of s. 12(3) was held to be to define the time at which the challenge is to be made rather than to require the public announcement of jurors' names. (See further on this case, para 7.5.5.4.)

The prosecution, too, can challenge for cause, and, in addition, has the right to 'stand by' any would-be juror and, if necessary, the entire panel. 'Stand by' is meant to be only a provisional challenge on the part of the prosecution. In theory, the prosecution can only challenge for cause and, when a juror is stood by, the inquiry into the cause is postponed until a later date. In practice, there will usually be sufficient other qualified persons in attendance from whom to form a jury, and to allow the trial to proceed, so that the prosecution's 'cause' is never investigated.

The Attorney-General's 'Guidelines on the Exercise by the Crown of its Right to Stand By' (published as a *Practice Note* [1988] 3 All ER 1086) reaffirm the principles of random jury selection and that no one should be treated as disqualified or ineligible except as provided for by Parliament in the Juries Act 1974, as amended.

According to the 'Guidelines', it is accepted practice that the prosecution's right of stand by will be exercised sparingly, and only in exceptional circumstances, and that it will not be used in order to influence the overall composition of a jury or to obtain a tactical advantage over the defence. The circumstances in which it would be proper to exercise the right are listed as:

(a) where a jury check (see para 6.5.2 below) has revealed information which strongly suggests that, given the facts of the case and the offences to be tried, a particular juror might be a security risk, be susceptible to improper approaches, or be influenced by improper motives in arriving at a verdict, and the Attorney-General has personally authorised use of the right of stand by;

(b) where a juror is about to be sworn in who is 'manifestly unsuitable' and the defence agrees that the exercise of the right of stand by would be appropriate; the 'Guidelines' give the example of an illiterate juror about to try a complex case.

When challenging for cause, counsel (whether for the prosecution or the defence) should not ask questions of the juror in order to establish the cause; any questioning should be left to the trial judge. Irregularities at the Angry Brigade trial in 1972 led to changes in procedure. At the suggestion of counsel for the defence, the trial judge asked would-be jurors to exclude themselves on a number of grounds; for example, if they were members of the Conservative Party, or if they had relatives in the police or armed forces in Northern Ireland. *In toto*, 29 persons were challenged and another 19 admitted belonging to one or other of the judge's categories.

After the trial was concluded, the Lord Chief Justice issued a *Practice Direction* in which he emphasised that jurors should not be excused on account of race, religion, politics, or occupation. By implication, the trial judge should not ask questions about such matters. A juror should be excused only on the more traditional grounds of personal connection with the facts of the case (or with a party or a witness), or personal hardship, or conscientious objection to jury service (*Practice Direction* [1973] 1 All ER 240; see also *R v Broderick* [1970] Crim LR 155, CA; the 1973 *Practice Direction* was revoked and replaced in 1988 by another *Practice Direction*, which is now incorporated in *Practice Direction (Criminal Proceedings: Consolidation), Part IV* [2002] 1 WLR 2870). In 1973, the Lord Chancellor issued a directive that jury panels would no longer include the occupations of prospective jurors.

In *R v Andrews* (1998) *The Times*, 15 October, CA, the Court of Appeal held that the questioning of potential jurors by means of a questionnaire (or, *obiter*, orally), with a view to discovering whether they are biased, is of doubtful efficacy and could be counterproductive in that it might bring to the attention of, or remind, jurors of some matter which the parties and the court would wish them to disregard. Such questioning should be avoided unless the circumstances are most exceptional. In all other cases, jurors must be trusted to be faithful to their oaths to return true verdicts in accordance with the evidence given in court.

6.5.2 Jury vetting

The practice of *jury vetting* (i.e., checking the background of jurors) appears to conflict with the Lord Chief Justice's *Practice Direction* of 1973 in which it was reaffirmed that a jury 'consists of 12 individuals chosen at random from the appropriate panel' ([1973] 1 All ER 240 *per* Lord Widgery CJ). It is also in defiance of Blackstone's caveat that the liberties of England depend on the jury remaining 'sacred and inviolate; not only from all open attacks (which none will be so hardy as to make), but also from all secret machinations, which may sap and undermine it' (4 Bl Comm 350).

In 1978, at the trial of Messrs Aubrey, Berry, and Campbell (the so-called 'ABC' official secrets trial), two journalists and a soldier were charged with collecting secret information. Counsel for one of the defendants discovered that some weeks before the trial began the prosecution had been supplied with a list of potential jurors. Those on the list had been scrutinised so as to check their 'loyalty'. As a result of this discovery the trial was stopped and a new trial ordered before a fresh jury. All three defendants were ultimately acquitted. In October 1978, pressure from MPs and the press led the Attorney-General to make public his 'Guidelines on Jury Checks', which, it was then admitted, had been in existence since 1974 following on the Angry Brigade trial.

The Attorney-General's Guidelines have since been revised and the current version is open to inspection in the law reports (see *Practice Note (jury: stand by: jury checks)* [1988] 3 All ER 1086).

The Guidelines reaffirm three basic principles—that a jury should be chosen by random selection; that only those matters specified in the Juries Act 1974, as amended, should disqualify a person from jury service; and that the correct way for the prosecution to exclude a prospective juror is to stand him by, or challenge him for cause, in open court. However, the Guidelines highlight certain exceptional types of case of public importance in which the statutory safeguards of majority verdicts and the imposition of criminal liability on a disqualified person 'may not be sufficient to ensure the proper administration of justice'. These are identified as cases involving:

(a) national security, where part of the evidence is likely to be heard in camera;

(b) terrorism.

In these exceptional cases it is felt that a potential juror's political beliefs may be so biased as to 'interfere with his fair assessment of the facts of the case or lead him to exert improper pressure on his fellow jurors'. And, in security cases, there is a danger that a juror, either voluntarily or under pressure, may reveal evidence given *in camera*.

The actual vetting is done by means of a check on police criminal records and police Special Branch records. In cases falling under (a) above, the security services may also be involved. The involvement of Special Branches and the security services requires the personal authority of the Attorney-General on the application of the Director of Public Prosecutions. The questioning of family, neighbours, and friends is not in general permissible except where it is necessary to confirm the identity of a juror about whom doubts have been raised following the initial check on criminal or Special Branch records.

There is no statutory basis for the Attorney-General's Guidelines on Jury Checks, and the authority of a judge is not required before vetting can take place. The prosecution can (though only with the personal authority of the Attorney-General) stand by a prospective juror who falls foul of the Guidelines without, in practice, any reason being given in open court. However, it is stated that a person should not be stood by unless the check has revealed a *strong* reason for believing that he might be a security risk, or susceptible to improper approaches, or be influenced in arriving at a verdict.

There is no *duty* to communicate the reason for the stand by to the defence, although counsel may use his discretion to disclose it if its nature and source permit. If the prosecution do not stand by a juror but vetting has revealed that he may be biased against the accused, the Guidelines advise that defence counsel should be warned, even if only in a general way.

The question of jury vetting is controversial, and has given rise to a conflict of judicial opinion within the Court of Appeal. It came before the Court for the first time in *R v Crown Court at Sheffield (ex parte Brownlow)* [1980] 2 All ER 444, CA, and the practice was condemned as unconstitutional by Lord Denning MR and Shaw LJ sitting in the civil division. Both said that jury vetting is unconstitutional because it is not sanctioned by the Juries Act 1974 ([1980] 2 All ER 444 at pp. 453 and 455, respectively). They regarded it as a serious invasion of a person's privacy. The third judge in the appeal, Brandon LJ, while not directly condemning the practice, said (at p. 456) that if jury vetting is permitted then the defence should be entitled to receive any resulting information.

It is clear that, since the appeal in the *Sheffield* case was dismissed, what was said about jury vetting was strictly *obiter*. This fact was seized on in the second case to reach the Court of Appeal (this time the criminal division), *R v Mason* [1980] 3 All ER 777, CA. Here, the Court said that some vetting of jurors was necessary in order to eliminate disqualified persons. There was some evidence that prospective jurors were not disclosing that they were disqualified, although since the Juries Act 1974 there had been only two prosecutions for serving on a jury while disqualified. One resulted in an acquittal and the other in a fine of £10.

Lawton LJ, delivering the judgment of the court, said that the principle of random selection of the jury had been qualified for centuries: certain persons were disqualified by statute; the defence has the right of peremptory challenge (since abolished in 1989); the prosecution has the right of stand by; and the judge can intervene to excuse a juror who, for example, cannot read or hear. His Lordship said that the only authority competent to check the criminal convictions of jurors are the police, and, since it is a criminal offence to serve on a jury while disqualified, the police are only performing their usual duty to prevent crime when they engage in jury vetting. He made it clear that his judgment was only concerned with the vetting of the jury panel for criminal convictions. He refused to comment on any information other than convictions or on the desirability of making other enquiries about jurors. He said that the comments about jury vetting made in the *Sheffield* case were *obiter* and not binding, and that in *Mason* the court had been able to examine the issue in greater depth.

In *R v McCann and others* (1990) 92 Cr App R 239, CA, the Court of Appeal, criminal division, preferring *Mason* to the *Sheffield* case, went further and declared that the practice of jury vetting in accordance with the Attorney-General's Guidelines, and the exercise by the Attorney-General of his right to 'stand by' prospective jurors, are constitutionally proper. (*R v McCann and others* is the case of the 'Winchester Three'; for the full story, see para 7.6.)

The legal position on jury vetting appears, then, to be this. The checking of criminal records for convictions is permissible and desirable in order to eliminate disqualified persons (*R v Mason* and *R v McCann*, above). The checking of police Special Branch records, and the making of other enquiries, in accordance with the Attorney-General's Guidelines, as outlined above, is also lawful (*R v McCann*, above).

Jury vetting does not necessarily always work to the advantage of the prosecution. While Michael Bettaney, an M15 officer, was convicted in 1984 of charges under the Official Secrets Act by a jury which had been vetted by M15 and Special Branch officers, Clive Ponting, a senior civil servant with the Ministry of Defence, was *acquitted* of similar charges by a vetted jury in the following year.

6.5.3 Death or discharge of jurors

In a criminal case, the trial (even if it is for murder) may continue if a juror *dies or is discharged* because of illness or any other reason, provided that the number of jurors does not fall below nine (Juries Act 1974, s. 16(1)).

The judge may discharge a juror otherwise than in open court if there is a good reason (*R v Richardson* [1979] 3 All ER 247, CA). The Court of Appeal has jurisdiction to review the trial judge's exercise of his discretion to excuse a juror under the Juries Act 1974, but will only interfere where the judge has been capricious or where injustice has resulted. Thus, in *R v Hambery* [1977] QB 924, CA, the Court of Appeal refused to interfere where the trial judge had discharged a juror who was due to go on holiday the next day. It made no difference that the judge had not enquired of the juror whether it was important for her holiday to start on that day; the importance could be inferred. The judge had not been capricious, and no injustice had resulted, because a trial can continue with less than the maximum number of jurors present. Jurors are entitled to some consideration and if justice can be administered properly without undue inconvenience to jurors then it should be ([1977] QB 924 *per* Lawton LJ at p. 930).

In a *criminal* case, no appeal lies against the trial judge's decision to discharge the *whole jury*. This is because the Court of Appeal, criminal division, can only hear an appeal against a conviction on indictment (Criminal Appeal Act 1968, s. 2(1), as substituted by the Criminal Appeal Act 1995), and if the entire jury is discharged before a verdict is delivered there is no conviction against which to appeal (*R v Hambery*, CA, above, *per* Lawton LJ at p. 929; *R v Gorman* [1987] 2 All ER 435, CA; *Gladding* v *Channel 4 TV Corporation* [1999] EMLR 475, CA).

In a *civil* case, however, an appeal does lie against the trial judge's decision to discharge the whole jury because the civil division of the Court of Appeal is not subject to the same restriction as the criminal division, having jurisdiction, under s. 16(1) of the Supreme Court Act 1981, 'to hear and determine appeals from any judgment or order of the High Court' (*Gladding* v *Channel 4 TV Corporation*, above).

6.5.4 Majority verdicts

Majority verdicts, which were first introduced into English law by the Criminal Justice Act 1967, are now dealt with in s. 17 of the Juries Act 1974. In 2004, of those convicted in the Crown Court after a plea of not guilty 23 per cent were convicted on majority verdicts (*Judicial Statistics 2004*, Cm 6565, 2005, p. 91).

In the Crown Court, a majority verdict is acceptable if:

(a) there are not less than eleven jurors, and ten of them agree; or

(b) there are ten jurors, and nine of them agree (Juries Act 1974, s. 17(1)(a)–(b)).

In either case, the foreman of a jury which has reached a verdict of *guilty* must state in open court the number of jurors who agreed *and* disagreed with the verdict (without, of course, giving their names) (ibid., s. 17(3)).

Compliance with this requirement is mandatory (rather than merely directory) before the judge can accept a majority verdict of guilty. If the requirement is not observed at all, the defendant's conviction may be quashed. Thus, in *R v Barry* [1975] 2 All ER 760, CA, where the judge accepted a majority verdict of guilty without the foreman of the jury stating, or even being asked to state, the number of jurors who agreed to, or dissented from, the verdict, the Court of Appeal allowed the defendant's appeal and quashed his conviction.

In *R v Reynolds* [1981] 3 All ER 849, CA, in answer to a question by the clerk of the court asking how many agreed, and how many disagreed, with the verdict, the foreman said, 'Ten agreed'. The Court of Appeal held that the defendant's conviction would be quashed. *R v Reynolds* was followed reluctantly in *R v Pigg* [1982] 2 All ER 591, CA, in which the Court of Appeal, while accepting itself as bound by *Reynolds*, expressed the view (at p. 595) that it was absurd that it should be mandatory for the foreman to state how many jurors dissented when all that was required to find out was a simple arithmetical calculation.

The Crown appealed to the House of Lords in *R v Pigg*, and the House ([1983] 1 All ER 56) reversed the decision of the Court of Appeal, overruled *R v Reynolds*, and, at the same time, approved *R v Barry*, above. The House of Lords held that, although compliance with the requirement of s. 17(3) is mandatory, the precise form of words used by the clerk of the court when asking the foreman of the jury the number who agreed and dissented, and by the foreman in his reply, is not an essential part of that requirement. Section 17(3) is satisfied as long as the words used by the clerk and the foreman make it clear to an ordinary person how the jury was divided. If, in the case of a jury of *twelve*, the foreman, on being asked how many jurors agreed with the verdict, replies that *ten* agreed, s. 17(3) is satisfied because it is a necessary and inevitable inference which would be obvious to any ordinary person that *two* disagreed.

At the same time, it was made clear in *R v Pigg* that if the requirement of s. 17(3) is not observed at all, as in *R v Barry*, the judge cannot lawfully accept the jury's verdict. For example, in *R v Mendy* [1992] Crim LR 313, CA, the foreman of the jury at the end of a lengthy trial of two counts of conspiracy to defraud answered a question put to him by the clerk of the court by saying (ambiguously) that, in respect of one count, the defendant had been found guilty 'by a majority of us all'. The professionals involved in the case (the clerk of the court, the trial judge, counsel and their instructing solicitors) seem to have simply assumed that this was a unanimous verdict.

On appeal, however, the issue was raised as to whether the foreman was returning a majority verdict or a unanimous one. If his answer was meant to indicate a majority verdict, then s. 17(3) had not been complied with because there was no statement about how many jurors agreed and disagreed with the verdict. Without needing to apply *R v Barry*, it was held that since the case was one involving the liberty of the subject the benefit of the ambiguity should be given to the defendant and his conviction on that count was quashed. The Court of Appeal stressed how important it was for counsel and solicitors to pay the closest attention to procedural matters in order to avoid for the future the 'horrifying possibility' of a verdict obtained in a long and expensive City fraud trial being set aside on a technicality.

Where the foreman of the jury fails to state the number of jurors who agreed and disagreed with a majority verdict of guilty, the verdict is not unlawful if the omission is corrected quickly by the foreman after the entire jury has been reassembled (*R v Maloney* [1996] 2 Cr App R 303, CA, in which the defective verdict was delivered on a Friday and corrected on the following Monday, the next working day).

In *R v Millward* [1999] 1 Cr App R 61, CA, the clerk of the court asked whether a verdict had been reached on which at least ten were agreed, to which the foreman said: 'Yes'. The clerk asked the further question: 'Is that the verdict of you all or by a majority?', to which the foreman replied: 'All of us'. The clerk then said: 'You find the defendant guilty and that is the verdict of you all?', and the foreman answered: 'Yes'. A verdict of guilty was recorded. The next day the foreman wrote to the judge stating that she had made a mistake and that, in fact, only ten jurors had agreed and two had dissented. On appeal against conviction, it was argued that s. 17(3) had not been complied with.

The Court of Appeal dismissed the appeal, holding that the clerk had asked the right questions, the jury's answer was not ambiguous, and there was a clear statement that the verdict was unanimous. It was further held that, even if that conclusion was wrong, there was an acceptable majority verdict and there was nothing to indicate that the conviction was unsafe.

The cases of *Barry*, *Mendy*, and *Maloney* were distinguished in *Millward* because, unlike *Barry* and *Maloney*, the clerk had asked the right questions, and, unlike *Mendy*, the jury's answer was not ambiguous. Unlike all three earlier cases, there was in *Millward* a clear and explicit statement of a unanimous verdict. (See also *R v Lewis* (2001) *The Times*, 26 April, CA, in which *Millward* was applied.)

There is no requirement to state the number of jurors who agreed and disagreed in the case of a verdict of *not guilty*.

There are provisions designed to ensure that the jury does not resort to a majority verdict too hastily. The trial judge must encourage the jury to reach a unanimous verdict. In the Crown Court, at least two hours' deliberation must be allowed to the jury (Juries Act 1974, s. 17(4)). In addition, *Practice Directions* first issued in 1967 and 1970 by the Court of Appeal state that if the members of the jury are not agreed after two hours they should be sent back to the jury room at least once with a view to reaching *unanimity*. Then, if necessary, they should be sent back at least once to see if they can reach the required *majority*. (The two *Practice Directions* are now incorporated in *Practice Direction (Criminal Proceedings: Consolidation), Part IV* [2002] 1 WLR 2870.) If all these efforts fail, the jury will be discharged and the case may be retried before a fresh jury.

Although judges are meant to observe the Court of Appeal's guidance, it does not have the force of law. Accordingly, the acceptance of a majority verdict in contravention of it does not necessarily render the defendant's conviction unsafe so as to entitle him to have it quashed on appeal (*R v Trickett* [1991] Crim LR 59, CA; *R v S (a juvenile)*, 10 February 1997, CA, unreported).

It may be mentioned here that in *civil* cases in the High Court the rules about majority verdicts are the same as in the Crown Court (Juries Act 1974, s. 17(1)). In the county courts, where there can be a jury of eight (County Courts Act 1984, s. 67), the verdict need not be unanimous if seven jurors agree (Juries Act 1974, s. 17(2)). The jury must, however, be given reasonable time for deliberation, although the two-hour rule does not apply in the civil courts (ibid., s. 17(4)).

6.5.5 Judicial pressure on the jury

It is generally accepted that a jury must be allowed to consider its verdict free from all external pressures, such as violence, threats, intimidation, or bribery. It is also important that a jury should not be subjected to undue pressure from the trial judge. If it is so subjected, this may lead to the defendant's conviction being quashed on appeal on the ground that it is unsafe.

One of the most extreme examples of improper judicial pressure occurred in

R v *McKenna* [1960] 1 QB 411, CCA, in which the trial judge, Stable J, threatened the jury members that if they did not return a verdict within another ten minutes they would be locked up all night. Inside six minutes the jury brought in verdicts of guilty against the defendants, having spent the previous two and a quarter hours unable to agree. On appeal, the convictions were quashed because of a material irregularity in the course of the trial. The Court of Criminal Appeal said ([1960] 1 QB 411 *per* Cassels J at p. 422):

It is a cardinal principle of our criminal law that in considering their verdict, concerning, as it does, the liberty of the subject, a jury shall deliberate in complete freedom, uninfluenced by any promise, unintimidated by any threat. They still stand between the Crown and the subject, and they are still one of the main defences of personal liberty. To say to such a tribunal in the course of its deliberations that it must reach a conclusion within ten minutes or else undergo hours of personal inconvenience and discomfort, is a disservice to the cause of justice . . .

Two points should be noted about the facts of *R* v *McKenna*, above. First, they arose in the days before majority verdicts were permitted. Now, if the trial judge considers that the jury is taking too long to reach a decision he can accept a majority verdict as long as the jury has been deliberating for at least two hours. Secondly, although until 1994 a jury was not allowed to separate once it had retired to consider its verdict, it became the usual practice for the jury to be accommodated in an hotel, rather than to be locked up in the jury room, if an overnight stay was necessary.

When a jury is sent to an hotel for the night, the object is to give its members a break from the case they are trying (*R* v *Young* [1995] QB 324, CA). The jury is not meant to continue its deliberations at the hotel, and it should be directed against doing this by the trial judge. A failure on the judge's part so to do may lead to a successful appeal against conviction (*R* v *Tharakan* [1995] 2 Cr App R 368, CA, where a medical practitioner, who had been convicted of obtaining and attempting to obtain property by deception after a 24-day trial, had his convictions quashed because the judge had failed to direct the jury not to deliberate at the hotel).

A jury may now be allowed by the trial judge to separate at any time, whether before or after retiring to consider its verdict (Juries Act 1974, s. 13, as substituted by the Criminal Justice and Public Order Act 1994). If the jury is unable to reach a verdict during a day's sitting, an overnight stay in an hotel away from family is no longer inevitable since under s. 13, at the discretion of the judge, it may be permitted to separate and go home.

The Court of Appeal has provided guidance to judges on how to direct a jury which is to be allowed to separate before reaching its verdict. The direction should remind members of the jury of the following matters:

(a) that the case must only be decided on the evidence and arguments presented in court;

(b) that the evidence has been completed and it would be wrong to seek or receive further information of any sort about the case;

(c) that the case must not be discussed with anyone except other members of the jury in the jury room;

(d) that when they leave court they should try to put the case on one side until they return to court and get to the jury room to begin or continue their deliberations.

The direction should be given in full before the jury separates for the first time, and a brief reminder should be given before each subsequent dispersal (*R* v *Oliver* [1996] 2 Cr App R 514, CA).

It is not to be regarded as improper pressure for the judge simply to ask the jury, as a practical administrative matter, whether a verdict is likely to be reached that evening (*R* v *Bean* (1991) *The Times*, 1 May, CA). On the other hand, in a case where the trial judge had sent a message to the jury after 6 pm asking for deliberations to be continued, and within minutes thereafter the jury had convicted the defendant by a majority verdict of 10–2, the judge's action was construed as improper pressure amounting to a material irregularity in the course of the trial. Instead of sending the offending message, he should have brought the jury back into court and enquired about the prospects of reaching a verdict (*R* v *Wharton* [1990] Crim LR 877, CA; see also *R* v *Duggan* [1992] Crim LR 513, CA).

In *R* v *Walhein* (1952) 36 Cr App R 167, CCA, the trial judge said to a jury who were unable to agree:

. . . it makes for *great public inconvenience and expense* if jurors cannot agree owing to the unwillingness of one of their number to listen to the arguments of the rest . . . (Emphasis added.)

On appeal, this direction was upheld by the Court of Criminal Appeal. However, it has since been disapproved on the basis that reference to 'public inconvenience and expense' may amount to pressure on jurors to express agreement with a view they do not really hold. If this is the case, the defendant's conviction might be quashed on appeal on the ground that it is unsafe. In *R* v *Watson* [1988] 1 All ER 897, CA, Lord Lane CJ said (at p. 903):

One starts from the proposition that a jury must be free to deliberate without any form of pressure being imposed on them, whether by way of promise or of threat or otherwise. They must not be made to feel that it is incumbent on them to express agreement with a view they do not truly hold simply because it might be inconvenient or tiresome or expensive for the prosecution, the defendant, the victim or the public in general if they do not do so . . .

It is at the trial judge's discretion to decide whether to give a specific direction, and at what stage of the trial, to a jury which is, or might be, unable to agree. According to the Court of Appeal in *R* v *Watson*, such a direction is best given in the following form (*per* Lord Lane CJ at p. 903):

Each of you has taken an oath to return a true verdict according to the evidence. No one must be false to that oath, but you have a duty not only as individuals but collectively. That is the strength of the jury system. Each of you takes into the jury box with you your individual experience and wisdom. Your task is to pool that experience and wisdom. You do that by giving your views and listening to the views of the others. There must necessarily be discussion, argument and give and take within the scope of your oath. That is the way in which agreement is reached. If, unhappily, [ten of] you cannot reach agreement you must say so.

6.5.6 Communications between the jury and the judge

The trial judge must at all times be in a position to give proper and accurate advice to jurors on any matter of law or fact which is troubling them. At the same time it is important that, as a general rule and for the sake of impartiality and openness, there should be no private or secret communication between the jury and the judge.

During the jury's deliberations any communication from the jury should normally be dealt with by the judge in open court in the presence of the entire jury and of the defendant and his counsel. Failure to observe this practice may result on appeal in the quashing of the defendant's conviction on the ground that it is unsafe (see, e.g., *R v Lamb* (1974) 59 Cr App R 196, CA, *R v Townsend* [1982] 1 All ER 509, CA, *R v Rose* [1982] 2 All ER 536, CA, *R v Woods* (1988) 87 Cr App R 60, CA, *R v Sipson Coachworks Ltd* [1992] Crim LR 666, CA, and *R v Obellim* [1997] 1 Cr App R 355, CA, in all of which the conviction was quashed).

In *R v Gorman* [1987] 2 All ER 435, CA (in which the conviction was not quashed), the Court of Appeal laid down the following guidelines:

(a) if the note from the jury is about some matter unconnected with the trial (e.g., a request to pass on a message to a juror's relative), it can be handled by the judge without involving counsel or bringing the jury back into court;

(b) in other cases, the judge should state in open court, in the presence of the defendant and his counsel, the nature and content of the note and, if he considers it helpful, may seek the assistance of counsel before bringing the jury back into court in order to deal with the note;

(c) if the note contains information which the jury need not and should not have disclosed (such as details of voting figures), the judge should not read out that particular information.

The *Gorman* approach was applied by the majority of the Judicial Committee of the Privy Council in *Ramstead* v *The Queen* [1999] 2 AC 92, PC, and by the Court of Appeal in *R v Tantram* [2001] EWCA Crim 1364, CA.

6.5.7 Jurors' expenses and compensation for loss of earnings

A juror is not paid for his services but he is entitled to travelling expenses, subsistence expenses (including vouchers and other benefits which may be used to pay for subsistence), and compensation for loss of earnings. The amounts are fixed by the Lord Chancellor by statutory instrument, and the money is provided by Parliament out of taxation (Juries Act 1974, s. 19(1), (3)–(4), and s. 19(1A) (as inserted by the Criminal Justice Act 2003)). All of these allowances are tax-free. They are paid not only to jurors but to all those who perform unpaid public duties, such as justices of the peace and witnesses in criminal trials.

6.5.8 The representativeness of the criminal jury

The jury in criminal cases is frequently under attack. Its representativeness was called in question by some research carried out in Birmingham, where it used to be the practice (since discontinued) to summon for jury service only half as many women as men (see Baldwin and McConville, *Jury Trials*, 1979, p. 97). The jury is probably no longer 'predominantly male, middle-aged, middle-minded, and middle-class'. No evidence was found that younger juries are less likely to convict (ibid., pp. 101–2). But it was found that racial minorities were not adequately represented on juries in Birmingham. In the period investigated, only 0.7 per cent of jurors were of Asian or Afro-Caribbean origin. Again, although 10 per cent of the population of Birmingham is of Irish extraction, only 3.6 per cent of jurors were from there (ibid., p. 98).

Subsequent research has confirmed suspicions about the representativeness of the criminal jury. Work done in 1993 on behalf of the Royal Commission on Criminal Justice found that men were over-represented on juries, as were age groups from 35 to 65, while people over 65 were under-represented. Professional/managerial workers were represented in almost the same proportion as in the general population, but service-industry and skilled manual workers were found to be significantly under-represented (Zander and Henderson, *Crown Court Study*, Royal Commission on Criminal Justice Research Study No. 19, HMSO, 1993).

The electoral register, which is updated once a year and is the starting-point for jury summoning, is known to be incomplete and, therefore, potentially unrepresentative. Home Office research revealed that in 1999 about 8 per cent of those eligible to be on the electoral register were not. Those not registered included 21 per cent of people aged 20 to 24, 38 per cent of those living in privately rented furnished accommodation, and a disproportionate number from ethnic minorities. About 24 per cent of black voters, 15 per cent of those from the Indian sub-continent, and 24 per cent of voters from other ethnic minorities, were not registered (*Jury Excusal and Deferral*, Home Office Research Findings No. 102, Home Office, 1999).

6.5.9 Trials on indictment without a jury

Sections 43–50 of the Criminal Justice Act 2003 provide for certain cases which are presently tried on indictment at the Crown Court by judge and jury to be tried instead by a judge sitting alone. These provisions are not yet in force.

Auld LJ in his *Review of the Criminal Courts of England and Wales* (Lord Chancellor's Department, 2001) proposed that in *serious or complex fraud* cases the trial judge should have the power to direct trial by himself and two lay members drawn from a panel of experts. The government decided that these cases (estimated at 15–20 a year) should instead be tried by a judge sitting alone. It was also decided to introduce measures for trial by judge alone in some circumstances involving *jury tampering* (*Justice For All*, Cm 5563, 2002).

While the Criminal Justice Bill of 2003 was before Parliament, the clause dealing with the trial of serious or complex fraud cases by a judge sitting alone was rejected by the House of Lords. In order to save the Bill, the government agreed that the offending clause (now s. 43 of the Criminal Justice Act 2003, below) would not be implemented except after an affirmative resolution of both Houses of Parliament. Given the troubled legislative history of the fraud provisions, it is most improbable that such a resolution will ever be passed. In the meantime, it was further agreed to explore other ways of dealing with these fraud cases, including the use of specialist juries.

Section 43 of the Criminal Justice Act 2003 would allow the prosecution to apply to a judge of the Crown Court for a serious or complex fraud trial to be conducted without a jury. If the judge is satisfied that the length or complexity (or both) of the trial is likely to make it so burdensome on the jury that the interests of justice require serious consideration to be given to conducting the trial without a jury, he would have a discretion to order trial without a jury. In deciding whether or not he is satisfied, he must have regard to any steps which might reasonably be taken to reduce the complexity or length of the trial. The judge's order would have to be approved by the Lord Chief Justice or a judge nominated by him.

Section 44 would allow the prosecution to apply for the trial of *any* indictable offence to be conducted without a jury where there is a danger of *jury tampering*. The judge *must* order trial without a jury if two conditions are satisfied. First, that there is evidence of a 'real and present danger' that jury tampering would take place. Secondly, that the likelihood of tampering is so substantial as to make it necessary in the interests of justice for the trial to take place without a jury, in spite of any steps (including the provision of police protection) which might reasonably be taken to prevent tampering. Examples of what might constitute evidence of a real and present danger of jury tampering are set out in the section.

Section 45 provides for an appeal to the Court of Appeal for both prosecution and defendant against the decision of a judge made under ss. 43 and 44.

Section 46 deals with trials already under way where jury tampering appears to have taken place. If, in the exercise of his common law powers, the trial judge is minded to discharge the jury, he must notify the prosecution and defence, explain his reasons, and allow the parties an opportunity to make representations.

If the judge, after considering the representations, discharges the jury, he can make an order that the trial should continue without a jury if he is satisfied that (a) tampering has taken place, and (b) continuing without a jury would be fair to the defendant.

Alternatively, the judge must terminate the trial completely if this is necessary in the interests of justice. If he does so, he can order a retrial and has a discretion to order that the retrial should take place without a jury if he is satisfied that both the conditions in s. 44 (above) are likely to be fulfilled in respect of the retrial.

Section 47 provides for an appeal to the Court of Appeal against the decision of a judge to continue a trial in the absence of a jury, and against a decision that a retrial be conducted in the absence of a jury.

6.6 Juries in civil cases

6.6.1 Introduction

In the county courts, a jury of eight may be called (County Courts Act 1984, ss. 66–67), although trial by jury is practically obsolete there. In the High Court, a jury of 12 may be called. In the Chancery Division and the Family Division, trial by jury is practically obsolete. In the Queen's Bench Division, a jury is used in perhaps only 1 per cent of all cases. (For the coroner's jury, see paras 2.5.4 and 2.5.6.)

The use of the jury in civil cases has declined considerably over the past 150 years. Before 1854, all cases tried in the common law courts were heard by a judge sitting with a jury. The Common Law Procedure Act 1854 then laid down that common law actions could be tried without a jury if the parties agreed. There was a further decline after the reorganisation of the structure of the courts effected by the Judicature Acts in 1875, and s. 6 of the Administration of Justice (Miscellaneous Provisions) Act 1933 further curtailed the use of the jury in civil cases. The provisions of the 1933 Act were repealed and replaced by the Supreme Court Act 1981, s. 69.

6.6.2 The right to a jury in civil cases

The Supreme Court Act 1981, s. 69, provides that there is only a *right* to trial by jury where there is a charge of fraud against a party, and in claims for libel, slander, malicious prosecution, or false imprisonment. A similar provision is contained in s. 66 of the County Courts Act 1984.

Juries today are most commonly used in defamation actions, although their virtual abolition in such actions was recommended in 1975 by the Faulks Committee on Defamation (*Report*, Cmnd 5909, 1975), which proposed that jury trial in defamation actions should be at the discretion of the judge. The proposal was not implemented.

When Parliament was debating the Bill which became the Supreme Court Act 1981, an amendment proposed in the House of Lords would have implemented the Faulks Committee recommendation. The proposed amendment, which was defeated, was a result of *Orme* v *Associated Newspapers Group Ltd* (31 March 1981, unreported), a libel action brought by the Unification Church (the 'Moonies') against the *Daily Mail*. The action was tried with a jury on more than 100 days between October 1980 and the end of March 1981. In December 1982, the Unification Church lost a two-week appeal before the Court of Appeal, based on alleged misdirections by the trial judge, and was left to pay the costs of the trial and the appellate proceedings estimated at some £800,000.

Even in the four exceptional cases mentioned above, the right to a jury is *qualified*. The judge can refuse to allow trial by jury if the case 'requires any prolonged examination of documents or accounts or any scientific or local investigation which cannot conveniently be made with a jury' (Supreme Court Act 1981, s. 69(1); County Courts Act 1984, s. 66(3)).

It has been held that the word 'conveniently' refers to the efficient administration of justice. If the administration of justice is likely to suffer by the presence of a jury, the judge should consider ordering a trial by judge alone. (See, e.g., *Goldsmith* v *Pressdram Ltd* [1987] 3 All ER 485, CA—order for trial by judge alone upheld; *Viscount De L'Isle* v *Times Newspapers Ltd* [1987] 3 All ER 499, CA—order for trial by judge alone reversed; *Beta Construction Ltd* v *Channel Four TV Co. Ltd* [1990] 2 All ER 1012, CA—order for trial by judge alone upheld; *Taylor* v *Anderton* [1995] 2 All ER 420, CA—order for trial by judge alone upheld; *Aitken* v *Preston and others* (1997) *The Times*, 21 May, CA—order for trial by judge alone upheld; *Oliver* v *Calderdale Metropolitan Borough Council* (1999) *The Times*, 7 July, CA—order for trial by judge alone upheld.)

During the passage through the House of Commons of the Bill which became the

Supreme Court Act 1981, the government proposed an amendment that jury trial should not be available in civil cases where the probable length of the trial made the action one which could not conveniently be tried with a jury. The proposed amendment, which was not accepted, was again a direct result of *Orme* v *Associated Newspapers Group Ltd.*, above.

It has, in any event, been made clear by subsequent case law that the likely prolongation of a trial is a relevant factor, among others, to be taken into account under s. 69 of the Supreme Court Act 1981 as it stands. In *Goldsmith* v *Pressdram Ltd* [1987] 3 All ER 485, CA, Lawton LJ said (at p. 492) that:

[a] trial with a jury inevitably takes longer than a trial by a judge alone. If the trial is made much longer because of the time taken up by the jury examining documents, then an element of inconvenience arises.

In *Beta Construction Ltd* v *Channel Four TV Co. Ltd* [1990] 2 All ER 1012, CA, the defendants had admitted liability in a libel action and the only issue to be tried was the quantum of damages. This would involve a consideration of actual and estimated losses of net profits, which could only be worked out by a long investigation of the claimants' accounts and other documents. The claimants wanted damages to be assessed by a jury while the defendants wanted them assessed by a judge alone. At first instance, it was decided that damages should be quantified by a judge alone, and, on appeal, that decision was upheld. The Court of Appeal said that whether a libel action involving 'prolonged examination of documents or accounts' can conveniently be tried by a jury, so as to be consistent with the efficient administration of justice, depends on a consideration of such factors as:

(a) the extent to which the presence of a jury might add to the length of the trial;

(b) the extent to which the presence of a jury might add to the cost of the trial by reason of its increased length and the necessity of photocopying a multitude of documents for use by the jury;

(c) any practical difficulties which a jury trial might cause, such as the physical problem of handling in the confines of a jury box large bundles of bulky documents;

(d) any special complexities in the documents or accounts which might lead a jury to misunderstand the issues in the case.

6.6.3 The judge's discretion to order trial by jury in civil cases

In all other civil cases in the Queen's Bench Division there is no *right* to trial by jury but the judge has a *discretion* to allow it (Supreme Court Act 1981, s. 69(3); a similar provision relating to trial in a county court is contained in the County Courts Act 1984, s. 66(2)).

The guidelines for the proper exercise of this discretion were laid down by a full Court of Appeal of five members in *Ward* v *James* [1966] 1 QB 273, CA. This was an action for personal injuries sustained in a motor accident. The claimant sued the defendant in negligence and applied for trial by jury. The judge ordered a jury, and the defendant appealed against this order to the Court of Appeal which held that, in the circumstances, there should be a jury to hear the case. The Court of Appeal laid down the following guidelines:

(a) the judge's discretion is not absolute but must be exercised judicially;

(b) normally, personal injury cases should be tried by a judge sitting alone so as to achieve the three basic objectives of assessability, uniformity, and predictability of awards of damages;

(c) a jury in civil cases should be used only in exceptional circumstances, such as where there is a substantial dispute about the facts.

Ward v *James* [1966] 1 QB 273, CA, had a significant effect in the Queen's Bench Division of the High Court. Before it, about two per cent of personal injury cases were tried by jury ([1966] 1 QB 273 *per* Lord Denning MR at p. 290). After it, the proportion dropped to less than one per cent. The full effect of the decision on the trial of civil cases was always difficult to predict, especially because no comprehensive guidance was given on what are 'exceptional circumstances'. But at least the Court of Appeal made it very clear that it was not seeking to abolish trial by jury. Lord Denning MR, in particular, stressed the point in the following well-known passage (ibid., at p. 295):

Let it not be supposed that this court is in any way opposed to trial by jury. It has been the bulwark of our liberties too long for any of us to seek to alter it. Whenever a man is on trial for serious crime, or when in a civil case a man's honour or integrity is at stake, or when one or other party must be deliberately lying, then trial by jury has no equal.

The question of whether a jury should be used in a personal injury case arose before the Court of Appeal again in *Hodges* v *Harland & Wolff Ltd* [1965] 1 All ER 1086, CA. The claimant was injured at work by a spindle which caught his trousers and injured his penis. One result of the injury was that the claimant could no longer perform the sex act although he still retained his sexual urge. The judge ordered trial by jury after taking into account all the relevant factors, including the desirability of uniformity of awards. The defendant appealed. The Court of Appeal held that the judge was right and that there should be a jury as the case was an exceptional one. On this occasion, Lord Denning MR (at p. 1087) emphasised that the decision in *Ward* v *James*, above, had not taken away the right to trial by jury in civil cases:

It was not this court but Parliament itself which years ago took away any absolute right to trial by jury and left it to the discretion of the judges. This court in *Ward* v *James* affirmed that discretion and . . . laid down the considerations which should be borne in mind by a judge when exercising his discretion.

In *Singh* v *London Underground* (1990) *The Independent*, 25 April, an application for trial by jury of a claim arising from the King's Cross Underground fire disaster of November 1987 was refused on the ground that a case involving such wide issues and technical topics was unsuitable for a jury.

In another personal injury case, *H* v *Ministry of Defence* [1991] 2 All ER 834, CA, the claimant was a regular soldier diagnosed by the Army Medical Service as suffering from Peyronie's disease. This is a congenital disease resulting in curvature of the penis and making sexual intercourse difficult. He submitted to a test designed to assess the extent of the abnormality. The test involved injecting the penis with a saline solution; as a result of the test the claimant's penis became infected at the site of the injection. He was then advised to have a skin graft, but while he was under the anaesthetic awaiting surgery it was discovered that a skin graft was impossible and, instead, a major part of his penis was amputated.

On discovering what had happened to him, the claimant suffered severe psychological trauma and brought an action for damages for personal injuries caused by the defendants' negligent medical treatment. The defendants admitted liability and the only issue remaining between the parties was the assessment of damages. On the claimant's application, the judge ordered trial by jury on the ground that the circumstances of the claimant's claim were exceptional.

The Court of Appeal allowed the defendants' appeal and held that, although the claimant's injuries were most unusual and distressing, the circumstances of his claim were not exceptional and did not, therefore, displace the legislative presumption contained in s. 69(3) of the Supreme Court Act 1981 against trial by jury in those civil cases not specified in s. 69(1). Applying *Ward* v *James* [1966] 1 QB 273, CA, the court said it was appropriate for there to be trial by a judge alone since it was important that an award of compensatory damages to the claimant should be compatible with the conventional scale of personal injury awards (see [1991] 2 All ER 834, CA, *per* Lord Donaldson MR, delivering the judgment of the court, at p. 840).

6.7 Confidentiality of jury deliberations and finality of verdicts

6.7.1 Introduction: the common law rule on confidentiality

It has been a rule of the common law for over 200 years that in both civil and criminal cases the deliberations of a jury take place in secret, and that, after the trial, they remain confidential. In essence, this judge-made rule is about the admissibility of evidence since, as strictly applied, it insists that, after the conviction of the defendant in a criminal case or after the delivery of the verdict in a civil case, the court will not admit evidence about, or otherwise inquire into, what went on in the jury room.

The rule is based, first, on the need to guarantee that a jury is able to engage in full and frank exchanges without fear of repercussions, and, secondly, on the desire to achieve finality in legal proceedings. It is in the interests of the public, and of the efficient administration of justice, that a jury's verdict should be final and unchangeable. The principle of finality also protects jurors themselves from pressure to explain or alter their verdicts, thus minimising the risk of criticism, ridicule, and intimidation. The principle of finality does not, of course, preclude an *appeal* against a verdict to a higher court as long as pursuing the appeal does not involve breaching the confidentiality of the jury's deliberations.

As long ago as 1785, Lord Mansfield held that the court could not receive affidavit evidence from two jurors alleging that the jury had reached its verdict by tossing a coin (*Vaise* v *Delaval* (1785) 1 Durn & E 11). Similarly, in the civil case of *Boston* v

W.S. Bagshaw & Sons [1966] 1 WLR 1135n, CA, the Court of Appeal declined to admit affidavit evidence from all 12 jurors after the trial. Here, the claimant had sued the defendants for libel and the question of malice was raised. The jury found there was no malice and the claimant's case failed. The jurors were surprised at this result, and each swore an affidavit that he meant to find malice and wished to change his answer. The claimant moved for a new trial. The Court of Appeal refused the motion, Lord Denning MR saying (at p. 1136):

> To my mind it is settled as well as anything can be that it is not open to the court to receive any such evidence as this. Once a jury have given their verdict, and it has been accepted by the judge, and they have been discharged, they are not at liberty to say they meant something different.

Lord Denning's dictum was applied by the Judicial Committee of the Privy Council in *Nanan v The State* [1986] 3 All ER 248, PC, where it was held that affidavit evidence from four jurors that they had been under a misapprehension in agreeing to a guilty verdict in a criminal trial was inadmissible.

In *R v Miah* [1997] 2 Cr App R 12, CA, the Court of Appeal refused to order an investigation into complaints contained in a document emanating from a juror which suggested that some of the jurors had not approached their task as they ought to have done. *R v Miah* was applied in *R v Qureshi* [2002] 1 WLR 518, CA, in which the Court of Appeal declined to enquire any further into allegations made by a juror after the defendant's conviction that some members of the jury had been racially prejudiced against him and had decided that he was guilty at the start of the trial. The Court of Appeal had gone so far as to instruct a senior police officer to take a statement from the juror, avoiding any enquiry about what had taken place during the jury's deliberations, but then held that it could not have regard to the statement nor authorise any further enquiries of the juror.

R v Qureshi, above, was, in turn, followed in *R v Mirza* [2002] Crim LR 921, CA, and *R v Connor and Rollock* [2002] EWCA Crim 1236, CA. In both of these cases, a juror, after conviction of the defendants, alleged that there had been misconduct during deliberations in the jury room. In one case, the allegation was that the other jurors had been racially prejudiced against the defendant. In the second case, it was alleged that the other jurors had not considered the evidence against each co-defendant separately but had decided to convict them both in order to save time. The Court of Appeal held in each case that it was prevented from investigating the allegations, both by the common law confidentiality rule and by s. 8 of the Contempt of Court Act 1981. The two cases subsequently went on appeal to the House of Lords (see para 6.7.4 below).

6.7.2 Exceptions

Over the years, exceptions have been developed to the common law rule on confidentiality. In the main, the exceptions relate only to situations where it is alleged that the jury has been affected by extraneous matters, i.e., influences extrinsic to its deliberations, the investigation of which does not involve invading the privacy of the jury room. For example, in *R v Wooller* (1817) 2 Stark 111, not all of the jurors were able to get into the court room for delivery of the verdict because the jury box was occupied by a jury trying another case. Consequently, the verdict was delivered out of the hearing of some jurors (who did not, in fact, agree with it). Although the court declined to accept affidavits from the jurors, it did admit a statement about what had gone wrong from the trial judge. The court ordered a new trial.

R v Wooller, above, was followed on similar facts in the *civil* case of *Ellis v Deheer* [1922] 2 KB 113, CA. On this occasion, however, affidavit evidence was admitted from the three jurors who alleged that they were unable to hear what the foreman said, and who did not, in fact, agree with the verdict. A new trial was ordered.

A similar result would ensue if it was alleged, and proved, that a juror was not competent to understand the proceedings due to, for example, insufficiency of English (*Ras Behari Lal v King-Emperor* (1933) 50 TLR 1, PC).

In *R v Young* [1995] QB 324, CA, the Court of Appeal admitted affidavit evidence which it had ordered to be taken from each of the 12 jurors, and from the two bailiffs who looked after them in the hotel at which they had stayed, in deciding that the use of a makeshift ouija board by four members of the jury at the hotel constituted a material irregularity in the trial.

Several cases have arisen in connection with verdicts mistakenly given or not given. In *R v Froud* [1990] Crim LR 197, CA, the jury foreman mistakenly announced a verdict of not guilty. He realised his mistake immediately and corrected the verdict to one of guilty. The judge accepted this verdict and refused later to reopen the matter. The Court of Appeal held that the jury was entitled to rectify its verdict before it was recorded, or promptly thereafter.

In *R v Aylott* [1996] 2 Cr App R 169, CA, owing to a misunderstanding on the part of the judge, the jury was discharged before it could deliver the verdict it had reached. Shortly thereafter, the judge realised his mistake, reassembled the jury, and accepted its guilty verdict. The Court of Appeal held that the judge had acted correctly in the interests of fairness and justice.

In the somewhat unusual circumstances of *R v Bills* [1995] 2 Cr App R 643, CA, the judge was held to have been wrong to accept a changed verdict. The jury had found the defendant not guilty of wounding with intent, but guilty of the lesser offence of unlawful wounding. The members of the jury remained in the jury box while the defendant's previous convictions were read out and then they were discharged. Later the same day they purported to change their verdict, saying they had meant to find the defendant guilty of the more serious offence. The judge accepted the changed verdict.

The Court of Appeal held that the changed verdict was unsafe since the jury had been influenced by hearing about the defendant's previous convictions, and to allow that verdict to stand would be unfair to the defendant. The judge's summing-up had been clear, there was no evidence of any misunderstanding by the jury, their original verdict was plain, and there was no indication of dissent when it was announced. The Court of Appeal quashed the changed verdict and substituted the original verdict.

R v Bills was distinguished in *Igwemma v Chief Constable of the Greater Manchester Police* [2002] 2 WLR 204, CA, a *civil* case, where the changed verdict was allowed to stand in place of the original verdict (which had been based on a misunderstanding). The jury, unlike the one sitting in *Bills*, had heard nothing prejudicial to the claimant after giving its original verdict.

6.7.3 Section 8 of the Contempt of Court Act 1981

A deficiency in the common law rule on confidentiality was identified in *Attorney-General v New Statesman* [1980] 1 All ER 644, DC. This case arose out of the acquittal by a jury in 1979 of a well-known politician, Jeremy Thorpe, on charges of conspiracy to murder and incitement to murder (*R v Holmes and others*, 22 June 1979, unreported). In view of the massive publicity surrounding the case and the consequent difficulty of finding an unbiased jury, the verdicts of not guilty were seen by many as a remarkable vindication of trial by jury. It soon became clear, however, that extraneous factors had been at work on the minds of the jurors, particularly the widely reported fact that the press had 'bought' some prosecution witnesses.

Soon after the acquittal of Mr Thorpe, the *New Statesman* published an interview with one of the jurors, who remained anonymous and was not paid for his revelations (see 'Thorpe's trial: how the jury saw it', *New Statesman*, 27 July 1979). It was said in the published interview that the members of the jury had made up their minds to acquit on the first day of the trial.

In contempt proceedings brought by the Attorney-General against the *New Statesman*, it was held that the mere disclosure of the secrets of the jury room was not contempt at common law unless it interfered with the finality of jury verdicts or with the attitude of future jurors towards their responsibilities. On the facts, the *New Statesman* was held not to be in contempt. It was said that each case of disclosure must be judged in the light of the circumstances in which the disclosure took place (*Attorney-General v New Statesman*, above).

Section 8 of the Contempt of Court Act 1981 was enacted to fill the lacuna revealed in *Attorney-General v New Statesman* by allowing contempt proceedings to be brought in circumstances not provided for by the common law.

Section 8 provides that it is a contempt to obtain, disclose, or solicit any particulars of statements made, opinions expressed, arguments advanced, or votes cast by members of a jury in the course of their deliberations in any legal proceedings, whether criminal or civil (Contempt of Court Act 1981, s. 8(1)). Proceedings for this sort of contempt can only be brought by, or with the consent of, the Attorney-General, or on the motion of a court having jurisdiction to deal with it, such as the trial court in the case of an attempt by a journalist to obtain particulars from a juror (ibid., s. 8(3)).

Section 8(1) prohibits disclosures both by a member of the jury and by a non-juror who has acquired information directly or indirectly from a juror (*Attorney-General v Associated Newspapers Ltd* [1994] 1 All ER 556, HL; see further on this case, para 8.2.1.5.2).

It is not a contempt under s. 8(1) to disclose particulars in the legal proceedings in question for the purpose of enabling the jury to arrive at its verdict, or in connection with the delivery of the verdict (ibid., s. 8(2)(a)), or to disclose them in evidence in subsequent legal proceedings for an offence alleged to have been committed in relation to the jury in the earlier proceedings (ibid., s. 8(2)(b)). However, there is no exception permitting disclosure of jury room deliberations for the purpose of an appeal by the defendant against his conviction by the jury.

6.7.4 Relationship between the common law rule and section 8

In *R* v *Young* [1995] QB 324, CA (the 'ouija board case', para 6.7.2 above), it was said that *judges* are bound by s. 8 like everyone else, and the Court of Appeal cannot, therefore, investigate what took place in the jury room without committing contempt of court. This view of s. 8 was held to be mistaken by the House of Lords in *R* v *Mirza* [2004] 2 WLR 201, HL.

In this case, the appeals in *R* v *Mirza* and *R* v *Connor and Rollock* (see para 6.7.1 above) were heard together. By a majority, the House of Lords held that the allegations of the juror in each case were not admissible as evidence, and that, on the facts, all the defendants had had a fair trial. The decisions of the Court of Appeal were affirmed.

The House examined the relationship between the common law rule on con- fidentiality, s. 8 of the Contempt of Court Act 1981, and art. 6 of the European Convention on Human Rights (right to a fair trial).

It was made clear that it is the common law rule which regulates the admission of evidence about jury deliberations, and not s. 8 of the 1981 Act, which does not apply to the courts since a court (meaning the judge) cannot be in contempt of itself. Section 8 is directed at persons other than judges. It applies to members of a jury, and the particular mischief which it seeks to remedy is the communication of information to, and in, the media which ought to be kept secret. It was further pointed out that the common law rule on the secrecy of jury deliberations had already been held to be compatible with art. 6 by the European Court of Human Rights itself in *Gregory* v *United Kingdom* (1997) 25 EHRR 577, ECHR, where it was described as a 'crucial and legitimate feature of English trial law'.

Clearly, the most appropriate time for a juror to make allegations of impropriety is *before* the verdict is given. If this is done (anonymously by means of a note, if neces- sary), then the matter can be dealt with by the trial judge (see para 6.5.6 above). Section 8 of the Contempt of Court Act 1981 would not be an impediment since it does not forbid disclosures for the purpose of enabling a jury to arrive at its verdict. In *R* v *Mirza* [2004] 2 WLR 201, HL, it was recommended that jurors should be reminded before the trial begins (and, possibly, during the trial) that they are under a duty to inform the court immediately of any irregularity (internal as well as external) arising during their deliberations (*per* Lords Slynn, Hope, Hobhouse, and Rodger at [51], [126], [148], and [165], respectively).

It was suggested, *obiter*, by three of their Lordships in *R* v *Mirza*, above, that the common law rule might be flexible enough to allow evidence to be admitted in support of an allegation that the jury *did not deliberate properly at all*, but reached its verdict in the jury room by an illegitimate and irrational process, such as tossing a coin, drawing lots, or using a ouija board (see *per* Lords Slynn, Hope, and Rodger at [55], [123], and [166], respectively).

R v *Mirza* [2004] 2 WLR 201, HL, was followed quickly by *Practice Direction (Crown Court: Guidance to Jurors)* [2004] 1 WLR 665, which added a new section to *Practice Direction (Criminal Proceedings: Consolidation), Part IV* [2002] 1 WLR 2870. Under the new arrangements, the trial judge should warn jurors that it is their duty to bring to his attention promptly any concerns about the behaviour of fellow jurors or about the behaviour of others which affects the jurors. He should stress that, unless this is done at the time, it may be impossible to put things right later. The judge should consider whether the warning should be repeated (especially in a long trial) before the jury retires to consider its verdict.

R v Mirza, above, was applied in *R v Smith* [2005] 1 WLR 704, HL, in which the House of Lords had to consider what steps the trial judge can take in order to deal with allegations of impropriety made by a juror. In the *Smith* case, the allegations communicated in a letter to the judge were to the effect that other jury members were ignoring the judge's directions on the law, that they were engaging in speculation, and were bargaining over the verdicts to be reached. After consulting counsel, the judge dealt with the allegations by giving the jury further directions. The jury eventually convicted the defendants on majority verdicts.

On the defendants' appeal, it was argued that the judge should have questioned the members of the jury about the criticisms made in the juror's letter. The House of Lords held that this would not have been appropriate. Their Lordships refused to list the circumstances (if any) in which it might be permissible for the trial judge to question the jury. The two remaining courses of action open to the judge were to discharge the jury or to give them further directions. It was held that, on the facts, he had been entitled to prefer the latter course to the former. However, it was decided that his further instructions were not sufficiently comprehensive and emphatic. He had failed to give the jury a stern warning that they must follow his directions on the law, stick to the evidence without speculation, and decide on their verdicts without bargaining. The defendants' convictions were quashed as unsafe.

R v Mirza [2004] 2 WLR 201, HL, was applied again in *Attorney-General v Scotcher* [2005] 1 WLR 1867, HL. Here the appellant had served on a jury which, by a majority verdict, convicted two defendants who were brothers. He wrote anonymously to the defendants' mother stating that he had voted for acquittal and discussing in detail his views on the unsatisfactory nature of the jury's deliberations. The letter was referred to the police, the appellant's identity was discovered, and the Attorney-General gave his consent to the institution of proceedings against him for contempt of court under s. 8(1) of the Contempt of Court Act 1981. The House of Lords held, first, that the appellant was guilty of contempt since, by writing to the defendants' mother, he had created all the risks to the confidentiality of the jury's deliberations which s. 8(1) was designed to prevent, and, secondly, that his desire to expose a miscarriage of justice was no defence.

The *Scotcher* case is also important for indicating some circumstances in which the appellant would *not* have been in contempt. For example, if, instead of writing to the defendants' mother, he had communicated his misgivings directly to the Crown Court or the Court of Appeal, or indirectly to the Crown Court (via the jury bailiff or the clerk of the court), or in a sealed letter to the defendants' solicitors or counsel, or to a citizens' advice bureau, with a request that the letter be passed on unopened to the appropriate court authorities, there would have been no contempt provided that he acted without malice, dishonesty, or other improper motive.

With a view to identifying how the support and protection provided to juries might be enhanced, the Department for Constitutional Affairs issued a Consultation Paper in January 2005 seeking views on two issues. First, whether there should continue to be a complete bar on all research into a jury's deliberations, and, secondly, whether the circumstances should be broadened in which those deliberations can be investigated following the making of allegations of improper behaviour in the jury room (*Jury Research and Impropriety*, Department for Constitutional Affairs, 2005). Responses received during the consultation period were published in November 2005 (*Jury Research and Impropriety: Response to Consultation*, Department for Constitutional Affairs, 2005). In the light of these responses, the government decided that more research into the jury decision-making process should be allowed—initially within the confines of the existing law without amending s. 8 of the Contempt of Court Act 1981.

On the question of jury impropriety, the government's view is that the law should be allowed to develop on a case-by-case basis rather than by means of amendments to the statute law.

6.8 Future of the jury

6.8.1 Some advantages of trial by jury

Several advantages have long been claimed for trial by jury. In words which have become something of a cliché, the jury is a bastion of liberty against the state. It is, however, ironic that today some offences which involve the state are triable without a jury, i.e., summarily in the magistrates' court. Examples are assaulting the police and offences arising out of demonstrations and strikes.

Trial by jury involves ordinary people in the day-to-day administration of justice and thus prevents the domination of the system by professional judges. But the jury is used only in a very small proportion of cases, and the value of lay participation in those cases where there is a jury can be effectively diminished by the influential position of the judge in a Crown Court trial.

For instance, a significant proportion of acquittals by juries are on the direction of the trial judge. The figures show that in 2004, 66.7 per cent of defendants who pleaded not guilty to all charges were totally acquitted. Of those acquitted, 31 per cent were acquitted by the jury alone after a full trial, 11 per cent were acquitted by the jury on the direction of the judge because, for example, the prosecution case was weak, and 58 per cent were discharged by the judge alone because, for example, the prosecution had offered no evidence (*Judicial Statistics 2004*, Cm 6565, 2005, p. 91).

The attitude of juries has been known to mitigate the harshness of the criminal law and to lead to its reform. In the early decades of the nineteenth century all felonies (some 146 of them, except petty larceny and mayhem) were, in theory, punishable by death. One of them was theft of goods or money above the value of one shilling. Juries would often find the value of the property to be less than one shilling and so avoid the imposition of the death penalty.

In the twentieth century, jurors who were motorists themselves were reluctant to convict persons accused of motor-manslaughter, which carried a sentence of life imprisonment. In 1956, Parliament was moved to create the separate, less serious offence of causing death by reckless driving, which carried a sentence of up to five years' imprisonment (Road Traffic Act 1956, s. 8; later contained in the Road Traffic

Act 1988, s. 1, by an amendment to which, made by the Road Traffic Act 1991, the offence of causing death by *reckless* driving was abolished and replaced by the offence of causing death by *dangerous* driving for which the sentence is, under the Road Traffic Offenders Act 1988, sch. 2, pt 1, as most recently amended by the Criminal Justice Act 2003, imprisonment for up to 14 years).

The introduction in 1967 of the so-called 'breathalyser law', which is based on the presence in the bloodstream of a prescribed level of alcohol, was due in part to the reluctance of juries to convict motorists of drunken driving (Road Safety Act 1967; the relevant provisions are now contained in the Road Traffic Act 1988).

The acquittals of Jeremy Thorpe in 1979 (para 6.7.3 above), Clive Ponting in 1985 (para 6.5.2 above), and of Patrick Pottle and Michael Randle in 1991—described in some quarters as perverse verdicts—have been explained as the reaction of a lay jury to an unsatisfactory law or to unfair treatment of the defendant.

Messrs Pottle and Randle were prosecuted 25 years after the event for helping George Blake, the convicted MI6 double agent, to escape from Wormwood Scrubs prison. Despite admitting their guilt publicly, discussing the episode in the media, and publishing a book about it (*The Blake Escape: How We Freed George Blake, and Why*), they pleaded not guilty at their trial and were acquitted by a jury which was not only sympathetic (and no doubt unimpressed by the bunglings of the security services and the prosecuting authorities over the years) but was also prepared to ignore the direction of the trial judge that the defendants had no defence in law (see *The Independent*, 27 June and 5 July 1991).

6.8.2 Some disadvantages of trial by jury

6.8.2.1 Introduction

There are disadvantages to trial by jury. Jurors can be intimidated, and this factor, *inter alia*, led to the suspension of jury trial for terrorist offences in Northern Ireland. The law on contempt of court, and the existence of criminal offences relating to interference with jurors, are not necessarily deterrents to determined and unscrupulous people.

The Royal Commission on Criminal Justice (*Report*, Cm 2263, 1993) found it necessary to recommend that every effort should be made to protect jurors from intimidation, that 'sensitive' cases be tried in courtrooms where the positioning of the public gallery does not facilitate intimidation of jurors by members of the public, that in new courtrooms jury boxes should not be situated opposite the public gallery, and that, so far as practicable, jurors be given waiting and eating areas separate from the ordinary public. There is evidence that the intimidation, or attempted bribery, of jurors and witnesses participating in trials involving armed robbery or drugs offences is a highly organised activity carried on by criminal gangs (and see the comments of Lord Bingham CJ on jury-nobbling in *R v Comerford* [1998] 1 All ER 823, CA, at p. 827).

The law was strengthened by the Criminal Justice and Public Order Act 1994, s. 51(1)–(2), as substituted by the Youth Justice and Criminal Evidence Act 1999, which created new offences of intimidating or harming, or threatening to harm, jurors, witnesses, or persons assisting in the investigation of offences. In addition, s. 54 of the Criminal Procedure and Investigations Act 1996 was enacted to deal with the mischief of 'tainted' acquittals obtained by criminals who use money, threats, or violence to evade conviction by means of interference with, or intimidation of, jurors or witnesses (see para 7.5.2). The Criminal Justice Act 2003, s. 44 (not yet in force) will allow the prosecution to apply for the trial of an indictable offence to be conducted without a jury where there is a danger of jury tampering (para 6.5.9 above).

The danger of intimidation in the *workplace* is dealt with by the Employment Rights Act 1996, as amended by the Employment Relations Act 2004. With effect from April 2005, it is unlawful for an employer to dismiss, or discriminate against, an employee who has been summoned for jury service or who has been absent from work on jury service.

Jurors may not always be able to bring to bear that degree of objectivity which is essential to the impartial administration of justice. They may, for example, be prejudiced, albeit subconsciously, in favour of attractive members of the opposite sex, or against newspapers in libel suits (para 6.8.2.2 below), or against the police in actions for false imprisonment and malicious prosecution (para 6.8.2.3 below).

The inexperience and ignorance of jurors may mean that they are too easily influenced by what counsel says to them at the expense of the real issues in the case. They may be incapable of following the evidence in a long and complicated trial of libel or commercial fraud. In 1986, the Committee on Fraud Trials under the chairmanship of Lord Roskill, a Lord of Appeal in Ordinary, recommended abolition of jury trial in complex criminal fraud trials in favour of trial before a tribunal consisting of a judge sitting with two laymen with business, City, or accounting skills (*Report*, HMSO, 1986).

The government of the day rejected this central recommendation and decided instead to introduce statutory changes designed to make it easier for juries to follow the proceedings. These changes are contained in the Criminal Justice Act 1987.

In a case of *serious or complex fraud*, a judge of the Crown Court can order the holding of a 'preparatory hearing' before the time when the jury is sworn if he considers that such a hearing would produce substantial benefits for the purpose of:

(a) identifying issues which are likely to be material to the determinations and findings which are likely to be required during the trial;

(b) if there is to be a jury, assisting its comprehension of those issues and expediting the proceedings before it;

(c) determining a prosecution application for trial without a jury in a case of serious or complex fraud;

(d) assisting the judge's management of the trial; or

(e) considering questions as to the severance or joinder of charges (Criminal Justice Act 1987, s. 7(1), as amended by the Criminal Justice Act 2003).

At the preparatory hearing the judge may, *inter alia*, decide any question of law relating to the case, including any question about the admissibility of evidence (ibid., s. 9(3), as amended by the Criminal Justice Act 2003). He may order the prosecution to prepare its evidence in a form which is likely to aid comprehension by a jury (ibid., s. 9(4), as amended by the Criminal Justice Act 2003).

Auld LJ in his *Review of the Criminal Courts of England and Wales* (Lord Chancellor's Department, 2001) made a proposal similar to that of Lord Roskill 15 years earlier, namely, that in *serious or complex fraud* cases the trial judge should have the power to order trial by himself and two lay members drawn from a panel of experts. This proposal was again unacceptable to the government, which decided that serious or complex fraud cases should instead be tried by a judge sitting alone. Section 43 of the Criminal Justice Act 2003 was enacted to achieve this change, but it is highly unlikely that the section will ever be brought into force (see para 6.5.9 above).

In the meantime, preparatory hearing measures, similar to those available in fraud cases since 1987, were introduced, as from January 1999, by the Criminal Procedure and Investigations Act 1996 for use in *non-fraud* cases of such complexity, likely length, or seriousness (the latter word added by the Criminal Justice Act 2003) that a judge of the Crown Court believes that a preparatory hearing would produce substantial benefits.

Jury awards of damages in civil cases may be unrealistic and too sympathetic towards the claimant. Appealing against the quantum of damages assessed by a jury is usually fruitless since the Court of Appeal will only interfere where the assessment is so high (or so small) as to be perverse (para 6.1 above).

The generosity of juries in running-down cases led to the principles laid down in *Ward v James* [1966] 1 QB 273, CA (para 6.6.3 above), and exaggerated awards of damages were responsible, in part, for the abolition of the action for breach of promise in 1970 (Law Reform (Miscellaneous Provisions) Act 1970, s. 1, following recommendations of the Law Commission in Law Com. No. 26, 1969). The problem of over-generous awards in libel cases remains.

6.8.2.2 Jury awards in libel cases

Jury awards in libel cases continue to give concern, in spite of judicial and other efforts to control them. It is always difficult to put a value on a person's reputation, but awards have got out of hand and seem often to contain an element of punishment which, in general, is not justified in law (see para 9.2.2 for the circumstances in which punitive damages can be awarded).

Enormous awards were made in the late 1980s and throughout the 1990s. In 1987, a jury awarded what were then record libel damages of £500,000 to the novelist and politician, Jeffrey Archer, against whom it was alleged by *The Star* newspaper that he had paid a prostitute for sex. Costs (also payable by the unsuccessful defendants) were estimated at a further £670,000. In 1988, the actress Koo Stark received damages of £300,000 from the publishers of the *Sunday People* following allegations that she had resumed her relationship with the Duke of York after her marriage. Sonia Sutcliffe, the estranged wife of Peter Sutcliffe (the so-called 'Yorkshire Ripper'), was awarded libel damages of £600,000 in 1989. She had successfully sued the publishers of the satirical magazine, *Private Eye*, in respect of assertions that she was prepared to capitalise on her notoriety as the wife of a multiple murderer by selling her story to the Press.

The jury's award in this case was, however, later set aside on appeal (*Sutcliffe v Pressdram Ltd* [1990] 1 All ER 269, CA). The Court of Appeal took the view that an award of £600,000 was so high as to be irrational. The action was settled by the parties for damages of £60,000, one-tenth of the jury's original award.

Lord Donaldson MR (at pp. 283–4) called for juries to be given improved guidance by the judge which was aimed at helping them to appreciate the real value of large sums of money. The guidance could possibly take the form of inviting them to consider what the result of their awards would be in terms of weekly, monthly, or annual income if the damages were invested in a building society deposit account without touching the capital sum, or, in the case of a smaller award, to consider what could be bought with it.

Any hopes that the comments of the Court of Appeal in *Sutcliffe* v *Pressdram Ltd*, above, would in themselves be enough to bring about a return to a saner level of awards were dashed just six weeks later when the jury in Lord Aldington's libel case awarded damages of £1.5m, following a 41-day trial the costs of which were estimated as approaching a further £1m (*Aldington* v *Watts and Tolstoy*, 30 November 1989, unreported, but see *The Times*, 1 December 1989).

The unprecedented sum of £1.5m was awarded despite the trial judge's efforts to advise the jury in accordance with *Sutcliffe* v *Pressdram Ltd*. Thus, he had warned them against giving 'grossly too much' (what he described as 'Mickey Mouse money') because to do so could simply lead to an appeal, had reminded them to pay no attention to any other case they may have heard about, and had pointed out that it was not their function to punish or 'fine' the defendants.

The *Aldington* case had a number of peculiar features. The defendants were not national newspaper proprietors but private individuals. The libel was contained in a pamphlet and did not, therefore, have the mass circulation of a newspaper, although 10,000 copies of the pamphlet were distributed to Members of Parliament, peers, staff, parents and boys at Winchester College where Lord Aldington was Warden, and to his neighbours. It has to be admitted that the libel was particularly grave, involving as it did allegations that Lord Aldington, as an officer in the Second World War, had been responsible for the repatriation, torture, and massacre of some 70,000 Cossacks and Yugoslavs.

Nevertheless, an award of such massive proportions as £1.5m cannot be justified, and, if taken to appeal, would almost certainly have resulted in a retrial or a substantial reduction of the damages if the parties had invited the Court of Appeal itself to reassess them. As it was, Count Tolstoy was ordered to pay into court the sum of £124,900 as security for costs if he wished to pursue an appeal (*Aldington* v *Watts and Tolstoy (No. 2)* (1990) *The Independent*, 20 July, CA). On his being unable to do so, no appeal was pursued in the case. Both defendants later became bankrupt.

In 1995, the European Court of Human Rights unanimously condemned the jury's award as disproportionate and a violation of Count Tolstoy's right to freedom of expression under art. 10 of the European Convention on Human Rights. By a majority decision, the Court decided that the Court of Appeal's order for security for costs did not violate Count Tolstoy's right of access to court under art. 6 (*Tolstoy Miloslavsky* v *United Kingdom* (1995) 20 EHRR 442, ECHR).

In 1991, a jury awarded record damages of £150,000 for *slander* (defamation in temporary form, such as the spoken word). This was in a case where a female doctor was found to have slandered a male doctor by making, in front of patients, allegations of sexual harassment (*Smith* v *Houston*, 25 October 1991; the case is unreported but see *The Independent*, 26 October 1991). On appeal, the damages were reduced to

£50,000 by the Court of Appeal (see *The Times*, 17 December 1993). The defendant became bankrupt and the claimant received very little in respect of damages or costs. Indeed, he ended up being sued by his solicitor for unpaid fees (see *The Times*, 13 January 1995).

William Roache, the actor who plays the part of Ken Barlow in the television series *Coronation Street*, was awarded libel damages of £50,000 in 1991 after he was found to have been defamed in the *Sun* newspaper by a description that he was 'self-satisfied, smug and boring' (*Roache* v *News Group Newspapers Ltd*, 4 November 1991, unreported; see *The Times*, 5 November 1991).

Libel damages of £250,000 were awarded to Esther Rantzen, the television presenter and founder of 'Child Line' (a charity established to help victims of abuse), as a result of allegations in *The People* newspaper that she had deliberately and improperly kept quiet about the activities of a suspected child abuser (*Rantzen* v *Mirror Group Newspapers*, 16 December 1991, unreported; see *The Times*, 17 December 1991). The damages were later described by the Court of Appeal as excessive and 'disproportionately large' in comparison to the harm suffered by the claimant, and were reduced to £110,000 (*Rantzen* v *Mirror Group Newspapers (1986) Ltd* [1993] 4 All ER 975, CA; see further below).

In 1992, the Australian 'pop' star and actor, Jason Donovan, was awarded £200,000 against the publishers of *The Face* magazine in which it had wrongly been suggested that he was 'queer' (*Donovan* v *Wagadon Ltd*, 3 April 1992, unreported; see *The Times*, 4 April 1992).

In 1993, a jury awarded damages totalling £350,000 (including an exemplary sum of £275,000) to Elton John against the publishers of the *Sunday Mirror* newspaper in which it had been alleged that at a Hollywood party he had spat out partially chewed food into a napkin and recommended the practice to guests as a good way to lose weight. The fictitious story was found to be libellous because it wrongly suggested that Mr John had relapsed after successfully overcoming drug and alcohol addiction and the eating disorder, *bulimia nervosa* (*John* v *Mirror Group Newspapers*, 4 November 1993, unreported; see *The Times*, 5 November 1993). The damages were later reduced to £75,000 by the Court of Appeal (see below).

In 1994, libel damages totalling £1,485,000 were awarded in respect of an editorial and review in *Yachting World* which contained a scathing attack on a revolutionary trimaran design (*Walker* v *Sheahan and others*, 8 July 1994, unreported; see *The Times*, 9 July 1994). The damages were later reduced by agreement to £160,000 (see *The Times*, 23 August 1995). Later that year, a retired police superintendent was awarded £375,000 after being libelled by media allegations of sexual abuse at a children's home (*Anglesea* v *Newspaper Publishing plc and others*, 6 December 1994, unreported; see *The Times*, 7 December 1994).

The football manager, Graeme Souness, was awarded damages of £750,000 against the publishers of *The People* newspaper over an article in which his ex-wife made unflattering allegations about his parsimonious attitude towards money and the maintenance of their children (*Souness* v *Mirror Group Newspapers*, 15 June 1995, unreported; see *The Times*, 16 June 1995). The damages were later reduced by agreement to £100,000 (see *The Times*, 24 February 1996).

In 1996, damages of £625,000 were awarded to a consultant orthopaedic surgeon against the publishers of the *Daily Mirror* after the newspaper had described him as 'Dr Dolittle' for allegedly failing to do more to find a hospital bed for a patient with serious head injuries who later died (*Percy* v *Mirror Group Newspapers*, 23 February 1996, unreported; see *The Times*, 24 February 1996).

Section 8 of the Courts and Legal Services Act 1990 authorised the making of rules of court by which the Court of Appeal was given a new statutory power to reassess of its own motion a jury's award of damages. Under the former law, the Court of Appeal itself could only reassess an irrational award made by a jury if both parties agreed to its doing so. If the parties did not agree, the Court of Appeal had no other alternative but to order a retrial—with consequent expense and delay.

By s. 8 of the 1990 Act, and rules made thereunder (currently CPR, r. 52.10(3)), if it is shown on appeal that the jury's award is excessive (or inadequate), the Court of Appeal can reduce (or increase) the damages (without being invited to do so by the parties) to 'such sum as appears to the court to be proper'.

Gorman v *Mudd*, 15 October 1992, unreported, CA, appears to be the first case in which the Court of Appeal exercised its new power under s. 8. Here, libel damages of £150,000 awarded by a jury to the Conservative Member of Parliament, Teresa Gorman, in respect of allegations in a 'mock press release' that she was vain and spiteful were reduced to £50,000 on appeal (see *The Times*, 16 October 1992).

In *Clark* v *Chief Constable of Cleveland Constabulary* (1999) *The Times*, 13 May, CA, the Court of Appeal used its s. 8 power to *increase* from £500 to £2,000 compensatory damages awarded by a jury in a malicious prosecution action against the police.

Since the House of Lords has an inherent jurisdiction to exercise any power conferred on the Court of Appeal, it follows that the House has the same power to substitute a jury's award of damages (*Grobbelaar* v *News Group Newspapers Ltd* [2002] 1 WLR 3024, HL, para 6.1 above).

Given the terms and purpose of s. 8, it always seemed likely that the Court of Appeal would be expected in its judgments to give general guidance on levels of damages for use by trial judges in addressing juries in future cases. The opportunity to provide such guidance was taken by Neill LJ in delivering the judgment of the court in *Rantzen* v *Mirror Group Newspapers (1986) Ltd* [1993] 4 All ER 975, CA (see above), where the Court of Appeal stated its readiness to interfere more frequently with excessive jury awards. It was held that the award to the claimant of £250,000 libel damages could not be justified on any objective standard of reasonable compensation, or necessity, or proportionality.

On the question of guidance for jurors, the Court of Appeal decided that it would not be right to allow reference to be made to jury awards in previous libel cases since they do not establish any meaningful norm for use in future cases.

While recognising the unsatisfactory nature of the present practice whereby a libel claimant can recover much more by way of damages for a transient injury to his reputation than is awarded to the victim of a serious accident for a permanent injury, the suggestion that libel juries should be referred to awards made in personal injuries actions was also rejected as unsatisfactory.

It was further decided, however, that awards made or confirmed by the Court of Appeal in the exercise of its power under s. 8 of the Courts and Legal Services Act 1990 were different and could be relied on as establishing the correct norm. This conclusion was reached by inferring that one of the purposes of s. 8 must have been to allow the Court of Appeal, over a period of time, to provide guidance for use in subsequent cases.

Pending the development of a body of guidance under its new power, the Court of Appeal laid down that juries should be asked by the trial judge to consider:

(a) the purchasing power of the sum they are minded to award;

(b) whether the sum is proportionate to the damage suffered by the claimant; and

(c) whether the sum is necessary in order to provide adequate compensation for, and to re-establish the reputation of, the claimant ([1993] 4 All ER 975 *per* Neill LJ at p. 997).

Within three years the Court of Appeal was persuaded to review *Rantzen v Mirror Group Newspapers (1986) Ltd* [1993] 4 All ER 975, CA, in view of the fact that, notwithstanding what was laid down in that case, juries continued to make libel awards so large as to be divorced from reality.

The review of the *Rantzen* case was carried out in *John v MGN Ltd* [1996] 2 All ER 35, CA, where libel damages of £350,000 awarded to Elton John in November 1993 (see above) were reduced to £75,000, and what were described as 'modest but important' changes of practice were approved.

The Court of Appeal, while endorsing the *Rantzen* view that juries should not be referred to awards made by other juries in previous libel cases but that they could be referred to awards approved or substituted by the Court of Appeal, departed from *Rantzen* in two important respects. First, it decided to allow juries to be referred to conventional personal injury awards as a means of checking the reasonableness of their proposed libel awards. Secondly, it decided to permit counsel and the judge to indicate levels of award which, in their opinion, would be appropriate for the libel in question.

The decision of the Court of Appeal in *John v MGN Ltd* [1996] 2 All ER 35, CA, to look again at the question of referring the jury to other awards of damages was influenced by a number of considerations: the body of guidance to be built up in the Court of Appeal, and referred to in *Rantzen*, had scarcely developed in practice; juries had continued to make what appeared to be 'grossly excessive' awards; the High Court of Australia was now in favour of allowing libel juries to be referred to personal injury damages (see *Carson v John Fairfax and Sons Ltd* (1993) 67 ALJR 634, High Court of Australia); and, since *Rantzen*, the European Court of Human Rights had held in *Tolstoy Miloslavsky v United Kingdom* (1995) 20 EHRR 442, ECHR (see above) that Count Tolstoy's rights under art. 10 of the European Convention on Human Rights had been violated.

The *compensatory* damages of £75,000 awarded to Elton John were held to be excessive because the newspaper article, though it was untrue, offensive and distressing, had not attacked his integrity or damaged his reputation as a performer. A compensatory award of £25,000 was substituted.

The Court of Appeal further held that the case was a suitable one for the award of *exemplary* damages because the defendants had calculated to make a profit out of their story which would exceed any compensation paid to the claimant (see para 9.2.2), and, in the circumstances, a compensatory award by itself was not adequate to mark the gravity of the defendants' conduct, to punish them, or to deter them and others from behaving in the same way again. However, the jury's award of £275,000 as exemplary damages was considered excessive and a sum of £50,000 was substituted—sufficient, according to the Court of Appeal, to do justice to both parties.

Further measures to tackle the problem of excessive jury libel awards became law as from 28 February 2000 by virtue of the Defamation Act 1996, which reduced the limitation period for commencing defamation proceedings from three years to one year, and introduced a summary procedure for disposing of libel and slander claims.

Under this summary procedure, a judge *sitting without a jury* can:

(a) dismiss a claim if it appears that it has no realistic prospect of success and there is no reason why it should be tried; or

(b) give judgment for the claimant, and grant him summary relief, if it appears that there is no defence to the claim which has a realistic prospect of success and that there is no other reason why the claim should be tried.

'Summary relief' for this purpose means:

(a) a declaration that the defendant's offending statement was false and defamatory of the claimant;

(b) an order that the defendant publish or cause to be published a suitable correction and apology;

(c) damages not exceeding £10,000, or such other amount as the Lord Chancellor may prescribe by order;

(d) an order restraining the defendant from publishing or further publishing the matter complained of.

It seems clear from subsequent developments that *John* v *MGN Ltd* [1996] 2 All ER 35, CA, above, has not resolved the problem of excessive jury awards. In *Kiam* v *MGN Ltd* [2002] 3 WLR 1036, CA, damages of £105,000 were awarded despite the fact that the trial judge, adopting the approach laid down in the *John* case, had referred the jury to awards in personal injury cases, and had indicated that an award exceeding £75,000–£80,000 in the present case might be regarded as excessive.

However, on this occasion the Court of Appeal, by a majority, declined to interfere with the jury's award because, although it exceeded the judge's indication, it did not substantially exceed the most that any jury could reasonably have thought appropriate in the circumstances. In his dissenting judgment, Sedley LJ said that *John* had not succeeded in its avowed purpose because jury awards in libel cases were escalating once more ([2002] 3 WLR 1036 at [66]–[67]: '. . . it seems to me that the train has left the station again and is now accelerating').

6.8.2.3 Jury awards against the police

Following a number of cases in which juries had awarded very large sums of exemplary damages in actions against the Metropolitan Police for malicious prosecution, false imprisonment, wrongful arrest, and assault, the Metropolitan Police Commissioner in *Thompson* v *Commissioner of Police of the Metropolis* [1997] 2 All ER

762, CA, sought clarification from the Court of Appeal on the directions which the trial judge should include in his summing up to help the jury decide what to award by way of damages (including exemplary damages) in such actions.

The Court of Appeal (*per* Lord Woolf MR, delivering the judgment of the court, at pp. 775–8) laid down that, where exemplary damages are claimed (e.g., because there has been oppressive or arbitrary behaviour; see para 9.2.2), the jury should be told that:

(a) they should be awarded only if the jury considers that the compensatory and aggravated damages they have already awarded are, in the circumstances, an inadequate punishment for the defendants;

(b) they are a windfall for the claimant and, if payable out of public funds, the sum awarded may no longer be available to the police to spend for the benefit of the public; and

(c) the amount awarded should be adequate to indicate the jury's disapproval of the oppressive or arbitrary behaviour of the police, but no more.

The sum is unlikely to be less than £5,000, and might be as much as £25,000. Exemplary damages of more than £25,000 should not be awarded, except in cases where officers of at least the rank of superintendent have been directly involved in the police misconduct. The absolute maximum is £50,000.

These figures will be adjusted to take account of future inflation. In the two cases with which the Court of Appeal was concerned in *Thompson*, jury awards of exemplary damages were reduced from £50,000 to £25,000, and from £200,000 to £15,000, respectively.

6.8.3 Some alternatives to trial by jury

The question of whether the jury should be retained must be considered in the light of possible alternatives. Trial by a single professional judge is the norm in *civil* cases. *Summary criminal offences* are triable by a judge sitting alone in those areas of the country where there is a district judge (magistrates' courts). However, in the more serious criminal cases heard in the Crown Court it may be considered undesirable to have guilt or innocence determined by one individual with his in-built prejudices and with no opportunity for discussion. Where there is trial by jury, at least ten persons have to be satisfied instead of just one.

Another alternative is trial by a *bench* of professional judges. This method is used when criminal *appeals* are heard in the Divisional Court of the Queen's Bench Division, the Court of Appeal, and the House of Lords. But to resort to a plurality of judges to try criminal cases at first instance would necessitate the appointment of many more judges. This would be expensive financially and would leave the legal profession depleted of some of its best practising talent. Moreover, the idea is open to the objection that it would allow guilt to be decided solely by lawyers.

Trial by a *combined bench* of professional judges and laymen is already a reality in the Crown Court. Appeals from magistrates' courts to the Crown Court *must* be heard by two to four justices in addition to the judge or recorder (Supreme Court Act 1981,

s. 74(1)). On appeals in licensing and gaming cases, four justices *must* be present, and, on appeals from youth courts, two justices authorised to sit in a youth court by the Lord Chancellor *must* sit, and the court must include a man and a woman (Criminal Procedure Rules 2005 (SI 2005, No. 384), r. 63.7). Theoretically all members of the court are of equal status, but when the voting is tied the judge or recorder has a casting vote (Supreme Court Act 1981, s. 73(3)).

Up to four justices *may* sit at a trial on indictment, except for cases listed for plea of not guilty (ibid., ss. 8(1) and 75(2); *Practice Direction (Criminal Proceedings: Consolidation), Part IV* [2002] 1 WLR 2870). Thus their presence is effectively confined to the hearing of guilty pleas.

7

Appeals and the correction of miscarriages of justice

7.1 Appeals in civil cases

7.1.1 From the magistrates' courts

Appeals from the magistrates' courts in licensing matters go to the Crown Court (Supreme Court Act 1981, s. 45). Appeals from the magistrates' courts in family proceedings lie to the Family Division of the High Court (ibid., s. 61 and sch. 1).

7.1.2 From the county courts

Prior to 2 May 2000, most appeals from the county courts lay direct to the Court of Appeal, civil division. On that date, new provisions for appeals from the county courts (and the High Court) were introduced by the Access to Justice Act 1999, Part 52 of the Civil Procedure Rules 1998, the *Practice Direction* to CPR Part 52, and the Access to Justice Act 1999 (Destination of Appeals) Order 2000 (SI 2000, No. 1071). The effect of the new provisions is fully explained in the judgments of Brooke LJ in *Tanfern Ltd* v *Cameron-MacDonald* [2000] 1 WLR 1311, CA, and *Clark* v *Perks* [2001] 1 WLR 17, CA.

The general rule now is that a *first* appeal lies to the next level of judge in the court hierarchy. Thus, in a county court an appeal normally lies from a first instance decision of a district judge to a circuit judge, and from a first instance decision of a circuit judge to a High Court judge (Access to Justice Act 1999 (Destination of Appeals) Order 2000, art. 3).

Exceptionally, however, an appeal lies from a district judge or a circuit judge *direct to the Court of Appeal* from a final decision given in some multi-track cases, and from a final decision taken in certain specialist proceedings, which include proceedings in admiralty, arbitration, and company law, along with commercial and mercantile actions, and patents court and technology and construction court business (ibid., art. 4, as amended).

In limited circumstances, a circuit judge or a High Court judge can order an appeal to be *transferred to the Court of Appeal* instead of hearing it himself. This can only be done where the appeal would raise an important point of principle or practice, or there is some other compelling reason for the Court of Appeal to hear it (CPR, r. 52.14). In addition, the Master of the Rolls has the power (under the Access to Justice Act 1999, s. 57) to direct that an appeal which would normally be heard by a county court or the High Court should be heard instead by the Court of Appeal.

An appeal lies on a question of law or fact. In most cases, permission to appeal is required since an appeal rarely lies as of right. Permission to appeal is not required in the three cases involving the liberty of the subject mentioned in para 7.1.4.1 below (CPR, r. 52.3).

In addition, and somewhat anomalously, permission is not required at present to appeal from a district judge to a circuit judge, or from a circuit judge to a High Court judge, against a decision made at first instance in the *small claims track* (see further, para 1.6.4.6).

Permission to appeal can be obtained either from 'the lower court' or from 'the appeal court'. The lower court is the court from whose decision it is sought to appeal, and the appeal court for this purpose is the court which will hear a first appeal if permission to appeal is granted. Thus, in order to appeal from a first instance decision of a circuit judge (the lower court) to a High Court judge (the appeal court), permission of either of those judges is required. By virtue of the *Practice Direction* to CPR Part 52, the application for permission to appeal should be made orally at the hearing before the lower court. Where no such application is made, or the lower court refuses permission to appeal, an application for permission can be made to the appeal court.

The conditions on which permission to appeal may be granted, and the circumstances in which the appeal court will allow an appeal, are the same as for appeals taken from the High Court to the Court of Appeal (see para 7.1.4.1 below).

All *second* appeals from the county courts lie to the Court of Appeal (Access to Justice Act 1999 (Destination of Appeals) Order 2000, art. 5). Thus, if a circuit judge has heard an appeal from a first instance decision of a district judge, any further appeal from the circuit judge will be heard by the Court of Appeal and not by a High Court judge. The circumstances in which the Court of Appeal will hear a second appeal are severely restricted (see para 7.1.4.1 below).

7.1.3 From the Employment Appeal Tribunal

Appeals from the Employment Appeal Tribunal lie to the Court of Appeal, civil division, on points of *law*. Permission to appeal, granted either by the Employment Appeal Tribunal or by the Court of Appeal, is required. On questions of *fact*, the decision of the Tribunal is final (except in a case of committal for contempt of court) and is not subject to appeal to the Court of Appeal (Employment Tribunals Act 1996, s. 37). Nor, since the Employment Appeal Tribunal is a superior court of record, is its decision on a question of fact subject to judicial review in the High Court.

7.1.4 From the High Court

7.1.4.1 Introduction

Unless the leapfrog procedure (described in para 7.1.4.3 below) is used, an appeal normally lies from a master or district judge of the High Court to a High Court judge, and from a High Court judge to the Court of Appeal, civil division (Access to Justice Act 1999 (Destination of Appeals) Order 2000, art. 2). However, an appeal lies *direct to the Court of Appeal* from a master or district judge of the High Court in the same two exceptional cases mentioned in para 7.1.2, above (final decisions taken in multi-track claims or in specialist proceedings).

By CPR, r. 52.3, permission to appeal is required except where the appeal is against:

(a) the making of a committal order;

(b) a refusal to grant habeas corpus; or

(c) an order made under s. 25 of the Children Act 1989 (secure accommodation orders).

In these three cases, where the liberty of the subject is at stake, an appeal lies as of right.

A *suspended* committal order is a committal order for the purposes of CPR, r. 52.3, and can, therefore, be appealed without permission (*Wilkinson* v *S* [2003] 1 WLR 1254, CA).

Permission to appeal can be given either by the lower court (master, district judge of the High Court, or High Court judge) or by the appeal court (High Court judge or Court of Appeal) (ibid.). No *appeal* lies against a decision to give or refuse permission, but this does not affect any right under rules of court to make a *further application* for permission to the same or another court (Access to Justice Act 1999, s. 54(4)). Where, for example, a High Court judge refuses an application for permission to appeal, a further application for permission can be made to the Court of Appeal, and where the Court of Appeal, without a hearing, refuses permission to appeal, the person seeking permission can request the decision to be reconsidered at a hearing (CPR, r. 52.3).

Permission to appeal will only be given where:

(a) the court considers that the appeal would have a real prospect of success; or

(b) there is some other compelling reason why the appeal should be heard (ibid.).

An order giving permission can limit the issues to be heard and can be made subject to conditions (ibid.). If the court grants permission to appeal on some issues only, it will expressly refuse permission to appeal on any remaining issues. Those other issues cannot then be raised at the appeal hearing without the permission of the appeal court (*Practice Direction* to CPR Part 52).

In effect, every appeal court (whether a circuit judge in the county court, or a High Court judge, or the Court of Appeal) now has the same powers in relation to appeals governed by Part 52 of the CPR. When hearing an appeal it has all the powers of the lower court. It also has power, *inter alia*, to:

(a) affirm, set aside, or vary any order or judgment made or given by the lower court;

(b) refer any claim or issue for determination by the lower court;

(c) order a new trial or hearing (CPR, r. 52.10).

An appeal is not usually a rehearing of the case, but is limited to a review of the decision of the lower court unless:

(a) a practice direction makes different provision for a particular category of appeal; or

(b) the court considers that in the circumstances of an individual appeal it would be in the interests of justice to hold a rehearing (ibid., r. 52.11).

The appeal court will not usually allow oral evidence, or evidence which was not before the lower court, to be presented. It does, however, have power to order otherwise (ibid.)

The appeal court will only allow an appeal where the decision of the lower court was:

(a) wrong; or

(b) unjust because of a serious procedural or other irregularity in the proceedings (ibid.).

Severe limitations are imposed on *second* appeals. Where there has already been one appeal against the decision in question, for example from a district judge to a circuit judge in the county court, or from a master to a judge in the High Court, any second appeal lies to the Court of Appeal (Access to Justice Act 1999 (Destination of Appeals) Order 2000, art. 5), and requires the permission of the Court of Appeal itself (CPR, r. 52.13). By s. 55(1) of the Access to Justice Act 1999, such permission cannot be given unless:

(a) the appeal would raise an important point of principle or practice; or

(b) there is some other compelling reason for the Court of Appeal to hear the appeal.

7.1.4.2 Appeals from the High Court in criminal cases

In a *criminal* cause or matter, not even the criminal division of the Court of Appeal has jurisdiction to hear an appeal from a judgment of the High Court, with the exception of an appeal from a single judge of the High Court in proceedings for criminal contempt of court (Supreme Court Act 1981, s. 18(1)(a)). Criminal appeals from the High Court will normally lie direct to the House of Lords (para 7.2.1.3 below).

A judgment of the High Court is regarded as being given in a criminal cause or matter if that judgment could lead to a criminal trial or punishment. In *R v Stipendiary Magistrate at Lambeth (ex parte McComb)* [1983] 1 All ER 321, CA, the civil division of the Court of Appeal held that it had jurisdiction to hear an appeal from a decision of the Divisional Court of the Queen's Bench Division in judicial review proceedings relating to exhibits to be used in a criminal prosecution. The decision of the Divisional Court was not itself one which could lead to a trial or punishment and so was not a judgment of the High Court in 'a criminal cause or matter'. It followed that the Court of Appeal was not precluded from hearing an appeal against it.

In *Re O* [1991] 1 All ER 330, CA, it was held that restraint and charging orders made by the High Court under part VI of the Criminal Justice Act 1988, with the object of preserving assets for later availability as the potential targets of any confiscation order made in criminal proceedings, are merely collateral to those criminal proceedings and are civil in character. They are not made 'in a criminal cause or matter' and, accordingly, the Court of Appeal has jurisdiction to hear an appeal against them. This decision was approved by the House of Lords in *Government of the United States of America* v *Montgomery* [2001] 1 WLR 196, HL.

It has been held that an anti-social behaviour order made by a magistrates' court under s. 1 of the Crime and Disorder Act 1998 is not made 'in a criminal cause or matter'. It follows that any appeal in judicial review proceedings arising from the making of such an order lies properly to the Court of Appeal (*R* (*McCann*) v *Crown Court at Manchester* [2002] 3 WLR 1313, HL).

On the other hand, an order of a High Court judge in criminal proceedings allowing inspection of bank accounts is made in 'a criminal cause or matter', and, therefore, no appeal lies to the Court of Appeal against the making of the order (*Bonalumi* v *Secretary of State for the Home Department* [1985] 1 All ER 797, CA).

7.1.4.3 The leapfrog procedure

There has existed since 1969 a procedure for missing out, or 'leapfrogging', the Court of Appeal so as to enable an appeal to be taken from the High Court direct to the House of Lords (Administration of Justice Act 1969, ss. 12–15). The procedure is little used.

The conditions which must be satisfied before such a direct appeal can be taken are that:

(a) the trial judge has granted a certificate of satisfaction (Administration of Justice Act 1969, s. 12), and

(b) the House of Lords has given leave to appeal (ibid., s. 13).

As to (a), a trial judge can only grant a certificate if all the parties consent *and* the case involves a point of law of general public importance which is *either* concerned wholly or mainly with the construction of a statute or of a statutory instrument, *or* is one where the trial judge is bound by a previous decision of the Court of Appeal or the House of Lords (ibid., s. 12(1) and (3)). The granting of a certificate by the trial judge is discretionary. No appeal is possible against the granting or refusal of a certificate (ibid., s. 12(5)).

As to condition (b), the application for leave to appeal is determined by the House of Lords without a hearing (ibid., s. 13(3)). If leave is granted, any appeal to the Court of Appeal from the decision of the trial judge is precluded (ibid., s. 13(2)(a)).

7.1.5 From the Court of Appeal

An appeal may be taken from the Court of Appeal to the House of Lords. Leave of either court is required (Administration of Justice (Appeals) Act 1934, s. 1(1)). If the Court of Appeal refuses leave to appeal, a party may nevertheless apply to the Appeal Committee of the House of Lords for leave to appeal. The Appeal Committee has existed since 1934 in order to consider, in private, petitions to the House for leave to appeal. Each petition must be heard by at least three judges (Administration of Justice (Appeals) Act 1934, s. 1(2)).

The fact that the Appeal Committee grants or refuses leave to appeal in any particular case does not indicate either disapproval or approval of the decision of the court below (*Wilson* v *Colchester Justices* [1985] 2 All ER 97, HL). This is so in the case of *criminal* as well as *civil* appeals. The House of Lords is able to deal only with a limited number of cases each year and it is important for the proper development of the law that those cases should be chosen carefully (ibid., *per* Lord Roskill at p. 100).

The *Appeal* Committee must not be confused with the *Appellate* Committee, which has existed since 1948 and which hears the actual appeal sitting in a committee room, and not the chamber, of the House.

As the ultimate appellate court, the House of Lords has an inherent jurisdiction (albeit rarely exercised) to rehear an appeal, and to rescind or vary an order made earlier in the proceedings, with a view to correcting any injustice that may have occurred. The House is reluctant to reopen an appeal, and will certainly not do so merely on the ground that the earlier decision is now thought to be wrong. It seems that an appeal will only be reopened where a party has been the victim of an unfair procedure through no fault of his own (*R* v *Bow Street Metropolitan Stipendiary Magistrate (ex parte Pinochet Ugarte) (No. 2)* [1999] 1 All ER 577, HL, *per* Lord Browne-Wilkinson at pp. 585–6, the other Law Lords agreeing with him).

In the case of *Cassell & Co. Ltd* v *Broome (No. 2)* [1972] AC 1136, HL, the House later varied a costs order made in the main proceedings because the parties had not had a fair opportunity to present argument on the question of costs. (See further on this litigation, para 9.2.2.) In *Pinochet (No. 2)*, above, Senator Pinochet was successful in his petition to have an earlier decision of the House set aside on the ground of bias on the part of one of the Law Lords involved in the case. A rehearing before a differently constituted Appellate Committee of the House was ordered. (See further on the *Pinochet* litigation, para 10.5.3.2.1.)

A civil appeal is usually taken to the House of Lords on a question of *law*, although the appeal can be on a question of fact, such as the quantum of damages. In a civil appeal to the House which involves a question of law, there is no statutory requirement that the question must necessarily be one of general public importance (cf. criminal appeals, para 7.2.2.5 below), although as a matter of *practice*, leave to appeal is liable to be refused by the Appeal Committee if the petition does not raise an 'arguable point of law of general public importance' (*Procedure Direction* [1988] 2 All ER 831).

In some instances the decision of the Court of Appeal is declared by statute to be final. It follows that no further appeal is possible in these instances to the House of Lords. The decision of the Court of Appeal is final, for example, in any appeal from a county court in probate proceedings (County Courts Act 1984, s. 82) and in any appeal from the Chancery Division of the High Court in insolvency proceedings (Insolvency Act 1986, s. 375(2)).

7.2 Appeals in criminal cases

7.2.1 Appeal following summary trial

An appeal may lie from the magistrates' court to the Crown Court or, by way of case stated, to the Queen's Bench Division of the High Court.

7.2.1.1 Appeal to the Crown Court

If the defendant pleaded guilty, an appeal lies from the magistrates' court to the Crown Court *against sentence only*. There is no appeal to the Crown Court against *conviction* if the defendant pleaded guilty before the justices. If, on the other hand, the defendant pleaded not guilty, he can appeal to the Crown Court against either conviction or sentence or both (Magistrates' Courts Act 1980, s. 108(1)).

An appeal to the Crown Court takes the form of a complete rehearing of the case with witnesses but without a jury. The Crown Court has power to confirm, reverse, or vary the decision under appeal, and may impose any sentence which the magistrates' court could have imposed, whether more or less severe than that actually inflicted (Supreme Court Act 1981, s. 48(2) and (4), as amended).

It is anomalous that the Crown Court can increase sentence whereas the Court of Appeal, which is superior to the Crown Court, cannot do so on an appeal by the defendant (see para 7.2.2.2 below). Where the case has been referred to the Crown Court by the Criminal Cases Review Commission under s. 11 of the Criminal Appeal Act 1995, the Crown Court cannot increase the sentence (Criminal Appeal Act 1995, s. 11(6)).

A further appeal lies, at the instance of either the defence or the prosecution, from the decision of the Crown Court by way of case stated to the High Court, but only on the grounds that the decision is wrong in law or was given in excess of jurisdiction (Supreme Court Act 1981, s. 28(1)). In a few betting and gaming matters the decision of the Crown Court is final and cannot be appealed against (ibid., s. 28(2), as amended).

7.2.1.2 Appeal to the Queen's Bench Division of the High Court

As an alternative to appealing to the Crown Court, an appeal may sometimes be taken direct to the High Court *by way of case stated* from the magistrates' court. Either the defence or the prosecution may take advantage of this procedure, which is available only where it is alleged that the justices' decision is wrong in law or was given in excess of jurisdiction (Magistrates' Courts Act 1980, s. 111(1); note that a decision may be wrong in *law* because of a perverse finding on the *facts*). The party so appealing will lose his right to appeal to the Crown Court (ibid., s. 111(4)).

On a case stated, the High Court may affirm, reverse, or vary the decision under appeal, or remit the case to the magistrates' court with its opinion, or 'make such other order . . . as it thinks fit' (Supreme Court Act 1981, s. 28A, as substituted by the Access to Justice Act 1999; this provision was formerly contained in s. 6 of the Summary Jurisdiction Act 1857). It has been held that the words, 'make such other order . . . as it thinks fit', empower the High Court to order a rehearing before the same or a different bench of justices provided that a fair trial is still possible (*Griffith* v *Jenkins* [1992] 1 All ER 65, HL).

The procedure by way of case stated is set in motion when either the defendant or the prosecutor makes written application to the magistrates' court to 'state a case' for the opinion of the High Court. The application must include a statement of the question of law or jurisdiction on which the opinion of the High Court is required (Criminal Procedure Rules 2005 (SI 2005, No. 384), r. 64.1). The case stated by the magistrates will usually include a statement of the original information laid against the defendant, the facts as found by the justices, the submissions of the parties, the cases cited (if any), the decision of the justices, and the question for decision by the High Court.

The magistrates can only refuse to state a case if, in their opinion, the application is frivolous (Magistrates' Courts Act 1980, s. 111(5)). Refusal to state a case may enable the applicant to seek, by way of judicial review, a mandatory order to compel the magistrates to state a case (ibid., s. 111(6)). Thus, in *R v West Midlands Magistrates (ex parte PMS International Group plc* [1993] COD 455, DC, a mandatory order was issued in circumstances where there was no basis on which the magistrates could reasonably have concluded that the application to state a case was frivolous. They had given no reasons for their decision, and the case involved a substantial legal point about the admissibility of evidence which clearly affected the defence's conduct of its case.

A further appeal from the High Court in a *criminal* case lies direct to the House of Lords (see para 7.2.1.3 below). If, however, the appeal to the High Court by way of case stated under s. 111 of the Magistrates' Courts Act 1980 is on a *non-criminal* matter, the decision of the High Court is declared to be 'final' (Supreme Court Act 1981, s. 28A(4), as substituted by the Access to Justice Act 1999). It follows that no appeal direct to the House of Lords is possible since the case is not a criminal one. Moreover, by virtue of s. 18 of the Supreme Court Act 1981, the Court of Appeal has no jurisdiction to hear an appeal (see, e.g., *Westminster City Council v O'Reilly* [2004] 1 WLR 195, CA, where the Court of Appeal declined jurisdiction in a licensing case).

7.2.1.3 Appeal to the House of Lords from the Queen's Bench Division of the High Court

A further appeal may be taken by either party from the High Court to the House of Lords. This is so whether the case reached the High Court via the Crown Court or direct from the magistrates' court. But the appeal to the House of Lords will only lie if:

(a) the High Court has certified that the case involves a point of law of general public importance; *and*

(b) the High Court or the House of Lords has given leave to appeal on it appearing that the point is one which ought to be considered by the House of Lords (Administration of Justice Act 1960, s. 1, as amended by the Access to Justice Act 1999).

7.2.2 Appeal following trial on indictment

7.2.2.1 Appeal against conviction

Against *conviction* in the Crown Court, the defendant can appeal to the Court of Appeal, criminal division, but only with leave of the Court of Appeal or if the trial judge grants a certificate that the case is fit for appeal (Criminal Appeal Act 1968, s. 1(2), as substituted by the Criminal Appeal Act 1995).

This provision, which applies whether the appeal is on a question of law, or of fact, or on a question of mixed law and fact, makes it more difficult to appeal against a Crown Court conviction than was the case before 1996, when an appeal against conviction lay as of right (i.e., without leave) on a question of *law*.

Although the trial judge has power to facilitate an appeal by certifying that the case is fit for appeal, it has been held that he should not exercise the power unless the case has 'exceptional features' (*R* v *Bansal* (1998) *The Times*, 29 December, CA). It follows that most applications for leave will be made to the Court of Appeal itself. Leave can be granted by a single judge of the Court of Appeal. If he refuses leave, the appellant can have his application considered by the full Court (Criminal Appeal Act 1968, s. 31(2)–(3), as amended). If the single judge grants leave on some grounds but refuses it on other grounds, the leave of the full Court is required in order to pursue those grounds on which the single judge refused leave (*R* v *Jackson* [1999] 1 All ER 572, CA; *R* v *Cox* [1999] 2 Cr App R 6, CA).

The actual appeal against conviction must be heard by at least *three* judges in the Court of Appeal (Supreme Court Act 1981, s. 55(4)).

It should be noted that, unlike the case of summary trial, a defendant may appeal against conviction even if he pleaded guilty at his trial on indictment in the Crown Court. On an appeal against conviction only, the Court of Appeal has no power to interfere with the sentence imposed on the defendant by the Crown Court.

The defendant is allowed only one appeal. If that is unsuccessful, he cannot bring a second appeal even if he wishes to call fresh evidence under s. 23 of the Criminal Appeal Act 1968 (*R* v *Pinfold* [1988] 2 All ER 217, CA). Nor can he go back to the Court of Appeal where the House of Lords has decided his appeal, and either dismissed it or directed that his conviction be restored, albeit that he now wishes to pursue grounds of appeal which were not considered by the House of Lords (*R* v *Berry (No. 2)* [1991] 2 All ER 789, CA, approved by the House of Lords in *R* v *Mandair* [1994] 2 All ER 715, HL).

7.2.2.2 Appeal against sentence

Against the *sentence* of the Crown Court, the defendant may appeal to the Court of Appeal with leave of the Court of Appeal (Criminal Appeal Act 1968, ss. 9(1) and 11(1), as amended). Leave may be granted by a single judge of that Court or, if he refuses it, by the full Court (ibid., s. 31(2)–(3), as amended). Leave of the Court of Appeal to appeal against sentence is not required if the Crown Court judge who passed the sentence grants a certificate that the case is fit for appeal against sentence (ibid., s. 11(1A), as inserted by the Criminal Justice Act 1982).

The actual appeal against sentence can be heard by *two* judges (Supreme Court Act 1981, s. 55(4)).

No appeal is possible against a sentence which is fixed by law (Criminal Appeal Act 1968, s. 9(1), as amended), such as life imprisonment for murder. Notwithstanding this prohibition, the Court of Appeal in the unusual circumstances of *R (Lichniak)* v *Secretary of State for the Home Department* [2001] 3 WLR 933, CA, granted permission to appeal against a mandatory life sentence imposed for murder in order that the Court could investigate, under s. 3 of the Human Rights Act 1998, whether such a sentence is compatible with the European Convention on Human Rights. (On hearing the substantive appeal, it was held that the sentence violated neither art. 3 (inhuman or degrading treatment) nor art. 5 (liberty and security) of the Convention. The House of Lords ([2002] 3 WLR 1834) later affirmed the decision of the Court of Appeal.)

On an appeal by the defendant against sentence, the Court of Appeal cannot *increase* the sentence but is limited to confirming it, or reducing it, or varying it from one form of detention to another (Criminal Appeal Act 1968, s. 11(3); cf. the wider power of the Crown Court under the Supreme Court Act 1981, s. 48(4), as amended, para 7.2.1.1 above).

By s. 172 of the Criminal Justice Act 2003, every court when sentencing a defendant must take into account any relevant sentencing guidelines issued by the Sentencing Guidelines Council which was established under that Act. These guidelines are intended to assist in achieving consistency in sentencing.

The function of the Sentencing Guidelines Council is to consider whether to frame new sentencing guidelines or to review existing guidelines. It is under a *duty* to consider whether to do so when requested by the Secretary of State or by the Sentencing Advisory Panel which was set up under the Crime and Disorder Act 1998 (Criminal Justice Act 2003, ss. 170–171).

If the Council decides to frame or revise guidelines it must have regard to:

(a) the need to promote consistency in sentencing;

(b) the sentences imposed by courts in England and Wales for offences to which the guidelines relate;

(c) the cost of different sentences and their relative effectiveness in preventing re-offending;

(d) the need to promote public confidence in the criminal justice system; and

(e) the views of the Sentencing Advisory Panel (ibid., s. 170(5)).

To enable the defendant and other interested parties to understand why a particular sentence was chosen, the court when passing sentence must in almost every case state in open court its reasons for deciding on that sentence. It must also explain to the defendant in ordinary language the effect of the sentence (ibid., s. 174(1)). If the court departs from any relevant guidelines issued by the Sentencing Guidelines Council, it must give reasons for doing so (ibid., s. 174(2)).

7.2.2.3 Deterrent against appealing: section 29(1) of the Criminal Appeal Act 1968

A deterrent against appealing to the Court of Appeal is contained in s. 29(1) of the Criminal Appeal Act 1968, which provides that:

The time during which an appellant is in custody pending the determination of his appeal shall, subject to any direction which the Court of Appeal may give to the contrary, be reckoned as part of the term of any sentence to which he is for the time being subject.

Two former Lord Chief Justices have said that appellants with no merit in their appeals cannot expect to be given any credit under this provision (*Practice Note* [1970] 1 All ER 1119, Lord Parker CJ; *Practice Note* [1980] 1 All ER 555, Lord Widgery

CJ; these Practice Notes are now incorporated in *Practice Direction (Criminal Proceedings: Consolidation)* [2002] 1 WLR 2870). Thus, a convicted defendant may find that, on the direction of the Court of Appeal, the time he spent in prison awaiting the inevitable outcome of his unmeritorious appeal will have to be served all over again.

7.2.2.4 Appeals in cases of death: section 44A of the Criminal Appeal Act 1968

Before 1996, if the defendant died before an appeal could be started, or before a remitted appeal could be determined, the right of appeal died with him, and the Court of Appeal had no jurisdiction to hear, or to continue to hear, the appeal (*R v Kearley (No. 2)* [1994] 3 All ER 246, HL). This could cause hardship to the family of a deceased defendant on whom a financial penalty had been imposed because there was no procedure for challenging the penalty and recovering any money paid under it.

In *R v Kearley (No. 2)*, above, Parliament was called on to consider remedying this unjust situation (ibid., *per* Lord Jauncey at pp. 253–4, the other Law Lords agreeing with him). A response was provided in the Criminal Appeal Act 1995, which inserted a new s. 44A into the Criminal Appeal Act 1968 allowing a person approved by the Court of Appeal to begin or continue an appeal. The provision also applies to criminal appeals lying from the Court of Appeal to the House of Lords. The first case under s. 44A was *R v W (crime: pursuing deceased's appeal)* (1997) *The Times*, 8 January, CA. Here, a widow was allowed to pursue her deceased husband's appeal against conviction for indecently assaulting his daughter. However, the appeal was dismissed on the merits.

7.2.2.5 Appeal to the House of Lords from the Court of Appeal

A further appeal may be taken by either the defendant or the prosecutor from the Court of Appeal, criminal division, to the House of Lords. But this appeal will only lie if:

(a) the Court of Appeal has certified that the case involves a point of law of general public importance; *and*

(b) the Court of Appeal or the House of Lords has given leave to appeal on it appearing that the point is one which ought to be considered by the House (Criminal Appeal Act 1968, s. 33(2)).

An application to the Court of Appeal for leave to appeal to the House of Lords must be made within 28 days of the decision of the Court of Appeal against which it is sought to appeal, or, if later, the date on which that Court gives reasons for its decision. An application to the House of Lords for leave to appeal must be made within 28 days of the Court of Appeal's refusal to give leave (ibid., s. 34(1), as amended by, and s. 34(1A), as inserted by, the Courts Act 2003).

Pending the outcome of the appeal in the House of Lords, the Court of Appeal has power to order that the defendant be detained or not be released except on bail (Criminal Appeal Act 1968, s. 37(1)–(2)). If no order for detention is made, and the defendant is allowed to go free, he cannot be detained again if the appeal goes against him in the House of Lords (ibid., s. 37(5); *R v Hollinshead* [1985] 2 All ER 769, HL, where the Court of Appeal was criticised for refusing to make an order under s. 37

which meant that the defendants went completely unpunished for conspiracy to defraud electricity boards when their convictions were restored by the House of Lords).

For the purpose of disposing of an appeal, the House of Lords can exercise any powers of the Court of Appeal or remit the case to that Court (ibid., s. 35(3)). For example, if the Court of Appeal does not deal with a ground of appeal which is relevant to whether a conviction should be allowed to stand, the House can either deal with that ground itself or send the matter back to the Court of Appeal to be dealt with (*R v Mandair* [1994] 2 All ER 715, HL).

7.3 The Attorney-General's reference of points of law: section 36 of the Criminal Justice Act 1972

The prosecution has no right of appeal to the Court of Appeal following an acquittal on indictment, not even on a point of law. There is, however, provision for a 'reference' to be made by the Attorney-General in such a case. By s. 36(1) of the Criminal Justice Act 1972:

The Attorney-General may, if he desires the opinion of the Court of Appeal on a point of law which has arisen in the case, refer that point to the court, and the court shall, in accordance with this section, consider the point and give their opinion on it.

This provision merely confers on the Attorney-General a *discretion;* he cannot be compelled to make a reference. The acquittal of the defendant is not affected by the opinion of the Court of Appeal. His identity must not be revealed without his consent. He may appear before the Court to present argument, or be represented by counsel with his costs paid out of central funds (Criminal Justice Act 1972, s. 36(2), (5), and (7)). The Court of Appeal can refer the point of law to the House of Lords if of the opinion that it ought to be considered there (ibid., s. 36(3)).

The reference procedure is an exception to the practice of the English courts not to give advisory opinions on academic points of law. It is, however, little used—perhaps because it is thought only to be appropriate for determining exceptionally difficult points of law. The Criminal Justice Act 1972 contains no such limitation, and, as Lord Widgery CJ said in *Attorney-General's Reference (No. 1 of 1975)* [1975] QB 773, CA (at p. 778):

It would be a mistake to think, and we hope people will not think, that references by the Attorney-General are confined to cases where very heavy questions of law arise and that they should not be used in other cases. On the contrary, we hope to see this procedure used extensively for short but important points which require a quick ruling of this court before a potentially false decision of law has too wide a circulation in the courts.

The Attorney-General's reference is a useful procedure for providing future guidance to the Crown Court by clearing up doubtful points in the criminal law, a branch of English law where certainty is particularly important. It was used in *Attorney-General's Reference (No. 3 of 1994)* [1997] 3 All ER 936, HL, in which it was held that a person can be charged with the manslaughter of a child who, though born alive, subsequently dies where that person has deliberately and unlawfully injured either the child while a foetus *in utero* or the mother carrying the child.

7.4 The Attorney-General's reference of unduly lenient sentences: section 36 of the Criminal Justice Act 1988

The prosecution cannot appeal to the Court of Appeal against the sentence imposed on the defendant. However, by s. 36(1) of the Criminal Justice Act 1988:

If it appears to the Attorney-General . . . that the sentencing of a person in a proceeding in the Crown Court has been unduly lenient . . . he may, with the leave of the Court of Appeal, refer the case to them for them to review the sentencing of that person; and on such a reference the Court of Appeal may—

(i)   quash any sentence passed on him in the proceeding; and

(ii) in place of it pass such sentence as they think appropriate for the case and as the court below had power to pass when dealing with him.

The Attorney-General thus has a *discretion* to refer unduly lenient sentences; he cannot be compelled to make a reference. He requires leave of the Court of Appeal to refer the case. Leave must be applied for within 28 days from the day when the original sentence was passed (Criminal Justice Act 1988, sch. 3, para 1).

'Sentence' has the same meaning as in the Criminal Appeal Act 1968, namely, 'any order made by a court when dealing with an offender', except that it does not include an interim hospital order under the Mental Health Act 1983 (Criminal Justice Act 1988, s. 35(6)). Accordingly, absolute and conditional discharges can be referred under s. 36 as well as sentences by way of imprisonment and fines. There is no power to review a sentence which was passed after a retrial ordered by the Court of Appeal, and is the same sentence as that imposed when the defendant was originally convicted, since the judge at the retrial is bound by the Criminal Appeal Act 1968 not to impose a sentence of greater severity than that passed on the original conviction (*Attorney-General's Reference (No. 82a of 2000) (2002) The Times*, 28 February, CA).

Section 36 of the Criminal Justice Act 1988 was enacted as a result of the view taken by the government that public confidence in the criminal justice system was being undermined by what it regarded as the unduly lenient sentences imposed for some serious offences. It should be noted, however, that the power of the Court of Appeal under s. 36 is not limited to *increasing* a referred sentence. It may, instead, *confirm* or even *reduce* a sentence if it is not satisfied that it is unduly lenient (s. 36(1)(ii), above, and see *Attorney-General's Reference (No. 4 of 1989)* [1990] 1 WLR 41, CA, where probation was substituted for two concurrent 18-month suspended prison sentences imposed for incest and indecent assault).

When the new procedure first came into force it was limited to sentences imposed for *indictable* offences. It has since been extended, by Orders made by the Secretary of

State under s. 35(4) of the Act, to cover sentences imposed by the Crown Court for some *either-way* offences, including indecent assault, threats to kill, cruelty to persons under the age of 16 (1994), certain drugs offences and unlawful sexual intercourse (2000), and certain racially or religiously aggravated offences (2003).

The hearing at which the sentence is reviewed under s. 36 must be conducted before *three* judges of the Court of Appeal (Supreme Court Act 1981, s. 55(4)(aa), as inserted by the Criminal Justice Act 1988). The defendant is entitled to be present at the hearing (Criminal Justice Act 1988, sch. 3, para 6). If he is represented by counsel, he is entitled to payment of his costs out of central funds (ibid., para 11).

In conducting its review, the question which the Court of Appeal must ask itself is whether the sentence imposed by the trial judge is 'outside the range of sentences which the judge, applying his mind to all the relevant factors, could reasonably consider appropriate' (*Attorney-General's Reference (No. 4 of 1989)* [1990] 1 WLR 41, CA).

The term of any sentence of imprisonment imposed by the Court of Appeal as a result of the review begins to run, unless otherwise directed, from the time when it would have begun to run if passed originally by the Crown Court (Criminal Justice Act 1988, sch. 3, para 10).

When the Court of Appeal has concluded its review, any point of law involved in the sentence may be referred, by either the Attorney-General or the defendant, to the House of Lords for its opinion (ibid., s. 36(5)). A reference to the House of Lords requires the leave of the Court of Appeal or the House itself, and leave must not be granted unless the Court of Appeal has certified that the point of law involved is one of general public importance *and* it appears to the Court of Appeal or the House of Lords that the point is one which ought to be considered by the House (ibid., s. 36(6)).

The s. 36 procedure was first used in *Attorney-General's Reference (No. 1 of 1989)* [1989] 3 All ER 571, CA, in which the Court of Appeal quashed a sentence of three years' imprisonment, imposed on the defendant after his conviction for incest, and substituted a sentence of six years' imprisonment.

In *Attorney-General's Reference (No. 3 of 1989)* and *Attorney-General's Reference (No. 5 of 1989)* (1989) 11 Cr App R(S) 486, CA, the Court of Appeal made legal history when, for the first time, it substituted custodial sentences for the non-custodial penalties imposed by the Crown Court.

In a much-publicised case in 1993, a 15 year-old youth on his conviction for rape had been made the subject of a three year supervision order with a condition that he attend a special activity programme, and his parents were ordered to pay £500 compensation to the victim so that she could have a holiday. The Court of Appeal held that the sentence was too lenient. It was set aside and replaced by a sentence that the youth serve two years' detention in a young offender institution (*Attorney-General's Reference (No. 3 of 1993)* [1993] Crim LR 472, CA).

In 2002, in an effort to curb the dramatic escalation of robbery involving the theft of mobile telephones, the sentences imposed on two defendants were increased in *Attorney-General's Reference (Nos. 4 & 7 of 2002)* [2002] Cr App R(S) 77, CA. The Court of Appeal indicated that a custodial sentence is inevitable for this sort of crime unless there are exceptional circumstances.

7.5 Correction of miscarriages of justice

7.5.1 Introduction

Most of the publicity surrounding miscarriages of justice is concerned with the wrongful conviction of *defendants*. However, it must be appreciated at the outset that the *prosecution* can be the victim of miscarriages of justice, and that, since the prosecution is representative of the public interest, these are miscarriages which affect the wider public. As Lord Hobhouse put it in *R v Mirza* [2004] 2 WLR 201, HL, at [131]–[132]:

[I]t is fundamentally wrong to use the phrase 'miscarriage of justice' selectively as if it only related to perverse convictions. This presents a false picture . . . [A] perverse verdict of not guilty . . . is also a miscarriage of justice . . . A failure to convict a defendant whose guilt has been proved is a breach of the social contract between the state and its citizens. It is a failure of the state to provide to citizens the protection to which they are entitled against the criminal activities of others. It leads to a failure of public confidence in the justice system and the ability of the state to maintain 'the Queen's peace'. It encourages vigilantes. It encourages intimidation and discourages witnesses from giving evidence. These are, to coin a phrase, 'clear and present dangers'. In simple Human Rights terms, they are a breach of the victims' human rights . . .

Various procedures are now available to deal with miscarriages of justice suffered by the prosecution.

There is the Attorney-General's reference of points of law to the Court of Appeal under s. 36 of the Criminal Justice Act 1972. However, as noted in para 7.3 above, an opinion of the court which is favourable to the prosecution does not affect the acquittal of the defendant.

The Court of Appeal has power, under s. 36 of the Criminal Justice Act 1988, to correct a miscarriage of justice occurring in the sentencing process by increasing an unduly lenient sentence brought to its attention by the Attorney-General (para 7.4 above).

In appropriate cases, the High Court can quash tainted acquittals under s. 54 of the Criminal Procedure and Investigations Act 1996 (para 7.5.2 below).

In addition, and of more recent origin, are the powers conferred by the Criminal Justice Act 2003 to apply for a retrial of serious offences (para 7.5.3 below) and to exercise a right of appeal against rulings made by a Crown Court judge during the course of a trial (para 7.5.4 below).

A procedure for correcting miscarriages of justice suffered by defendants convicted at the Crown Court is provided by s. 2(1) of the Criminal Appeal Act 1968 (para 7.5.5 below). For defendants convicted and sentenced by a magistrates' court, there is s. 142 of the Magistrates' Courts Act 1980 (para 7.5.6 below). Other mechanisms available are the Criminal Cases Review Commission's reference under the Criminal Appeal Act 1995 (para 7.5.7 below), pardon (para 7.5.8 below), and compensation (para 7.5.9 below).

7.5.2 Quashing of tainted acquittals: section 54 of the Criminal Procedure and Investigations Act 1996

The Royal Commission on Criminal Justice (*Report*, Cm 2263, 1993) recommended that where a person is convicted of conspiracy to pervert the course of justice by interfering with the jury in a trial which resulted in the defendant's acquittal it should be possible for the defendant to be tried again for the same offence.

Section 54 of the Criminal Procedure and Investigations Act 1996 implements this proposal in an extended form. It deals with the mischief of 'tainted' acquittals obtained by interfering with, or intimidating, a juror or witness, and is targeted mainly at criminals who use money, threats, or violence to evade conviction. The section applies to acquittals in respect of offences alleged to have been committed on or after 15 April 1997.

Where a person has been acquitted of an offence, and the same or another person has been convicted of an administration of justice offence involving interference with, or intimidation of, a juror or witness (or potential witness) in any proceedings which led to the acquittal, and it appears to the convicting court that there is a real possibility that, but for the interference or intimidation, the acquitted person would not have been acquitted, the court must certify that it so appears (Criminal Procedure and Investigations Act 1996, s. 54(1)–(2)).

The court must not so certify if, because of lapse of time or for any other reason, it would be contrary to the interests of justice to take proceedings against the acquitted person for the offence of which he was acquitted (ibid., s. 54(5)).

If a court does so certify, an application can be made to the High Court for an order quashing the acquittal, and, if the High Court makes such an order, proceedings can then be taken against the acquitted person for the offence of which he was acquitted (ibid., s. 54(3)–(4)).

For the purposes of s. 54, 'administration of justice offences' are the offence of perverting the course of justice, the offence under s. 51 of the Criminal Justice and Public Order Act 1994 of intimidating or harming, or threatening to harm, jurors, witnesses, or persons assisting in the investigation of offences, and an offence of aiding, abetting, counselling, procuring, suborning, or inciting another person to commit perjury under s. 1 of the Perjury Act 1911 (Criminal Procedure and Investigations Act 1996, s. 54(6)).

The High Court cannot quash an acquittal under s. 54 of the Criminal Procedure and Investigations Act 1996 unless four conditions are satisfied (ibid., s. 54(3)). These conditions are as follows (ibid., s. 55):

(a) it is likely that, but for the interference or intimidation, the acquitted person would not have been acquitted;

(b) it would not be contrary to the interests of justice, because of lapse of time or for any other reason, to take proceedings against the acquitted person for the offence of which he was acquitted;

(c) the acquitted person has been given a reasonable opportunity to make written representations to the court; and

(d) it appears that the conviction for the administration of justice offence will stand.

The usefulness of the procedure to quash tainted acquittals, and the prevalence of the mischief at which it is aimed, may be questioned. Although the procedure has been available since April 1997, it appears never to have been used.

7.5.3 Retrial for serious offences

The Criminal Justice Act 2003 has changed the double jeopardy rule by allowing retrials in respect of a number of serious offences where 'new and compelling evidence' has come to light since a defendant's acquittal. These provisions were brought into force on 4 April 2005.

The double jeopardy rule prevents the risk of a retrial for the same offence hanging over a defendant who has already been either acquitted or convicted of it. It allows the defendant to plead *autrefois acquit* ('formerly acquitted') or *autrefois convict* ('formerly convicted') if charged later with the same offence. The government, while acknowledging the importance of the rule to acquitted defendants, was of the opinion that it is capable of causing grave injustice to victims and the community in certain cases where compelling fresh evidence comes to light after an acquittal. (White Paper, *Justice For All*, Cm 5563, 2002. The opinion was expressed in the light of the Stephen Lawrence Inquiry Report.)

The future of the double jeopardy rule was considered by the Law Commission in *Double Jeopardy and Prosecution Appeals* (Law Com. No. 267, Cm 5048, 2001), and by Auld LJ in his *Review of the Criminal Courts of England and Wales* (Lord Chancellor's Department, 2001). However, as foreshadowed in *Justice For All*, above, the changes contained in the Criminal Justice Act 2003 go further than the recommendations of the Law Commission and Auld LJ.

It is provided that if a person has been acquitted (whether before or after the passing of the Act) of a qualifying offence, a prosecutor can apply to the Court of Appeal for an order quashing the acquittal and authorising a retrial for the offence (Criminal Justice Act 2003, ss. 75–76). The application can only be made with the written consent of the Director of Public Prosecutions (ibid., s. 76(3)). 'Qualifying offences' are those listed in the Act, and include murder, manslaughter, and rape (ibid., s. 75(8) and sch. 5, pt 1).

The Court of Appeal *must* order the acquittal to be quashed, and a retrial to be held, if satisfied that there is new and compelling evidence in the case, and that it is in the interests of justice for the order to be made. If not so satisfied, the prosecutor's application must be dismissed (ibid., ss. 77–79).

Evidence is 'new' if it was not adduced at the original trial of the acquitted person. It is 'compelling' if it is reliable and substantial and, when considered in the context of the outstanding issues, appears to be highly probative of the case against the acquitted person (ibid., s. 78(2)–(3)).

Whether it is 'in the interests of justice' for the order to be made is decided by considering, in particular, any existing factors which make a fair trial unlikely, the length of time since the alleged offence was committed, and whether the police and prosecution have acted with due diligence (ibid., s. 79).

An appeal lies on a point of law to the House of Lords from decisions made by the

Court of Appeal on an application for a retrial (Criminal Appeal Act 1968, s. 33(1B), as inserted by the Criminal Justice Act 2003).

With a view to safeguarding the fairness of any retrial, the Court of Appeal has power to make an order imposing reporting restrictions in respect of matters connected with the application if it believes this to be necessary in order to avoid a substantial risk of prejudice to the administration of justice in a retrial (Criminal Justice Act 2003, s. 82). The order can be made either of the Court's own motion or on the application of the Director of Public Prosecutions. The reporting restrictions cease to have effect at the end of the retrial (or earlier if it becomes clear that the acquitted person can no longer be retried).

If an acquittal is quashed, and a retrial ordered, the acquitted person can only be retried on an indictment preferred by the direction of the Court of Appeal. He must be arraigned on this indictment within two months of the date on which the Court of Appeal ordered a retrial, unless the Court allows a longer period. The Court can only extend this period if satisfied that the prosecutor has acted with due expedition since the order was made, and that there is still a good and sufficient cause for a retrial despite the lapse of time. If the arraignment for the retrial does not take place within the two-month period or any further time allowed, the acquitted person can apply to the Court of Appeal to have the order for retrial set aside and the previous acquittal restored (ibid., s. 84).

In a press release in November 2005, the Director of Public Prosecutions announced that he had given his consent for the first application under the reformed double jeopardy law to be made to the Court of Appeal. It is the case of a man who has already been tried twice for the same murder. On each occasion, the jury failed to reach a verdict and he was formally acquitted in 1991. In February 2006, the Director of Public Prosecutions (with the agreement of the acquitted person) applied for, and was granted, an order under s. 82 of the Criminal Justice Act 2003 (above) restricting publicity in relation to the application. The Court of Appeal said that reporting restrictions were necessary in the interests of justice in a case like this since, if a new trial were ordered, publicity of the very fact that the Court had concluded that there was 'compelling' evidence for ordering a retrial might itself be regarded as prejudicial to the subsequent trial. On the question of press releases, the Court of Appeal said that, although that of November 2005 in the instant case might have been appropriate (since it related to the first ever application under the new provisions), it was doubtful whether a press release would be appropriate in any future case (*In re D (Acquitted Person: Retrial)* [2006] WLR (D) 51, 27 February 2006, CA).

7.5.4 Prosecution appeals against rulings

7.5.4.1 Introduction

As already noted, a defendant convicted at the Crown Court has a qualified right of appeal against conviction and sentence. The prosecution has no corresponding right of appeal against an *acquittal*, whether as a result of the jury's verdict or a ruling of the trial judge which has the effect of terminating the trial prematurely.

Jury acquittals remain sacrosanct, but ss. 57–74 of the Criminal Justice Act 2003 provides for an interim prosecution right of appeal against two categories of ruling made by a Crown Court judge during a criminal trial.

First, there is a 'general right of appeal in respect of rulings' (Criminal Justice Act 2003, s. 58), and, secondly, there will be a 'right of appeal in respect of evidentiary rulings' (ibid., s. 62). In each instance, the appeal only lies if the prosecution has leave to appeal. This can be granted either by a Crown Court judge or the Court of Appeal (ibid., s. 57(4)). If, in the case of the general right of appeal, the prosecution fails to obtain leave to appeal, or abandons the appeal, it must agree that the defendant should be acquitted (ibid., s. 58(8)). This rule does not apply to evidentiary appeals.

If leave to appeal has been granted, a judge decides whether the appeal will be expedited or non-expedited. If he decides on expedition, the trial is adjourned pending the outcome of the appeal. If the appeal is to be non-expedited, any jury which has been empanelled may be discharged (ibid., ss. 59 and 64).

It is expressly provided that the new provisions do not confer a right of appeal against a ruling that a jury be discharged, or against a ruling from which the prosecution can already appeal under other legislation (ibid., s. 57(2)), such as a ruling given by the judge at a preparatory hearing on the admissibility of evidence and other questions of law relating to the case (see para 6.8.2.1).

With a view to safeguarding the fairness of any resumed trial or retrial, a general prohibition on the reporting of appeals is imposed (ibid., s. 71). The prohibition, which does not apply to certain matters relating to the court, the defendant, the offence charged, witnesses, and legal representatives, ends at the conclusion of the trial. The prohibition can be lifted before then, either completely or to a specified extent, by order of a judge, the Court of Appeal, or the House of Lords. If the defendant objects to the early lifting of the prohibition, it can only be lifted if it would be in the interests of justice to do so. Reporting restrictions imposed by other legislation continue to apply in addition to those in s. 71.

An appeal lies on a point of law to the House of Lords from a decision made by the Court of Appeal on a prosecution appeal against a ruling (Criminal Appeal Act 1968, s. 33(1), as amended by the Criminal Justice Act 2003). The right of appeal is available to both the prosecution and the defence.

7.5.4.2 General right of appeal

The general right of appeal applies to rulings which are made by the judge before the time when he starts his summing up to the jury (Criminal Justice Act 2003, s. 58(1) and (13), as amended), and which, in effect, terminate the trial.

The statutory provisions relating to the general right of appeal were brought into force on 4 April 2005.

On the appeal, the Court of Appeal can confirm, reverse, or vary the ruling appealed against (ibid., s. 61(1)). It can only *reverse* the ruling on three grounds, namely, that it (a) was wrong in law; (b) involved an error of law or principle; or (c) was a ruling which it was not reasonable for the judge to have made (ibid., s. 67).

If it *confirms* the ruling, it must order the acquittal of the defendant for the offence which is the subject of the appeal (ibid., s. 61(3)). If it *reverses or varies* the ruling, it must do one of three things: order the Crown Court proceedings to be resumed; order a fresh trial; or order the defendant to be acquitted of the offence which is the subject of the appeal (ibid., s. 61(4)). It can only order the resumption of the Crown Court proceedings or a fresh trial where it is necessary in the interests of justice to do so (ibid., s. 61(5)).

7.5.4.3 Right of appeal in respect of evidentiary rulings

The provisions relating to this right of appeal are not yet in force.

The right of appeal will apply to evidentiary rulings made by the trial judge in relation to a qualifying offence (whether made before or after the start of the trial) at any time before the opening of the defence case (ibid., s. 62(2)). An 'evidentiary ruling' is a ruling which relates to the admissibility or exclusion of any prosecution evidence. 'Qualifying offences' are those serious offences listed in the Act (ibid., s. 62(9) and sch. 4, pt 1).

Leave to appeal against an evidentiary ruling cannot be granted unless a judge or the Court of Appeal is satisfied that the ruling significantly weakens the prosecution case (ibid., s. 63).

On the appeal, the Court of Appeal can confirm, reverse, or vary the ruling (ibid., s. 66(1)). It can only *reverse* the ruling on the three grounds mentioned above in relation to the general right of appeal (ibid., s. 67).

If the Court of Appeal confirms, reverses, or varies the evidentiary ruling, it must do one of three things: order the Crown Court proceedings to be resumed; order a fresh trial; or order the defendant to be acquitted of the offence which is the subject of the appeal (ibid., s. 66(2)). It can only order an acquittal if the prosecution has indicated that it does not intend to continue with the prosecution of that offence (ibid., s. 66(3)).

7.5.5 Section 2(1) of the Criminal Appeal Act 1968

7.5.5.1 Introduction

Section 2(1) of the Criminal Appeal Act 1968 (as substituted by the Criminal Appeal Act 1995) provides as follows:

Subject to the provisions of this Act, the Court of Appeal—

(a) shall allow an appeal against conviction if they think that the conviction is unsafe; and

(b) shall dismiss such an appeal in any other case.

Under the amended s. 2(1), there is now only one ground for allowing an appeal—the unsafeness of the conviction. Two other grounds (available before 1996) of a wrong decision of any question of law, and a material irregularity in the trial, have been abolished as separate grounds. In *R v Chalkley* [1998] 2 All ER 155, CA, the Court of Appeal held that it has no power under the amended s. 2(1) to allow an appeal where it is dissatisfied in some way with what happened before, or at, the trial (for example, because there has been a wrong decision of a question of law or a material irregularity), but does not also think that the conviction is unsafe. On this narrow approach, procedural irregularities could be ignored if there was no doubt as to the defendant's guilt.

In *R* v *Mullen* [1999] 3 WLR 777, CA, the Court of Appeal preferred a broader approach to 'unsafeness' than that laid down in *Chalkley* [1998] 2 All ER 155, CA. Having found the meaning of 'unsafe' to be ambiguous, the Court of Appeal in *Mullen* referred to *Hansard* and concluded that the amended form of s. 2(1) was intended to restate the practice of the Court of Appeal prior to the amendment. Accordingly, it was held that a conviction can be unsafe because of abuse of process prior to trial, even if there is no doubt as to the defendant's guilt. (Mr Mullen's conviction was quashed. The rejection by the Home Secretary of his subsequent application for compensation led to further litigation; see para 7.5.9 below.)

Chalkley was disapproved, and *Mullen* approved, in *R* v *Togher* [2001] 3 All ER 463, CA. Preferring the broader approach to the court's power to intervene supported by *Mullen* to the narrower approach adopted in *Chalkley*, the Court of Appeal in *Togher* said that if a defendant has been denied a fair trial it will almost be inevitable that the conviction will be regarded as unsafe. If *Mullen* and *Togher* are correct, it means that under the amended s. 2(1) there is a comprehensive test of lack of safety which applies across the board and is sufficiently wide to cover all the situations previously included in s. 2(1) before it was amended.

Guidance given by the House of Lords emphasises that 'unsafe' is an ordinary word of the English language, and that no particular thought process is involved in deciding whether a conviction is unsafe. It is, therefore, unhelpful and wrong in principle to attempt to put a gloss on the words of s. 2(1). In a case where fresh evidence is admitted, the Court of Appeal should test its provisional view on the safety of the conviction by asking whether the evidence, if it had been given at the trial, might reasonably have affected the jury's decision to convict. There is no connection between the safety or otherwise of a conviction and the possibility of ordering a retrial. If a conviction is considered to be unsafe, it is the clear statutory duty of the Court of Appeal under s. 2(1) to allow the appeal whether or not a retrial is possible (*R* v *Pendleton* [2002] 1 WLR 72, HL, reversing the decision of the Court of Appeal, especially *per* Lords Bingham and Hobhouse at [19]–[20] and [38]–[39], respectively).

In deciding on the safety of a conviction imposed many years earlier, the standards of fairness to be applied are those considered appropriate at the time of the hearing before the Court of Appeal, and not those in force at the time of the original conviction (*R* v *Bentley* (1998) *The Times*, 31 July, CA, where a 46-year-old conviction was quashed posthumously; *R* v *Johnson* (2000) *The Times*, 21 November, CA, in which a 32-year-old conviction was quashed). It is thus open to the Court of Appeal to quash as unsafe a conviction which was regarded as perfectly proper when it was originally imposed.

If the Court of Appeal allows an appeal under s. 2(1), and does not order a retrial, it must quash the defendant's conviction. The quashing of the conviction operates as a direction to the trial court to enter an acquittal (Criminal Appeal Act 1968, s. 2(3)). Section 2(1) does not empower the Court of Appeal to declare the innocence of the defendant. That is a function exerciseable only by a jury; the Court of Appeal is solely concerned with whether a conviction can be allowed to stand (*R* v *McIlkenny and others* [1992] 2 All ER 417, CA; this is the case of the 'Birmingham Six'—for the full story, see para 7.6 below).

7.5.5.2 Receiving fresh evidence: section 23 of the Criminal Appeal Act 1968

At the hearing of an appeal, the Court of Appeal has power to admit fresh evidence (even though it was available but not called at the trial) if to do so is necessary or expedient in the interests of justice (Criminal Appeal Act 1968, s. 23(1), as amended by the Criminal Appeal Act 1995). The discretion to admit fresh evidence may be exercised (but only in very exceptional cases) where the reason that the evidence was not given at the trial was the defendant's unequivocal plea of guilty (*R* v *Lee* [1984] 1 All ER 1080, CA).

The Court of Appeal, in deciding whether to admit any evidence, must consider in particular whether that evidence appears to be capable of belief, whether it might render the defendant's conviction unsafe, whether it would have been admissible at the trial, and whether there is a reasonable explanation for the failure to adduce it at the trial (Criminal Appeal Act 1968, s. 23(2), as substituted by the Criminal Appeal Act 1995).

These four matters are not intended to be an exhaustive list of preconditions which must be satisfied before fresh evidence can be admitted. They must be taken into account 'in particular', and, as long as they are considered, it is possible for the Court of Appeal to receive fresh evidence when all four matters are not satisfied. For example, credible admissible evidence affecting the safety of a conviction can be admitted even though no reasonable explanation is provided for the failure to adduce it at the trial (*R* v *Sale* (2000) *The Times*, 14 June, CA).

Any fresh evidence will usually be supplied in written form, but the court commonly hears witnesses orally in order to determine whether the evidence is capable of belief (*R* v *Sale*, above, disapproving *R* v *McLoughlin*, 30 November 1999, CA, unreported).

The Court of Appeal, having received and considered the fresh evidence, can allow the appeal against conviction and, if the interests of justice so require, can order a retrial. Alternatively, it may simply allow the appeal and quash the conviction without ordering a retrial (see para 7.5.5.3 below).

By s. 23A of the Criminal Appeal Act 1968 (as inserted by the Criminal Appeal Act 1995 and amended by the Criminal Justice Act 2003), on an appeal against conviction or an application for leave to appeal against conviction, the full Court of Appeal (but not a single judge thereof) can direct the Criminal Cases Review Commission to investigate and report on any matter if:

(a) the matter is relevant and ought, if possible, to be resolved before the appeal or application is determined;

(b) the matter is likely to be resolved by the court as a result of the investigation; and

(c) the matter cannot be resolved without an investigation.

7.5.5.3 Ordering a retrial: section 7 of the Criminal Appeal Act 1968

Before the summer of 1989, the Court of Appeal, on allowing an appeal, had statutory power to order a retrial instead of quashing the defendant's conviction in only one specific instance, namely, where the appeal was allowed by reason of fresh evidence received by the Court on appeal. In other cases, the defendant who successfully demonstrated that his conviction was unsafe was permitted to go free even though there may have been overwhelming evidence of his guilt.

The law was changed by the Criminal Justice Act 1988 so as to provide the Court of Appeal with a *general* power to order a retrial. This discretion is not limited to fresh evidence cases, but is exercisable whenever the Court of Appeal allows an appeal against conviction under s. 2(1) of the Criminal Appeal Act 1968 and the interests of justice require the defendant to be retried (Criminal Appeal Act 1968, s. 7(1), as amended by the Criminal Justice Act 1988).

The new trial must not be for a completely different offence, but only for (a) the offence of which the defendant was convicted at the original trial; or (b) an offence of which he could have been convicted at the original trial (for example, manslaughter instead of murder); or (c) an offence charged in an alternative count of the indictment in respect of which no verdict was given in consequence of the defendant being convicted of another offence (Criminal Appeal Act 1968, s. 7(2), as amended by the Criminal Justice Act 2003).

Where a retrial has been ordered and the new trial has not begun within two months, leave of the Court of Appeal is required for it to take place (Criminal Appeal Act 1968, s. 8(1), as amended by the Criminal Justice Act 1988). After two months the defendant can apply to the Court of Appeal to set aside the retrial order and to direct the trial court to enter a verdict of acquittal (ibid., s. 8(1A), as inserted by the Criminal Justice Act 1988). On an application by the prosecution under s. 8(1), or by the defendant under s. 8(1A), the Court of Appeal can only grant leave for the retrial to take place if satisfied that there has been no undue delay (ibid., s. 8(1B), as inserted by the Criminal Justice Act 1988).

The Court of Appeal is not under a duty to order a retrial on allowing an appeal against conviction. Instead, it may simply quash the conviction without ordering a retrial (ibid., s. 2(1)–(2)). This course might be adopted where the alleged crime was committed many years earlier and the passage of time would make it difficult to arrive at the truth in a new trial, or where sensational press coverage of a trial has created a real risk of prejudice against the defendant.

It was for this latter reason that the Taylor sisters were not ordered to be retried (para 7.6 below). On the other hand, a retrial was ordered by the Court of Appeal in the case of Susan Whybrow and Dennis Saunders whose convictions were quashed on the ground that the trial judge's frequent interventions (sometimes sarcastic and scornful) during their evidence had gone so far beyond the bounds of legitimate judicial conduct as to deprive the defendants of a fair trial (*R v Whybrow; R v Saunders* (1994) *The Times*, 14 February, CA). They had been convicted at the Crown Court at Norwich of conspiracy to murder Mrs Whybrow's husband by making it look as though his sit-on lawnmower had accidentally toppled over and either crushed him or caused him to fall into a pond and drown. At the second trial, held at the Old Bailey, they were acquitted by the jury (see *The Times*, 22 March 1994). They had already served prison sentences for conspiracy to cause grievous bodily harm to Mr Whybrow, a charge to which they had pleaded guilty at the original trial.

7.5.5.4 *Venire de novo*

Most appeals against conviction come before the Court of Appeal under s. 2(1) of the Criminal Appeal Act 1968 (para 7.5.5.1 above). The ground for allowing an appeal mentioned in s. 2(1) presupposes that the trial itself was perfectly valid. This ground is not appropriate where the basis of the defendant's appeal is that his trial was a *nullity*.

Where the trial was a nullity, the Court of Appeal has an inherent common law power to order a *venire de novo* (in effect, a new trial or, perhaps more accurately, a *proper* trial), but is not bound to do so. Where the Court of Appeal decides that the trial was a nullity it does not 'quash' the 'conviction' since, technically, there is no 'conviction' to 'quash'. Instead, it orders the purported conviction to be 'set aside and annulled' (*R* v *Booth* (1998) *The Times*, 26 November, CA).

The Court of Appeal has no jurisdiction to order a *venire de novo* where a material irregularity occurred during a trial which was validly commenced and which was validly concluded by a conviction following an unequivocal verdict of guilty from the jury (*R* v *Rose and others* [1982] 2 All ER 731, HL). Such a trial is not a nullity.

If, however, the first 'trial' was never validly commenced, but was void *ab initio* by reason of a material irregularity, the Court of Appeal can order a *venire de novo*. An example of this would be where the first 'trial' takes place before a court which has no jurisdiction to try the offence with which the defendant is charged. Similarly, where the first 'trial', although validly commenced, has concluded without a properly constituted jury returning a valid verdict, a *venire de novo* can be ordered. Examples are where the defendant changes his plea of not guilty to one of guilty during the course of the hearing, and the judge discharges the jury without obtaining a verdict of guilty from them, or where the jury's verdict is ambiguous (ibid.).

It seems that the irregularity must be one of *procedure*. A *venire de novo* will not be ordered where, for instance, the judge's direction to the jury contains a mistake, or the judge has left some issue to the jury which should not have been left. (These circumstances may, however, render a conviction 'unsafe' within the meaning of s. 2(1) of the Criminal Appeal Act 1968.) It is also clear that the procedural irregularity must be *fundamental* in the sense of being so serious that the hearing can properly be described as a mistrial or nullity.

In *R* v *Morais* [1988] 3 All ER 161, CA, the Court of Appeal ordered a *venire de novo* in a case where the defendant's trial was a nullity by reason of the fact that the bill of indictment was not signed by 'the proper officer of the court' (as required by s. 2(1) of the Administration of Justice (Miscellaneous Provisions) Act 1933), but merely initialled by a judge. On the other hand, in *R* v *Laming* (1990) 90 Cr App R 450, CA, the trial was held not to be a nullity by reason only of the fact that a two-page indictment had not been signed by the court clerk immediately after the last count but only on the first page.

In *R* v *Comerford* [1998] 1 All ER 823, CA, the Court of Appeal refused to order a *venire de novo* in a case where the defendant had been convicted by an anonymous jury, the members of which had been given police protection by the trial judge and allowed to be called and sworn by numbers rather than by names. It was held that if it is reasonably thought to be desirable to withhold jurors' names in order to thwart suspected jury-nobbling, the trial is valid provided that the defendant's right to challenge jurors is preserved. On the facts, it was held that the defendant had not been deprived of his right of challenge; accordingly, his trial was a valid one. (See further on *R* v *Comerford*, para 6.5.1.)

7.5.6 Section 142 of the Magistrates' Courts Act 1980

Under s. 142 of the Magistrates' Courts Act 1980, the magistrates' courts have power to reopen cases in order to rectify mistakes. That power was extended when s. 142 was amended by the Criminal Appeal Act 1995. It is still, however, of somewhat limited scope.

A magistrates' court can at any time vary or rescind a sentence or other order imposed or made by it if it appears to be in the interests of justice to do so, unless the Crown Court (by way of an appeal) or the High Court (by way of case stated) has already dealt with the matter (Magistrates' Courts Act 1980, s. 142(1)–(1A)).

After convicting a person, a magistrates' court can at any time direct that the case should be heard again by different justices (whether the defendant pleaded guilty or not guilty) if it is in the interests of justice to do so, unless the Crown Court (by way of an appeal) or the High Court (by way of case stated) has already dealt with the matter (ibid., s. 142(2)–(2A)). Where a magistrates' court gives a direction that the case should be heard again by different justices, the conviction (and any consequent sentence or other order) is treated as a nullity and has no effect (ibid., s. 142(3)).

The decision of a magistrates' court under s. 142 can be challenged in the High Court in judicial review proceedings.

The ambit of s. 142 has been explained and limited by judicial decision. The purpose of the section is to enable a magistrates' court to rectify mistakes. It does not extend to allowing a defendant to obtain a rehearing in circumstances where he pleaded guilty before the magistrates and is, therefore, unable to appeal to the Crown Court. It has been held that it is not 'in the interests of justice' to allow s. 142 to be so used since the 'interests of justice' include the interests of the courts and the public that there should be certainty and an end to litigation (*R* v *Croydon Youth Court (ex parte Director of Public Prosecutions)* [1997] 2 Cr App R 411, DC).

7.5.7 The Criminal Cases Review Commission's reference: the Criminal Appeal Act 1995

Before April 1997, the Home Secretary could refer the case of a person convicted on indictment to the Court of Appeal for review under s. 17 of the Criminal Appeal Act 1968. This provision was repealed with effect from 31 March 1997 by the Criminal Appeal Act 1995, thereby abolishing the Home Secretary's long-standing power of referral. The responsibility for referring possible miscarriages of justice back to the Court of Appeal was taken over by a new body created by the 1995 Act, the Criminal Cases Review Commission. The Commission came into existence on 1 January 1997, and took on its casework responsibilities at the end of March 1997.

The Criminal Cases Review Commission can at any time refer to the *Court of Appeal* the conviction and/or sentence (except a sentence fixed by law) of any person tried in the Crown Court (Criminal Appeal Act 1995, s. 9). The Criminal Cases

Review (Insanity) Act 1999 makes it clear that the Commission has the power to refer to the Court of Appeal a case in which the verdict returned was the now obsolete 'guilty but insane'.

In addition, the Commission can at any time refer to the *Crown Court* the conviction and/or sentence of any person tried by a magistrates' court (Criminal Appeal Act 1995, s. 11).

The reference of a case back to the Court of Appeal, or to the Crown Court, can be made on the Commission's own initiative or after an application made by, or on behalf of, the convicted person (ibid., s. 14). The reference is treated for all purposes as an appeal by the person against the conviction and/or sentence (ibid., ss. 9(2)–(3) and 11(2)–(3)). When a case is referred back to the *Crown Court*, the original *sentence* imposed by the magistrates' court cannot be increased (ibid., s. 11(6)).

On a reference back to the *Court of Appeal*, the appellant is not allowed to rely on any ground which is not related to a reason given by the Commission for making the reference unless the Court of Appeal gives leave (ibid., s. 14(4A)–(4B), as inserted by the Criminal Justice Act 2003). This provision is designed to overrule the effect of *R v Smith (No. 3)* [2003] 1 WLR 1647, CA, in which the Court of Appeal held reluctantly that, under the law as it then was, the appellant, on a reference back by the Commission, was allowed to raise before the Court any grounds which were considered and rejected by the Commission, those which were not considered by the Commission, and even those which were rejected by the Court during any normal appeal against conviction.

In the case of a reference back to the *Crown Court*, it remains possible for the appellant to rely on any ground relating to the conviction whether or not it is related to any reason given by the Commission for making the reference (ibid., s. 14(5), as amended by the Criminal Justice Act 2003).

The Commission must not make a reference unless:

(a) an appeal against conviction or sentence has been decided, or leave to appeal against it has been refused; and

(b) in the case of a *conviction*, it considers there is a real possibility that the conviction would not be upheld because of an argument, or evidence, not raised at the trial or on appeal; or

(c) in the case of a *sentence*, it considers there is a real possibility that the sentence would not be upheld because of an argument on a point of law, or information, not raised at the trial or on appeal (ibid., s. 13).

In considering whether to refer a case to the Court of Appeal, or to the Crown Court, the Commission must take into account any application or representations made to the Commission by, or on behalf of, the defendant, any other representations made to the Commission in relation to the application, and any other matters which appear to the Commission to be relevant (ibid., s. 14(2)). In considering whether to refer a case to the *Court of Appeal*, the Commission can at any time refer any point arising in the case to the Court of Appeal for its opinion. The Court is obliged to consider the point referred and to provide the Commission with its opinion thereon (ibid., s. 14(3)).

When the Commission makes a reference it must give to the Court of Appeal, or to the Crown Court, as appropriate, a statement of its reasons for making the reference, and must send a copy of that statement to every person who is likely to be a party to any proceedings on the appeal arising from the reference (ibid., s. 14(4)). When the Commission decides not to make a reference following an application made by, or on behalf of, the defendant, it must give a statement of the reasons for its decision to the person who made the application (ibid., s. 14(6)).

A decision by the Commission not to refer a case to the Court of Appeal is susceptible to challenge in an application for judicial review (*R* v *Criminal Cases Review Commission (ex parte Pearson)* [1999] 3 All ER 498, DC, where, however, the challenge failed).

No time limit is imposed by the Criminal Appeal Act 1995 for references by the Commission. Accordingly, the Commission has power to refer a conviction no matter how long ago the trial took place. In *R* v *Kansal (No. 2)* [2001] 3 WLR 751, CA, which was decided after the Human Rights Act 1998 came into force, the Court of Appeal reluctantly held that it had no option but to declare a conviction unsafe if it resulted from the admission of evidence obtained in breach of art. 6 of the European Convention on Human Rights (right to a fair trial), even though the original trial took place before the 1998 Act became operative.

This conclusion was reached by reason of the Court of Appeal's interpretation of the retrospectivity provision in s. 22(4) of the 1998 Act. Section 22(4) provides that s. 7(1)(b) of the 1998 Act 'applies to proceedings brought by or at the instigation of a public authority whenever the act in question took place', and s. 7(1)(b) provides that 'a person who claims that a public authority has acted (or proposes to act) in a way which is made unlawful by section 6(1) may . . . rely on the Convention right or rights concerned in any legal proceedings'. A court is a 'public authority' (Human Rights Act 1998, s. 6(3)).

Delivering the judgment of the Court of Appeal in *Kansal (No. 2)*, above, Rose LJ (at [24]) expressed concern over the ambit of the Commission's discretion, and predicted an alarming workload for the Commission and for the Court of Appeal. He thought that if Parliament had not intended this to be a consequence of the Human Rights Act 1998, the matter should be looked at urgently either by Parliament or by the House of Lords on appeal from the Court of Appeal.

Less than two months after *Kansal (No. 2)* was decided, a majority of the House of Lords held in *R* v *Lambert* [2001] 3 WLR 206, HL, that the correct interpretation of ss. 7(1)(b) and 22(4) of the Human Rights Act 1998 is that Parliament did not intend the Act to have retrospective effect in such a situation, and so 'did not intend to bring about the situation which was a cause of concern to Rose LJ' in *Kansal (No. 2)* (*R* v *Lambert* [2001] 3 WLR 206, HL, *per* Lord Hutton at [173]).

According to *Lambert*, therefore, a defendant cannot rely on the Human Rights Act 1998 in an appeal against a conviction which took place before the Act came into force. Lord Steyn dissented on the ground that s. 6(1) clearly makes it unlawful for an appellate court to act incompatibly with a convention right. If, he said, the majority view in *Lambert* is correct then the Court of Appeal and the House of Lords will (contrary to the wording of s. 6(1)) have 'to act in a way which is incompatible with a convention right' ([2001] 3 WLR 206 at [29]).

Four months after *Lambert* was decided, the House heard an appeal from the Court of Appeal in the case of *R v Kansal (No. 2)* [2001] 3 WLR 1562, HL, and the question of retrospectivity (on which fresh arguments were presented before the House) was raised again. A majority of three to two of their Lordships held that *Lambert* was wrongly decided, and that the Human Rights Act should extend retrospectively to appeals since this would be consistent with the purpose of s. 22(4), namely to give effect to the state's treaty obligations under art. 13 of the Convention. However, a majority of four to one held that there was no compelling reason not to continue applying *Lambert* even though it was wrong. The decision of the Court of Appeal in *Kansal (No. 2)* was reversed and the defendant's conviction restored.

The conflict of judicial opinion in the House of Lords remains unresolved since *Lambert* was not overruled in *Kansal (No. 2)*, and the two cases are equally authoritative. In the latter case, Lord Lloyd suggested that the only satisfactory solution was to have the appeal reargued before a panel of seven Law Lords ([2001] 3 WLR 1562, HL, at [21]). This, however, was always unlikely to happen. *Kansal (No. 2)* was later applied by the House of Lords in *R v Benjafield* [2002] 2 WLR 235, HL, *R v Lyons* [2002] 3 WLR 1562, HL, and *Bellinger v Bellinger* [2003] 2 WLR 1174, HL.

7.5.8 Pardon

Miscarriages of criminal justice do still occur in spite of the existence of procedures designed to prevent or remedy them. It is not unknown for a miscarriage of justice to go undetected, or unremedied, during a normal appeal against conviction under s. 2(1) of the Criminal Appeal Act 1968, or on a reference out of time to the Court of Appeal.

The commonest causes of miscarriages of justice are said to be wrongful identification of the defendant, false confessions, perjury by a co-defendant or witness, police misconduct, and bad trial tactics on the part of defence lawyers. It is worth noting here that the Court of Appeal will not treat incompetent advocacy as rendering a conviction 'unsafe', within the meaning of s. 2(1) of the Criminal Appeal Act 1968 (para 7.5.5.1 above), unless the advocacy is 'flagrantly incompetent' (*R v Gautam* (1987) *The Times*, 4 March, CA, applied in *R v Ensor* [1989] 2 All ER 586, CA).

When a wrongful conviction comes to light, the Home Secretary can pardon the defendant by exercising the royal prerogative of mercy. A *free* pardon relieves the defendant from all penalties and punishments resulting from the conviction. But the conviction itself cannot be quashed by the Crown and still stands. It can only be quashed by the Court of Appeal under the Criminal Appeal Act 1968 (*R v Foster* [1984] 2 All ER 679, CA; but note that the quashing of a conviction does not operate as a declaration of the defendant's innocence—*R v McIlkenny and others* [1992] 2 All ER 417, CA, para 7.5.5.1 above).

The royal prerogative of mercy is a flexible power capable of being exercised in many different circumstances. Accordingly, it is within the Home Secretary's powers to grant a *partial* or *conditional* pardon, or a *posthumous* pardon, in recognition of the fact that a mistake has been made in the administration of the criminal justice system. This is illustrated by *R v Secretary of State for the Home Department (ex parte Bentley)* [1993] 4 All ER 442, DC, the case in which the sister of Derek Bentley challenged the decision taken in 1992 by the then Home Secretary, Kenneth Clarke, not to grant a posthumous pardon to her brother. (The public law aspects of this case are discussed more fully in para 10.2.2.1.)

Derek Bentley was convicted with Christopher Craig in 1952 of the murder by shooting of PC Sidney George Miles. Bentley and Craig were together on the roof of a warehouse in Croydon when they were disturbed by police officers, and Craig, who was carrying a gun, fired the fatal shot. Bentley, aged 19 with a mental age of 11, was sentenced to death (the mandatory sentence for murder at that time), but Craig, then aged 16, was too young for the death penalty and was sentenced to be detained during Her Majesty's pleasure. (He was released from custody in 1963 and maintains that Bentley never uttered the famous ambiguous words 'Let him have it, Chris'.)

The jury had recommended mercy for Bentley, and two senior civil servants at the Home Office advised the Home Secretary of the day, Sir David Maxwell Fyfe (who later became Lord Chancellor as Viscount Kilmuir LC), to act on this recommendation and to reprieve Bentley and commute the death sentence to one of life imprisonment. However, the Home Secretary decided to ignore the recommendation and advice. Bentley was hanged at Wandsworth prison on 28 January 1953, his appeal against conviction having been dismissed earlier.

Kenneth Clarke's refusal to pardon Bentley was based on what emerged as a long established policy of successive Home Secretaries that a *free* pardon should be granted only if the moral, as well as the technical, innocence of the convicted person can be established. Mr Clarke did not believe that Derek Bentley was either morally or technically innocent, although he made it clear that if he had been the Home Secretary in 1953 he would have reprieved Bentley.

On an application for judicial review of Mr Clarke's decision, it was held in effect that in considering the case as one for a *free* (i.e., 'unconditional' or 'full') pardon he had directed his mind to the wrong question. He had not given any, or sufficient, consideration to the possibility of granting some other form of pardon which would be suitable to the circumstances of the particular case, such as a *conditional* pardon whereby the penalty is removed on condition that a lesser sentence is served.

If Bentley had been reprieved in 1953, the substitution of a sentence of life imprisonment would have constituted a conditional pardon, and the court in 1993 could see no objection in principle to the grant of a *posthumous conditional* pardon where a death sentence had already been carried out. It was said that the grant of such a pardon would be a recognition by the state that a mistake had been made, and that a reprieve should have been granted, and that it was wrong to regard the prerogative of mercy as a power only to be exercised in cases falling into specific categories, or as no more than an 'arbitrary monarchical right of grace and favour', when, in truth, it is now a constitutional safeguard against mistakes in the criminal justice system (see *R v Secretary of State for the Home Department (ex parte Bentley)* [1993] 4 All ER 442, DC, *per* Watkins LJ at pp. 454–5).

The court was further of the opinion that there was a compelling argument that even by the standards of 1953 the then Home Secretary's decision was clearly wrong (ibid., *per* Watkins LJ at p. 455). Although it was not considered appropriate to make any formal order, the court took the unusual step of inviting the Home Secretary to look at the matter again, and to examine whether it would be just to exercise the royal prerogative of mercy in a way which would give recognition to the generally accepted view that Bentley should have been reprieved 40 years earlier.

Less than four weeks after this decision, a new Home Secretary (Michael Howard) granted to Derek Bentley a partial posthumous pardon, which recognised that he should not have been executed. Bentley's conviction was, of course, left intact at this stage (see *R v Foster* [1984] 2 All ER 679, CA, above).

Subsequently, the question of Bentley's conviction was referred back to the Court of Appeal by the Criminal Cases Review Commission in the light of new medical evidence and submissions about irregularities at the trial. The conviction was quashed by the Court of Appeal in July 1998 (*R v Bentley* (1998) *The Times*, 31 July, CA). At first, Bentley's family was refused compensation (see *The Times*, 13 April 1999), but the Home Secretary changed his mind after four weeks (see *The Times*, 13 May 1999) in the light of fresh legal advice on the impact of *R v Secretary of State for the Home Department (ex parte Garner)* (1999) *The Times*, 4 May, DC (see para 7.5.9 below).

As a result of changes made by the Criminal Appeal Act 1995, the Home Secretary is able to seek the assistance of the Criminal Cases Review Commission in connection with the prerogative of mercy. He can refer to the Commission any matter arising in his consideration of whether to recommend the exercise of the prerogative of mercy in relation to a conviction and on which he desires its assistance. The Commission is then obliged to consider the referred matter and to give a statement of its conclusions on it to the Home Secretary, who, in considering whether to recommend mercy, must treat the Commission's statement as conclusive of the matter referred (Criminal Appeal Act 1995, s. 16(1)). In addition, the Commission can in any case express its opinion to the Home Secretary that he should consider whether to recommend the exercise of the prerogative of mercy in relation to that case. If it does so, the Commission must supply the Home Secretary with the reasons for its opinion (ibid., s. 16(2)).

7.5.9 Compensation

Prior to 1988, a person who was pardoned, or whose conviction was quashed, had no *right* to compensation even though he may have spent many years in prison before the miscarriage of justice came to light. He might, however, have been granted an *ex gratia* payment in recognition of hardship suffered, although such payments were extremely rare. The *ex gratia* compensation scheme is part of the royal prerogative.

With effect from October 1988, the Criminal Justice Act 1988 introduced a statutory scheme of compensation for the victims of miscarriages of justice. The statutory scheme confers a *right* to compensation, but only in the limited circumstances laid down in the Act. Cases falling outside the scope of the statutory scheme will continue to be dealt with under the *ex gratia* scheme, which, by its very nature, involves no right to compensation (see below).

As amended by the Criminal Appeal Act 1995, s. 133 of the Criminal Justice Act 1988, which was passed in order to give effect to art. 14.6 of the International Covenant on Civil and Political Rights 1966, provides as follows:

(1) Subject to subsection (2) below, when a person has been convicted of a criminal offence and when subsequently his conviction has been reversed or he has

been pardoned on the ground that a new or newly discovered fact shows beyond reasonable doubt that there has been a miscarriage of justice, the Secretary of State shall pay compensation for the miscarriage of justice to the person who has suffered punishment as a result of such conviction or, if he is dead, to his personal representatives, unless the non-disclosure of the unknown fact was wholly or partly attributable to the person convicted.

(2) No payment of compensation under this section shall be made unless an application for such compensation has been made to the Secretary of State.

(3) The question whether there is a right to compensation under this section shall be determined by the Secretary of State.

(4) If the Secretary of State determines that there is a right to such compensation, the amount of the compensation shall be assessed by an assessor appointed by the Secretary of State.

(5) In this section 'reversed' shall be construed as referring to a conviction having been quashed—

(a) on an appeal out of time; or

(b) on a reference . . . under the Criminal Appeal Act 1995 . . .

(6) For the purposes of this section a person suffers punishment as a result of a conviction when sentence is passed on him for the offence of which he was convicted . . .

Thus, under the statutory scheme the Home Secretary has a *duty* (and not merely a *discretion*) to pay compensation. That duty only arises, however, in the limited circumstances outlined in s. 133. In particular,

(a) there must have been a conviction; and

(b) the conviction must have been reversed outside the normal appeal process, or the convicted person pardoned, on the basis of a new or newly discovered fact; and

(c) that fact must show beyond reasonable doubt that there has been a miscarriage of justice; and

(d) a positive application for compensation under s. 133 must be made to the Home Secretary.

For the purposes of s. 133 of the Criminal Justice Act 1988, a conviction is only 'quashed' when an appeal against it has been allowed in full under s. 2 of the Criminal Appeal Act 1968, and not when a lesser verdict has been substituted for it under s. 3 of that Act. Where, therefore, nine years after the event the Court of Appeal substituted for a jury's verdict of guilty of murder a verdict of guilty of attempted grievous bodily harm, it was subsequently held in an application for judicial review of the Home Secretary's refusal of compensation under s. 133 that the claimant had no entitlement under the section (*R (Christofides)* v *Secretary of State for the Home Department* [2002] 1 WLR 2769, DC).

Since the phrase 'new or newly discovered fact' in s. 133 qualifies the words 'his conviction has been reversed' in subsection (1) as well as the word 'pardoned', it follows that there is no right to compensation under s. 133 where a conviction has been quashed on the ground of:

(a) wrongful admission of evidence at the trial (*R* v *Secretary of State for the Home Department (ex parte Bateman)* [1993] COD 493, DC); or

(b) invalidity of a by-law (*R* v *Secretary of State for the Home Department (ex parte Howse)* [1993] COD 494, DC).

Section 133 is intended to apply to natural persons only and not to corporate bodies. Accordingly, a company, not being a 'person' within the meaning of s. 133(1), has no right to compensation under the statutory scheme (*R* v *Secretary of State for the Home Department (ex parte Atlantic Commercial (UK) Ltd)* (1997) *The Times*, 10 March).

The 'new or newly discovered fact' referred to in s. 133 has to be newly discovered since the date of the final decision in the case (i.e., after the end of any appeal brought within the ordinary time limit, or after the expiry of the time limit for bringing an appeal where none has been brought), and it has to be the *principal*, if not the only, reason for the quashing of a conviction (*R (Murphy)* v *Secretary of State for the Home Department* [2005] 1 WLR 3516, DC). In this case, the claimants unsuccessfully sought judicial review of the Home Secretary's refusal to pay compensation under s. 133 following the quashing of their murder convictions by the Court of Appeal in 2002 on a reference by the Criminal Cases Review Commission. Their first appeals against conviction (brought within the ordinary time limit) had been unsuccessful in 1993. The challenge to the Home Secretary's decision failed because the material which the claimants relied on as a newly discovered fact had been disclosed before the hearing of their first appeals in 1993, and the alleged newly discovered fact was not, in any case, the *principal* reason why the convictions were quashed in 2002. It was just one of a number of factors which led the Court of Appeal to conclude that the convictions were unsafe.

Not all those whose convictions are subsequently quashed as unsafe by the Court of Appeal are victims of a 'miscarriage of justice' within the meaning of s. 133. In *R (Mullen)* v *Secretary of State for the Home Department* [2004] UKHL 18, HL, the claimant had been convicted of conspiracy to cause explosions likely to endanger life or cause serious injury to property, and was sentenced to 30 years' imprisonment. Although entertaining no doubt about the guilt of the claimant (who did not even argue that he was innocent), the Court of Appeal had quashed the conviction on the ground that his deportation from Zimbabwe to the United Kingdom to stand trial involved serious abuses of process on the part of the authorities (*R* v *Mullen* [1999] 3 WLR 777, CA, para 7.5.5.1 above).

The claimant, who had spent ten years in prison, applied to the Home Secretary for compensation under s. 133. The Home Secretary decided to reject the application, and the claimant sought judicial review of that decision. The House of Lords held that he was not entitled to compensation. A majority of their Lordships held that there had been no 'miscarriage of justice' within the meaning of s. 133 because there had been no failure in the trial process itself. The claimant had been properly convicted on the evidence. His conviction had later been quashed because of something which had happened *before* the commencement of the trial process and which had enabled his trial to take place, namely a serious abuse of executive power which had led to his removal from Zimbabwe to the United Kingdom. This may have constituted a 'newly discovered fact', but the claimant was not the victim of a 'miscarriage of justice'.

Lord Steyn held that there had been no 'miscarriage of justice' because the claimant was not innocent. According to his Lordship, proof of innocence is required before it can be said that there has been a 'miscarriage of justice' within the meaning of s. 133 (see [2004] UKHL 18, HL, at [56]). If this view is correct, compensation under s. 133 is payable only to those ultimately proved innocent. In the opinion of Lord Bingham, the phrase 'miscarriage of justice' in s. 133 extends further than this and covers those who, *whether guilty or not*, should clearly not have been convicted at their trials (ibid., at [4]). The other three Law Lords involved in the appeal found it unnecessary to reach any conclusion on these competing views. (Mr Mullen had also applied for an *ex gratia* payment, as to which see below.)

The Home Secretary is solely responsible for deciding whether there is a right to compensation in individual cases (Criminal Justice Act 1988, s. 133(3)). If his decision is in the affirmative, the actual compensation payable is not assessed by the Home Secretary but by an assessor appointed by him (ibid., s. 133(4)). Those eligible for appointment are lawyers practising in the United Kingdom, persons who hold or have held judicial office in the United Kingdom, and members of the Criminal Injuries Compensation Board (ibid., sch. 12).

The amount of the award is calculated using the principles applied in the assessment of damages in the law of tort. The relevant factors include loss of earnings during any period of imprisonment, loss of future earning capacity, damage to reputation, hardship (including mental suffering, injury to feelings, and inconvenience), costs incurred in the original proceedings and in establishing innocence, and expenses incurred as a result of imprisonment (including family expenses). In assessing compensation attributable to suffering, damage to reputation or similar damage, the assessor must have regard in particular to the seriousness of the offence of which the applicant was convicted and the severity of the sentence imposed on him, the way in which the offence was investigated and prosecuted, and any other convictions of, and sentences imposed on, the applicant (ibid., s. 133(4A), as inserted by the Criminal Appeal Act 1995).

The Court of Appeal has held that it is correct for the assessor to deduct from the amount of compensation awarded for loss of earnings the living expenses that the claimant would have had to meet had he not been in prison. This deduction was said to be necessary in order to avoid compensating the claimant for a loss that he has not suffered (*R (O'Brien and others)* v *Independent Assessor* (2004) *The Times*, 7 September, CA).

The *ex gratia* scheme of payments continues to operate for those miscarriages of justice not covered by the statutory scheme. The Home Secretary has a discretion in the matter of *ex gratia* payments, and, as long as he has followed his own stated policy and has acted fairly, his refusal to make an award cannot be successfully challenged in judicial review proceedings (*R* v *Secretary of State for the Home Office (ex parte Chubb)* [1986] Crim LR 809, DC; *R* v *Home Secretary (ex parte Weeks)* (1988) *The Guardian*,

23 February, DC; *R v Secretary of State for the Home Department (ex parte Harrison)* [1988] 3 All ER 86, DC; *In Re McFarland* [2004] UKHL 17, HL, below; *R (Mullen) v Secretary of State for the Home Department* [2004] UKHL 18, HL, below).

The Home Secretary's policy (as outlined in *R v Secretary of State for the Home Department (ex parte Bateman)* [1993] COD 494, DC, above) is to make an *ex gratia* payment provided that:

(a) the claimant has spent time in custody following a wrongful conviction or charge which resulted from serious default on the part of the police or some other public authority; or

(b) there are exceptional circumstances, such as the emergence of facts which completely exonerate the claimant.

In addition to the police, the Crown Prosecution Service is clearly a 'public authority' for the purposes of (a) above, but the trial judge is not. That was the opinion of the Divisional Court in *R v Secretary of State for the Home Department (ex parte Harrison)*, above, *R v Secretary of State for the Home Department (ex parte Bateman)*, above, and *R v Secretary of State for the Home Department (ex parte Garner)*, below. The matter was raised before the House of Lords in *In Re McFarland* [2004] UKHL 17, HL, where the majority agreed with the Divisional Court's approach (see, e.g., *per* Lord Bingham at [15]). Lord Steyn dissented because he found the exclusion of judges 'strange', 'curious', and 'unjust', and was prepared to treat judges and juries as 'public authorities' for the purposes of the *ex gratia* scheme (at [28]).

Although the trial judge is not a public authority within (a) above, it has been suggested that his errors or misconduct may constitute 'exceptional circumstances' within the meaning of (b) above, and that the Home Secretary should at least consider in every case where judicial misconduct is alleged whether it was so gross as to give rise to exceptional circumstances. Failure to do this may amount to an improper fetter on the exercise of his discretion (*R v Secretary of State for the Home Department (ex parte Garner)* (1999) *The Times*, 4 May, DC).

Since it is impossible for a company to have 'spent time in custody', it follows that a corporate body is not covered by the *ex gratia* scheme (*R v Secretary of State for the Home Department (ex parte Atlantic Commercial (UK) Ltd)* (1997) *The Times*, 10 March).

It seems that an *ex gratia* payment will not be made unless, *inter alia*, on a balance of probabilities the claimant was more likely than not to have been innocent of the crime for which he was convicted (see *R v Secretary of State for the Home Department (ex parte Harrison)*, above, at p. 88). This aspect of the *ex gratia* scheme has been criticised on the ground that it enables the Home Secretary to make a decision on the guilt or innocence of the claimant.

The claimant in *R (Mullen) v Secretary of State for the Home Department* [2004] UKHL 18, HL, above, had alternatively claimed an *ex gratia* payment of compensation. This was refused by the Home Secretary on the ground that, although the claimant's case fell within (a) above, it was right to depart from his usual policy having regard to the exceptional circumstances of the case, in particular the fact that the claimant had been properly convicted and that it would be an 'affront to justice' to pay compensation to him. On the claimant's application for judicial review, the House of Lords held that the Home Secretary's departure from his stated policy was justified and was not irrational in the circumstances.

The amount of an *ex gratia* award is assessed by a leading barrister on the same principles as those mentioned above in relation to the statutory scheme. The level of compensation offered in these cases is not over-generous. The value of liberty is often put at a lower figure than the value of a person's reputation.

There are two situations in which the victim of an injustice has no prospect of even an *ex gratia* award of compensation, and may also be left without any legally enforceable remedy.

The first is the case of the person who has been prosecuted and acquitted either by a magistrates' court or the Crown Court. Acquittals affect a significant proportion of defendants, especially in the Crown Court. For example, in 2004, the proportion of defendants acquitted in the Crown Court who pleaded not guilty to all charges was 66.7 per cent (*Judicial Statistics 2004*, Cm 6565, 2005, p. 90). It goes without saying that not all acquitted defendants are victims of injustice. However, those who *are* such victims have no legal remedy unless they can succeed against the police in a civil action for damages for malicious prosecution, a tort which is notoriously difficult to prove.

The second concerns a person who has been kept in custody and then released later without being prosecuted. He has no legal remedy unless he can succeed against the police in a civil action for the tort of false imprisonment. In recent years there appears to have been an increase in the number of out-of-court settlements agreed by police authorities in actions against them for false imprisonment and/or malicious prosecution.

7.6 Some notable miscarriages of justice

In October 1989, on a reference by the Home Secretary under s. 17 of the Criminal Appeal Act 1968 (since repealed by the Criminal Appeal Act 1995: see para 7.5.7 above), the Court of Appeal quashed the convictions of four people who had been deprived of their liberty for over 14 years. The case was that involving the so-called 'Guildford Four', perhaps the most serious (and most widely-publicised) miscarriage of justice.

Patrick Armstrong, Gerard Conlon, Paul Hill, and Carole Richardson were tried before Donaldson J and a jury in October 1975. They were convicted of conspiracy to cause explosions and of five murders arising from the bombing of a public house in Guildford a year earlier, and also of causing an explosion likely to endanger life at another public house in Guildford on the same day. In addition, Patrick Armstrong and Paul Hill were convicted of two murders arising from the bombing of a public house in Woolwich. The 'Guildford Four' were convicted solely on the basis of confessions which they retracted at the trial. There was no corroborating scientific or identification evidence.

Messrs Armstrong, Conlon, and Hill were sentenced to life imprisonment. Carole

Richardson, who was under 18 at the time of the alleged offences, was ordered to be detained during Her Majesty's pleasure. Their applications for leave to appeal against conviction were refused in October 1977 by the Court of Appeal, and in January 1987 the Home Secretary refused to refer the case back to that Court. He did, however, order an inquiry to be carried out by the Avon and Somerset police into the way in which the case had been investigated in 1974–5.

A reference to the Court of Appeal was finally made in January 1989 in the face of mounting pressure from the media, senior churchmen, politicians, and two retired Law Lords. The grounds on which the reference was made related particularly to new alibi and medical evidence in the case of Carole Richardson.

Before the reference came on for hearing, the Director of Public Prosecutions announced publicly that, in the light of further new evidence discovered by the Avon and Somerset police, it 'would be wrong for the Crown to seek to sustain the convictions' (see *The Times*, 18 October 1989). An expedited hearing of the reference was arranged, and it emerged that some of the original investigating police officers had perjured themselves and presented fabricated evidence at the trial. This revelation rendered the convictions unsafe and they were quashed (see *R v Richardson and others* (1989) *The Times*, 20 October, CA).

After the case of the 'Guildford Four', other disturbing examples of actual or potential miscarriages of justice came to light. In April 1990, the convictions of the 'Winchester Three' were quashed as unsafe in *R v McCann and others* (1990) 92 Cr App R 239, CA, following prejudicial comments in the media by, *inter alia*, the then Secretary of State for Northern Ireland (Mr Tom King) and a retired senior judge (Lord Denning).

In 1987, the three defendants, John McCann, Finbar Cullen, and Martina Shanahan, had been discovered close to the home of the Secretary of State for Northern Ireland and were arrested under anti-terrorism legislation. They were tried in 1988 for conspiracy to murder the Secretary of State and conspiracy to murder persons unknown. Each defendant chose not to give evidence. While counsel were making their closing speeches to the jury, the Home Secretary announced the government's intention to change the law on an accused's right to silence. This announcement received wide publicity in the media. Tom King and Lord Denning, in separate television interviews, used words which conveyed the impression that in terrorist cases a failure to answer questions or to give evidence was tantamount to an admission of guilt.

An application to discharge the jury, and for a retrial, was dismissed by the trial judge, Swinton Thomas J, who instead merely warned the jury, in his summing-up, to disregard any broadcasts on the right to silence. After deliberating for 15 hours, the jury returned a verdict of guilty on each count by a majority of ten to two. On appeal against conviction, the Court of Appeal held that the impact of the media coverage on the fairness of the trial could not be overcome by any direction to the jury, and that the judge ought to have discharged the jury and ordered a retrial. The defendants' appeals were allowed and their convictions set aside on the ground that they were unsafe. (Note that the Court of Appeal itself was unable to order a retrial because this was not a fresh evidence case, and the relevant part of the Criminal Justice Act 1988 which conferred a more general power to order a retrial (para 7.5.5.3 above) was not in force at the relevant time.)

Despite spending many months in prison awaiting trial and the outcome of their appeal, the 'Winchester Three' did not qualify for compensation under s. 133 of the Criminal Justice Act 1988. Their convictions were neither reversed on the ground of a 'new or newly discovered fact', nor on an appeal out of time or on a reference by the Home Secretary under what was then s. 17 of the Criminal Appeal Act 1968.

The story of the 'Maguire Seven' began at the time of the arrest of two of the 'Guildford Four', which had led the police to visit the Maguire home in December 1974 where seven people were apprehended. They were charged with unlawful possession of an explosive substance. No explosives were found on the premises, but there were traces of nitroglycerine on the hands of the male defendants and on plastic gloves used by Mrs Maguire.

The 'Maguire Seven'—Anne Maguire and her husband Patrick Maguire, two of their children (Patrick, aged 13 and Vincent, aged 16), Sean Smyth (Mrs Maguire's brother), Giuseppe Conlon (Mrs Maguire's brother-in-law and the father of Gerard Conlon, one of the 'Guildford Four'), and Patrick O'Neill (a family friend)—were convicted in 1975 after a trial before Donaldson J and a jury. They were given custodial sentences of between four and 14 years. Their appeals against conviction were dismissed by the Court of Appeal in 1977. One of them, Giuseppe Conlon, died in prison in 1980.

In June 1990, several years after the last of them had been released from prison, the Director of Public Prosecutions and the Home Secretary announced that the convictions of the 'Maguire Seven' could not be allowed to stand (see *The Independent*, 15 June 1990). Their case was referred to the Court of Appeal, and eventually their convictions were quashed as unsafe on the ground that their hands could have been *innocently* contaminated by nitroglycerine (*R v Maguire and others* [1992] 2 All ER 433, CA).

In February 2005, the Prime Minister, Tony Blair, made a public apology to the 'Maguire Seven' and the 'Guildford Four' for the trauma of their wrongful convictions and 'the stigma which wrongly attaches to them to this day'. He said that they 'deserve to be completely and publicly exonerated' (see *The Times*, 10 February 2005).

The 'Birmingham Six'—Richard McIlkenny, Hugh Callaghan, Gerard Hunter, Patrick Hill, Billy Power, and John Walker—were tried by Bridge J (later Lord Bridge, a Lord of Appeal in Ordinary) and a jury in 1975. They were convicted of murder and sentenced to imprisonment for life. The trial judge said that they had been convicted on the 'clearest and most overwhelming evidence' he had ever heard.

They were allegedly the Birmingham pub bombers of 1974 responsible for the deaths of 21 people. They were refused leave to appeal against their convictions in 1976. Lord Widgery CJ said there was no evidence that they had been knocked about in custody 'beyond the ordinary'. In 1987, the Home Secretary referred the case to the Court of Appeal, but the appeal was dismissed at the end of the hearing in the following year. Lord Lane CJ described as 'wholly incredible' the defendants' contention that the police had conspired to present a false case against them (see *The Independent*, 29 January 1988).

In 1990, the Home Secretary referred their case back to the Court of Appeal for a second time. After spending more than 16 years in prison, their convictions were quashed as unsafe in 1991 in the light of fresh scientific evidence (including electrostatic document analysis (ESDA) of police interview notes), and fresh evidence of police ill-treatment of them. The defendants were released from prison in March 1991, and the Court of Appeal's reasons for quashing their convictions are reported in *R v McIlkenny and others* [1992] 2 All ER 417, CA.

In September 1991, the Home Secretary referred the cases of Winston Silcott, Mark Braithwaite, and Judith Ward to the Court of Appeal.

Messrs Silcott and Braithwaite had received life sentences as a result of their convictions for the murder of PC Keith Blakelock during the riot at Broadwater Farm, Tottenham, in 1985. The case of the third man convicted for the Broadwater Farm murder, Engin Raghip, had already been referred to the Court of Appeal.

The 'Tottenham Three' referrals were made because of doubts over the genuineness of their confessions to the police. Their convictions were quashed on the grounds that police notes of an interview with Silcott had been tampered with (demonstrated by electrostatic document analysis (ESDA)), the police officer in charge of the case had lied under oath at their trial in 1987, fresh psychological evidence about Raghip's level of intelligence had been wrongly excluded at an earlier Court of Appeal hearing presided over by Lord Lane CJ in 1988 at which leave to appeal against conviction was refused, and that Braithwaite had been wrongly denied access to legal advice during police interviews.

For what is believed to be the first time ever, the Court of Appeal apologised for the fact that the defendants had suffered as a result of the shortcomings of the criminal justice system (*R v Silcott and others* (1991) *The Independent*, 6 December).

Judith Ward had been sentenced to imprisonment for life in 1974 following the deaths of 12 people caused by the bombing of a coach on the M62 in Yorkshire. The reference back to the Court of Appeal was made because the cases of the 'Maguire Seven' and the 'Birmingham Six' had cast doubt on the scientific evidence relied on by the prosecution at her trial.

She was released on bail in May 1992. Her convictions were quashed in the following month on the grounds that there had been a material irregularity in the course of the trial, and that the convictions were unsafe and unsatisfactory. The material irregularity consisted of the prosecution's failure to disclose evidence to the defence which cast doubt on the prosecution case. The convictions were unsafe and unsatisfactory for two reasons. First, the fresh forensic evidence available before the Court of Appeal rendered the scientific case against Miss Ward insupportable. Secondly, in view of her mental state at the time of the trial no reliance should have been placed on the confessions she had made to the police. (See *R v Ward* [1993] 2 All ER 577, CA.)

The Darvell brothers, Wayne and Paul, from Swansea, were convicted in 1986 of murdering the manageress of a sex-shop. Their convictions were quashed in July 1992 after the Home Secretary had referred their case to the Court of Appeal and it was found, *inter alia*, that police records of interviews with Wayne Darvell had been altered so as to produce false admissions that his brother had committed the murder. Lord Taylor CJ, on behalf of the Court of Appeal, apologised to the brothers for the ordeal they had been put through (see *The Independent*, 1 August 1992).

In December 1992, the convictions of the 'Cardiff Three' (Stephen Miller, Yusef Abdullahi, and Tony Paris) were quashed. They had stood trial in 1990 for the murder of a prostitute in Cardiff and were found guilty. Two other men, John Actie and his cousin, Ronnie Actie, who stood trial with them were acquitted.

Police interviews of Stephen Miller were central to the prosecution's case against all three convicted men. These interviews had been conducted in a bullying fashion. Miller, who was on the borderline of mental handicap, denied any participation in the murder more than 300 times but eventually 'confessed' after 13 hours of questioning spread over five days. The interviews were recorded on 19 tapes.

The Court of Appeal held that the trial judge had been wrong to allow the interviews to be introduced in evidence since they were oppressive and the confession obtained was unreliable. The convictions of all three men were decided to be unsafe. In delivering the judgment of the Court, Lord Taylor CJ was severely critical both of the interviewing police officers and of Miller's former solicitor, who was present throughout much of the interviewing but did not intervene to prevent the police asking questions, or making comments, which were improper (i.e., oppressive, threatening, or insulting), or to advise his client to remain silent (R v *Miller and others* (1992) *The Independent,* 17 December, CA). A new investigation into the murder uncovered fresh DNA evidence from the crime scene and led to the conviction of the real murderer in 2003 on a plea of guilty.

Michelle and Lisa, the Taylor sisters, were found guilty in July 1992 of murdering the wife of a man with whom Michelle had allegedly had an affair. Their convictions were quashed on appeal in June 1993 for two reasons. First, there had been a failure by the police detective sergeant in charge of the case to disclose (even to the prosecution) the existence of a statement by a witness that the second woman he saw leaving the victim's home around the time of the murder might have been black. The Taylor sisters are white, and the witness's later statement made for the purposes of the trial described the two women he had seen as 'blonde'. The police also failed to disclose that this witness had written to Barclays Bank (the victim's employers) before the trial claiming the £25,000 reward offered for information about the murder.

Secondly, the press coverage of the trial (described as 'unremitting, extensive, sensational, inaccurate, and misleading') had created a real risk of prejudice against the defendants. It was for this reason that the Court of Appeal did not think it appropriate to order a retrial.

The papers in the case were ordered to be sent to the Attorney-General for him to consider whether to institute proceedings for contempt of court against any newspaper (R v *Taylor* (1993) *The Times,* 15 June, CA). When, after a lengthy delay, the Attorney-General decided that it was not appropriate to start proceedings, the Taylor sisters were granted leave to apply for judicial review of his decision (see *The Times,* 7 December 1994). It was ultimately held, however, that it was not reviewable because of the well-established principle (exemplified in *Gouriet* v *Union of Post Office Workers* [1978] AC 435, HL) that the courts will not review the exercise of the Attorney-General's discretion in relation to decisions taken in the execution of his public office (*R* v *Attorney-General (ex parte Taylor)* (1995) *The Times,* 14 August, DC).

In June 1993, the Court of Appeal when quashing the conviction of Ivan Fergus, by then aged 16, took the unusual step of declaring not only that the conviction was unsafe but also that Mr Fergus was 'wholly innocent'. He had been convicted in November 1991 of assault with intent to rob—largely on the basis of wrongful identification evidence provided by a short-sighted bank clerk. He was sentenced to 15 months' youth custody but was released pending his appeal after serving six months of the sentence.

The conviction was quashed for a number of reasons. The main ground was the failure of the police to disclose, first, a photograph of the defendant taken at the time of his arrest at the age of 13, and, secondly, a crime report containing the first details given to the police by the victim. These documents, which were disclosable under a common law duty of disclosure, would have revealed that Mr Fergus was not the perpetrator of the crime.

Other grounds were that the trial judge should have withdrawn the case from the jury; that the judge's summing up amounted to a material irregularity in its failure to put adequately to the jury specific weaknesses in the identification evidence; that the defendant's former counsel and solicitor had failed to prepare and conduct his defence properly; that the defendant had not had a fair trial due to the failure of the police to take statements from alibi witnesses. (See *R v Fergus* (1993) *The Times*, 30 June, CA.)

In 1995, the convictions of four men—Colin Phillips, Stuart Blackledge, Bryan Mason, and Paul Grecian—were set aside in controversial circumstances. They had been convicted in 1992 of breaking an arms embargo by selling 300,000 artillery fuses to Iraq under an export licence which listed Jordan as the end-user. Three were given suspended one-year prison sentences, and the fourth was fined. The Court of Appeal subsequently decided that the withholding of government documents had prevented the men from mounting a proper defence at their trial. Indeed, it was the non-availability of these documents which had pressurised the men into pleading guilty. The documents, for which government ministers had successfully invoked public interest immunity from disclosure, would have shown that within government circles there was widespread knowledge of, and complicity in, an arms trade with Iraq (see *The Times*, 8 November 1995).

The political pressure generated by this case prompted the government to establish an inquiry under Sir Richard Scott, Vice-Chancellor of the Chancery Division of the High Court, into the so-called 'arms to Iraq affair'. The *Scott Report* was published in February 1996 and was highly critical of the behaviour of some ministers and civil servants (*Report of the Inquiry into the Export of Defence Equipment and Dual-Use Goods to Iraq and Related Prosecutions*, 15 February 1996). There were, however, no resignations in its aftermath.

The case of Mahmood Hussein Mattan was the first to be referred to the Court of Appeal by the Criminal Cases Review Commission under s. 9 of the Criminal Appeal Act 1995.

Mr Mattan was hanged at Cardiff prison in 1952 following his conviction for murder. In 1998, his conviction was quashed as unsafe after the evidence of the main prosecution witness was found to be unreliable. In its judgment, the Court of Appeal welcomed the creation of the Criminal Cases Review Commission and expressed profound regret that it had taken 46 years for Mr Mattan's conviction to be shown to be unsafe. It further suggested that capital punishment is not an appropriate ultimate sanction in a criminal justice system which is human and, therefore, fallible, and that no one associated with the system can afford to be complacent (see *The Times*, 25 February 1998 and *R v Mattan* (1998) *The Times*, 5 March, CA).

Patrick Nicholls and Ryan James had their murder convictions quashed in 1998 when it transpired that the victim in each case had not been murdered at all.

Mr Nicholls had spent 23 years in jail following conviction for the murder of an elderly widowed friend who, according to the evidence of two pathologists, had been suffocated and beaten about the face. She had actually died from a heart attack. The Court of Appeal expressed its 'great regret' that Mr Nicholls had been wrongly convicted on flawed evidence and had spent such a long time in jail (see *The Times*, 13 June 1998).

Mr James was convicted in 1995 of poisoning his wife with a horse sedative, but the later discovery of a suicide note rendered his conviction unsafe. It appeared that his wife killed herself but made her death look like murder in order to gain revenge on Mr James for having an affair with another woman (see *The Times*, 29 July 1998).

Paul Blackburn was 15 years of age when he was convicted of attempting to murder a nine year-old boy. His case was referred to the Court of Appeal by the Criminal Cases Review Commission, and his conviction was quashed in 2005 after he had spent 25 years in detention. It emerged that when he was interviewed by the police before his trial he had not been told of his right to consult a solicitor and neither a parent nor his social worker was present. In addition, fresh linguistic evidence suggested that there had been substantial police involvement in the wording of his written admissions (see *The Times*, 26 May 2005).

8

Contempt of court

8.1 Introduction

Contempt of court has traditionally been classified as either criminal or civil, although for some time this classification has been criticised as 'unhelpful and almost meaningless' (*Jennison* v *Baker* [1972] 2 QB 52, CA, *per* Salmon LJ at p. 61; see also *Home Office* v *Harman* [1982] 1 All ER 532, HL, *per* Lord Scarman at p. 542; *Attorney-General* v *Newspaper Publishing plc* [1987] 3 All ER 276, CA, *per* Sir John Donaldson MR at p. 294 and Lloyd J at p. 306). The abolition of the distinction between criminal and civil contempt was recommended in 1974 by the *Phillimore Committee on Contempt of Court* (Cmnd 5794, 1974), but so far there has been no legislative move to abandon it.

Criminal contempt is a non-arrestable common law offence punishable by imprisonment and/or a fine, or an order to give security for good behaviour. Alternatively, the court may grant an injunction to restrain a repetition of the contempt, as in *Attorney-General* v *Times Newspapers Ltd* [1974] AC 273, HL (the 'thalidomide case'; see para 8.2.2.3 below). Criminal contempt extends to such matters as insulting behaviour in the face of the court and conduct which interferes with the proper administration of justice, both criminal and civil. The offence of criminal contempt is triable summarily without a jury. Trial of the offence on indictment is obsolete.

In *Attorney-General* v *Times Newspapers Ltd* [1991] 2 All ER 398, HL, the term 'contempt of court' was criticised by the House of Lords as 'inaccurate and misleading' and as 'a less than happy description' of a concept designed to promote the effective administration of justice rather than (as is commonly supposed) the dignity of judges (see especially [1991] 2 All ER 398 *per* Lords Ackner, Oliver, and Jauncey at pp. 406–7, 413, and 423, respectively). This criticism is consistent with earlier *dicta*, and their Lordships re-emphasised Lord Diplock's statement in *Attorney-General* v *Leveller Magazine Ltd* [1979] AC 440, HL, at p. 449 that

. . . although criminal contempt of court may take a variety of forms they all share a common characteristic: they involve an interference with the due administration of justice, either in a particular case or more generally as a continuing process. *It is justice itself that is flouted by contempt of court* . . .

(Emphasis added. See to the same effect *Re Johnson* (1888) 20 QBD 68, CA, *per* Bowen LJ at p. 74; *Jennison* v *Baker* [1972] 2 QB 52, CA, *per* Salmon LJ at p. 61; *Attorney-General* v *Times Newspapers Ltd* [1974] AC 273, HL, *per* Lord Cross at p. 322.)

Civil contempt, sometimes called 'contempt in procedure', consists of disobedience to the judgments or orders of the court, or a breach of an undertaking given to the court. In essence, civil contempt is a *private wrong* to the person entitled to the benefit of the order or undertaking, although contempt proceedings are not an action in tort even where the contempt has incidentally been committed during an action in tort (*Express and Star* v *National Graphical Association (1982)* [1986] ICR 589, CA).

Civil contempt is also a *criminal offence* punishable by imprisonment and/or a fine under the inherent jurisdiction of the court. Alternatively, an injunction may be granted against the contemnor. Because proceedings for civil contempt are criminal in nature the standard of proof required to establish the contempt is the criminal standard, i.e., proof beyond reasonable doubt (*Dean* v *Dean* [1987] 1 FLR 517, CA).

The High Court has power to impose a fine of an unlimited amount for civil contempt. The power has been used to inflict swingeing fines on trade unions which have defied court orders made during industrial disputes. Thus, in 1983 the NGA was fined a total of £675,000 for contempt, and, in 1984, the NUM and the TGWU were each fined £200,000.

A further remedy, not available in cases of criminal contempt, is a writ of sequestration under which the contemnor's property is placed temporarily in the hands of sequestrators, who manage the property and receive the rents and profits from it until the contempt is purged.

Sequestration is regarded as a drastic remedy and one to be resorted to sparingly. It is a particularly appropriate method of enforcing court orders against a company or its directors, or against a trade union, especially where the contempt consists of a refusal to pay a sum of money. The assets of the NUM were sequestrated during the miners' strike of 1984–85, and, in 1986, SOGAT '82 was fined £25,000, and had its assets sequestrated, for contempt of orders made in legal proceedings arising out of the disputed move of Times Newspapers from Fleet Street to Wapping.

Normally, proceedings for civil contempt will be commenced either by the litigant in whose favour the court order was made or by the Attorney-General. It seems, however, that where neither the litigant nor the Attorney-General commences proceedings, the court itself may act of its own motion in exceptional cases of clear contempt where it is urgent and imperative for the contempt to be punished immediately (*Clarke* v *Chadburn* [1985] 1 All ER 211, *per* Sir Robert Megarry V-C at p. 215; *Re M and others (minors) (breach of contact order: committal)* [1999] 2 All ER 56, CA).

8.2 Criminal contempt of court

8.2.1 Interference with the proper administration of justice in the courts

8.2.1.1 Interrupting court proceedings

Any words or acts which obstruct or interfere with the proper administration of justice in the courts, and are calculated to do so, are punishable as criminal contempts. Thus, in a much-publicised case in May 1993 a young man was sentenced to 14 days' imprisonment for interrupting the proceedings during a trial in the Crown Court at Cardiff. His contempt consisted of wolf-whistling at a female juror from the public gallery (*R v Powell* (1993) *The Times*, 3 June, CA; he was released on bail after 24 hours on application to the High Court and, within six days, the Court of Appeal, while confirming that he was guilty of contempt, quashed the sentence of imprisonment on the ground that it was an inappropriate punishment in the circumstances).

It is usual, though not essential, for the words or acts to be spoken or done in the courtroom itself (*Balogh v Crown Court at St Alban's* [1974] 3 All ER 283, CA, *per* Lord Denning MR, Stephenson LJ, and Lawton LJ at pp. 287, 292, and 294, respectively). Assaulting, insulting or abusing the judge, whether in court or outside, is a contempt, as is interference with jurors, witnesses, or the parties to, or the lawyers appearing in, a case. (See further on interference with jurors, para 8.2.1.5 below.) The prohibition on interfering with witnesses extends to interference with proper and reasonable attempts by a party's lawyers to identify and interview potential witnesses (*Connolly v Dale* [1996] 1 All ER 224, DC, where a detective superintendent of police was held guilty of contempt).

In *Morris v Crown Office* [1970] 2 QB 114, CA, some students supporting the Welsh Language Society staged a demonstration in the High Court in London and interrupted the proceedings there. They were found to be in contempt of court, and some of them were sent to prison for three months while others were fined. On appeal, the Court of Appeal affirmed that their conduct amounted to contempt, and that the judge had power to deal with them summarily as he did. However, those imprisoned were released and were instead bound over to be of good behaviour and to keep the peace.

In *Balogh v Crown Court at St Alban's* [1974] 3 All ER 283, CA, Balogh, a bored solicitor's clerk, made preparations to introduce laughing gas into a courtroom through a ventilation duct. For this purpose he stole a cylinder of nitrous oxide from a hospital. He was apprehended before he could carry out his plan and was charged with the theft of the cylinder. He was taken before the senior judge in the court building, Melford Stevenson J, who was the judge presiding in the courtroom next to the courtroom in which Balogh was engaged. The judge regarded Balogh's conduct as serious and sentenced him to six months' imprisonment for contempt of court. Balogh appealed, contending, *inter alia*, that he was not in contempt of the judge's court because it was the proceedings in the court next door that he had intended to subvert.

The Court of Appeal reaffirmed the inherent jurisdiction of superior court judges to punish contempt of court summarily, of their own motion, whenever there had been a gross interference with the course of justice in a case which was being tried or was about to be tried. This power exists whether the judge has seen the contempt with his own eyes, or it has merely been reported to him; it is not limited to cases of contempt committed in the face of the court. It was emphasised, however, that the power should only be used where the contempt had been proved beyond reasonable doubt, and where it was urgent and imperative for the judge to act immediately to prevent justice being obstructed or undermined.

On the facts, Balogh was held not to be guilty of contempt of court because he had not gone far enough to commit it. His acts were preparatory acts falling short of contempt. In addition, the Court of Appeal made it clear that, even if Balogh had been in contempt, the judge should not have exercised his summary jurisdiction to punish it. There was no sufficient urgency and it was not imperative to act immediately as Balogh was already in custody on a charge of theft.

8.2.1.2 Scandalising the court

Conduct which scandalises the court or otherwise lowers its authority is a contempt. Thus, scurrilous abuse of a judge, attacks on his personal character, or imputations of partiality, are all punishable. In 1900, the writer of an article in a local newspaper, the *Birmingham Daily Argus*, was proceeded against for publishing comments which attacked the personal qualities of Darling J as a judge. The article was held to be a scurrilous personal abuse of the judge and a contempt of court. The writer was fined £100 and ordered to pay costs of £25 (*R v Gray* [1900] 2 QB 36, DC). In 1928, the editor of the *New Statesman* was proceeded against after making allegations of bias against Avory J. The editor was found to be in contempt, although he was not imprisoned or fined but merely ordered to pay the costs of the proceedings (*R v New Statesman (Editor)* (1928) 44 TLR 301, DC. See also *R v Wilkinson* (1930) *The Times*, 16 July, in which the *Daily Worker* was punished for contempt, having described Swift J as a 'bewigged puppet exhibiting a strong class bias').

It is not contempt to criticise judges for their judicial conduct as long as the comments are fair and made in good faith (see, e.g., *R v Commissioner of Police of the Metropolis (ex parte Blackburn) (No. 2)* [1968] 2 QB 150, CA, where it was held that an article in *Punch* written by Quintin Hogg (who eventually became Lord Chancellor as Lord Hailsham LC), which was highly critical of a certain judicial decision on the gaming laws, was inaccurate and perhaps in bad taste but did not amount to a contempt of court).

The law of contempt applies only to courts of justice properly so-called and to the judges of such courts. It follows that scandalous comments made about a judge otherwise than in his capacity as a member of such a court cannot amount to contempt (*Attorney-General v British Broadcasting Corporation* [1980] 3 All ER 161, HL; *Badry v Director of Public Prosecutions of Mauritius* [1982] 3 All ER 973, PC).

8.2.1.3 Impeding access to the courts

It is a contempt to impede a person's right of access to the courts. In *Raymond v Honey* [1982] 1 All ER 756, HL, the governor of Albany Prison stopped a letter written by an inmate to his solicitor because the letter contained an allegation that an assistant governor had committed theft. The inmate thereupon prepared an application to the High Court for leave to commit the governor to prison for contempt arising out of the stopping of his letter. The governor also stopped the application.

The House of Lords held that the governor was in contempt of court in stopping the application since his conduct amounted to a denial of the inmate's right of access to the courts. The governor was not, however, subjected to any punishment for his contempt. It was further held that the stopping of the inmate's letter to his solicitor was *not* a contempt since there was not enough evidence to show that it had effectively obstructed the proper administration of justice.

8.2.1.4 Refusing to answer questions in court or to disclose sources of information

8.2.1.4.1 *Introduction* It is a contempt committed in the face of the court for a witness in legal proceedings to refuse to be sworn or to refuse to answer a question without lawful excuse.

In *R* v *Samuda* (1989) 11 Cr App R(S) 471, CA, the defendant was found guilty of contempt for refusing to give evidence against a person whom he alleged had attempted to murder him. The Court of Appeal regarded the contempt as serious in the circumstances, and decided that the appropriate sentence was one of six months' imprisonment. A victim of violence and a witness to the attack, who both refused to give evidence at the attacker's trial (which consequently collapsed), were found to be in contempt in *R* v *Holt* (1996) *The Times*, 31 October, CA, although their custodial sentences were reduced on appeal. It was pointed out in this case that difficulties might be avoided if more use was made of what was then s. 23 of the Criminal Justice Act 1988, which provided that a written statement made by a witness may be admissible as evidence where the witness was too afraid to give oral evidence. (This provision has since been repealed and replaced by a similar provision contained in s. 116 of the Criminal Justice Act 2003.)

A lawful excuse for failing to give evidence may be based on the fear of self-incrimination, on privilege, on the irrelevance of the question, or on a genuine fear of reprisal if the fear is so real and compelling that the witness could not reasonably be expected to ignore it (*R* v *K* (1983) 78 Cr App R 82, CA; *R* v *Lewis* (1992) *The Times*, 19 November, CA, where the judge's finding of contempt was set aside and the sentence of nine months' imprisonment imposed on the defendant for the 'contempt' was quashed).

Journalists have no special privilege entitling them not to reveal in court the source of their information. In 1963, three journalists refused to divulge to a tribunal of inquiry the source of their information relating to William Vassall, the Admiralty spy. An application was made to the High Court for an order that the journalists were in contempt of the tribunal. It was held that they were. One of them received a six months' suspended prison sentence (*Attorney-General* v *Clough* [1963] 1 QB 773). The sentence was suspended in the hope that he would change his mind and identify his informant. He did not. However, the informant identified himself and the journalist did not serve his sentence. The other two journalists were sent to prison for terms of six months and three months, respectively (*Attorney-General* v *Mulholland; Attorney-General* v *Foster* [1963] 2 QB 477, CA).

8.2.1.4.2 *Section 10 of the Contempt of Court Act 1981 and article 10 of the European Convention on Human Rights* Since the case of the three journalists, above, it has been provided by s. 10 of the Contempt of Court Act 1981 that no court has power to order a person to disclose, nor is any person guilty of contempt for refusing to disclose, the source of any information contained in a publication for which he is responsible, unless the court is satisfied that disclosure is necessary in the interests of justice, or of national security, or for the prevention of disorder or crime.

Section 10 thus expressly recognises the public interest in allowing newspapers to protect their sources, subject only to the demand for disclosure made by the four overriding public interests mentioned in the section (*Re an inquiry under the Company Securities (Insider Dealing) Act 1985* [1988] 1 All ER 203, HL, *per* Lord Griffiths at p. 207; *X Ltd v Morgan-Grampian (Publishers) Ltd* [1990] 2 All ER 1, HL, *per* Lords Bridge, Oliver and Lowry at pp. 7, 16, and 18, respectively). However, s. 10 would have made no difference in a case like that involving the three journalists, above, since it was 'in the interests of . . . national security' that they should reveal their sources.

The scope of s. 10 was considered by the House of Lords in *Secretary of State for Defence v Guardian Newspapers Ltd* [1984] 3 All ER 601, HL, a case in which the request of the government for the return of a copy of a 'leaked' secret memorandum was resisted by *The Guardian* on the ground that the markings on the document would be likely to identify the 'leaker'. The House of Lords, by a majority of three to two, decided that disclosure of the source of information was necessary in the interests of national security. The risk to national security lay in the possibility that, unless identified, the 'leaker' might in future leak other classified documents with more serious consequences. Their Lordships were unanimous in the view that the burden of proving that one or more of the four exceptions applied (the interests of justice, the interests of national security, the prevention of disorder, the prevention of crime) is on the party seeking the disclosure order. The standard of proof required is that applicable in civil proceedings generally, namely, the balance of probabilities.

The House of Lords has since held that the phrase 'prevention of . . . crime' in s. 10 is to be construed in a wide rather than a narrow fashion. It is a reference to the prevention of a particular identifiable future crime (*Re an inquiry under the Company Securities (Insider Dealing) Act 1985* [1988] 1 All ER 203, HL).

In *X Ltd v Morgan-Grampian (Publishers) Ltd* [1990] 2 All ER 1, HL, the House of Lords was called on to consider directly for the first time the phrase 'in the interests of justice' in s. 10. William Goodwin, a trainee journalist employed by the publishers of a weekly journal, *The Engineer*, intended to write an article based on confidential information about a company's financial affairs which had come into his possession. He refused to hand over to the company the notes from which the source of his information could be identified.

The House of Lords held unanimously that disclosure was necessary in the interests of justice, even though the information was not 'contained in a publication' as required literally by s. 10. Their Lordships were of the opinion that s. 10 applies equally to a source of information contained in an *intended* publication since the underlying purpose of the statutory protection of sources is as valid *before* publication as *after* it (see especially [1990] 2 All ER 1 *per* Lords Bridge and Lowry at pp. 7 and 17, respectively).

Their Lordships were further of the opinion that the phrase 'in the interests of justice' was not to be confined as referring only to the administration of justice in the course of legal proceedings in a court of law. Lord Diplock's *dictum* to the contrary in *Secretary of State for Defence v Guardian Newspapers Ltd* [1984] 3 All ER 601, HL, at p. 607, was expressly disapproved. The House in *X Ltd v Morgan-Grampian (Publishers) Ltd*, above, said it was in the interests of justice that persons should be able to exercise their legal rights, and to protect themselves against legal wrongs, with or without commencing proceedings (see especially [1990] 2 All ER 1 *per* Lords Bridge and Oliver at pp. 9 and 17, respectively).

When Mr Goodwin continued to refuse to disclose his source, he was fined £5,000 for contempt by Hoffmann J in the High Court (see *The Times*, 11 April 1990). The order requiring him to reveal his source, and the fine imposed on him for declining to do so, were later held by a majority of the European Court of Human Rights to be violations of his right to freedom of expression under art. 10 of the European Convention on Human Rights (*Goodwin* v *United Kingdom* (1996) 22 EHRR 123, ECHR).

Article 10 has featured in several subsequent cases. As scheduled to the Human Rights Act 1998, it provides that:

(a) Everyone has the right to freedom of expression. This right shall include freedom to hold opinions and to receive and impart information and ideas without interference by public authority . . .

(b) The exercise of these freedoms, since it carries with it duties and responsibilities, may be subject to such formalities, conditions, restrictions or penalties as are prescribed by law and are necessary in a democratic society, in the interests of national security, territorial integrity or public safety, for the prevention of disorder or crime, for the protection of health or morals, for the protection of the reputation or rights of others, for preventing the disclosure of information received in confidence, or for maintaining the authority and impartiality of the judiciary.

It follows from the second paragraph of art. 10 that the right of freedom of expression conferred by the first paragraph is not an unqualified right.

8.2.1.4.3 *When is disclosure 'necessary'?* Even if the case falls within one of the four exceptional categories mentioned in s. 10 of the Contempt of Court Act 1981, disclosure of a source of information will not be ordered by the court unless it is 'necessary' to do so. 'Necessary' means something more than merely 'useful' or 'expedient' and something less than 'indispensable'. It has been suggested that the nearest paraphrase is 'really needed' (*Re an inquiry under the Company Securities (Insider Dealing) Act 1985* [1988] 1 All ER 203, HL, *per* Lord Griffiths at p. 209; *X Ltd* v *Morgan-Grampian (Publishers) Ltd* [1990] 2 All ER 1, HL, *per* Lord Oliver at p. 16).

Thus, in *Maxwell* v *Pressdram Ltd* [1987] 1 All ER 656, CA, the Court of Appeal upheld the trial judge's decision not to order disclosure because, although it may have been 'relevant', 'important', or 'desirable', disclosure was not 'necessary' in the interests of justice in the context of the claimant's libel action.

In *X* v *Y* [1988] 2 All ER 648, a health authority was refused an order against a journalist and his newspaper to compel disclosure of the identities of health authority employees who had supplied information from hospital records about two doctors who were continuing in general practice despite having contracted AIDS. Disclosure was held not to be 'necessary' for the prevention of crime since it was not the health authority's function to prevent crime and a criminal investigation was unlikely to ensue even if the source of the information were disclosed.

In *Saunders* v *Punch Ltd* [1998] 1 All ER 234, it was held not to be necessary in the interests of justice to order disclosure of a source of information protected by legal professional privilege. An anonymous article in the defendants' magazine led the claimant, the former chairman and chief executive of Guinness plc, to believe that the defendants were in possession of unpublished records of meetings between him and his solicitors. At these meetings, a Department of Trade and Industry inquiry into alleged fraudulent activities at Guinness plc was discussed. The claimant obtained an injunction to prevent further publication, and applied for an order compelling the defendants to divulge the source of their information. It was decided that the public interest in receiving the information was not outweighed by the interests of justice, and that the claimant was already sufficiently protected by the injunction already obtained.

On the facts of *Camelot Group plc* v *Centaur Communications Ltd* [1998] 1 All ER 251, CA, the opposite conclusion was reached. It was held that the public interest in enabling the claimants (organisers of the National Lottery) to identify a disloyal employee who had 'leaked' a copy of the claimants' draft accounts to the defendants (who revealed them in their publication *Marketing Week*) outweighed the public interest in enabling the defendants to protect their sources of information, and that a disclosure order was necessary in the interests of justice.

Applying *X Ltd* v *Morgan-Grampian (Publishers) Ltd* [1990] 2 All ER 1, HL, and the *Camelot* case, above, it was held in *Michael O'Mara Books Ltd* v *Express Newspapers plc* (1998) *The Times*, 6 March, that it was in the interests of justice that the deputy editor of *The Express* newspaper should disclose the name of the person who had supplied him with a photocopied typescript of a book entitled, *Fergie—Her Secret Life.*

In determining whether it is 'necessary' to compel the revelation of a source of information, the court may also consider whether the applicant for the order has himself taken steps to discover the source of the leak—apart from applying to the court—and whether the leaked information is of great importance (*Broadmoor Hospital* v *Hyde* (1994) *The Times*, 18 March). A similar conclusion was reached in *Saunders* v *Punch Ltd* [1998] 1 All ER 234, above. It is clear from these two cases that an application to the court under s. 10 of the Contempt of Court Act 1981 is not to be treated as a remedy of first resort. In *John* v *Express Newspapers plc* [2000] 3 All ER 257, CA, one reason why the Court of Appeal refused to order the defendants to disclose the source of leaked information was that other means of identifying the source existed but had not been explored.

The *John* case concerned the leaking to the press of draft advice prepared by counsel in the course of litigation between Elton John and a firm of accountants, PricewaterhouseCoopers. There was no internal investigation at counsel's chambers. Such an investigation might have identified the source of the leak or, at least, narrowed down the number of suspects. At first instance, the judge had made an order compelling disclosure of the source. The Court of Appeal held that his decision was wrong since he had not attached sufficient importance to the failure to conduct an inquiry at counsel's chambers, and had attached too much importance to the threat posed to legal confidentiality by an individual making use of a barrister's discarded draft advice. Failure to conduct an inquiry meant that a court order for disclosure of the source was not 'necessary' in the interests of justice. Approving the approach adopted in *Saunders* v *Punch Ltd*, above, Lord Woolf MR said at [27]:

Before the courts require journalists to break what a journalist regards as a most important professional obligation to protect a source, the minimum requirement is that other avenues should be explored.

Any other decision, he said (at [29]),

would be wrongly interpreted as an example of lawyers attaching a disproportionate significance to the danger to their privilege while undervaluing the interests of journalists and thus the public.

In the exceptional circumstances of *Ashworth Hospital Authority* v *MGN Ltd* [2002] 1 WLR 2033, HL, the House of Lords upheld a s. 10 order and decided that it did not violate art. 10 of the European Convention on Human Rights. Here, an employee at a secure hospital supplied an intermediary with medical information held on the hospital computer database about the convicted 'Moors murderer', Ian Brady. The intermediary passed the information to a journalist who wrote an article in the *Daily Mirror* containing verbatim extracts from the confidential information. An internal inquiry at the hospital failed to identify the employee responsible for the leak. The hospital applied for an order requiring the defendants to identify their sources.

It was held that a disclosure order was justified in the 'interests of justice' since the hospital's interest in preserving confidentiality came within one or more of the legitimate aims in art. 10 (para 8.2.1.4.2 above) for restricting the defendants' right to freedom of expression, that the hospital had used all other reasonable means to identify the source, and that identification was necessary to minimise the significant risk of future sales of confidential information. The public interest in the protection of journalists' confidential sources was overridden by the hospital's interest in preserving the confidentiality of medical records which ought to be protected in any democratic society.

Their Lordships further confirmed, first, that a s. 10 order can be made against a person who is involved in the informant's wrongdoing but has not committed any civil or criminal wrong; secondly, that disclosure can be ordered even though the claimant does not intend to bring legal proceedings against the informant; and thirdly, that the court will not order disclosure of a journalist's source of information 'in the interests of justice' within the meaning of s. 10 unless that is a necessary and proportionate response in the circumstances of the case.

8.2.1.5 Contempt in relation to juries

8.2.1.5.1 *Contempt at common law and under the Juries Act 1974* A person may be punished for contempt committed in the face of the court if he fails to attend the court in answer to a jury summons or if, having answered the summons, he is not available to serve as a juror or is unfit for service by reason of drink or drugs. Alternatively, he may be tried summarily for an offence under the Juries Act 1974 (Juries Act 1974, s. 20(1)–(2)). He is not liable to be punished if he can show some reasonable cause for his failure to attend, or for not being available, when called on to serve as a juror (ibid., s. 20(4)).

A juror may also be punished for contempt if he refuses to be sworn or if he misbehaves in court.

It is a contempt for a juror to refuse to give a verdict with the intention of interfering with the administration of justice. In *R* v *Schot* [1997] 2 Cr App R 383, CA, two jurors who had been given prison sentences by the trial judge for refusing, on conscientious grounds, to return a verdict were held on appeal not to have been guilty of contempt. Although they had probably committed the *actus reus* of contempt, there was no evidence that they had done so with the requisite *mens rea* (i.e., with the intention of interfering with the administration of justice).

Quite apart from the statutory contempt created by s. 8 of the Contempt of Court Act 1981 (para 8.2.1.5.2 below), any interference, or attempted interference, with a juror for what he has done in the discharge of his duty is an interference with the proper administration of justice and, therefore, a contempt at common law (*R* v *Martin* (1848) 5 Cox CC 356).

Thus, in *R* v *Palache* (1993) 14 Cr App R(S) 294, CA, the defendant was held to be guilty of common law contempt by threatening the jury after it had returned a guilty verdict against her mother. The defendant's conduct (making remarks in the presence of the jury about the effect of their verdict, banging a water carafe against a glass screen, and moving towards the jury with the carafe in her hand before being stopped by the clerk of the court) was described as a 'grave contempt'.

In *Attorney-General* v *Judd* (1994) *The Times*, 15 August, DC, the defendant's deliberate harassment at a car-boot sale of a member of the jury which had tried him was held to be a common law contempt. He was also found to have violated s. 8 of the Contempt of Court Act 1981 by asking the juror how many jurors had found him guilty.

8.2.1.5.2 *Contempt under section 8 of the Contempt of Court Act 1981* The statutory form of contempt created by s. 8 of the Contempt of Court Act 1981 affects both jurors and outsiders, and is designed to preserve the secrets of the jury-room in the interests of the proper administration of justice.

It is a contempt to obtain, disclose, or solicit any particulars of statements made, opinions expressed, arguments advanced, or votes cast, by members of a jury in the course of their deliberations in any legal proceedings, whether criminal, civil, or in a coroner's court (ibid., s. 8(1)).

Proceedings for contempt under s. 8 can only be instituted by, or with the consent of, the Attorney-General, or on the motion of a court having jurisdiction to deal with the contempt (ibid., s. 8(3)).

There are two exceptional cases where it is not a contempt to disclose particulars (or to publish them once disclosed). First, where the particulars are disclosed in the legal proceedings in question for the purpose of enabling the jury to arrive at their verdict or in connection with the delivery of that verdict (ibid., s. 8(2)(a)). Secondly, where the particulars are disclosed in evidence in any subsequent legal proceedings for an offence alleged to have been committed in relation to the jury in the earlier proceedings (ibid., s. 8(2)(b)). There is, however, no exception allowing disclosure of jury-room deliberations for the purpose of an appeal against conviction, or for the purpose of research into the practical workings of the jury system.

Since there is no exception to jury secrecy in relation to appeals, lawyers and others must take care not to fall foul of s. 8 by interviewing or taking statements from jurors after the trial (*R* v *McCluskey (Kevin)* (1994) 98 Cr App R 216, CA; *R* v *Mickleburgh* (1994) *The Times*, 26 July, CA; *R* v *Young* [1995] QB 324, CA; *R* v *Khan* (1995) *The Times*, 14 April, CA; *R* v *Schot* [1997] 2 Cr App R 383, CA, above). Furthermore, the trial judge should not be approached with enquiries relating to the jury because, after verdict and sentence, he is regarded as *functus officio* ('having performed his function' and, therefore, no longer having jurisdiction). In such cases, the guidance of the Court of Appeal should be sought instead (*McCluskey, Mickleburgh, Young,* and *Khan,* above).

It was held by the House of Lords in *R* v *Mirza* [2004] 2 WLR 201, HL, disapproving dicta of Lord Taylor CJ in *R* v *Young* [1995] QB 324, CA (the 'ouija board case'), that the Court of Appeal is not bound by s. 8 since it cannot be in contempt of itself. (See further on *Mirza* and *Young*, para 6.7.)

The scope of s. 8(1) was considered in *Attorney-General* v *Associated Newspapers Ltd* [1994] 1 All ER 556, HL. Here the *Mail on Sunday* had published an article referring to statements, opinions, and arguments made during their deliberations by some members of the jury in the much-publicised 'Blue Arrow' fraud trial of 1991–92. The newspaper had not obtained its information directly from the jurors but from transcripts of interviews with them conducted by persons who had advertised in another newspaper offering a reward to jurors who contacted a box number.

The Attorney-General brought proceedings for contempt under s. 8(1) against the publishers of the *Mail on Sunday*, its editor, and one of its journalists. Their somewhat feeble defence was that the prohibition against disclosure of jury deliberations was limited to disclosure by members of the jury, and did not apply where the information had been obtained indirectly from another source. The defendants were found to be in contempt and were fined £30,000, £20,000, and £10,000, respectively.

On appeal to the House of Lords, it was held that the meaning of s. 8(1) is plain and unambiguous, and that the word 'disclose' in the subsection applies not only to disclosure of jury deliberations by a member of the jury to a friend or neighbour but also to disclosure to the public at large by a non-juror who has acquired the information directly or indirectly. This interpretation is confirmed by reference to the mischief which the subsection was intended to remedy. That mischief comprises not only disclosure by individual jurors but also publication of the forbidden particulars to the general public ([1994] 1 All ER 556 *per* Lord Lowry at p. 564).

8.2.1.6 Tape recorders, photography, and drawing in court

By s. 9 of the Contempt of Court Act 1981, it is a contempt to use in court, or bring into court for use, a tape recorder or other instrument for recording sound, except with leave of the court, which may be granted subject to conditions.

This form of contempt is probably one of strict liability and not dependent on proof of *mens rea* (*Re Hooker* [1993] COD 190, DC; see further on this case, para 8.4.4 below).

It is also a contempt under s. 9 to publish a recording of legal proceedings by playing it in the hearing of the public, or any section of the public, or to dispose of the recording with a view to such publication.

For contempts under s. 9 the court may, in addition to punishing the contemnor in the normal way, order the tape recorder or any recording made with it, or both, to be forfeited. They may then be sold or otherwise disposed of in such manner as the court directs (ibid., s. 9(3)). Section 9 does not apply to the making or use of sound recordings as official transcripts of legal proceedings (ibid., s. 9(4)).

In the light of s. 9 of the Contempt of Court Act 1981, a Practice Direction on the use of tape recorders was issued jointly by the Lord Chief Justice, the Master of the Rolls, the President of the Family Division, and the Vice-Chancellor of the Chancery Division in 1981 (now incorporated in *Practice Direction (Criminal Proceedings: Consolidation), Part I* [2002] 1 WLR 2870).

Taking (or attempting to take) photographs, and making (or attempting to make) portraits or sketches, in court (which includes not only the court room, but also the building and the precincts of the building in which the court is held) are criminal offences under s. 41 of the Criminal Justice Act 1925, as amended by the Constitutional Reform Act 2005 so as to exclude the proposed Supreme Court of the United Kingdom from the prohibition. The restrictions on recording, photographic, and drawing activities are based essentially on the desire to uphold the dignity of the judicial process.

8.2.2 Prejudicing a fair trial in particular proceedings: the *sub judice* or strict liability rule

8.2.2.1 Introduction

It is of the first importance that legal proceedings should be tried fairly in our courts and that they should not be 'tried' in advance on television or in the press. There is an ever-present danger that unrestrained public discussion of the issues arising in a case may prejudice a subsequent fair trial by reason of the effect that discussion might have on the minds of prospective jurors and witnesses.

The *sub judice* rule is designed to deter prejudicial comment so as to secure a fair trial in both criminal and civil proceedings. At common law it was a contempt, *irrespective of intention*, to interfere with the outcome of particular legal proceedings by publishing prejudicial comments. The rule was made the subject of important changes by the Contempt of Court Act 1981, which refers to it as the 'strict liability rule', so-called because breach of it does not depend on proof of any intention to interfere with the trial. It is sometimes referred to as 'statutory contempt' (see, e.g., *Re Lonrho plc* [1989] 2 All ER 1100, HL, *per* Lord Bridge at p. 1113) in order to distinguish it from common law contempt, which continues to exist in this area but for which a person cannot be liable unless there was an intention to interfere with the trial. (See further on the post-1981 common law, para 8.2.2.4.8 below.)

8.2.2.2 The position at common law

The common law position is well illustrated by some of the cases involving the press, which, along with the broadcasting media, are particularly affected by this aspect of contempt of court.

In 1949, the *Daily Mirror* was fined £10,000 and its editor sent to prison for three months. Their contempt was in suggesting the names of persons who may have been murdered by John George Haigh, the so-called 'acid bath murderer', who at the time had been charged with only one murder (*R* v *Bolam (ex parte Haigh)* (1949) 93 SJ 220, DC). In 1957, W.H. Smith & Son Ltd were fined £50 for contempt in distributing the American magazine, *Newsweek*, containing highly prejudicial material which had not been given in evidence, during the trial for murder of Dr John Bodkin Adams (*R* v *Griffiths (ex parte Attorney-General)* [1957] 2 QB 192, DC). (This case was decided before the defence of innocent publication or distribution was first introduced in 1960 (para 8.2.2.4.4 below).

In 1974, the *Socialist Worker* and its editor, Paul Foot, were each fined £250 for publishing the names of two blackmail victims after the trial judge had directed that they should be referred to only as Mr Y and Mr Z (*R* v *Socialist Worker (ex parte Attorney-General)* [1975] QB 637, DC). However, in *Attorney-General* v *Leveller Magazine Ltd* [1979] AC 440, HL, the House of Lords found that the defendants were not in contempt for disclosing the name of a witness whom the magistrates' court had allowed to be referred to only as Colonel B. The magistrates had given no positive direction to the press not to publish his name, and Colonel B himself had given evidence from which his identity could be discovered.

This type of disclosure is now dealt with by s. 11 of the Contempt of Court Act 1981, which bears a close resemblance to some of the speeches in *Attorney-General* v *Leveller Magazine Ltd*, above. Section 11 appears to confirm rather than confer a power to withhold matters from publication. It provides that where a court allows a name or other matter to be withheld from the public in proceedings before the court, the court may give such directions prohibiting the publication of that name or matter in connection with the proceedings as appear to be necessary for the purpose for which it was withheld.

In the *Leveller* case [1979] AC 440, HL, the majority of the House of Lords was of the opinion that for the purposes of liability at common law there had to be a clear direction or order restricting publication since it was not contempt to ignore a mere request not to publish. In view of this it seems reasonable to suppose that the court, when exercising its discretion under s. 11, must be seen clearly to be giving directions and not merely making a request.

Section 11 does not impose strict liability and so a person cannot be guilty under it unless he knew of the court's order prohibiting publication. At the same time, as long as the court has allowed a name to be withheld and has given directions under s. 11, it would now seem to be immaterial that the defendant, as in the *Leveller* case, has discovered the name by deduction from the evidence given in the proceedings. However, if the court has not first allowed the name to be withheld from the public in the particular proceedings it has no jurisdiction to give a direction under s. 11 prohibiting publication of the name. Any such direction can be ignored by the press and others (*R* v *Arundel Justices (ex parte Westminster Press Ltd)* [1985] 2 All ER 390, DC).

The power contained in s. 11 to authorise the withholding of information from the public is an exception to the principle of open justice. Accordingly, the court must exercise the power very sparingly, and only in exceptional circumstances where it can be shown to be necessary. It is, for example, inappropriate for a magistrates' court to make an order under s. 11 prohibiting publication of the defendant's name and business addresses and the allegations made against him, in order to protect the defendant's business from damage and possible closure (*R* v *Dover Justices (ex parte Dover District Council)* (1991) 156 JP 433, DC). Similarly, in the absence of special circumstances, there is no justification for using s. 11 to give special treatment to the legal profession with regard to anonymity (*R* v *Legal Aid Board (ex parte Kaim Todner)* [1998] 3 All ER 541, CA).

Although the *sub judice* rule is capable of applying to *all* legal proceedings, whether criminal or civil, original or appellate, it was considered most unlikely that the courts would treat as contempt, either at common law or under the statutory strict liability rule, a public discussion of the merits of a pending civil action which was to be tried by a judge sitting without a jury, or a public discussion of a civil or criminal case which was pending before the Court of Appeal or House of Lords (*R* v *Duffy (ex parte Nash)* [1960] 2 QB 188, DC; *Vine Products Ltd* v *Green* [1966] Ch 484; *Attorney-General* v *British Broadcasting Corporation* [1979] 3 All ER 45, CA, reversed on other grounds at [1980] 3 All ER 161, HL).

A professional judge, it was said, can be relied on to be completely uninfluenced by anything he reads in the newspapers or sees on television (*Attorney-General* v *British Broadcasting Corporation*, CA, above, *per* Lord Denning MR at p. 52; in the House of Lords, Lord Salmon (at p. 169) agreed with this statement but Viscount Dilhorne (at p. 163) did not). As Lord Bridge pointed out in the *Lonrho* contempt case ([1989] 2 All ER 1100, at p. 1117, and see below), it is common to find in legal journals discussion and criticism of decisions which are subject to pending appeals, and, in the newspapers, criticism of criminal convictions, sentences, and awards of damages in libel cases.

The view shared by Lords Denning and Salmon was vindicated by the subsequent decision of the House of Lords in the unusual case of *Re Lonrho plc* [1989] 2 All ER 1100, HL, which arose in the aftermath of Lonrho plc's unsuccessful bid to acquire House of Fraser plc (and with it the Harrods department store), which had instead been taken over by the Al Fayed brothers. It was alleged that there had been an attempt by Lonrho and its advisers to influence the decision of the Law Lords in pending proceedings by means of material published in *The Observer* newspaper.

The House of Lords took a serious view of the matter and, of its own motion as a superior court within the meaning of s. 19 of the Contempt of Court Act 1981 (see para 8.4.2 below), instituted contempt proceedings. It was held, however, that there had been no contempt of court because the possibility that the Law Lords would be influenced by what appeared in *The Observer* was too remote. Lord Bridge said ([1989] 2 All ER 1100, HL, at pp. 1116–7):

The possibility that a professional judge will be influenced by anything he has read about the issues in a case which he has to try is very ... remote ... So far as the appellate tribunal is concerned, it is difficult to visualise circumstances in which any court in the United Kingdom exercising appellate jurisdiction would be in the least likely to be influenced by public discussion of the merits of a decision appealed against or of the parties' conduct in the proceedings.

8.2.2.3 The 'thalidomide case'

The cases on the *sub judice* rule described at the beginning of para 8.2.2.2 are eclipsed by the litigation in the 1970s in the 'thalidomide case', *Attorney-General* v *Times Newspapers Ltd* [1974] AC 273, HL.

Between 1958 and 1961, the Distillers Co. Ltd manufactured and marketed a sedative which contained the drug thalidomide. The sedative was said to be safe for pregnant women but, in fact, many of those who used it gave birth to babies with serious deformities. Some 460 such children were born. Writs were issued against Distillers claiming damages for negligence. Some 70 actions were settled out of court, but by the middle of 1972 there were 389 claims outstanding.

In September 1972, the *Sunday Times* published the first of what was intended to be a series of articles to draw attention to the plight of the children. The article urged Distillers to make a larger out-of-court settlement than they had offered hitherto. Distillers complained to the Attorney-General that the article was a contempt of court because some claims were still pending, but the Attorney decided to take no action. However, the editor of the *Sunday Times* made preparations to publish a second article intended to show that Distillers had not exercised due care to ensure that thalidomide was safe for pregnant women before they put it on the market.

The editor sent a copy of the second article to the Attorney-General, who commenced contempt proceedings for an injunction to restrain the proprietors of the newspaper from publishing the second article. The Queen's Bench Divisional Court granted the injunction but the Court of Appeal refused it. The House of Lords allowed the Attorney-General's appeal and sent the case back to the Divisional Court with a direction to grant the injunction.

The reasons given by the Law Lords for their decision were, first, that it was a contempt to publish comments on a specific issue which was before the court for determination where those comments gave rise to a real risk that the fair trial of the action would be prejudiced. Secondly, it was a contempt to use improper pressure to induce a litigant to settle a case on terms which were not agreeable to him.

The matter did not rest there, however, for Times Newspapers Ltd took the issue before the European Court of Human Rights, alleging that the House of Lords' decision violated art. 10 of the European Convention on Human Rights (freedom of expression; see para 8.2.1.4.2 above). The case was before the European Court of Human Rights for over four years pending judgment in April 1979. Meanwhile, in 1976, the Attorney-General had applied successfully to the Divisional Court to have the injunction discharged on the ground that by then most of the children's actions against Distillers had been settled. This was a tactical blunder because the Human Rights Court at Strasbourg was able to question whether the injunction had been necessary in the first place.

The European Court of Human Rights eventually decided, by 11 votes to nine, that the House of Lords' decision in the thalidomide case was a violation of the right of freedom of expression. The injunction had not been necessary in a democratic society for maintaining the authority of the courts since there was no social need which outweighed the public interest in freedom of expression.

8.2.2.4 Contempt of Court Act 1981

8.2.2.4.1 *Introduction* The Contempt of Court Act 1981 was enacted to reform the law, partly to take account of the decision of the European Court of Human Rights in the thalidomide case in 1979 and partly to implement some of the recommendations contained in the *Report of the Committee on Contempt of Court* (the *Phillimore* Committee, Cmnd 5794, 1974).

The changes effected by the Act relate particularly, although not exclusively, to the *sub judice* or strict liability rule. It should be noted that, by s. 7 of the 1981 Act, proceedings for contempt under the strict liability rule can only be commenced by, or with the consent of, the Attorney-General or on the motion of a court having jurisdiction to deal with the contempt. This includes the House of Lords, which, in *Re Lonrho plc* [1989] 2 All ER 1100, of its own motion commenced contempt proceedings against a multinational corporation (see para 8.2.2.2 above).

Although it is a matter of great importance to newspaper publishers and the like, no guidelines are laid down for the exercise of the Attorney-General's discretion to prosecute under the strict liability rule. Thus, although no proceedings were taken at common law (the 1981 Act was not then in force) in respect of almost universally irresponsible press speculation following the arrest in January 1981 of Peter Sutcliffe, the so-called 'Yorkshire Ripper', the publishers and editor of the *Daily Mail* were proceeded against (unsuccessfully) for publishing an article at the time of the trial of Dr Arthur (*Attorney-General* v *English* [1982] 2 All ER 903, HL, para 8.2.2.4.7 below), as were the publishers of five newspapers for reports concerning Michael Fagan, the so-called 'Buckingham Palace intruder' (*Attorney-General* v *Times Newspapers Ltd and others* (1983) *The Times*, 12 February, DC).

8.2.2.4.2 *Definition of 'strict liability rule' and 'publication': sections 1 and 2* Section 1 of the Contempt of Court Act 1981 defines the strict liability rule as 'the rule of law whereby conduct may be treated as a contempt of court as tending to interfere with the course of justice in particular legal proceedings regardless of intent to do so'.

It is provided that the strict liability rule applies only to 'publications' and, moreover, only to publications which create a *substantial* risk that the course of justice in the proceedings in question will be *seriously* impeded or prejudiced (ibid., s. 2(2)). Whether this is so is essentially a question of fact. This provision has, it seems, relaxed the law since, according to the House of Lords in *Attorney-General* v *Times Newspapers Ltd* [1974] AC 273, HL, the common law position was that the risk need only have been *real* (not *substantial*) that the proceedings would be impeded or prejudiced (not necessarily *seriously* so).

'Publication' includes any speech, writing, broadcast, or other communication in whatever form, which is addressed to the public at large or any section of the public (ibid., s. 2(1)).

The expression 'substantial risk' in s. 2(2) is intended to exclude a risk which is remote. A slight or trivial risk of serious prejudice is not enough, nor is a substantial risk of slight prejudice (*Attorney-General* v *English* [1982] 2 All ER 903, HL, *per* Lord Diplock at p. 919). The 'risk' must be a practical, and not a theoretical, risk (*Attorney-General* v *Guardian Newspapers Ltd* [1992] 3 All ER 38, DC).

8.2.2.4.3 *'Active' proceedings: section 2(3) and schedule 1* The strict liability rule will not apply to a publication unless the proceedings in question are 'active' at the time of the publication (Contempt of Court Act 1981, s. 2(3) and sch. 1).

Briefly, *criminal* proceedings are active, as appropriate, from the time of (a) arrest without warrant; or (b) the issue of a warrant for arrest; or (c) the issue of a summons to appear; or (d) the service of an indictment or other document specifying the charge; or (e) oral charge. They are concluded by acquittal, sentence, discontinuance, or by operation of law (ibid., sch. 1).

Civil proceedings are active from the time the case is set down for hearing or when a date for the trial or hearing is fixed, and they cease to be active when the proceedings are disposed of, discontinued, or withdrawn (ibid.).

Appellate proceedings are active only from the time when they are commenced. There is thus freedom to comment on a case between the conclusion of the proceedings at first instance and the initiation of an appeal. Depending on their type, appellate proceedings are usually commenced (a) by application for leave to appeal, or to apply for judicial review, or by notice of such an application, or (b) by notice of appeal. Appellate proceedings cease to be active when they are disposed of, abandoned, discontinued, or withdrawn (ibid.).

However, where the court in appellate *criminal* proceedings remits the case to the court below, or orders a new trial or a *venire de novo*, any further or new proceedings which result are treated as active from the conclusion of the appellate proceedings (ibid., para 16). Since the strict liability rule applies to such a case in the interval between the conclusion of the appellate proceedings and the start of the fresh proceedings, there is no unfettered freedom of comment.

8.2.2.4.4 *Innocent publication or distribution: section 3* By s. 3 of the Contempt of Court Act 1981, innocent publication or distribution is a defence to a charge of criminal contempt under the strict liability rule. This statutory defence was first introduced by s. 11 of the Administration of Justice Act 1960, which is now repealed and replaced by s. 3 of the 1981 Act.

A publisher is not in contempt if at the time of publication, and having taken all reasonable care, he does not know and has no reason to suspect that relevant proceedings are active (Contempt of Court Act 1981, s. 3(1)). A distributor is not in contempt if at the time of distribution, and having taken all reasonable care, he does not know that the publication contains prejudicial material and has no reason to suspect that it is likely to do so (ibid., s. 3(2)). The burden of proving this defence is on the publisher or distributor (ibid., s. 3(3)).

8.2.2.4.5 *Postponement of publication of reports: section 4* By s. 4(1) of the Contempt of Court Act 1981, it is not a contempt under the strict liability rule to publish contemporaneously and in good faith a fair and accurate report of legal proceedings held in public. However, under s. 4(2), a court has a discretion to order that publication of reports of a case before it be postponed.

Some judges had claimed the power to order postponement at *common law*. Thus, at the time of the Cato Street conspiracy trials, publication of reports was postponed in *R v Clement* (1821) 4 B & Ald 218. Postponement of publication was also ordered at the time of the Poulson trials in the 1970s. *R v Clement*, above, was overruled by the Judicial Committee of the Privy Council in *Independent Publishing Co. Ltd v Attorney-General of Trinidad and Tobago* [2004] 3 WLR 611, PC. This case was concerned with the laws of a country which has no equivalent of our Contempt of Court Act 1981, s. 4(2), and it was held that there is no *common law* power to make an order postponing the publication of reports of legal proceedings held in public. Such a power would need to be conferred by *statute*. At the same time, it was emphasised that, even without legislation, it is open to the judge to warn the media in an appropriate case that they run the risk of contempt proceedings if they publish matter which is likely to prejudice the fair administration of justice.

The statutory judicial discretion of postponement contained in s. 4(2) of the Contempt of Court Act 1981 is not unfettered. A court can only order postponement of publication 'where it appears to be necessary for avoiding a substantial risk of prejudice' in the proceedings or in any other proceedings which are pending or imminent. Moreover, the postponement order can only last for 'such period as the court thinks necessary' for the purpose of avoiding the risk of prejudice.

Section 4(2) thus lays down two requirements for the making of a postponement order. The court must consider, first, whether publication would create 'a substantial risk of prejudice', and, secondly, whether the order 'appears to be necessary' for the avoidance of that risk. If there is no substantial risk of prejudice, then no order should be made. If there *is* a substantial risk, an order should still not be made unless it is necessary to do so. If it is necessary to make an order, its terms should be no wider than is consistent with avoiding the risk of prejudice (*Ex parte Telegraph plc* [1993] 2 All ER 971, CA).

Section 4(2) applies only to a 'report of legal proceedings held in public'. A film of a person being arrested for drug offences, taken by a television camera crew for a programme about drug trafficking, is not a report of 'legal proceedings held in public'. Accordingly, magistrates have no jurisdiction to order that publication of the film be postponed. If in such a case a breach of the strict liability rule is feared, the appropriate remedy is an application to the High Court for an injunction to prevent publication (*R v Rhuddlan Justices (ex parte HTV Ltd)* [1986] Crim LR 329, DC).

It is important that postponement orders made under s. 4(2) are formulated in precise terms. To that end, a Practice Note issued in 1983 provides that such an order must be in writing and must state (a) its precise scope; (b) the time at which it shall cease to have effect, if appropriate; and (c) the specific purpose of making the order. (Now incorporated in *Practice Direction (Criminal Proceedings: Consolidation), Part I* [2002] 1 WLR 2870. For a case where the Practice Note was not properly observed, see *Re Central Independent Television plc and others* [1991] 1 All ER 347, CA, para 8.2.2.4.6 below.)

The same requirements were extended by the Practice Note (see now the *Practice Direction* of 2002, above) to orders made under s. 11 of the 1981 Act prohibiting the publication of any name or other matter in connection with proceedings before the court (para 8.2.2.2 above). It is envisaged that a court will give notice to the press that an order has been made under s. 4(2), or under s. 11, and that the court staff will be prepared to answer enquiries about specific cases. But it is emphasised that reporters of cases and their editors remain responsible for ensuring that there is no breach of any order, and that *they* should take the initiative in making enquiries if there is doubt in any particular case.

The court has an inherent discretion to hear representations from the news media both before and after making a postponement order under s. 4(2) (*R v Beck (ex parte Daily Telegraph plc)* (1991) [1993] 2 All ER 177, CA; *Attorney-General v Guardian*

Newspapers Ltd [1992] 3 All ER 38, DC; *R v Clerkenwell Magistrates' Court (ex parte Telegraph plc)* [1993] 2 All ER 183, DC).

8.2.2.4.6 *Appeals against postponement orders: section 159 of the Criminal Justice Act 1988* The principle of open justice requires that legal proceedings are held in open court to which the press and public are admitted, that evidence given in the proceedings is communicated publicly, and that dissemination of fair and accurate reports of the proceedings to the public at large is not discouraged (*Attorney-General v Leveller Magazine Ltd* [1979] AC 440, HL, *per* Lord Diplock at p. 450). While recognising that there are exceptions (both by statute and at common law) to the basic principle, the courts have stressed that any departure from it in an individual case rests on necessity and not on what the judge in his discretion regards as expedient or convenient (*Scott v Scott* [1913] AC 417, HL, *per* Viscount Haldane LC at p. 437).

Maintenance of the principle of open justice makes it essential to take seriously the disquiet expressed in the media over recent years about, first, the number of postponement orders being made under ss. 4 and 11 of the Contempt of Court Act 1981, and, secondly, the increasing number of occasions on which members of the public are excluded from criminal trials.

Postponement and exclusion orders made by *magistrates' courts* can be challenged in an application to the High Court for judicial review. However, this procedure is not available where the order is made by the *Crown Court* in relation to a trial on indictment (Supreme Court Act 1981, s. 29(3), as amended; see para 10.5.2.3). Instead, a right of appeal has been provided which enables 'a person aggrieved' to appeal against orders made by the Crown Court postponing reports of a trial on indictment, or postponing publication of details emerging in it, or excluding the public from the trial (Criminal Justice Act 1988, s. 159).

The appeal lies to, and only with the leave of, the Court of Appeal, criminal division (ibid., s. 159(1)), which may confirm, reverse, or vary the order appealed against (ibid., s. 159(5)). The decision of the Court of Appeal is final, there being no further appeal to the House of Lords (ibid., s. 159(1)).

When hearing an appeal under s. 159, the function of the Court of Appeal is to form its own view of what is the best order to make based on the material placed before it. Its task is not limited simply to reviewing the judge's order made at first instance (*R v Beck (ex parte Daily Telegraph plc)* (1991) [1993] 2 All ER 177, CA, *per* Farquharson LJ at p. 180; *Ex Parte Telegraph plc* [1993] 2 All ER 971, CA, *per* Lord Taylor CJ at p. 977; *Ex Parte The Telegraph Group plc* [2001] 1 WLR 1983, CA, *per* Longmore LJ at [3]).

The first appeals under the procedure provided by s. 159 of the Criminal Justice Act 1988 were brought by Mr Tim Crook, a freelance journalist whose job involved supplying national news organisations with reports of legal proceedings (*Re Crook* [1992] 2 All ER 687, CA). One of the arguments of the appellant in *Re Crook*, above, was that, even if it is right to exclude the *public*, the *Press* should not be excluded. The argument was rejected. The Court of Appeal was of the view that if exclusion of the public is necessary, it would not usually be right to make an exception in favour of the Press because to do so would genuinely aggrieve other members of the public with as much interest in the proceedings, such as the victim's family or the defendant's family.

In *Re Crook,* above, the Court of Appeal took the opportunity to offer guidance to judges on the exclusion of the public from proceedings. Each application for exclusion must be considered on its individual merits and the public must only be excluded when it is strictly necessary. A judge should not adjourn into chambers as a matter of course, but only if he believes that something might be said which would be more appropriately heard in private than in public. If he does decide to sit in chambers, he should return to open court if, and as soon as, it becomes apparent that it was not necessary, after all, to exclude the public.

In *Re Central Independent Television plc and others* [1991] 1 All ER 347, CA, the Court of Appeal held that s. 159 of the Criminal Justice Act 1988 can be used to appeal against an order which is spent and where, therefore, the challenge to it is largely academic. To hold that an order wrongly made cannot be reversed on appeal simply because it has ceased to operate would undesirably lessen the effectiveness of s. 159 (ibid., *per* Lord Lane CJ at p. 351).

There is no right of appeal under s. 159 against a judge's decision *not* to exclude the public from the proceedings. This is because s. 159(1)(b) refers only to 'any order restricting the access of the public to the whole or any part of a trial on indictment or to any proceedings ancillary to such a trial', and not to an order *refusing* to restrict such access (*R* v *Salih* (1994) *The Times,* 31 December, CA).

8.2.2.4.7 *Bona fide discussion: section 5* Section 5 of the Contempt of Court Act 1981 provides that a publication made as part of 'a discussion in good faith of public affairs or other matters of general public interest is not to be treated as a contempt of court under the strict liability rule if the risk of impediment or prejudice to particular legal proceedings is merely incidental to the discussion'.

This section is a particularly liberalising measure since it recognises that it is in the public interest that there should be no suspension of public debate of public issues merely because there is an incidental risk of prejudice to particular legal proceedings. If the terms of s. 5 are satisfied, the strict liability rule has no application so that the publication in question will be a contempt only if an *intention* to impede or prejudice particular legal proceedings can be proved.

Section 5 of the 1981 Act was considered by the House of Lords in *Attorney-General* v *English* [1982] 2 All ER 903, HL. The *Daily Mail* published an article written in support of someone who was seeking election to Parliament as a pro-life candidate. The article was concerned with the preservation of the sanctity of human life, and alleged that handicapped babies had been, and were likely to be, allowed to die of starvation and by other means. The article was published in the same week that Dr Leonard Arthur, a well-known consultant paediatrician, went on trial for the murder of a Down's Syndrome baby by starvation, although the article made no mention of Dr Arthur's trial.

The Attorney-General commenced contempt proceedings under the 1981 Act against the proprietors of the *Daily Mail* and its editor, Mr David English. It was not suggested that the article was intended to influence Dr Arthur's trial. Instead, the Attorney-General relied on the strict liability rule contained in s. 1 of the Act. The editor and publishers relied on s. 5 of the Act.

The House of Lords held that neither the editor nor the publishers were in contempt. Although publication of the article created the risk of serious prejudice to Dr Arthur's trial within the meaning of s. 2(2) of the Act, the trial had not been mentioned in the article, and the risk of the jury reading the article and allowing it to prejudice their minds against Dr Arthur on evidence which did not justify a finding of guilty was 'merely incidental' to the discussion contained in the article.

In a judgment with which the other Law Lords expressly agreed, Lord Diplock said (at p. 919) that the word 'discussion' in s. 5 is not limited to the airing of views and the debating of principles and arguments. It also includes the making of accusations.

There is a suggestion in Lord Diplock's judgment that s. 5 may not provide protection in circumstances like those in the thalidomide case, *Attorney-General v Times Newspapers Ltd* [1974] AC 273, HL, para 8.2.2.3 above. In *English*, above, he said (at p. 920) that the offending article was in nearly all respects the antithesis of the offending article in the thalidomide case where the whole subject of the article was the pending civil actions against Distillers, and the whole purpose of it was to put pressure on Distillers in the lawful conduct of their defence in those actions. He concluded that it would, therefore, be difficult to argue that the risk of prejudice to the proceedings was 'merely incidental' to the discussion in the article. In *English*, on the other hand, the article did not mention Dr Arthur's trial and the risk of prejudice to it could properly be described as merely incidental to the discussion of the wider issues contained in the article.

If Lord Diplock's interpretation of s. 5 is correct, it would seem that the 1981 Act has failed to take account of the majority judgment of the European Court of Human Rights in the thalidomide case.

8.2.2.4.8 *Preservation of the common law: section 6* Section 6 of the Contempt of Court Act 1981 provides that the Act does not:

(a) prejudice any defence available at common law to a charge of contempt under the strict liability rule;

(b) extend the strict liability rule beyond the limits previously set by the common law;

(c) restrict liability for contempt in respect of conduct which is *intended* to impede or prejudice the administration of justice.

Section 6(c) is particularly important because it preserves the common law of contempt in those cases where the strict liability rule cannot apply; as, for example, where the proceedings alleged to have been prejudiced were not active at the time of publication (for 'active' proceedings see para 8.2.2.4.3 above).

Common law contempt is of potentially wider application than the strict liability rule. For example, while under s. 1 of the Contempt of Court Act 1981 the strict liability rule is limited to interference with 'particular legal proceedings', it has been held that common law contempt may be committed where there are no proceedings at all pending or imminent at the time of publication. It was so decided by a two-man Divisional Court in *Attorney-General v News Group Newspapers Ltd* [1988] 2 All ER 906, DC. This was a case in which the proprietors of the *Sun* newspaper were fined £75,000 for common law contempt in publishing allegations that a named doctor had raped an eight-year old girl, even though at the time of publication no proceedings against the doctor were pending or imminent (and, therefore, the strict liability rule did not apply). It was said that, although the need for a free press is axiomatic, 'the press cannot be allowed to charge about like a wild unbridled horse' (ibid., per Watkins LJ at p. 921).

What was thought to have been decided in *Attorney-General v News Group Newspapers Ltd*, above, was thrown into doubt when in the later case of *Attorney-General v Sport Newspapers Ltd* [1992] 1 All ER 503, DC, a differently constituted two-man Divisional Court was unable to agree on the point. Bingham LJ (at pp. 515–6) thought that the *News Group* decision, though admittedly extending the boundaries of common law contempt as previously understood, should be followed. Hodgson J (at p. 536) was of the opinion that the common law of contempt began to apply only from the time when relevant proceedings were pending and that, accordingly, the *News Group* case was wrongly decided. In his view, conduct committed before proceedings came into existence which was deliberately prejudicial to the administration of justice was more appropriately punished by the ordinary criminal law relating to perversion of the course of justice than by extending the scope of common law contempt.

Liability for common law contempt remains dependent on proof of *intention* to interfere with the administration of justice. Without such an intention there will be no liability. In *Attorney-General v Sport Newspapers Ltd* [1992] 1 All ER 503, DC, proceedings were brought against the publishers and editor of *The Sport* newspaper following the appearance of a front-page story in which a man wanted by the police for questioning about the disappearance of a 15-year old girl was described as a 'vicious rapist' and 'sex monster'. Bingham LJ and Hodgson J were unanimous in holding that, on the facts, the defendants were not guilty since they lacked the necessary *mens rea*. Proceedings for contempt under the strict liability rule were not possible in this case because criminal proceedings against the wanted man were not 'active' at the time of publication.

A case which illustrates clearly the relationship between common law contempt and the strict liability rule is *Attorney-General v Hislop* [1991] 1 All ER 911, CA. Here, contempt proceedings were brought against the publishers and editor of the satirical magazine, *Private Eye*. Shortly before the trial of Sonia Sutcliffe's libel action against *Private Eye* (see *Sutcliffe v Pressdram Ltd* [1990] 1 All ER 269, CA, para 6.8.2.2), two articles appeared in the magazine suggesting that Mrs Sutcliffe, wife of the 'Yorkshire Ripper', had known what her husband was doing at the time of the murders and either did nothing about it or gave him an alibi by lying to the police, and that she was defrauding the Department of Social Security. The Attorney-General alleged that publication of the articles had created a real and substantial risk of serious prejudice to the fair trial of the libel action in that the articles had been published with the intention of persuading Mrs Sutcliffe to drop her libel action or of prejudicing potential jurors in the action.

The Court of Appeal held that, in relation to the improper pressure put on Mrs Sutcliffe, the defendants were in contempt both at common law (since *intention* to deter Mrs Sutcliffe had been established), and under the strict liability rule. With regard to the prejudicing of potential jurors, however, there was contempt only under the strict liability rule since there had been no intention to influence prospective jurors. Each defendant was fined £10,000.

It is quite common for questions of *criminal* contempt to arise out of allegations of interference with the administration of *civil* justice (see, e.g., *Attorney-General* v *Hislop*, above, and *Attorney-General* v *Times Newspapers Ltd* [1991] 2 All ER 398, HL, where the issue was whether a person who knowingly interferes with the administration of justice in *civil* proceedings between two other persons is guilty of *criminal* contempt even though he is not named in any court order and has not assisted in the breach of any court order by a party against whom it was made).

It has been held to be a criminal contempt for a person who is not a party to legal proceedings, first, to inspect without leave of the court the documents on the court file if it is known that leave is required, or, secondly, to gain access to the court file by deceiving court officials or by subterfuge (*Dobson* v *Hastings* [1992] 2 All ER 94, where, on the facts, a *Daily Telegraph* journalist, and her editor, were found by Sir Donald Nicholls V-C not to be in contempt since the extracts from the court file had not been obtained by trickery or dishonesty on the part of the journalist, and their publication by the newspaper was not done with an intention to interfere with the administration of justice).

8.3 Civil contempt of court

It is a civil contempt, sometimes called a contempt in procedure, to disobey a judgment or order of the court, or to act in breach of an undertaking given, either expressly or by implication, to the court. Thus, it is a civil contempt to disobey a decree of specific performance, to act contrary to the terms of an injunction, to defy a mandatory, prohibiting, or quashing order, or an order of habeas corpus, or to fail to honour an implied undertaking given to the court not to make improper use of documents disclosed on discovery (as in *Home Office* v *Harman* [1982] 1 All ER 532, HL).

In the important case of *M* v *Home Office* [1993] 3 All ER 537, HL, the House of Lords held that government departments and ministers acting in their official capacity are subject to the law of civil contempt, and that, although the sanctions of imprisonment, fines, and sequestration of assets are not appropriate, an order for the payment of costs can be made against them to emphasise the seriousness of the contempt.

In *M* v *Home Office*, above, the Home Secretary (Mr Kenneth Baker) had disobeyed an order of a High Court judge requiring him to procure the return to this country of a citizen of Zaire, whose claim for political asylum he had rejected and who had been removed from the United Kingdom despite having made an application for leave to apply for judicial review of the Home Secretary's decision. At first instance, the judge, relying on *Factortame Ltd* v *Secretary of State for Transport (No. 1)* [1989] 2 All ER 692, HL (see para 10.8.6.4), had taken the view that neither the Crown nor its officers were subject to the coercive jurisdiction of the court.

A majority of the Court of Appeal, in reversing this decision and finding Mr Baker personally to be in contempt, pointed out that *Factortame (No. 1)* was only concerned with the granting of injunctions and did not lay down that ministers were immune from contempt proceedings. No penalty was imposed for the contempt. However, as a mark of the court's displeasure Mr Baker was ordered to pay the costs of the proceedings personally, although he was indemnified in this respect out of public funds. This was the first time that a minister of the Crown had been found to be in contempt of court.

The order made by the Court of Appeal was affirmed by a unanimous House of Lords, subject to a variation whereby 'Secretary of State for Home Affairs' was substituted for 'Kenneth William Baker' as the person against whom the finding of contempt had been made.

The earlier finding against Mr Baker personally had been made because the Court of Appeal was of the opinion that, while individual ministers of the Crown and civil servants, being natural persons with legal personality, could be liable for civil contempt of court in respect of their personal acts or omissions, the Crown and government departments could not be so liable since they lacked sufficient legal personality.

The House of Lords disagreed with this opinion, holding that the Crown and government departments *are* invested with adequate legal personality for the purposes of the law of contempt (see [1993] 3 All ER 537 *per* Lord Woolf at pp. 566–7). Their Lordships further held that ministers acting in their official capacity, and government departments, are subject to other coercive aspects of the court's jurisdiction and may, therefore, have interim injunctions granted against them. (This aspect of *M* v *Home Office* is dealt with more fully in para 10.8.6.4.)

The danger inherent in exempting ministers of the Crown from the coercive jurisdiction of the court was emphasised in the speech of Lord Templeman in *M* v *Home Office* when he said (at p. 541) that

. . . the argument that there is no power to enforce the law by injunction or contempt proceedings against a minister in his official capacity would, if upheld, establish the proposition that the executive obey the law as a matter of grace and not as a matter of necessity, a proposition which would reverse the result of the Civil War.

Although a civil contempt is punishable by committal to prison, it has been held that imprisonment should not be ordered if it is possible to ensure obedience to the order of the court in some other way (see, e.g., *Danchevsky* v *Danchevsky* [1975] Fam 17, CA).

8.4 Punishment for contempt of court

8.4.1 Introduction

At common law, superior courts had the power to commit a contemnor to prison for an unlimited time or to impose a fine unlimited in amount. It was not uncommon, for example, for a person who had disobeyed a court order, or committed a contempt in the face of the court, to be kept in prison until he had 'purged his contempt' by apology and an undertaking to obey the order of the court.

Section 14 of the Contempt of Court Act 1981, which applies to both civil and criminal contempts (*Linnett* v *Coles* [1986] 3 All ER 652, CA), abolished the unlimited prison sentence as a penalty for contempt. Moreover, a person who has already been committed to prison once for failing to comply with a court order (civil contempt) cannot be sent to prison again for a continuing breach of the same order (*Kumari* v *Jalal* [1996] 4 All ER 65, CA).

Sentencing guidelines for some criminal contempts were laid down by the Court of Appeal in *R* v *Montgomery* [1995] 2 All ER 28, CA, a case in which a sentence of 12 months' imprisonment for refusing, through fear, to give evidence against ten persons on charges of conspiracy to damage property was reduced on appeal to one of three months' imprisonment.

It has been held that, on an application by a contemnor to purge his contempt, the court has no jurisdiction to release him and at the same time suspend the unserved balance of his prison sentence. The court's powers are limited to either ordering immediate unconditional release, or ordering deferred release at a stated future date, or refusing the application (*Harris* v *Harris* [2002] 2 WLR 747, CA).

8.4.2 Superior courts

A superior court can now only commit a person to prison for contempt for a fixed term of up to two years on any one occasion (Contempt of Court Act 1981, s. 14(1)). However, it retains its power to impose an unlimited fine.

For the purpose of the Act, 'superior court' means the Supreme Court of the United Kingdom, the Court of Appeal, the High Court, the Crown Court, the Courts-Martial Appeal Court, the Employment Appeal Tribunal, and any other court exercising in relation to its proceedings powers equivalent to those of the High Court. It also includes the House of Lords when sitting as a court (ibid., s. 19, as amended by the Constitutional Reform Act 2005, and see *Re Lonrho plc* [1989] 2 All ER 1100, HL). In addition, for the purposes of s. 14 of the Act, a county court is to be treated as a superior court and not as an inferior court (ibid., s. 14(4A), added by the County Courts (Penalties for Contempt) Act 1983 and reversing the decision of the House of Lords in *Peart* v *Stewart* [1983] AC 109).

8.4.3 Inferior courts

Where an inferior court has power to commit for contempt, the committal must be for a fixed term not exceeding one month on any one occasion. Where it has power to impose a fine for contempt, the fine must not exceed £2,500 on any one occasion (Contempt of Court Act 1981, s. 14(1)–(2), as amended by the Criminal Justice Act 1991). This sum may be altered by the Home Secretary by statutory instrument (Magistrates' Courts Act 1980, s. 143, as amended, and sch. 6A, as substituted by the Criminal Justice Act 1991).

The powers of a county court to deal with *criminal* contempts, such as contempts committed in the face of the court or in the immediate vicinity, are less severe than those for civil contempts.

By s. 118(1) of the County Courts Act 1984, a circuit judge, district judge, assistant district judge, or deputy district judge has power to deal with any person who:

(a) wilfully insults the judge of a county court, or any juror or witness, or any officer of the court during his sitting or attendance in court, or in going to or returning from the court; or

(b) wilfully interrupts the proceedings of a county court or otherwise misbehaves in court.

The judge can order any officer of the court, with or without the assistance of any other person, to take the offender into custody and detain him until the rising of the court. In addition, the judge may, if he thinks fit, send the contemnor to prison for a specified period not exceeding one month, or impose a fine not exceeding £2,500, or he may do both (County Courts Act 1984, s. 118(1), as amended by the Criminal Justice Act 1991). This sum may be altered by the Home Secretary by statutory instrument (Magistrates' Courts Act 1980, s. 143, as amended by the County Courts Act 1984).

A *threat* has been held to be an 'insult' within the meaning of s. 118(1)(a) (*Manchester City Council* v *McCann* [1999] 2 WLR 590, CA, in which the Court of Appeal adopted a purposive approach to the statutory provision—so that insulting a witness includes making threats—in view of the need to deal summarily and immediately with threats to the proper conduct of county court proceedings).

8.4.4 Magistrates' courts

The Contempt of Court Act 1981 gave the magistrates' courts power to punish a contempt committed in the face of the court or in the immediate vicinity. By s. 12(1):

A magistrates' court has jurisdiction . . . to deal with any person who:

(a) wilfully insults the justice or justices, any witness before or officer of the court or any solicitor or counsel having business in the court, during his or their sitting or attendance in court or in going to or returning from the court; or

(b) wilfully interrupts the proceedings of the court or otherwise misbehaves in court.

In the event of such a contempt, the court can order any officer of the court, or any constable, to take the offender into custody and detain him until the court rises. In addition, the court can, if it thinks fit, commit the offender to custody for a specified period not exceeding one month, or impose a fine not exceeding £2,500, or both (Contempt of Court Act 1981, s. 12(2), as amended by the Criminal Justice Act 1991). This sum may be altered by the Home Secretary by statutory instrument (Magistrates' Courts Act 1980, s. 143, as amended, and sch. 6A, as substituted by the Criminal Justice Act 1991).

The justices' clerk is an 'officer of the court' for the purposes of s. 12(1)(a), and describing the clerk's case-listing system as 'ridiculous' may possibly amount to 'wilfully insulting' him (*R* v *Tamworth Magistrates' Court (ex parte Walsh)* [1994] COD 277, DC; the court, however, did not have to decide the point since the magistrates' finding of contempt was quashed on other grounds—see para 8.4.7 below).

It has been held that the word 'wilfully' in s. 12(1)(b) qualifies the phrase 'otherwise misbehaves in court' as well as the phrase 'interrupts the proceedings of the court'. It follows that where it is alleged that the defendant's use of a tape recorder amounts to 'otherwise misbehaving in court', there is no power to punish for contempt under s. 12(2) unless it can be established that the 'misbehaviour' is 'wilful' in the sense that it involves an element of defiance or is such that a court should not reasonably be expected to tolerate it (*Re Hooker* [1993] COD 190, DC).

The use of the word 'wilfully' makes it clear that the defendant cannot be punished without proof of *mens rea*. 'Wilfully' means that the defendant must either *intend* his act to be an interruption or *recklessly risk* that result (*Bodden* v *Commissioner of Police of the Metropolis* [1989] 3 All ER 833, CA, *per* Beldam LJ at p. 837).

In *R* v *Havant Magistrates' Court and Portsmouth Crown Court (ex parte Palmer)* [1985] Crim LR 658, DC, it was decided that a *threat* is not an 'insult' within the meaning of s. 12(1)(a). Accordingly, it was held that a witness in a magistrates' court case who, after giving evidence, threatened to 'get' the accused and his solicitor while they were awaiting the verdict could not be dealt with by the magistrates for contempt.

This decision was disapproved by the Court of Appeal in the later case of *Manchester City Council* v *McCann* [1999] 2 WLR 590, CA (para 8.4.3 above), and may be regarded as no longer good law. Section 12(1)(a) of the Contempt of Court Act 1981 is in similar terms to s. 118(1)(a) of the County Courts Act 1984, except that the category of protected persons also covers solicitors and counsel, and in the *McCann* case, above, it was held that a threat *is* an 'insult' within the meaning of s. 118(1)(a) of the 1984 Act.

The wording of s. 12(1)(b) seems to suggest that 'wilfully interrupting the proceedings of' a magistrates' court is only a contempt if the act causing the interruption takes place *inside* the court. It has been held, however, that this is not the case and that it may be contempt whether the act causing the interruption takes place *inside* or *outside* the court. Accordingly, the use of a loudhailer in the street outside the court building may constitute a contempt under s. 12(1)(b) if it prevents witnesses in court from being heard (*Bodden* v *Commissioner of Police of the Metropolis* [1989] 3 All ER 833, CA).

The wisdom of conferring on magistrates' courts the power to punish summarily contempt committed in the face of the court is open to question in view of some of the trivial incidents reported in the newspapers. In 1983, a man was sent to gaol for a week (but was released after four days) by a stipendiary magistrate for whispering in the public gallery. In the same year, a man was committed for seven days by lay justices for chuckling at the back of the court. In 1984, a teenager was fined £50 for contempt because he giggled in the public gallery when a prison officer fell off his chair.

In order to ensure that a contempt offender is treated fairly by magistrates' courts within art. 6 of the European Convention on Human Rights (as scheduled to the Human Rights Act 1998), a *Practice Direction (Magistrates' Courts: Contempt)* was issued by the Lord Chief Justice in 2001 (now incorporated in *Practice Direction (Criminal Proceedings: Consolidation), Part V* [2002] 1 WLR 2870). This covers such matters as the giving of warnings, willingness to accept apologies, the granting of legal representation, avoiding a situation where the justices are judges in their own cause, and, in cases where the contempt is not admitted, ensuring that the offender is able to call and examine witnesses where evidence is relevant to the trial and that any period of committal is for the shortest time consistent with the preservation of good order in the administration of justice.

8.4.5 Contempts punishable by the Divisional Court of the Queen's Bench Division

The magistrates' courts have no power to punish a contempt not falling within s. 12(1) of the Contempt of Court Act 1981 (para 8.4.4 above). However, other contempts of magistrates' courts, such as a breach of the strict liability rule, can be punished by the Queen's Bench Divisional Court.

The coroners' courts, as inferior courts of record, have power to punish contempt committed in the face of the court, but other contempts of coroners' courts can only be punished by the Queen's Bench Divisional Court.

By virtue of RSC, Ord. 52, r. 1(2), the Divisional Court can also punish contempts of other inferior courts which have no inherent power to do so themselves, such as consistory courts, courts martial, and contempts of tribunals of inquiry set up on an *ad hoc* basis under the Inquiries Act 2005 (repealing and replacing the Tribunals of Enquiry (Evidence) Act 1921).

If the body in question is not a court at all, then RSC, Ord. 52, r. 1(2), does not apply and the Divisional Court has no powers of punishment. The Professional Conduct Committee of the General Medical Council has been held not to be a court since, although its function is a judicial one, it does not exercise the judicial power of the state (*General Medical Council* v *British Broadcasting Corporation* [1998] 3 All ER 426, CA). In contrast, a mental health review tribunal has been held to be a court for the purposes of the law of contempt. Its functions are essentially judicial and, in discharging them, it is exercising the judicial power of the state within the meaning of s. 19 of the Contempt of Court Act 1981 (*Pickering* v *Liverpool Daily Post and Echo Newspapers plc* [1991] 1 All ER 622, HL, overruling on this point *Attorney-General* v *Associated Newspapers Group plc* [1989] 1 All ER 604, DC).

Furthermore, it has been held that an employment tribunal is an 'inferior court' for the purposes of the law of contempt since it has many of the characteristics of a court of law and discharges judicial, as opposed to administrative, functions. It follows that the Divisional Court has jurisdiction to punish contempt of an employment tribunal (*Peach Grey & Co.* v *Sommers* [1995] 2 All ER 513, DC).

The county courts have jurisdiction to punish contempts committed in the face of the court (County Courts Act 1984, s. 118, para 8.4.3 above), and civil contempts consisting of a breach of an order of, or of an undertaking given to, the court. However, by virtue of RSC, Ord. 52, r. 1(2), a contempt which falls outside these categories can only be punished by the Queen's Bench Divisional Court. In *In Re G (a child) (contempt: committal)* [2003] 1 WLR 2051, CA, it was held that a circuit judge sitting in a county court to deal with a father's application for a contact order in relation to his child had no power to commit the father to prison. The alleged contempt consisted of publishing, on a website, material which identified volunteers at a contact centre and the child who was the subject of the contact proceedings. Since the father's action was not committed in the face of the court, and it was not a breach of an order or undertaking, it was only punishable in the Queen's Bench Division of the High Court.

8.4.6 Representation for contempts committed in the face of the court

Public funding for representation in proceedings for contempt committed, or alleged to have been committed, in the face of the court, is included within the scope of the Criminal Defence Service (Access to Justice Act 1999, s. 12(2)(f)). This is a re-enactment of provisions contained in earlier statutes pre-dating the establishment of the Criminal Defence Service. However, as before, there is no *right* to public funding; the court has a *discretion* to grant it. The court's decision is based on the 'interests of justice' (ibid., sch. 3, para 5).

The availability of public funding for this sort of contempt goes some way towards meeting the criticisms of 'instant justice' expressed by the Court of Appeal in *Balogh* v *Crown Court at St Alban's* [1974] 3 All ER 283, CA (see para 8.4.7 below).

8.4.7 Summary punishment of contempts committed in the face of the court

A major criticism of the power to deal summarily with contempts committed in the face of the court has not been met by the Contempt of Court Act 1981. Where the same person is both victim and judge in his own cause the court is not impartial, and it is virtually impossible to be certain that justice is being done. There is no jury to determine independently the question of guilt or innocence.

In *Balogh* v *Crown Court at St Alban's* [1974] 3 All ER 283, CA, the Court of Appeal was of the opinion that the summary jurisdiction should be exercised sparingly and only in the clearest and gravest cases. Other cases should, in the main, be referred to the Attorney-General for him to decide whether to apply, under RSC, Ord. 52, to the Divisional Court of the Queen's Bench Division for an order committing the defendant to prison. (See also *R* v *Schot* [1997] 2 Cr App R 383, CA, para 8.2.1.5.1 above.)

It is of the utmost importance that wherever possible a defendant should be allowed legal representation before a finding of contempt is made against him (*Balogh* v *Crown Court at St Alban's*, CA, above; *R* v *Powell* (1993) *The Times*, 3 June, CA, the 'wolf-whistle case', para 8.2.1.1 above; *R* v *Bromell* (1995) *The Times*, 9 February, CA, in which a finding of contempt was quashed as being unsafe when the judge had refused to hear the defendant's counsel except on the question of sentence; *R* v *Tamworth Magistrates' Court (ex parte Walsh)* [1994] COD 277, DC, where a finding was quashed because magistrates had acted unreasonably in summarily committing a solicitor for criticising the court listing system without giving him the opportunity to seek legal representation).

In the case of magistrates' courts, some of the difficulties associated with summary punishment of contempts committed in the face of the court can be avoided by a strict adherence to *Practice Direction (Criminal Proceedings: Consolidation), Part V* [2002] 1 WLR 2870 (para 8.4.4 above).

8.5 Appeals in contempt cases

8.5.1 Introduction

Before 1960, there was no right of appeal to a higher court in a case of *criminal* contempt, although there was such a right in a case of *civil* contempt. Section 13 of the Administration of Justice Act 1960 introduced a right of appeal in cases of criminal contempt, and provided a uniform procedure for appeal in cases of both criminal and civil contempt.

The defendant has the right of appeal in all cases of contempt. In the case of criminal contempt, the appeal lies against both conviction and sentence. The person who originally applied to have the defendant punished can appeal if the application was to have the defendant committed to prison (Administration of Justice Act 1960, s. 13(2)).

The court hearing the appeal has power to reverse or vary the decision of the court below and to make such order as may be just (ibid., s. 13(3)). This includes the power to order a rehearing of the case before a different judge (*Duo* v *Duo* [1992] 3 All ER 121, CA, where a rehearing was ordered because justice had not been seen to be done at the first hearing in that the defendant had been denied the opportunity of properly presenting his case against an application to commit him to prison for contempt), and the power to substitute such other fine or sentence of imprisonment as the court considers just (*Linnett* v *Coles* [1986] 3 All ER 652, CA; *Mason* v *Lawton* [1991] 2 All ER 784, CA; *Delaney* v *Delaney* [1996] 1 All ER 367, CA). In exceptional circumstances a 'just' prison sentence could be a longer one than that imposed at first instance (*Linnett* v *Coles*, above; *Wilson* v *Webster* [1998] 1 FLR 1097, CA, in which the defendant's prison term was *increased* from 14 days to three months on an appeal by the *applicant*; these two cases were referred to in *Lomas* v *Parle* (*Practice Note*) [2004] 1 WLR 1642, CA, in which the Court of Appeal doubled the sentences imposed at first instance from two concurrent terms of four months' imprisonment to two concurrent terms of eight months' imprisonment).

The appellate court has power to release the defendant on bail pending the outcome of his appeal (ibid., s. 13(3)). This is an important power since there are no special provisions for expediting the hearing of appeals in contempt cases, although the Court of Appeal has said that it tries to hear such appeals within a day or two (*Balogh* v *Crown Court at St Alban's* [1974] 3 All ER 283, CA, *per* Lord Denning MR at p. 290).

8.5.2 Arrangements for the hearing of contempt appeals

(a) In cases of contempt of a *magistrates' court* under s. 12 of the Contempt of Court Act 1981, appeal lies to the Crown Court (Contempt of Court Act 1981, s. 12(5), applying s. 108 of the Magistrates' Courts Act 1980).

(b) For contempt of a *county court*, or any other inferior court from which appeals generally lie to the Court of Appeal, appeal lies to the Court of Appeal (Administration of Justice Act 1960, s. 13(2), as amended by the Access to

Justice Act 1999). Normally, permission is required in order to appeal against a decision of a county court in a contempt case. However, an appeal against a *committal order* made by the judge lies as of right and does not require permission (CPR, r. 52.3).

(c) In cases of contempt of an inferior court not falling under (a) or (b) above (e.g., a coroner's court), appeal lies to the High Court (ibid., s. 13(2), as amended by the Access to Justice Act 1999).

(d) For contempt of the *Crown Court*, appeal lies to the Court of Appeal (ibid.).

(e) From the decision of a single judge of the *High Court* in a contempt case (other than a decision on appeal), appeal lies to the Court of Appeal (ibid.). Permission to appeal is required unless the appeal is against a *committal order* made by the judge (CPR, r. 52.3).

(f) From the decision of a single judge of the High Court *on an appeal*, or of the Court of Appeal, appeal lies to the House of Lords (ibid.). Permission to appeal must first be obtained either from the High Court or the Court of Appeal, as appropriate, or from the House of Lords itself. If the appeal is against a decision given *on appeal* from a lower court, the High Court or Court of Appeal must, in addition, certify that the case involves a point of law of general public importance. If the appeal to the House of Lords is against a decision *at first instance* of the High Court or Court of Appeal in a contempt case, permission to appeal may be granted without such a certificate (ibid., s. 13(4)).

9

Remedies in private law

9.1 Distinction between common law and equitable remedies

The usual remedy awarded in a common law court for breach of contract was damages. This was so even for breach of a particular contract where damages might be inappropriate. Thus, if B contracted to sell to A a plot of land, and then B broke the contract by later refusing to transfer the land to A, the common law would award only damages to A. This was an unsatisfactory remedy in cases where A preferred ownership of the land to money compensation.

The expression 'common law' is used here to mean the principles of law common to the whole of England and developed from the twelfth century by judges sitting in the old common law courts. Ultimately, this body of law became too technical, rigid, and deficient in some of its content, and people who wanted discretionary or other relief not available at common law began to petition the King and his Council.

In time, the task of dealing with these petitions was delegated to the Chancellor, who was secretary of state for all departments, close to the King, and whose departmental office was the Chancery. Over the centuries, successive Lord Chancellors developed a body of principles, known as equity, with a view to countering the defects, and rectifying the injustices, of the common law. The Chancery itself emerged as a court in which equity was administered.

Equity was very inventive in providing a variety of new remedies: specific performance and the injunction, and also rescission and rectification. In this way, equity supplemented the common law. Thus, in the example given above, equity would be able to impose a just solution by means of an order of specific performance compelling B to transfer the land to A.

The common law and equity were at times in conflict, although the latter was never a self-sufficient body of rules since it pre-supposed the existence of the former. The common law courts and the Court of Chancery were abolished by the Judicature Act 1873 and replaced by the High Court (see para 1.4.1). Henceforth, law and equity were to be administered concurrently in all civil courts, from which it followed that both common law and equitable remedies were to be available in every civil court. At the same time, it was confirmed by statute that 'wherever there is any conflict or variance between the rules of equity and the rules of the common law with reference to the same matter, the rules of equity shall prevail'. (This rule is presently contained in the Supreme Court Act 1981, s. 49.)

The most important common feature of equitable remedies is that they are *discretionary*. At common law, if the claimant proved his case he was entitled to his remedy of damages and it mattered not that his own conduct had been bad, or that he had been dilatory in seeking relief, or that the outcome was unfair to the defendant. Equity, however, exercised a discretion over the granting of its remedies.

In particular, equity was interested in the conduct of the parties—*the conduct of the claimant* as well as that of the defendant. A remedy would be refused to those whose conduct was inequitable, or who had delayed in seeking relief, or whose claims would produce unfair results. Such claimants would be left to pursue whatever remedy they might have at common law. If an adequate remedy existed at common law, that in itself was a ground for denying an equitable remedy to the claimant.

The reforms of the late nineteenth century did not change this. Private law remedies are still referred to as belonging to the common law or to equity. Damages are still available as of right while equitable remedies are granted or withheld on the judicial exercise of discretion.

Only the major equitable remedies are dealt with in this book. Details of other remedies, such as account, appointment of a receiver, discovery, and delivery up and cancellation of documents, can be found in other works (see especially, *Snell's Equity*, 31st ed., 2005; Spry, *The Principles of Equitable Remedies*, 6th ed., 2001; Burrows, *Remedies for Torts and Breach of Contract*, 3rd ed., 2004).

9.2 Damages

An award of damages is a common law remedy available as of right. Damages are particularly important as a means of redress for breach of contract and the commission of a tort.

9.2.1 The object of damages

The object of damages for breach of contract is to place the claimant in the same position, as far as money can do it, as if the contract had been performed. The object of damages in the law of tort is to place the claimant in the same position, as far as money can do it, as if the tort had not been committed (i.e., *compensation* for the claimant). Thus, in contract and tort the common object of the remedy is *restitution*. In contract, the claimant is to be restored to the position he would have been in if performance had taken place. In tort, he is to be restored to the position he would have been in if no wrong had been done against him.

9.2.2 Types of damages

In both contract and tort the usual type of damages awarded is *compensatory* damages. But there also exist contemptuous, nominal, and exemplary damages.

Compensatory damages provide compensation for actual loss suffered, no more and no less. Most torts are actionable only on proof of loss or damage so that no action will lie, and no damages are recoverable, without such proof. Breach of contract is actionable *per se* (i.e., without proof of loss or damage).

Contemptuous damages usually consist of the lowest coin of legal tender, currently 1p. They are awarded where the claimant proves his case, but he has suffered no loss and the court has a low opinion of his claim.

In tort, contemptuous damages may be awarded because, morally at least, the claimant deserved what the defendant did to him. In *Dering* v *Uris* [1964] 2 QB 669, Dr Dering, an ex-prisoner at Auschwitz who had performed experimental surgical operations on inmates under Nazi pressure, sued Leon Uris for libel in respect of a reference to him in Uris's novel, *Exodus*. Libel is actionable without proof of damage. The trial of the action against the author and publisher of *Exodus* lasted for five weeks. The claimant's claim succeeded, but the jury, as an indication of their disapproval of his action, awarded him only a halfpenny in damages and he was ordered by the judge to pay the defence costs.

In a case decided towards the end of 1991, no damages at all were awarded to a successful libel claimant for what is believed to be the first time ever. A former teacher sued the publishers of *The People* newspaper over allegations that he was a sexual pervert and a danger to the boys he taught. The jury found that he had been libelled but awarded no damages, and the judge refused to order the claimant's costs to be paid by the unsuccessful defendants (*Standish* v *Mirror Group Newspapers*, 25 November 1991, unreported, but see *The Independent*, 26 November 1991; the founder of the 'ChildLine' charity, Esther Rantzen, fared considerably better when she later successfully sued the publishers of *The People* after it was alleged that she had kept quiet about Mr Standish's activities: see para 6.8.2.2).

If the damages awarded to the claimant are of the contemptuous type, the chances are that he will not be given his costs even though, technically, he has won the case. Costs usually 'follow the event' so that the winner has his costs paid by the loser. But the judge always has a discretion in the matter of costs, and the award of a derisory sum by way of damages is a material factor in exercising that discretion.

Nominal damages usually vary in amount between 5p and £2, although there is no conventional figure. They are awarded where the claimant proves his case but has suffered no loss and there is no moral blame attaching to him (although the claimant may still be deprived of his costs). Nominal damages are awarded, for example, in the case of torts which are actionable *per se* but where no loss has resulted, such as a technical trespass to land where no damage was done.

Nominal damages of £1 were awarded by the House of Lords in *Grobbelaar* v *News Group Newspapers Ltd* [2002] 1 WLR 3024, HL (see para 6.1), where the claimant won his libel action but the evidence showed that he had no reputation worthy of legal protection. The relatively large sum of five guineas was awarded as nominal damages in *Constantine* v *Imperial Hotels Ltd* [1944] KB 693 to the West Indian cricketer, Learie Constantine, who had been refused accommodation in one of the defendants' hotels. He was given accommodation in another of the defendants' hotels, but the defendants had broken their common law duty as common innkeepers and the claimant was entitled to damages.

Exemplary (sometimes called *punitive* or *vindictive*) *damages* are occasionally awarded to the claimant in order to punish, or to make an example of, the defendant for what he has done and to deter him (and others) from doing the same in the future. As Lord Nicholls put it in *Kuddus* v *Chief Constable of Leicestershire Constabulary* [2001] 2 WLR 1789, HL, at [51]:

Exemplary damages or punitive damages, the terms are synonymous, stand apart from awards of compensatory damages. They are additional to an award which is intended to compensate a [claimant] fully for the loss he has suffered. . . . They are intended to punish and deter.

Exemplary damages cannot be awarded in contract but they can be awarded in tort in the following three circumstances (*Rookes* v *Barnard* [1964] AC 1129, HL; *Cassell & Co. Ltd* v *Broome* [1972] AC 1027, HL).

(a) Where they are authorised by statute.

(b) For oppressive, arbitrary, or unconstitutional actions by servants of the government.

(c) Where the defendant calculated to make a profit out of the tort which would exceed any compensation payable to the claimant.

In *AB* v *South West Water Services Ltd* [1993] 1 All ER 609, CA, it was held by the Court of Appeal that an exemplary award could only be made for torts in respect of which exemplary damages were available prior to 1964, the year in which *Rookes* v *Barnard*, above, was decided. This conclusion was based on an analysis of some of the speeches in the two House of Lords' cases of *Rookes* v *Barnard* and *Cassell & Co. Ltd* v *Broome*, above.

There appeared to be no case prior to 1964 in which exemplary damages were awarded for public nuisance or misfeasance in public office. Accordingly, in *AB* v *South West Water Services Ltd*, above, a claim by 182 claimants for exemplary damages for public nuisance arising out of the contamination of their drinking water by large quantities of aluminium sulphate was struck out. And in *Kuddus* v *Chief Constable of Leicestershire Constabulary* (2000) *The Times*, 16 March, CA, a claim for exemplary damages for misfeasance in public office committed by a police constable was likewise struck out by the Court of Appeal.

When the *Kuddus* case went on appeal to the House of Lords ([2001] 2 WLR 1789), the decision of the Court of Appeal was reversed and *AB* v *South West Water Services Ltd* was overruled. A majority of their Lordships in *Kuddus* held that an analysis of the speeches in *Rookes* v *Barnard* and *Cassell & Co. Ltd* v *Broome* did not justify the conclusion reached by the Court of Appeal in *AB* v *South West Water Services Ltd* (see [2001] 2 WLR 1789 *per* Lords Slynn, Mackay, and Hutton at [21], [38], and [81], respectively).

It follows from the decision of the House of Lords in *Kuddus* that whether exemplary damages will be awarded for oppressive, arbitrary, or unconstitutional action by public officers ((b) above) depends entirely on the features of the allegedly objectionable behaviour. It is not a requirement that the tort in question must have been accepted before 1964 as justifying an exemplary award.

Exemplary damages are normally awarded *in addition to* compensatory damages. The decision of a jury to award only exemplary damages without at the same time awarding compensatory damages will usually be regarded as perverse. Thus, in *Cumber v Chief Constable of Hampshire* (1995) *The Times*, 28 January, CA, the successful claimant in an action for false imprisonment was awarded £50 exemplary damages— but no compensatory damages—by a jury which appears to have totally misunderstood its function. On appeal, it was held that the jury's verdict was perverse and the Court of Appeal substituted a global award of £350 to cover both compensatory and exemplary damages.

(a) *Authorised by statute.* What appears to be the only statute authorising in terms the award of exemplary damages is the Reserve and Auxiliary Forces (Protection of Civil Interests) Act 1951. By s. 13(2), in any action for damages for conversion or in other proceedings (which may be available in certain circumstances outlined in ss. 2 and 4) the court 'may take account of the conduct of the defendant with a view, if the court thinks fit, to awarding exemplary damages in respect of the wrong sustained by the [claimant]'.

There is no case law directly on s. 13(2), and it has been doubted whether the damages mentioned are exemplary in the true sense. In *Cassell & Co. Ltd v Broome* [1972] AC 1027, HL, Lord Kilbrandon expressed the opinion (at p. 1133) that 'exemplary' really meant 'aggravated'.

Section 97(2) of the Copyright, Designs and Patents Act 1988 allows the court in an action for infringement of copyright to 'award such *additional damages* as the justice of the case may require' having regard to all the circumstances and, in particular, to the flagrancy of the infringement and any benefit obtained by the defendant as a result of the infringement. It is not at all certain, however, that the 'additional damages' mentioned in s. 97(2) can properly be described as 'exemplary'.

According to Lord Clyde, *obiter*, in the Scottish appeal case of *Redrow Homes Ltd v Bett Brothers plc* [1998] 1 All ER 385, HL, they are 'more probably' aggravated damages of a compensatory nature (see also *Nottinghamshire Healthcare NHS Trust v News Group Newspapers Ltd* (2002) *The Times*, 1 April), and this seems to be confirmed by the history of the predecessor of s. 97(2).

This was s. 17(3) of the Copyright Act 1956. In *Williams v Settle* [1960] 2 All ER 806, CA, the Court of Appeal affirmed an exemplary award of £1,000 damages for breach of copyright. Although it is not clear from the report whether the damages were awarded at common law or under the statute, the 'additional damages' available under s. 17(3) of the 1956 Act were described in the Court of Appeal as 'exemplary'. In two subsequent cases in the House of Lords the view was advanced, *obiter*, that they are not true exemplary damages (*Rookes v Barnard* [1964] AC 1129, HL, *per* Lord Devlin at p. 1225; *Cassell & Co. Ltd v Broome* [1972] AC 1027, HL, *per* Lord Kilbrandon at p. 1133).

In *Nichols Advanced Vehicle Systems Inc. v Rees* [1979] RPC 127, Templeman J held that the defendant's breach of copyright was flagrant enough to attract an additional award under the statute. He ordered an enquiry as to the amount payable, indicating that it should be a moderate, but not excessive, sum. Without citing any of the case law, Templeman J refrained from describing the additional damages as 'exemplary'.

Additional damages in the modest sum of £200 were awarded under s. 17(3) of the

1956 Act by Mervyn Davies J for breach of copyright in a Ph.D. thesis in *Sushil Kumar Goswami* v *Hammons and others*, 29 October 1982, unreported.

'Additional damages' under s. 97(2) of the Copyright, Designs and Patents Act 1988 are only available as an addition to normal compensatory damages; they are not available as an addition to other remedies for breach of copyright, such as an account of profits (*Redrow Homes Ltd* v *Bett Brothers plc* [1998] 1 All ER 385, HL).

It seems highly unlikely that an award of exemplary damages for unlawful discrimination is justified on the wording of the Sex Discrimination Act 1975, the Race Relations Act 1976, or the Disability Discrimination Act 1995 (see below).

(b) *Oppressive, arbitrary, or unconstitutional actions.* The expression 'servants of the government' includes the police, local government, and statutory bodies. Thus, it has been held that exemplary damages may be awarded against a local authority in respect of unlawful discrimination in the selection of candidates for teaching posts in schools and colleges if the discriminatory acts are 'oppressive, arbitrary, or unconstitutional'. In *Bradford Metropolitan City Council* v *Arora* [1991] 3 All ER 545, CA, the Court of Appeal allowed an appeal from a decision of the Employment Appeal Tribunal and restored an employment tribunal's award of exemplary damages of £1,000 (made in addition to compensatory damages of £2,000) for the insulting manner in which a candidate had been treated.

It is doubtful whether the *Arora* case, above, was correctly decided since it seems likely that the anti-discrimination legislation does not authorise the award of exemplary damages. For example, s. 66(4) of the Sex Discrimination Act 1975, as amended, and s. 57(4) of the Race Relations Act 1976 (which deal with awards of damages in the county courts) are in almost identical terms:

For the avoidance of doubt it is hereby declared that damages in respect of an unlawful act of discrimination may include *compensation* for injury to feelings whether or not they include *compensation* under any other head. (Italics supplied.)

The wording of s. 25(2) of the Disability Discrimination Act 1995 is very similar. These three anti-discrimination statutes seem to contemplate the award of compensatory damages only; exemplary damages are not 'compensation'. This argument was not put forward in *Arora*.

The term 'servants of the government' may possibly include solicitors since they are officers of the Supreme Court. This would mean, for example, that clients could be liable in exemplary damages if their solicitors execute *Anton Piller* orders in an excessive and oppressive manner (*Columbia Picture Industries Inc.* v *Robinson* [1986] 3 All ER 338 *per* Scott J at p. 379; see para 9.4.2.5 below).

Actions against the police for wrongful arrest, false imprisonment, assault, and malicious prosecution are likely to raise allegations of oppressive or arbitrary behaviour. In *Thompson* v *Commissioner of Police of the Metropolis* [1997] 2 All ER 762, CA, the Court of Appeal laid down guidelines on how judges should instruct juries to help them decide what to award by way of damages in such cases (see para 6.8.2.3).

(c) *Calculated profit.* In *Cassell & Co. Ltd* v *Broome* [1972] AC 1027, HL, the defendants published a book knowing that it was probably defamatory of the claimant, Commander Broome. The book, entitled *The Destruction of Convoy PQ17*, accused the claimant of disobeying orders and of cowardly desertion of the convoy he was escorting during the Second World War. The book was published in spite of a warning from the claimant that he would sue the defendants if they published it.

The trial judge, Lawton J, was satisfied that the defendants had calculated that it was worth the risk of being sued because even the publicity value of litigation would help to sell more copies of the book. Accordingly, the jury was directed that exemplary damages might be awarded in such a case. The jury awarded compensatory damages of £15,000 and exemplary damages of £25,000. All seven judges who heard the appeal in the House of Lords thought that the amount awarded as exemplary damages was excessive, but, by a majority of four to three, they refused to disturb the jury's verdict.

The profit motive is not limited to money-making in the strict sense. It also covers cases where the defendant is intent on securing, at the expense of the claimant, some other advantage which can only be obtained by paying a price he is not prepared to pay, or by acting unlawfully (*Rookes* v *Barnard* [1964] AC 1129 *per* Lord Devlin at p. 1227). Thus, in *Drane* v *Evangelou* [1978] 2 All ER 437, CA, an exemplary award of £1,000 for trespass to land made against a landlord who had engaged in 'monstrous behaviour' to secure the unlawful eviction of his tenant was upheld, and in *Guppys (Bridport) Ltd* v *Brookling* (1983) 14 HLR 1, CA, an award of a similar amount was affirmed in a case of private nuisance where reconstruction works had caused serious disruption to two tenants. In *Design Progression Ltd* v *Thurloe Properties Ltd* [2005] 1 WLR 1, exemplary damages of £25,000 were awarded against a landlord of retail premises which, in breach of its duty under the landlord and tenant legislation, failed to decide within a reasonable time whether or not to consent to the tenant's proposed assignment of the remainder of a lease. The landlord had adopted a deliberately obstructive policy with the object of preventing the assignment, getting the premises back, and granting a new lease at a higher rent.

9.2.3 Difficulty of assessment of damages

Damages must still be awarded even though they are difficult to assess. For example, if an actor, through a breach of contract, loses the opportunity for publicity, damages must be awarded for this loss despite the fact that their quantification would be very difficult.

In *Chaplin* v *Hicks* [1911] 2 KB 786, CA, the claimant was prevented from taking part in a beauty contest because of the defendant's breach of contract. It was held that although it was impossible to assess the damages with any precision, a sum should be awarded in accordance with the value of the chance which the claimant had lost. She had been deprived of the opportunity of competing as one of 50 final candidates for 12 prizes in the form of theatrical engagements. The jury had awarded £100 damages, and the Court of Appeal refused to disturb this verdict.

In tort, damages are difficult to assess in defamation and personal injury cases but they must nevertheless be awarded.

9.2.4 **Damages for mental distress**

Damages may sometimes be awarded for mental distress as well as for the more common financial loss or physical injuries. Such damages are more likely to be awarded in tort than in contract.

The general rule in contract law is that damages are not recoverable for mental distress caused by a breach (*Addis v Gramophone Co. Ltd* [1909] AC 488, HL). For example, damages are not recoverable for mental distress, consisting of depression, frustration, anxiety, and injured feelings, arising incidentally out of dismissal or demotion at work (*Bliss v South East Thames Regional Health Authority* [1985] IRLR 308, CA, applying *Addis v Gramophone Co. Ltd*, above, and overruling *Cox v Philips Industries Ltd* [1976] 3 All ER 161). Nor are damages available for anguish and vexation arising from the breach of a purely commercial contract (*Hayes v James & Charles Dodd* [1990] 2 All ER 815, CA, applying *Bliss v South East Thames Regional Health Authority*, CA, above).

In some cases in recent years the courts have been prepared to compensate the claimant where the mental distress was a direct, as opposed to an incidental, result of the defendant's breach of contract. From these cases there has developed an exception to the general rule: where the contract is one whose very object is the provision of pleasure, relaxation, peace of mind, or freedom from harassment, damages for mental distress are recoverable on breach (see *Watts v Morrow* [1991] 4 All ER 937, CA, *per* Bingham LJ at pp. 959–60).

The county courts are regularly awarding modest damages for breaches of contracts to provide pleasure, relaxation, or peace of mind in such diverse contexts as ruined holidays or photographs, the supply of a wedding dress or double glazing, hire purchase transactions, landlord and tenant cases, building contracts, and the employment of estate agents and solicitors (see *Farley v Skinner* [2001] 3 WLR 899, HL, *per* Lord Steyn at [20]).

In *Jarvis v Swans Tours Ltd* [1973] 1 QB 233, CA, the claimant sued the defendants for failing to provide the houseparty holiday promised in their brochure. Among the claimant's disappointments in Switzerland were that there was no welcome party and no proper skiing facilities; the hotel owner did not speak English; the bar was open only on one evening; and the promised 'yodeller' consisted of a local man in his working clothes singing a few songs very quickly.

It was held that the defendants were in breach of contract and, in the county court, the claimant was awarded as damages half the price of the holiday. The claimant appealed and the Court of Appeal held that he was also entitled to damages for mental distress, which the Court defined as including inconvenience, disappointment, annoyance, and frustration. The claimant's damages were increased accordingly (see also *Jackson v Horizon Holidays Ltd* [1975] 3 All ER 92, CA).

In another case, the Court of Appeal held the claimant to be entitled to damages for mental distress caused by a breach of contract on the part of the defendant firm of solicitors in negligently failing to enforce a non-molestation injunction on her behalf (*Heywood v Wellers* [1976] QB 446, CA).

Jarvis v Swans Tours Ltd and *Jackson v Horizon Holidays Ltd*, above, were approved by the House of Lords in *Ruxley Electronics and Construction Ltd v Forsyth* [1995] 3 All

ER 268, HL. Here the builder of a swimming pool was in breach of contract in constructing a shallower pool than specified. The contract was treated as one to provide a 'pleasurable amenity', and a sum of £2,500 had been awarded as damages for loss of that amenity. Their Lordships thought that the sum was on the high side for a disappointed expectation but nevertheless refused to disturb it (see especially *per* Lord Lloyd at p. 289).

Damages for mental distress and inconvenience were refused in *Watts* v *Morrow* [1991] 4 All ER 937, CA, where the defendant's breach of an ordinary contract of survey had led the claimants to pay £15,000 more for their country house than it was really worth. The claimants were awarded compensatory damages of £15,000 on the diminution in value principle but were denied an additional sum by way of damages for mental distress and inconvenience because the 'very object' of a contract to survey the condition of a house for a prospective purchaser is not to provide 'pleasure, relaxation, peace of mind, or freedom from molestation' (*per* Bingham LJ at p. 960).

Watts v *Morrow*, above, was applied in *Alexander* v *Alpe Jack Rolls Royce Motor Cars Ltd* (1995) *The Times*, 4 May, CA, and in *Knott* v *Bolton* (1995) *The Independent*, 8 May, CA (which was, however, later overruled by the House of Lords: see below).

In the *Rolls Royce* case, it was argued that damages for mental distress should be available for breach of a contract to repair a motor car because it is akin to a contract to provide a relaxing holiday (as in *Jarvis* v *Swans Tours Ltd*, above) or to provide freedom from worry and anxiety (as in *Heywood* v *Wellers*, above). It was held, however, that breach of a contract to repair a car, even one so prestigious as a Rolls Royce, does not give rise to a liability for damages for distress and inconvenience or loss of enjoyment in the use of the car.

In *Knott* v *Bolton*, above, the claimants were unsuccessful in their attempt to obtain damages for mental distress from the defendant architect who had defectively designed their house so as to provide a much narrower staircase and smaller gallery than specified. The argument that the true nature of the contract entered into by the defendant was to provide pleasure for the claimants was rejected. The true nature of the contract, according to the Court of Appeal, was to design the house, and the provision of pleasure to the claimants was merely ancillary or incidental to that.

Knott v *Bolton* was overruled by the House of Lords in *Farley* v *Skinner* [2001] 3 WLR 899, HL, which concerned a contract to survey a house situated 15 miles from Gatwick Airport and which included an obligation to advise whether the property might be affected by aircraft noise. When the contract was broken by reason of the defendant's failure to ascertain this properly, the claimant sought damages for mental distress.

The House of Lords, applying *Ruxley Electronics and Construction Ltd* v *Forsyth* [1995] 3 All ER 268, HL, above, held that the defendant's obligation to advise on aircraft noise was a 'major or important' part of his contract with the claimant, the breach of which had caused significant interference with the claimant's enjoyment of his property for which he was entitled to be compensated. It was further held that the claimant had not forfeited his right to damages by deciding not to move house.

Although their Lordships thought that the award of £10,000 awarded at first instance was too high for this sort of case, they refused to set it aside as excessive. According to

Lord Steyn ([2001] 3 WLR 899 at [28]), awards of damages for mental distress should be 'restrained and modest' since it is 'important that logical and beneficial developments in this corner of the law should not contribute to the creation of a society bent on litigation'.

Knott v Bolton, CA, above, was held to have been wrongly decided because it placed too much emphasis on looking at the contract as a whole. It was also inconsistent with the later decision of the House of Lords in Ruxley, above. According to the House in Farley v Skinner, the claimants in Knott v Bolton, having been deprived of the benefit of the wider staircase and larger gallery to which they were entitled under the contract, should have been awarded damages.

It is clear from Farley v Skinner [2001] 3 WLR 899, HL, that in order to succeed in a claim for damages for mental distress it does not have to be shown that the provision of pleasure, relaxation, or peace of mind is the 'very' or 'sole' object of the contract, or that the defendant has guaranteed to achieve success. That would be too narrow an interpretation of the decision in Watts v Morrow [1991] 4 All ER 937, CA. It is sufficient that the provision of pleasure, relaxation, or peace of mind by means of the amenity in question is a major or important part of the contract and that the defendant has agreed (usually by implication) to exercise reasonable care. While it may safely be assumed that Watts v Morrow was correctly decided on its facts, the observations therein of Bingham LJ, although they provide 'a helpful point of departure for the examination of the issues' in this sort of case, were 'never intended to state more than broad principles' and should not be applied too literally (Farley v Skinner [2001] 3 WLR 899, HL, per Lord Steyn at [15]; see also per Lords Clyde and Hutton at [35] and [51], respectively).

In Malik v Bank of Credit and Commerce International SA (in liquidation) [1997] 3 All ER 1, HL, the House of Lords held that Addis v Gramophone Co. Ltd [1909] AC 488, HL, does not prevent recovery of damages for loss of reputation caused by a breach of contract, including a breach of a contract of employment. Thus, where an employer carries on his business dishonestly or corruptly, and is thereby in breach of his implied duty not to conduct his business in a way likely to do serious damage to the relationship of confidence and trust between employer and employee, an employee who suffers loss in the form of damage to his future employment prospects in consequence of such breach can sue for damages if the loss was reasonably foreseeable.

In the Malik case, two innocent former employees of BCCI, which ran its business fraudulently and collapsed in 1991, were accordingly held entitled to claim 'stigma compensation' arising from their difficulty in obtaining employment in the banking field because of their association with BCCI.

In Johnson v Gore Wood & Co. [2001] 2 WLR 72, HL, a majority of the House of Lords upheld the general principle in Addis v Gramophone Co. Ltd [1909] AC 488, HL, and refused to extend the exception to it, developed in cases like Ruxley [1995] 3 All ER 268, above, to a claim for damages brought against a firm of solicitors for mental distress caused to the claimant by its breach of contract in negligently handling his affairs. The alleged mental distress consisted of financial embarrassment and damage to family relationships. The claim was struck out. Lord Cooke (at p. 109), dissenting, would have allowed the claim for damages for mental distress to go to trial.

He described a changed way of life because of poverty, and damaged family relationships, as 'grievous forms of non-pecuniary harm', and he doubted 'the permanence of
Addis in English law'.

In the later case of *Johnson* v *Unisys Ltd* [2001] 2 WLR 1076, HL, further criticism of the *Addis* principle was expressed. *Johnson* v *Unisys Ltd* concerned an action
for wrongful dismissal in which the claimant, relying on the *Malik* case [1997] 3
All ER 1, HL, above, claimed damages for the manner of his dismissal, which, he
alleged, had caused a mental breakdown and inability to work. He argued that
breach by an employer of the implied contractual term of trust and confidence
would allow an employee to recover damages for loss arising out of the manner of
his dismissal.

The Law Lords held unanimously that the claim had been rightly struck out as
disclosing no reasonable cause of action, but they did so on different grounds. The
reason given by the majority was that it would be improper for judges so to extend the
implied term of trust and confidence, given the intention of Parliament (expressed in
the unfair dismissal legislation) that such claims should be heard by employment
tribunals, not by courts of law, and that the remedy provided should be restricted in
application and extent.

Malik, above, was distinguished because it was not a case on manner of dismissal.

Lord Steyn was critical of the *Addis* principle throughout his opinion, even to the
extent of suggesting ([2001] 2 WLR 1076 at [3]) that the headnote to *Addis* at [1909]
AC 488 is wrong in so far as it states that the case decided that a wrongfully dismissed
employee can never sue for damages for loss of employment prospects arising out of
the harsh and humiliating manner of his dismissal. (See also the comments of Lords
Nicholls and Hoffmann at [2] and [44], respectively.)

Farley v *Skinner* [2001] 3 WLR 899, HL, above, and Bingham LJ's dictum in *Watts* v
Morrow [2001] 3 WLR 899, CA, at pp. 959–60, above, were applied in the unusual case
of *Hamilton Jones* v *David & Snape* [2004] 1 All ER 657. On the claimant's instructions, the defendant firm of solicitors had obtained court orders prohibiting the
claimant's husband from removing the children from her care, and had sent copies
of the orders to the United Kingdom Passport Agency. However, in breach of their
contract with the claimant, they failed to renew the notification at the Passport
Agency, with the result that the claimant's husband was able to add the children to his
passport and remove them from the United Kingdom. It was held that a significant
purpose of the contract was to ensure, so far as possible, that the claimant kept
custody of the children for her own 'pleasure and peace of mind'. She was awarded
damages for mental distress.

9.2.5 Aggravated damages

In tort, *aggravated* damages (which are compensatory in nature and are not to be
confused with exemplary damages) may sometimes be awarded for injury to the
claimant's dignity and pride. Such cases would include an 'insolent and high-handed
trespass' to land (as in *Jolliffe* v *Willmett & Co.* [1971] 1 All ER 478, *per* Geoffrey Lane J
at p. 485; and note the award of aggravated damages in *Columbia Picture Industries
Inc.* v *Robinson* [1986] 3 All ER 338, para 9.4.2.5 below), or an attempted justification

of a libel where the defendant repeats his statements but fails to prove the truth of them.

In *W v Meah* [1986] 1 All ER 935, an action for trespass to the person (assault and battery), aggravated damages were awarded by Woolf J to two women, one of whom had been raped, and the other seriously sexually assaulted, in frightening and revolting circumstances.

Aggravated damages were awarded in *Appleton v Garrett* [1996] PIQR P1, an action for trespass to the person where a dentist, for personal profit, had performed a large amount of unnecessary dental work on the claimants' teeth.

Aggravated damages have been held to be available for the tort of malicious falsehood (*Khodaparast v Shad* [2000] 1 All ER 545, CA; this possibility had earlier been suggested, *obiter*, in *Joyce v Sengupta* [1993] 1 All ER 897, CA).

It seems that aggravated damages are not available in all torts. In *Kralj v McGrath* [1986] 1 All ER 54, the claimant contended that her compensation should include an award of aggravated damages for an obstetrician's 'horrific and wholly unacceptable' treatment of her while undergoing childbirth without an anaesthetic. However, Woolf J held that aggravated damages are not appropriate in medical negligence cases. Nor are they appropriate as compensation for anger and indignation in cases of public nuisance (*AB v South West Water Services Ltd* [1993] 1 All ER 609, CA).

9.2.6 Reform of aggravated and exemplary damages

In 1997, the Law Commission described the limited circumstances in which *exemplary* damages are available as 'plainly irrational' (*Aggravated, Exemplary and Restitutionary Damages*, Law Com No. 247, 1997). Under the Law Commission's recommendations, legislation would place exemplary damages on a 'clear, principled, but tightly controlled, footing'. At the same time, their availability would be extended to any legal wrong (other than breach of contract) where the defendant had 'deliberately and outrageously' disregarded the claimant's rights. Juries would no longer have the power to decide whether to award exemplary damages or to fix their amount; these matters would be for judges to determine. Exemplary damages would not be available where the defendant had been convicted of a criminal offence for the same conduct, or where another available remedy would be adequate punishment.

On *aggravated* damages, the Law Commission noted that they are really part of the law on damages for mental distress and referred to the confusion over whether aggravated damages contain a punitive, in addition to a compensatory, element. It was recommended that it should be made clear by legislation that they are intended only to provide *compensation* for the claimant's mental distress and not to punish the defendant for his conduct. To this end, it was proposed that they should go under the name of 'damages for mental distress' rather than the misleading name of aggravated damages.

In November 1999, following a period of consultation which resulted in no clear agreement on the way ahead, the government announced its intention not to introduce legislation to implement the Law Commission's proposals on exemplary damages.

Subsequently, in *Kuddus v Chief Constable of Leicestershire Constabulary* [2001]

2 WLR 1789, HL, the House of Lords considered the matter. As noted in para 9.2.2 above, their Lordships were unanimous in changing the common law so that an award of exemplary damages for oppressive, arbitrary, or unconstitutional action by public officers no longer depends on whether the tort in question was accepted before 1964 (the year in which *Rookes* v *Barnard* [1964] AC 1129 was decided) as justifying an exemplary award.

The wider question of whether exemplary damages should continue to be awarded at all was raised by their Lordships in *Kuddus* [2001] 2 WLR 1789, HL, although no argument on it had been presented by the parties. Lord Slynn (at [25]) did not think it an appropriate occasion to reopen the question. Lord Mackay (at [36]), in the absence of fuller argument and in view of the fact that the Law Commission after full consultation had recommended that exemplary damages should continue to be available, was content to decide the case on existing principles. Lord Nicholls (at [62]–[64]) emphasised the advantages of exemplary damages, while Lord Scott (at [107]) thought that the 'continuing need in the year 2001 for exemplary damages as a civil remedy in order to control, deter and punish . . . is not in the least obvious'.

Lord Hutton (a former Lord Chief Justice of Northern Ireland) preferred to express no concluded opinion on the question, but referred to some cases decided in Northern Ireland over the previous three decades which he thought indicated the value of exemplary awards in 'restraining the arbitrary and outrageous use of executive power and in vindicating the strength of the law' (at [75]). Lord Scott, on the other hand (at [108]), thought that 'the strength of the law' could have been vindicated just as well in those cases by an appropriate award of *aggravated* damages. He viewed the prospect of any extension of exemplary awards 'with regret' because 'victims of tortious conduct should receive due compensation for their injuries, not windfalls at public expense' (at [121]).

The question whether exemplary damages should in principle be awarded on the basis of *vicarious liability*, although not argued by the parties, was also raised by the House of Lords in *Kuddus*. It would appear that this question had never before been discussed in an English case (see *per* Lord Scott at [126]).

Vicarious liability arises where the defendant (usually an employer or public body) has personally committed no wrong, but is held to be liable for the torts of those under his control. Lord Slynn (at [27]) did not consider it right to rule on the question since it had not been argued, as did Lord Mackay (at [48]), Lord Nicholls (at [69]), and Lord Hutton (at [93]), although earlier (at [79]) he had referred to the value of being able to sue the Ministry of Defence or a chief constable vicariously in cases where it is not possible to identify the individual at fault. Lord Scott (at [128]–[137]) thought that an exemplary award based on vicarious liability was contrary to principle and should be rejected. He was critical (at [136]) of *Thompson* v *Commissioner of Police of the Metropolis* [1997] 2 All ER 762, CA (para 6.8.2.3), because it was decided on an assumption in the absence of any argument addressed to the court about whether exemplary damages could be awarded in a vicarious liability case.

9.3 Specific performance

9.3.1 Introduction

Specific performance is an order of the court directing the defendant to do what he has promised by contract to do. It is a creature of equity and is superior to the common law remedy of damages in that it provides for the specific enforcement of a contract instead of merely giving money compensation to the claimant.

Like other equitable remedies, specific performance is awarded or withheld on a discretionary basis. This means that the court is not bound to grant it merely because the conditions for it exist. Every breach of a binding contract entitles the injured party to the common law remedy of damages, but not every breach of contract entitles him to the equitable remedy of specific performance. Thus, in *Wood* v *Scarth* (1855) 2 K & J 33 the claimant's action for specific performance failed, but three years later his action for damages on the same facts succeeded (*Wood* v *Scarth* (1858) 1 F & F 293).

Although specific performance is a discretionary remedy, the discretion is exercised *judicially;* i.e., not according to the caprice of the individual judge, but according to rules which have been settled in decided cases over the centuries.

9.3.2 Circumstances in which specific performance will not usually be decreed

9.3.2.1 Where the claimant's injury would be sufficiently remedied by damages

The court will not order a superior remedy where an inferior one will do. For example, if a contract for the sale of goods is broken, and the goods are ordinary commercial articles, only damages will be awarded because similar goods can be obtained easily elsewhere.

However, specific performance may be granted where the goods are rare or unique or not easily obtainable elsewhere. Examples of such goods would be a rare jewel, an original painting by an old master, and even petrol during an oil shortage, as in *Sky Petroleum Ltd* v *VIP Petroleum Ltd* [1974] 1 All ER 954, in which an interlocutory injunction was granted to prevent an oil company from cutting off supplies of petrol to a garage when alternative supplies were not available because of the 1973 oil shortage. The injunction in these circumstances was tantamount to a decree of specific performance.

In *Behnke* v *Bede Shipping Co. Ltd* [1927] 1 KB 649, specific performance was ordered of a contract for the sale of a ship (approved in *Astro Exito Navegacion* SA v *Southland Enterprise Co. Ltd (No. 2)* [1982] 3 All ER 335, CA at p. 345). However, in *Cohen* v *Roche* [1927] 1 KB 169 specific performance was refused to the buyer of a set of Hepplewhite chairs on the ground that the chairs were not sufficiently rare or unique but were ordinary articles of commerce. The buyer was instead awarded damages for breach of contract.

There seems to be no doubt that, historically, the whole basis of the equitable jurisdiction to decree specific performance rested on the inadequacy of the damages remedy at common law (*Harnett* v *Yielding* (1805) 2 Sch & Lef 549, *per* Lord Redesdale LC at p. 556; *Adderley* v *Dixon* (1824) 1 Sim & St 607, *per* Leach V-C at p. 610). Specific performance was ideally suited to remedy the breach of a contract involving land because damages, even if substantial in amount, were not an adequate alternative since no single piece of land is exactly the same as any other. The claimant would much rather have the land he had contracted to buy than be fobbed off with a money judgment. Even so, specific performance is not always the appropriate remedy where the contract is for the sale of land. For example, in *Wroth* v *Tyler* [1974] Ch 30 specific performance of a contract for the sale of land was refused because, in the circumstances, it would have been unreasonable to decree it (see further, para 9.3.2.8 below).

At the same time, there seems to be no good reason why in modern law the availability of specific performance should be deliberately restricted to contracts involving land. In the case of all contracts, whether for the sale of land or otherwise, a principle based on what will produce the most just solution would be preferable to one based solely on the adequacy of damages as a remedy. Support for this approach can be found in *Beswick* v *Beswick* [1968] AC 58, HL, in which the House of Lords ordered specific performance of a contract to pay a sum of money because, in the circumstances, that was the more just and appropriate remedy.

9.3.2.2 Where constant supervision by the court would be necessary

If the court were to enforce specifically a contract to do continuous successive acts it would not have the machinery for ensuring that the contract was performed. The court will not embarrass itself by making an order which, in practice, it cannot enforce.

In *Ryan* v *Mutual Tontine Westminster Chambers Association* [1893] 1 Ch 116, CA, the lessor of a flat agreed in the lease that he would appoint a porter, who would be 'constantly in attendance' for the purposes of cleaning the common passages and stairs in the building, delivering letters, and accepting articles for safe custody. The porter appointed was absent from the premises for several hours each day in order to act as a chef at a nearby club. The Court of Appeal held that, although the lessor had committed a breach of contract, the only remedy for the disappointed lessee was an action for damages. Such a contractual term, which was of a continuing nature and involved the daily performance of a series of acts, would not be enforced by an order for specific performance.

It is difficult to see why the court would have to become involved at all in daily or any other periodical supervision in a case like *Ryan*. If the court had decreed specific performance and the lessor had still ignored his responsibilities, the lessee could have taken enforcement proceedings in the normal way against the lessor for breach of the court order.

The authority of *Ryan's* case has been weakened by subsequent developments. In particular, it was distinguished on the facts in *Posner* v *Scott-Lewis* [1986] 3 All ER 513, where specific performance was decreed of a covenant in a lease to employ a resident porter in a block of flats. Mervyn Davies J's judgment implies that difficulty of supervision is not by itself a ground for withholding specific performance. He said (at p. 521) that the decision to grant or refuse a decree depends on (a) whether what has to be done in order to comply with the court order is sufficiently defined; (b) whether supervision by the court would be involved to an unacceptable degree; and (c) the respective prejudices or hardships likely to be suffered by the parties. In the *Posner* case the decree of specific performance simply ordered the lessor to employ a resident porter within a specified period. The making of the order would not involve the court in any protracted supervisory role.

Contracts to build or repair will not normally be specifically enforced. The reason is not so much that constant supervision of the work would be required but that damages would usually be an adequate remedy for breach of such contracts. If the defendant builder refuses to build, the claimant can employ another builder and recover any loss as damages from the defendant.

Sometimes, however, building and repairing contracts have been specifically enforced. If the building work is clearly defined in the contract, if it is not possible for the claimant to employ another builder to build without committing a trespass because the defendant is in possession of the land, and if the work is not done the claimant will suffer a loss which cannot adequately be remedied by damages, then the contract may be enforced by means of an order for specific performance (*Wolverhampton Corporation* v *Emmons* [1901] 1 KB 515, CA; *Carpenters Estates Ltd* v *Davies* [1940] Ch 160). In *Rainbow Estates Ltd* v *Tokenhold Ltd* [1998] 2 All ER 860, a tenant's repairing covenant was specifically enforced because, on the facts, there was no adequate alternative remedy, and the problems of defining the work required to be done and the need for supervision could be overcome by sufficiently defining in the decree of specific performance what had to be done in order to comply with it.

It is the normal practice of the courts not to compel retail premises to remain open, or to reopen, under a 'keep open' covenant in a lease because of difficulty of supervision or because damages would usually be an adequate remedy for breach of such a covenant. In *Co-operative Insurance Society Ltd* v *Argyll Stores (Holdings) Ltd* [1996] 3 All ER 934, CA, the Court of Appeal held that, on the facts, damages would not be an adequate remedy and the defendants were ordered to reopen one of their *Safeway* supermarkets at a cost of £1m and to keep it open for the remaining 29 years of a lease.

On appeal to the House of Lords, this decision was reversed (see [1997] 3 All ER 297). Their Lordships held that the normal, settled practice of the courts not to make orders compelling a person to run a business is based on sound sense. Although it is not a binding rule, and the grant or refusal of specific performance rests with the judge's discretion, the normal practice should only be departed from in exceptional circumstances. Disagreeing with the Court of Appeal, it was held that this was not a suitable case for specific performance because the defendants' contractual obligation to 'keep . . . open for retail trade during the usual hours of business in the locality' was not sufficiently precise to be capable of specific performance since it left too much room for argument about whether the defendants were doing enough to comply with it. It would, therefore, require constant supervision by the court.

9.3.2.3 Where the contract is for personal work or services

The court is not a slave-driver and will not order a person to work against his will. In *Lumley* v *Wagner* (1852) 1 De GM & G 604, the defendant agreed to sing at the claimant's theatre for three months and 'not to use her talents' at any other theatre during that time. When the defendant broke her contract by singing elsewhere, Lord St Leonards LC refused to order her to sing for the claimant. He did, however, grant an injunction preventing her from singing elsewhere during the currency of the agreement (see further, para 9.4.2.2.3 below).

The judicial principle that specific performance will not be ordered of a contract for personal work or services has been given legislative force in the case of the contract of employment. By the Trade Union and Labour Relations (Consolidation) Act 1992, s. 236:

No court shall, whether by way of:

(a) an order for specific performance . . . of a contract of employment, or

(b) an injunction . . . restraining a breach or threatened breach of such a contract,

compel an employee to do any work or attend any place for the doing of any work.

The principle is reciprocal: the employee will not be compelled to work, and the employer will not be compelled to employ. An employee who is *wrongfully* dismissed can sue his employer for breach of contract, claiming damages or a declaration that the dismissal was wrongful. He cannot claim specific performance to compel the employer to take him back.

Under the code on *unfair* dismissal (which is not usually a breach of contract) contained in the Employment Rights Act 1996, an employment tribunal has power to recommend the reinstatement or re-engagement of the employee. But if the recommendation is not implemented there is no power of compulsion exercisable against the recalcitrant employer. He can only be ordered to pay extra compensation (Employment Rights Act 1996, ss. 113–117, as amended).

9.3.2.4 Where the contract is lacking in mutuality

The court will not grant specific performance to one party in a situation where it could not be granted to the other party. Thus, the court will not grant specific performance *to* a minor because it could not order specific performance *against* him (*Flight* v *Bolland* (1828) 4 Russ 298). It would be unfair to allow a minor to obtain specific performance against an adult in circumstances in which the contract is unenforceable by the adult against the minor.

As long as mutuality exists at the date of the trial, it does not matter that there was no mutuality at the date when the contract was made. For example, an obligation to repair on the part of the claimant is not normally specifically enforceable at the suit of the defendant who has agreed to grant a lease. There is, therefore, a lack of mutuality at the inception of the contract. However, if by the time of the hearing the repairs have been carried out by the defendant, and the claimant is willing to pay for them, there is mutuality of remedy and the claimant could be awarded specific performance of the agreement to grant the lease (*Price* v *Strange* [1978] Ch 337, CA).

There are a number of exceptions to the mutuality rule. It is, for instance, possible for the victim of a misrepresentation to obtain specific performance of a contract which could not be enforced against him because of his ability to rescind it (*Winch* v *Winchester* (1812) 1 Ves & B 375).

9.3.2.5 Where the contract is illegal or void

Equity will not aid the enforcement of a contract which is illegal or for some reason void according to English law. Illegal contracts include those which offend public policy. A contract which, or the performance of which, is contrary to public policy cannot be enforced either at common law, by an action for damages or for an amount due, or in equity, by a decree of specific performance (*Cleaver* v *Mutual Reserve Fund Life Association* [1892] 1 QB 147, CA, *per* Lord Esher MR at p. 151).

However, if the contract, or its performance, is not illegal there may be compelling reasons of public policy, such as the protection of fundamental freedoms, why it should be specifically enforced. In *Verrall* v *Great Yarmouth Borough Council* [1980] 1 All ER 839, CA, the Court of Appeal ordered specific performance of a contractual licence of short duration to enable the National Front to hold its annual conference of 1979 at Great Yarmouth, as agreed with the council and in spite of fears about the likelihood of public disorder caused by opponents of the National Front. Specific performance would not have been decreed if the National Front itself had intended to engage in violence or other unlawful behaviour (ibid., *per* Lord Denning MR at p. 845). In this event, it would have been against public policy specifically to enforce the contract.

A void contract is no contract at all so that, in reality, there is nothing to enforce. In *Webster* v *Cecil* (1861) 30 Beav 62, the defendant vendor offered to sell land for £1,250 by mistake instead of £2,250. The claimant purchaser must have known the figure of £1,250 was a mistake because earlier the defendant had refused to sell for £2,000. The claimant accepted the offer of £1,250 by return of post. The defendant realised his mistake and immediately informed the claimant of it. The claimant claimed specific performance of the 'contract', but it was held that he could not have it and he was left to whatever remedy he might have at common law.

In fact, there would be no remedy at common law either in the circumstances of *Webster* v *Cecil* since there was probably no binding contract due to operative mistake. Sir John Romilly MR said (ibid., at p. 64) that the court could not grant specific performance and compel a person to sell property for much less than its real value. In saying this, his Lordship must be taken to have assumed the existence of a mistake in addition to the low price because mere inadequacy of consideration does not of itself appear to be a ground for refusing specific performance. If there is no mistake on the part of the defendant, and no inequitable conduct on the part of the claimant, there is no reason why the court should not specifically enforce a contract to sell land at considerably less than its market value (*Collier* v *Brown* (1788) 1 Cox Eq Cas 428; *Coles* v *Trecothick* (1804) 9 Ves Jr 234, *per* Lord Eldon LC at p. 246).

9.3.2.6 When an agreement is made without any consideration at all

Equity will not aid a volunteer and so specific performance will not be granted of an agreement which is voluntary. This is so even if the agreement is binding at common law because it is contained in a deed (*Jefferys* v *Jefferys* (1841) Cr & Ph 138; *Re Pryce* [1917] 1 Ch 234).

9.3.2.7 Where the claimant's hands are not clean

This rule is based on the equitable maxim that 'he who comes into equity must come with clean hands'. It means that any conduct of the claimant which would make it inequitable to enforce the contract against the defendant will bar his action for specific performance. Such inequitable conduct covers not only actual fraud or innocent misrepresentation, but also trickiness and unfairness falling short of those extremes. In *Webster* v *Cecil* (para 9.3.2.5 above), for instance, the claimant's conduct in snapping at an offer which he must have known was made by mistake was inequitable and would of itself be a ground for refusing specific performance (see the explanation of *Webster* v *Cecil* given in *Tamplin* v *James* (1880) 15 ChD 215, CA, *per* James and Brett LJJ at pp. 221 and 222, respectively).

In *Redgrave* v *Hurd* (1881) 20 ChD 1, CA, the defendant was induced into making a contract to buy the claimant solicitor's house and a share in his practice by an innocent misrepresentation of the value of the practice. When the defendant discovered the truth he refused to complete the transaction and the claimant brought an action for specific performance. The defendant counterclaimed for rescission of the contract. The Court of Appeal dismissed the claimant's claim for specific performance and ordered the contract to be rescinded. It would have been inequitable to allow the claimant specifically to enforce a contract which had been induced by a misrepresentation, albeit an innocent one, even though the defendant had not taken the opportunity presented to him to check the accuracy of the claimant's statements.

Since the court's discretion to grant or refuse the remedy of specific performance cannot be fettered by agreement between the parties, any attempt by them in their contract to exclude the 'clean hands' principle will not be binding on the court. If the law were otherwise, the function of the court would be reduced to that of a rubber stamp (*Quadrant Visual Communications Ltd* v *Hutchison Telephone (UK) Ltd* (1991) [1993] BCLC 442, CA).

9.3.2.8 Where it would cause great hardship to the defendant

Specific performance as an equitable remedy is based on the desire to achieve fairness on both sides. It is not meant to be a punishment to the defendant, and so it will not be decreed if it would involve the defendant in doing something unlawful, or something which he has no power to do, or would otherwise inflict great hardship on him. In *Denne* v *Light* (1857) 8 De GM & G 774, specific performance of a contract to buy agricultural land was refused to the vendor on the ground that the land was completely surrounded by land belonging to other people over which there was no right of cartway for the purchaser.

In *Wroth* v *Tyler* [1974] Ch 30, the defendant contracted to sell his bungalow to the claimants. The defendant's title was registered as encumbered only by a building society mortgage but, on the day after exchange of contracts and unknown to the defendant, the defendant's wife entered in the charges register a notice of her rights of occupation under what was then the Matrimonial Homes Act 1967. She refused to withdraw her notice to enable the defendant to complete the sale with vacant possession and he withdrew from the contract.

The claimants sought specific performance, but it was refused by Megarry J on the ground that it would be highly unreasonable to compel the defendant to take legal proceedings against his own wife for an order terminating her rights of occupation under the 1967 Act. The claimants were instead awarded equitable damages under the Chancery Amendment Act 1858 (Lord Cairns's Act) (now s. 50 of the Supreme Court

Act 1981) based on the difference between the contract price and the market value of the bungalow at the date of judgment.

The hardship need not have been caused by the claimant or be connected with the subject-matter of the contract. Thus, in the extreme circumstances of *Patel* v *Ali* [1984] 1 All ER 978 the claimants were refused specific performance of a contract to sell a house where, during a long delay (which was not the fault of any of the parties) following the making of the contract, the defendant had had a leg amputated due to bone cancer and had given birth to two children. Specific performance was refused because of the hardship the defendant would suffer if she had to leave the house and move to an area where she would be without the friends and relations on whose daily assistance she relied to look after her home and children. The claimants were left instead to pursue their remedy in damages.

9.3.2.9 Where it would seriously prejudice the rights of a third party

In *Watts* v *Spence* [1976] Ch 165, the defendant and his wife were joint owners of the matrimonial home. The defendant wished to sell the house and began negotiations with the claimant, who was interested in buying it. The defendant's wife knew nothing of this and had not consented to the sale of her share. The defendant, thinking he could persuade his wife to join in the sale by the date fixed for completion, represented to the claimant that he (the defendant) was the sole owner. This induced the claimant to enter into a contract with the defendant to buy the house. On discovering what had happened, the defendant's wife refused to consent to the sale.

It was held that it would be unreasonable to grant specific performance against the defendant because a third party with an interest in the property (the defendant's wife) would be seriously prejudiced thereby. Instead, the claimant recovered damages for misrepresentation under s. 2(1) of the Misrepresentation Act 1967.

9.3.3 Damages in lieu of specific performance

When, before 1858, the Court of Chancery refused specific performance of a contract it normally had no power to award damages for the breach. The disappointed claimant was left to begin another action—this time in a common law court—to recover damages.

The law was changed by the Chancery Amendment Act 1858 (Lord Cairns's Act), s. 2, which gave the court a discretion to award damages in substitution for, or in addition to, specific performance. (It should be noted that damages will be awarded *in addition to* specific performance only where there has been some special damage (*Jacques* v *Millar* (1877) 6 ChD 153, in which the claimant, in addition to specific performance, was awarded damages of £250 in respect of loss of business profits).)

The 1858 Act itself was repealed in 1883, but the power it had conferred in s. 2 survived by virtue of the general provision made for the powers of the Supreme Court in the Judicature Act 1873 (later repealed and replaced by the Supreme Court of Judicature (Consolidation) Act 1925).

Despite the repeal in 1883, the 1858 Act has often been referred to as though it were still in force. The provision relating to damages in lieu of specific performance has been explicitly re-enacted in s. 50 of the Supreme Court Act 1981. The importance of the power is that a claimant whose claim for specific performance is rejected on some discretionary ground may, in the same action, be awarded damages instead.

Such damages are called *equitable damages* and must be distinguished from the *common law damages* discussed in para 9.2 above. Common law damages are available as of right to any claimant who establishes the breach of a binding contract. There is no such entitlement to equitable damages, which are awarded only on a discretionary basis and only under the jurisdiction originally conferred by the Chancery Amendment Act 1858. There is, however, no difference in the *measure* of common law and equitable damages. Damages at common law and in equity are measured in the same way in respect of both the basis and the date of assessment (*Johnson* v *Agnew* [1979] 1 All ER 883, HL).

It would appear that the court only has power to award equitable damages for breach of a contract which is specifically enforceable. If there is no jurisdiction to grant specific performance of a particular contract, then there is no power to substitute equitable damages and the claimant will be left to his common law remedy. Thus, equitable damages cannot be awarded in lieu of specific performance for breach of a contract for personal work or services (*Scott* v *Rayment* (1868) LR 7 Eq 112), or of a contract to sell or buy ordinary commercial goods (*Price* v *Strange* [1978] Ch 337, CA, *per* Buckley LJ at p. 369), or of a contract whose subject-matter no longer exists (*Hipgrave* v *Case* (1885) 28 ChD 356, CA; *Ferguson* v *Wilson* (1866) LR 2 Ch App 77).

These contracts are not specifically enforceable, and so the power to award equitable damages in substitution for specific performance cannot be exercised. If, however, the court had jurisdiction to grant specific performance when the proceedings were begun, it retains the power to substitute equitable damages, even though, by the date of judgment, specific performance has become impossible (*Cory* v *Thames Iron Works & Shipbuilding Co. Ltd* (1863) 8 LT 237).

If a claimant claims specific performance of a contract that is specifically enforceable, the court can substitute equitable damages whether or not they have been asked for in terms by the claimant (*Crawford* v *Hornsea Steam Brick & Tile Co. Ltd* (1876) 45 LJ Ch 432, CA). On the other hand, if a claimant does not claim specific performance it would seem that equitable damages cannot be awarded because there is nothing for which they can be a substitute (*Horsler* v *Zorro* [1975] Ch 302).

In most cases the claimant will gain no advantage in seeking equitable damages because of the availability, as of right, of common law damages. Since the Judicature Act 1873, all divisions of the High Court have had power to grant both common law and equitable relief in the same action. This means that if a court has no jurisdiction to decree specific performance in a particular case, or refuses to decree it in the exercise of its discretion, the court may nevertheless award *common law damages* (*Johnson* v *Agnew* [1979] 1 All ER 883, HL).

Sometimes, however, there is an advantage in seeking equitable as opposed to common law damages. There are some circumstances in which common law damages are not available because the claimant has no cause of action at common law. Yet, in the same circumstances, there may be a cause of action in equity justifying the exercise of the court's discretion to award equitable damages.

For instance, if a defendant commits an anticipatory breach of contract, by saying *now* that he will not perform when the time for performance comes round, the claimant cannot be awarded damages at common law if he refuses to accept the breach. He may, however, sue immediately for specific performance (*Hasham* v *Zenab* [1960] AC 316, PC), and the court may then, if it wishes, award equitable damages in substitution (*Oakacre Ltd* v *Claire Cleaners (Holdings) Ltd* [1981] 3 All ER 667, following a dictum of Turner LJ in *Phelps* v *Prothero* (1855) 7 De GM & G 722 at p. 733).

9.4 Injunctions

9.4.1 Introduction

The injunction is an equitable remedy and is, therefore, discretionary. It is a particularly useful remedy in contract to restrain a breach, in tort to prevent the continuation of a nuisance, in family law to control domestic violence, and in administrative law to prevent public authorities from acting unlawfully. Breach of an injunction is a contempt of court.

The power to grant injunctions, which was originally an inherent power, is regulated by statute. By s. 37 of the Supreme Court Act 1981, the High Court can grant an interlocutory or a perpetual injunction (either unconditionally or on terms) 'in all cases in which it appears to the court to be just and convenient to do so'.

The county courts can grant injunctions in actions within their jurisdiction by virtue of a statutory power to make any order which could be made in a similar case before the High Court (County Courts Act 1984, s. 38, as substituted by the Courts and Legal Services Act 1990). There are, however, strict limits on the power of the county courts to grant search orders (*Anton Piller* orders) and freezing injunctions (*Mareva* injunctions) (County Court Remedies Regulations 1991 (SI 1991, No. 1222); see further, para 1.6.3.1).

A pre-existing cause of action at the suit of the claimant against the defendant is a prerequisite to an application for an injunction under s. 37 of the Supreme Court Act 1981 (*The Siskina* [1979] AC 210, HL, endorsed in *Mercedes-Benz AG* v *Leiduck* [1996] 1 AC 284, PC; but note *Channel Tunnel Group Ltd* v *Balfour Beatty Construction Ltd* [1993] 1 All ER 664, HL, where it was suggested (*per* Lords Goff, Browne-Wilkinson, and Mustill at pp. 668, 670, and 685–6, respectively) that the case law may have unduly restricted the unfettered wording of s. 37).

Section 37 of the Supreme Court Act 1981, above, confers a very wide discretion to grant injunctive relief. That discretion must, however, be exercised judicially. As Jessel MR once put it (*Beddow* v *Beddow* (1878) ChD 89 at p. 93, commenting on s. 25 of the Judicature Act 1873, one of the predecessors of s. 37 of the Supreme Court Act 1981):

. . . I have unlimited power to grant an injunction in any case where it would be right or just to do so; and what is right or just must be decided, not by the caprice of the Judge, but according to sufficient legal reasons or on settled legal principles.

The wording of s. 37 authorises the granting of all the injunctions mentioned in the succeeding paragraphs, including search orders and freezing injunctions. It is also wide enough to enable the court to grant an injunction freezing money in a person's bank account where the court is satisfied that the money has been obtained from another in breach of the criminal law (*Chief Constable of Kent* v *V* [1983] QB 34, CA). An injunction will not be granted, however, where the tainted money has been mixed in the account with untainted money and it is not possible to separate the two (*Chief Constable of Hampshire* v *A* [1985] QB 132, CA), or where the money in issue cannot be identified as having been obtained in breach of the criminal law (*Chief Constable of Leicestershire* v *M* [1988] 3 All ER 1015).

The requirement of *The Siskina* [1979] AC 210, HL, above, that there must be a pre-existing cause of action to which an injunction can be attached appears to apply only in ordinary disputes concerning the alleged violation of *private* rights. As explained in *Morris* v *Murjani* [1996] 2 All ER 384, CA, *Chief Constable of Kent* v *V*, above, in which no pre-existing cause of action existed, was a case where the Chief Constable was acting under a general *public* duty to recover stolen property and restore it to the true owner.

In *Morris* v *Murjani* itself, it was held that a trustee in bankruptcy was entitled to an injunction restraining a bankrupt from leaving the country on the ground that the bankrupt would otherwise not comply with his duty under s. 333 of the Insolvency Act 1986 to 'give to the trustee such information ... as the trustee may for the purposes of carrying out his functions ... reasonably require'. This section does not expressly create a cause of action. Nevertheless, and *The Siskina* notwithstanding, the Court of Appeal decided that injunctive relief was available for violation of the section because it imposed a type of public duty on a bankrupt which was designed to help the trustee in bankruptcy to perform his statutory duties.

9.4.2 Types of injunctions

9.4.2.1 Prohibitory and mandatory injunctions

A *prohibitory* injunction is an order of the court to restrain the doing, continuance, or repetition of some wrongful act. Most injunctions issued by the courts are of this type.

A *mandatory* injunction is an order of the court to undo some wrongful act; for example, to pull down buildings which interfere with an established right to light and which constitute, therefore, a legal nuisance.

There are obvious similarities between a mandatory injunction and specific performance in that both require a positive act to be performed by the defendant. Specific performance, however, is available as a remedy only if there is a *contract* to enforce whereas, in most cases, the use of the mandatory injunction is confined to actions arising out of *tort*. However, there is no reason why a mandatory injunction should not be used to order the defendant to undo what he has already done in breach of a contract (*Lord Manners* v *Johnson* (1875) 1 ChD 673).

A mandatory injunction will not usually be issued where damages would be an adequate remedy, or where it would involve constant supervision by the court.

The reasonableness of the defendant's conduct is a relevant factor in deciding whether to order a mandatory injunction, as is the cost to the defendant of complying with it (*Redland Bricks Ltd* v *Morris* [1970] AC 652, HL, *per* Lord Upjohn at p. 665). The cost to the defendant is not such an important factor where he 'has tried to steal a march on' the claimant (ibid.).

If the court decides to grant a mandatory injunction, care must be taken to ensure that the defendant knows exactly what he has to do in order to comply with its terms (ibid.).

9.4.2.2 Interim and perpetual injunctions

9.4.2.2.1 *Introduction* An *interim* (formerly *interlocutory*) injunction can be granted before the trial of the action on the merits between the parties in order to preserve the *status quo*. It may last until the issue between the parties is finally decided. The object of an interim injunction is to prevent violation of the claimant's rights in a way which he could not be adequately compensated for in damages if his action succeeded at the full trial.

Quite often the parties treat the interim application as the trial of the action if their respective cases are ready for presentation. If the case is one of great urgency the claimant can obtain an interim injunction *ex parte* in the absence of the defendant. The injunction will operate until there can be a hearing with the defendant present a few days later.

A *perpetual* injunction may be granted after the claimant has proved his case at the actual trial of the action. It is called 'perpetual' because it is granted after the final determination of the parties' rights and not because it will necessarily operate for ever. It may, however, be expressed to last for ever.

9.4.2.2.2 *Interim injunctions* The decision whether to grant an interim injunction is usually a difficult one to make, involving a balancing of the parties' interests. The proceedings are conducted on affidavits only, without the benefit of oral evidence. To some extent the court must try to predict what is likely to happen later at the trial of the action because an interim injunction should not be granted on facts which would not justify the issue of a perpetual injunction.

The court will rarely grant a *mandatory* injunction on an interim application (*Locabail International Finance Ltd* v *Agroexport* [1986] 1 All ER 901, CA) because to do so could be premature and oppressive and, in a contract case, would be tantamount to decreeing specific performance. Nevertheless, the court has done so in the past in cases where the defendant has tried to steal a march by hurrying on with building work after being served with notice of the proceedings (*Daniel* v *Ferguson* [1891] 2 Ch 27, CA; *Von Joel* v *Hornsey* [1895] 2 Ch 774, CA), or where the circumstances are so exceptional that to refuse the interim mandatory injunction would produce a greater risk of injustice than to grant it (*Films Rover International Ltd* v *Cannon Film Sales Ltd* [1986] 3 All ER 772).

If a claimant is granted an interim injunction he must give an *undertaking in damages*. This is an undertaking to pay damages to the defendant if, at the trial, it turns out that the injunction was wrongly granted, either because at the trial the claimant cannot prove the case he alleged in the interim proceedings, or because the court, when it granted the interim relief, took a wrong view of the law (*Griffith* v *Blake*

(1884) 27 ChD 474, CA). These damages are to compensate the defendant for any loss he has sustained by being prevented from doing what he wanted to do. The claimant's undertaking is given to the court. An undertaking in damages will not necessarily be required where the claimant is the Crown, and the purpose of seeking the injunction is to enforce the law against the defendant (*F. Hoffman-La Roche & Co. AG* v *Secretary of State for Trade and Industry* [1975] AC 295, HL; *Director General of Fair Trading* v *Tobyward Ltd* [1989] 2 All ER 266).

This special privilege enjoyed by the Crown extends also to other public authorities in their role as law enforcers. It follows, therefore, that a local authority is not necessarily required to give an undertaking in damages if it wishes to obtain an interim injunction restraining a retailer from trading on Sundays (*Kirklees Metropolitan Borough Council* v *Wickes Building Supplies Ltd* [1992] 3 All ER 717, HL). Similarly, it has been held that a financial services regulator, acting in its role as a law enforcer on behalf of the Secretary of State for Trade and Industry, need not necessarily give an undertaking in damages when seeking a freezing injunction in support of claims for compensation against a defendant alleged to have infringed the financial services legislation (*Securities and Investments Board* v *Lloyd-Wright* [1993] 4 All ER 210, applying *F. Hoffman-La Roche & Co. AG* v *Secretary of State for Trade and Industry*, above, and *Kirklees Metropolitan Borough Council* v *Wickes Building Supplies Ltd*, above).

Whether an undertaking in damages is required from a public authority in any particular case is a decision for the court in the exercise of its discretion (*Kirklees Metropolitan Borough Council* v *Wickes Building Supplies Ltd*, above).

The principles to be taken into account by the court in deciding whether to grant an interim injunction are those laid down in 1975 by the House of Lords in *American Cyanamid Co.* v *Ethicon Ltd* [1975] AC 396, HL. Those principles have, however, been subjected to a process of refinement in recent years.

It is important to remember that the sole source of the jurisdiction of the High Court to grant injunctions is s. 37 of the Supreme Court Act 1981 (para 9.4.1 above), which confers a wide discretion and contemplates taking into account all the circumstances of a case, and that any 'principles' set out in the case law are merely guidelines which are not necessarily of universal application (*Factortame Ltd* v *Secretary of State for Transport (No. 2)* [1991] 1 All ER 70, HL, *per* Lord Goff at p. 118; *Lansing Linde Ltd* v *Kerr* [1991] 1 All ER 418, CA, *per* Butler-Sloss LJ at p. 434; in *Cambridge Nutrition Ltd* v *British Broadcasting Corporation* [1990] 3 All ER 523, CA, Kerr LJ said (at p. 535) that the '*American Cyanamid* case is no more than a set of useful guidelines which apply in many cases. It must never be used as a rule of thumb, let alone as a strait-jacket').

The *American Cyanamid* principles are as follows.

(a) The court must normally be satisfied that there is 'a serious question to be tried' between the parties, and that the claimant's claim is not frivolous or vexatious. In the event that there is no serious question to be tried, or if the claim is frivolous or vexatious, an interim injunction will be refused.

   In at least one exceptional case an interim injunction may be granted even though there is no serious question to be tried. It is established that a landowner whose title is not at issue is prima facie entitled to an injunction to restrain a trespass to his land, whether or not the trespass is harmful, subject to any defence that may be put forward by the defendant. In these circumstances, if there is no serious question to be tried between the parties because the evidence before the court shows that the defendant has no valid defence to the trespass claim, then the issues (mentioned below) of the balance of convenience, and the adequacy of damages as a remedy, do not arise for consideration, and the claimant may be granted his interim injunction (*Patel* v *WH Smith (Eziot) Ltd* [1987] 2 All ER 569, CA).

(b) If there is a serious question to be tried, the court will go on to consider 'the balance of convenience'. Whatever happens later at the trial of the action, one party will suffer a present disadvantage because of the grant or refusal of interim relief. With the interests of the parties and the public in mind, the court must consider the extent of the inconvenience likely to be suffered by each side. For example, in *Laws* v *Florinplace Ltd* [1981] 1 All ER 659, the balance of convenience was held to lie in favour of the claimants and, accordingly, an interim injunction was granted in the claimants' nuisance action restraining the defendants from carrying on the business of a sex shop.

In some situations, the existence of a serious question to be tried, and the fact that the balance of convenience lies in granting an injunction, will not necessarily be conclusive. If neither party is interested in damages, and the court's decision on the application for an interim injunction would be equivalent to giving final judgment in the dispute between the parties (for example, where timing is crucial because the case concerns the transmission of a broadcast or the publication of a newspaper article), the court should be prepared to assess the relative strength of the parties' cases before deciding whether the injunction should be granted (*Cambridge Nutrition Ltd* v *British Broadcasting Corporation* [1990] 3 All ER 523, CA, where, in the claimants' breach of contract action, an interim injunction restraining the transmission by the defendants of a television programme about very low calorie diets was refused; see also *Lansing Linde Ltd* v *Kerr* [1991] 1 All ER 418, CA, but cf *Lawrence David Ltd* v *Ashton* [1991] 1 All ER 385, CA).

The fact that an injunction would interfere with the public interest in freedom of speech is also a relevant consideration (*Cambridge Nutrition Ltd* v *British Broadcasting Corporation*, CA, above, *per* Kerr and Ralph Gibson LJJ at pp. 536 and 541, respectively; *Femis-Bank (Anguilla) Ltd* v *Lazar* [1991] 2 All ER 865; *Douglas* v *Hello! Ltd* [2001] 2 WLR 992, CA).

(c) If any damages awarded later to the claimant at the trial of the action would be an adequate remedy, an interim injunction should normally be refused.

Where, however, damages would not adequately compensate the claimant, or where it is doubtful whether a remedy in damages exists at all, the court may be prepared to take a wider view of the 'balance of convenience' and grant an interim injunction (*Factortame Ltd* v *Secretary of State for Transport (No. 2)* [1991] 1 All ER 70, HL, a most exceptional case involving law enforcement by a public authority in which, pending final judgment, the balance of convenience was held to lie in favour of granting an interim injunction against the Secretary of State restraining him from withholding, or withdrawing, registration in the register of British fishing vessels in respect of certain fishing boats owned by British companies whose directors and shareholders were mainly Spanish nationals).

(d) If damages would not be an adequate remedy for the claimant, an interim injunction may be granted if the claimant's undertaking in damages would give the defendant an adequate remedy should the claimant fail at the trial.

Where, however, the defendant would not be adequately compensated by damages the court may refuse an interim injunction (*Cambridge Nutrition Ltd v British Broadcasting Corporation* [1990] 3 All ER 523, CA, above, in which one of the reasons for refusing an injunction was that damages could not provide adequate compensation if the claimant's claim failed at trial; while the defendants would be able to recover their lost production costs, they would not be compensated for loss of their right to transmit a television programme on a topic of public interest in the form and at the time of their choice).

(e) If the inconvenience likely to be suffered by each party would not be widely different, the balance of convenience may be adjudged from the relative strengths of the parties' cases as revealed in the affidavits. But the claimant does not have to make out a prima facie case in order to obtain an interim injunction. At the interim stage the court will not embark on anything resembling a trial of the evidence on conflicting affidavits in order to evaluate the strength of either party's case.

Nor is it appropriate at that stage for the court to decide difficult questions of law which call for detailed argument and mature consideration. However, it may be appropriate at the interim stage to decide less difficult questions of law, especially if that course would finally and quickly determine the issue between the parties (*Factortame Ltd* v *Secretary of State for Transport (No. 2)* [1991] 1 All ER 70, HL, *per* Lord Jauncey at p. 123).

(f) In the particular circumstances of individual cases there may be many other special factors to be taken into account in addition to those already mentioned (see, e.g., the cases cited above: *Cambridge Nutrition Ltd* v *British Broadcasting Corporation*, CA; *Lansing Linde Ltd* v *Kerr*, CA; *Femis-Bank (Anguilla) Ltd* v *Lazar*; *Factortame Ltd* v *Secretary of State for Transport (No. 2)*, HL; *Douglas* v *Hello! Ltd*, CA).

In *R* v *Secretary of State for Health (ex parte Imperial Tobacco Ltd)* [2001] 1 WLR 127, HL, the question was raised whether domestic law or Community law applies when considering the grant of an interim injunction to restrain a minister from making regulations to implement a directive which is suspected to be invalid. If domestic law applies, then the *American Cyanamid* principles are relevant. If Community law governs the matter, the applicant will have to show that he has a strong case on the merits and that serious and irreparable damage would result if the injunction were not granted, and it appears that in Community law financial damage is not, in principle, regarded as 'irreparable' (*Zuckerfabrik Süderdithmarschen AG* v *Hauptzollampt Itzehoe* [1991] ECR I–415, CJEC).

The House of Lords held unanimously in the *Imperial Tobacco* case that the art. 234 reference procedure was not available since the offending directive had been held invalid by the European Court in the interval between the date of the hearing and the date of judgment in the *Imperial Tobacco* case. It was, therefore, not 'necessary' to decide any question in order to give judgment in the case before it. However, their Lordships were divided on the issue whether, had the procedure been available, a reference would have been necessary. The majority thought that it would, since in their view the applicable law is Community law and not domestic law.

It is not, in any event, entirely clear whether, and, if so, to what extent, domestic and Community law principles differ on this subject. The question was barely touched on in the *Imperial Tobacco* case since it was not necessary to decide it. Lord Slynn ([2001] 1 WLR 127 at p. 132) thought that in many respects the two sets of principles overlap—'but there may be differences, e.g., as to how far financial damage can be taken into account'.

An appellate court has only a limited function when reviewing the exercise of the judge's discretion to grant or refuse an interim injunction. The exercise of discretion cannot be interfered with, for example, merely because the members of the appellate court would have exercised the discretion differently if they had been sitting at first instance. It must be shown that the judge has misunderstood the law or the evidence before him, or that there has been a change of circumstances since the decision at first instance, or that the judge's decision is plainly wrong (*Hadmor Productions Ltd* v *Hamilton* [1982] 1 All ER 1042, HL, *per* Lord Diplock at p. 1046, with whom the other Law Lords expressly agreed; *Dimbleby & Sons Ltd* v *National Union of Journalists* [1984] 1 All ER 751, HL).

9.4.2.2.3 *Perpetual injunctions* The jurisdiction of a court of equity to grant a *perpetual* injunction is discretionary, but, as in the case of specific performance, the discretion is exercised judicially in accordance with rules laid down in decided cases. A perpetual injunction will not normally be granted if damages would sufficiently compensate the claimant (*Wood* v *Sutcliffe* (1851) 2 Sim NS 163), or if it would require constant supervision on the part of the court (*Attorney-General* v *Colney Hatch Lunatic Asylum* (1868) LR 4 Ch App 146). It may also be refused on account of the claimant's own inequitable conduct (*Measures Bros Ltd* v *Measures* [1910] 2 Ch 248, CA).

Although the grant of an injunction is a matter of discretion, the court will rarely refuse an injunction to prevent the breach of a negative contractual term—even where damages would be an adequate remedy and the contract is one for personal work or services. On the facts of *Lumley* v *Wagner* (1852) 1 De GM & G 604 (para 9.3.2.3 above), specific performance was refused. But the contractual obligation which the defendant had broken was expressed in negative terms, and the court granted an injunction to restrain her from singing elsewhere than in the claimant's theatre during the currency of the contract.

The court will not grant an injunction to achieve a purpose which could not be achieved by a decree of specific performance (but see the exceptional case of *Sky Petroleum Ltd* v *VIP Petroleum Ltd* [1974] 1 All ER 954, para 9.3.2.1 above). In *Page One Records Ltd* v *Britton* [1967] 3 All ER 822, the claimants, who were the managers of 'The Troggs' pop group, applied for an injunction to restrain the group from engaging another person as their manager in breach of contract. Stamp J refused to grant the injunction as its effect would have been the same as specific performance. It would have compelled the group to continue to employ the claimants as their managers, and in a situation where they had lost all confidence in the claimants. (This case was applied in *Warren* v *Mendy* [1989] 3 All ER 103, CA (boxer and manager); and see *Scandinavian Trading Tanker Co.* v *Flota Petrolera Ecuatoriana* [1983] 2 All ER 763, HL *per* Lord Diplock at p. 766.)

However, there is no objection to granting an injunction to restrain the breach of a negative contractual term where its effect will be to encourage, but not force, the defendant to perform the contract. Thus, in *Lumley* v *Wagner* (1852) 1 De GM & G 604, above, Lord St Leonards LC was aware that his order might tempt the defendant to fulfil her contract but he denied that it would compel her to do so.

In *Warner Brothers Pictures Inc.* v *Nelson* [1937] 1 KB 209, the defendant (Bette Davis) had an 'exclusive' contract with the claimants by which she agreed to act for them. She further agreed not to act for any third party, nor to engage in any other occupation, during the currency of the contract, without the claimants' written consent.

A dispute arose between the parties over money, and the defendant came to Britain to make a film in breach of contract. Branson J granted an injunction against her but limited it to stage and film activities. He said the injunction might 'tempt' her but would not 'drive' her to perform her contract with Warner Brothers because she could earn her living, although not as remuneratively, otherwise than by acting. The injunction was also limited to preventing breaches of the contract within the jurisdiction of the court, that is, England and Wales.

9.4.2.2.4 *Damages in lieu of an injunction* By virtue of s. 50 of the Supreme Court Act 1981 (para 9.3.3 above), the Court of Appeal or the High Court may award damages to a claimant either in substitution for, or in addition to, an injunction.

These damages are awarded in equity, and may be available when common law damages are not. In tort, for example, common law damages cannot be ordered for an injury which has not yet been inflicted but is only threatened or apprehended. However, an injunction *can* be ordered in such a situation, and, in the exercise of its equitable jurisdiction, the court may substitute damages for that injunction in a proper case.

Accordingly, in *Wrotham Park Estate Co. Ltd* v *Parkside Homes Ltd* [1974] 1 WLR 798 damages were awarded in lieu of a mandatory injunction in respect of the breach of a restrictive covenant; in *Bracewell* v *Appleby* [1975] Ch 408, damages were awarded instead of an injunction in respect of a trespass; in *Lyme Valley Squash Club Ltd* v *Newcastle under Lyme Borough Council* [1985] 2 All ER 405, damages were awarded instead of an injunction in relation to an interference with the claimant's right to light; and in *Jaggard* v *Sawyer* [1995] 2 All ER 189, CA, damages were awarded in lieu of an injunction in respect of both a continuing trespass and the breach of a restrictive covenant.

If the court grants damages in lieu of an injunction, the defendant is able to carry on as before with what he was doing on payment of a sum of money to the claimant. The court must take care that the substitution of damages for an injunction, and especially a *quia timet* injunction, does not become simply a device which allows the defendant to buy a licence to commit a wrong (*Shelfer* v *City of London Electric Lighting Co.* [1895] 1 Ch 287, CA; *Leeds Industrial Co-operative Society* v *Slack* [1924] AC 851, HL, *per* Lord Sumner, dissenting, at pp. 867–8 and 872).

In *Miller* v *Jackson* [1977] QB 966, CA, a majority of the Court of Appeal held that the activities of a village cricket club, which had been established for over seventy years, were a legal nuisance to the claimants by reason of cricket balls landing in their garden. It was also decided by a majority that an injunction should not be granted. Instead, damages of £400 were awarded in respect of past and future damage. The majority, clearly not wishing to forbid the game on this particular cricket ground, said that the case was one where the public interest should prevail over the private interest. (Lord Denning dissented on liability. On the question of the appropriate remedy, Geoffrey Lane LJ would have granted an injunction.)

In *Kennaway* v *Thompson* [1981] QB 88, CA, the trial judge held that the defendant club's motor-boat races on a lake were a nuisance to the claimant, who lived nearby. But he refused to grant an injunction on the ground that it would spoil the enjoyment of the large numbers of people who attended the races. Instead, he awarded damages of £15,000 under the equitable jurisdiction in order to compensate the claimant for damage likely to be suffered in the future.

The Court of Appeal set aside the judge's order and granted an injunction restricting the defendants' racing activities in each year and limiting the noise levels of boats using the lake at other times. The court said the claimant was entitled to an injunction despite the public interest, and tacitly disapproved of the refusal of an injunction by a differently constituted court in *Miller* v *Jackson*, CA, above, a case which, it was said, was not consistent with earlier Court of Appeal authority.

That authority was the case of *Shelfer* v *City of London Electric Lighting Co.* [1895] 1 Ch 287, CA, in which the Court of Appeal granted an injunction restraining the defendants from causing excessive vibration in spite of the fact that the defendants were performing a public service and the injunction might have deprived a section of the public of its electricity supply. It was said that the court should not allow itself to be used, by too often substituting damages for an injunction, as a tribunal for legalising unlawful acts. In addition, the fact that the defendant is a public benefactor was said not to be a sufficient reason for refusing an injunction and awarding damages instead. *Shelfer's* case was completely ignored in the judgments of two of the three members of the Court of Appeal in *Miller* v *Jackson*, CA, above.

Kennaway v *Thompson*, CA, above, was distinguished on the facts in *Tetley* v *Chitty* [1986] 1 All ER 663, where an injunction was granted to *stop altogether* (and not merely to restrict) noisy go-karting activities which were a nuisance to the claimants. Damages were also awarded as compensation for past inconvenience.

9.4.2.3 *Quia timet* injunctions

A court of equity can grant a *quia timet* ('because he fears') injunction where a claimant's rights have not yet been infringed but they are being threatened.

However, such an injunction is not commonly granted and a claimant must make out a strong case to obtain one. He must show that there is an imminent danger of very substantial damage, or further damage, occurring (*Crowder* v *Tinkler* (1816)

19 Ves Jr 617, *per* Lord Eldon LC at p. 622; *Attorney-General* v *Nottingham Corporation* [1904] 1 Ch 673).

If the court refuses a *quia timet* injunction, this does not prejudice the claimant's right to apply later for an interim or a perpetual injunction when the damage does occur.

9.4.2.4 *Ex parte* ('without notice') injunctions

A 'without notice' injunction is an interim order which is obtained by one party without notice to, and in the absence of, the other party. The court must be satisfied that the case is one of urgency; for example, that the delay in using the normal procedure would cause irreparable damage. Such an injunction only operates until both parties can be heard.

9.4.2.5 Search orders (*Anton Piller* orders)

9.4.2.5.1 *Introduction* A search order is an interim mandatory injunction of comparatively recent origin (see Denning, *The Due Process of Law*, 1980, pp. 123–30).

It is obtained without notice to the defendant so as to catch him off guard since its object is to prevent the defendant from removing, concealing, or destroying evidence in the form of documents or movable property (see *Crest Homes plc* v *Marks* [1987] 2 All ER 1074, HL, *per* Lord Oliver at pp. 1078–81).

A search order was originally called an *Anton Piller* order after the first case in which its use was sanctioned by the Court of Appeal—*Anton Piller KG* v *Manufacturing Processes Ltd* [1976] Ch 55, CA. The first reported case in the High Court appears to have been *EMI Ltd* v *Pandit* [1975] 1 All ER 418 in which Templeman J made an order and, in doing so, purported to follow a House of Lords' decision of 1821—*United Company of Merchants of England* v *Kynaston* (1821) 3 Bli 153. Before the *Anton Piller* case there were also a number of unreported High Court decisions (see *Anton Piller, per* Lord Denning MR at p. 58).

A search order will normally be obtained *before* final judgment in the action between the parties. It may, however, be granted *after* judgment to elicit documents which are essential to the claimant's efforts to execute the judgment (*Distributori Automatici Italia SpA* v *Holford General Trading Co. Ltd* [1985] 3 All ER 750).

Failure to comply with the order, either by refusing access or by destroying the evidence, is a contempt of court. (But note *Bhimji* v *Chatwani* [1991] 1 All ER 705—it is not contempt where the defendant, acting on legal advice, applies to the court for discharge or variation of a search order, and refuses access pending the hearing of the application, provided he makes a reasonable offer to protect the relevant documents in the meantime and there is no evidence of impropriety on his part.)

9.4.2.5.2 *Safeguards* It has been stressed that the order is not a search warrant authorising the claimant to enter the defendant's premises against the defendant's will (*Anton Piller* [1976] Ch 55, CA, *per* Lord Denning MR at p. 60). It is an order against the defendant *in personam* to permit the claimant to enter premises, or, if he refuses permission, to be in contempt of court (*Anton Piller per* Ormrod LJ at p. 62).

The courts are alive to the fact that the *ex parte* and draconian nature of a search order makes it a powerful weapon in the hands of claimants. To prevent the order from becoming too oppressive, it should be drawn so as to extend no further than is necessary to achieve the purpose for which it was sought, namely, to preserve evidence which might otherwise be concealed or destroyed. Thus, although documents may be retained for a short period for the purpose of taking copies, they should be returned to the defendant promptly. Furthermore, during the execution of a search order no material must be removed (whether with or without the defendant's consent) unless it is clearly covered by the terms of the order (*Columbia Picture Industries Inc.* v *Robinson* [1986] 3 All ER 338, where the defendant was awarded damages of £7,500 for the 'excessive and oppressive manner' in which a search order had been executed by the claimants' solicitors).

Since search orders have implications for civil liberties, the government decided that it was desirable that they should have the express approval of Parliament. Accordingly, they were placed on a statutory footing by s. 7 of the Civil Procedure Act 1997.

An order made by the High Court under s. 7 can direct that any person described in the order be permitted to enter premises (including any vehicle) in England and Wales, and to search for, or inspect, anything described in the order, and to make or obtain a copy, photograph, sample, or other record of anything so described. An order can also direct that any information or article described in the order be provided and that anything described in the order be retained for safe keeping.

Technically, s. 7 of the Civil Procedure Act 1997 *confers* jurisdiction on the High Court to grant search orders. In practice, it simply *confirms* what has been happening since the mid–1970s without imposing any new limits.

Concern that abuses of procedure might result in search orders falling into disrepute prompted Sir Donald Nicholls V-C to suggest, in *Universal Thermosensors Ltd* v *Hibben* [1992] 3 All ER 257 at p. 276, that stricter safeguards in the execution of a search order were required for the protection of the defendant's rights.

The *Universal Thermosensors* case provides a good example of a very badly executed search order. One of the defendants, alone in the house with her children in bed, had been called to the door in her nightclothes at 7.15 a.m. to be told by a strange man that he had a court order requiring her to permit him to enter her house, and that she was entitled to seek legal advice forthwith but otherwise she was not allowed to speak to anyone. No list was made of what documents were taken from which premises, and, at the defendants' business premises, a director of the claimants had been allowed to carry out a thorough search of all the documents of a competing company. The defendants were awarded £20,000 on the claimants' undertaking in damages.

Many of the procedural weaknesses identified in *Universal Thermosensors* (and earlier cases) were subsequently dealt with by *Practice Directions* issued in 1994 and 1997, which included new standard forms for use in search order cases. The matter is currently dealt with in the *Practice Direction* to CPR, Part 25, which lays down safeguards designed to prevent abuse.

9.4.2.5.3 *'Pirates'* and *'bootleggers'* The search order has been used often against so-called 'pirates' and 'bootleggers'. A 'pirate' includes a person who infringes the copyright in musical recordings and films by making and selling unauthorised copies, thus depriving the true owner of royalties.

Orders have been granted to protect the copyright in recorded Indian music (*EMI*

Ltd v *Pandit* [1975] 1 All ER 418), *Jaws* T-shirts (*Universal City Studios Inc.* v *Mukhtar & Sons* [1976] 2 All ER 330), and feature films which were being unlawfully repro- duced on video cassettes (*Rank Film Distributors Ltd* v *Video Information Centre* [1981] 2 All ER 76, HL).

A search order was made for the first time in a matrimonial case in *Emanuel* v *Emanuel* [1982] 2 All ER 342, in which the husband was ordered to permit the wife's solicitors to enter the husband's home to look for, inspect, photograph, and remove all documents relating to his earnings and capital.

A modern 'bootlegger' is a person who records a live musical performance from the audience, or from the radio, and then makes and sells unauthorised copies. The performers are thus deprived of royalties on the sale of recordings. But they have no copyright in the music they perform and no copyright in their own performance. It has been held that they are nevertheless entitled to the protection of a search order (*Ex parte Island Records Ltd and others* [1978] Ch 122, CA: *Lonrho Ltd* v *Shell Petroleum Co. Ltd (No. 2)* [1981] 2 All ER 456, HL).

A search order can be used to compel the disclosure of material information, such as the names and addresses of persons who have supplied, or been supplied with, illicit films and recordings. In *Rank Film Distributors Ltd* v *Video Information Centre*, above, the House of Lords held that a search order should not be made for the purpose of eliciting information from the defendant which would tend to incriminate him.

Their Lordships' decision would have had the effect of seriously diminishing the usefulness of the search order. It was quickly overruled by s. 72 of the Supreme Court Act 1981, which withdrew the privilege against incrimination in certain types of civil proceedings.

Section 72, as amended by the Civil Partnership Act 2004, provides that in civil proceedings in the High Court (and in any appeal thereon) for 'infringement of rights pertaining to any intellectual property', a person is not excused from answering questions under a search order or a freezing injunction, or from complying with any order made in those proceedings (such as an order for discovery of documents), on the ground that he might incriminate himself or his spouse or civil partner. At the same time, protection is given in any proceedings for a related criminal offence (such as conspiracy to defraud) in that the evidence obtained in the civil proceedings is not admissible in criminal proceedings, with the exception of proceedings for perjury or contempt of court (s. 72(3) and (4)).

It may sometimes happen that the disclosure of material information under a search order (such as the names and addresses of other persons involved in unlawful activity) will expose the defendant and his family to the risk of violence. That risk, although a relevant factor for the court in deciding whether to order disclosure, will not usually prevail over any pressing need the claimant has for the information, especially where the defendant is principally implicated in the unlawful conduct and is not merely a person who has become innocently involved in the wrongdoing of others. It is in the public interest that threats of violence should not prevent the exposure of unlawful activity, and a defendant in this situation may have to rely on police protection and the law of contempt if he is the recipient of such threats (*Coca-Cola Co.* v *Gilbey* [1995] 4 All ER 711).

9.4.2.6 Freezing injunctions (*Mareva* injunctions)

9.4.2.6.1 *Introduction* A freezing injunction is an interim injunction granted often on a 'without notice' application. Its object is to prevent the defendant from removing his assets out of the jurisdiction of the English courts (i.e., England and Wales), or from dissipating them within the jurisdiction. Breach of a freezing injunction by the defendant is a contempt of court.

Like the search order, the freezing injunction is a comparatively recent judicial development in English Law, being first used in 1975. It was originally called a *Mareva* injunction after one of the first cases in which it was used — *Mareva Compania Naviera SA* v *International Bulkcarriers SA* (1975) [1980] 1 All ER 213, CA, although *Nippon YK* v *Karageorgis* [1975] 3 All ER 282, CA, appears to have been the first reported decision in which the freezing type of injunction was granted.

The freezing injunction was invented in response to the situation where the claimant, having commenced proceedings against the defendant, believed there was a real danger that the defendant would remove his assets from the jurisdiction, or allow them to be dissipated, so that they would not be available to satisfy judgment should the claimant win the action. There was usually plenty of time for the defendant to act because of the delay of months, or even years, between the start of the proceedings and the date of judgment.

A freezing injunction will normally be obtained *before* final judgment in the action between the parties. It may, however, be granted *after* judgment to assist in execution of the judgment where there are grounds for believing that the defendant will dispose of his assets in order to frustrate execution (*Orwell Steel (Erection and Fabrication) Ltd* v *Asphalt and Tarmac (UK) Ltd* [1985] 3 All ER 747; *Babanaft International Co. SA* v *Bassatne* [1989] 1 All ER 433, CA).

A freezing injunction can be granted against any defendant and not just a foreign or foreign-based person or company (Supreme Court Act 1981, s. 37(3); this statutory provision was anticipated in *Prince Abdul Rahman* v *Abu-Taha* [1980] 3 All ER 409, CA, approving the decision of Sir Robert Megarry V-C in *Barclay-Johnson* v *Yuill* [1980] 3 All ER 190).

Although freezing injunctions have been granted mostly in commercial actions, their use extends to other causes of action. A freezing injunction has been granted, for instance, in a personal injury action (*Allen* v *Jambo Holdings Ltd* [1980] 2 All ER 502, CA), and in matrimonial proceedings (*Ghoth* v *Ghoth* [1992] 2 All ER 920, CA, below).

The right to apply for a freezing injunction is not a cause of action in itself. It cannot stand on its own. As in the case of other injunctions, the application must be ancillary or incidental to a pre-existing cause of action which the claimant has against the defendant (*The Siskina* [1979] AC 210, HL). A freezing injunction cannot be granted unless the claimant has a legal or equitable right which has been violated, or threatened with violation, by the defendant (*The Siskina*, above; the *Mareva* case, above; *Derby & Co. Ltd* v *Weldon (No. 2)* [1989] 1 All ER 1002, CA, *per* Neill LJ at p. 1019; *Zucker* v *Tyndall Holdings plc* [1993] 1 All ER 124, CA).

It follows from the rule requiring a pre-existing cause of action that, if no cause of action has yet arisen, the court has no jurisdiction to grant *conditional* relief in the form of a freezing injunction made subject to a condition that it does not come into effect unless and until the cause of action arises. This is so even though the court is presently satisfied on the evidence before it that the case is an appropriate one for ordering freezing-type relief when the cause of action does arise.

Although it is recognised that the denial of conditional relief in such circumstances might enable the defendant to take steps, in the interval between the cause of action arising and the granting of a full freezing injunction, to defeat the purpose of the remedy, it has been held that the practical convenience attendant on the invention of a conditional freezing injunction could only be achieved at the expense of inconsistency with the principle laid down by the House of Lords in *The Siskina* [1979] AC 210 (*Veracruz Transportation Inc.* v *VC Shipping Co. Inc.* [1992] 1 Lloyd's Rep 353, where the Court of Appeal, first, set aside a conditional freezing injunction granted at first instance restraining the defendants from dealing with a substantial part of the purchase price of a cruise ship pending delivery of the ship, which the claimants feared might not be in the state stipulated for in the contract, and, secondly, overruled *A* v *B* [1989] 2 Lloyd's Rep 423, the case in which the practice of granting conditional freezing injunctions appears to have been devised by Saville J).

9.4.2.6.2 *Issues for determination* There are three main issues for determination on an application for a freezing injunction (*Derby & Co. Ltd* v *Weldon (No. 1)* [1989] 1 All ER 469, CA).

(a) The claimant must have a good arguable case.

In an application for an interim injunction in an ordinary case, the claimant need only establish that there is a serious question to be tried (para 9.4.2.2.2 above), whereas the applicant for a freezing injunction must go further and show a good arguable case that money is due to him and that he needs the protection of a court order to preserve the fruits of any judgment he might obtain against the defendant (*Derby & Co. Ltd* v *Weldon (No. 1)*, above, *per* Parker LJ at p. 475; *Derby & Co. Ltd* v *Weldon (No. 2)* [1989] 1 All ER 1002, CA, *per* Neill and Butler-Sloss LJJ at pp. 1020 and 1022, respectively).

However, he need not show a probability that he will succeed on the merits in the main action (*Rasu Maritima SA* v *Perusahaan* [1978] QB 644, CA; *Etablissement Esefka International Anstalt* v *Central Bank of Nigeria* [1979] 1 Lloyd's Rep 445, CA; *Derby & Co. Ltd* v *Weldon (No. 1)*, above, *per* Parker LJ at p. 475).

(b) The claimant must satisfy the court that the defendant has assets within or outside the jurisdiction.

The term 'assets' includes cash and other property, such as an aeroplane (*Allen* v *Jambo Holdings Ltd* [1980] 2 All ER 502, CA). The term is wide enough to cover motor vehicles, jewellery, *objets d'art* and other valuables, as well as cash and securities (*CBS United Kingdom Ltd* v *Lambert* [1982] 3 All ER 237, CA, in which a husband and wife were ordered to deliver up, pending trial, the motor cars in their possession, to be kept in a garage chosen by the claimants' solicitors). Even a bank account which is overdrawn is sufficient evidence of assets within the jurisdiction, since the overdraft can be presumed to be secured by other assets (*Third Chandris Shipping Corporation* v *Unimarine SA* [1979] QB 645, CA).

The standard form of freezing injunction, which refers to 'the defendant's assets', applies only to assets owned by the defendant beneficially and which are available to satisfy the claim against him. It does not apply to assets of which the defendant is bare legal owner as a trustee and in which he has no beneficial interest. Thus, a defendant whose assets were the subject of a freezing injunction was held not to be in contempt of court by authorising the transfer of funds out of bank accounts in his name but in which he himself had no beneficial interest (*Federal Bank of the Middle East Ltd* v *Hadkinson* [2000] 2 All ER 395, CA).

(c) The claimant must show that there is a real risk that the defendant's assets will be removed from the jurisdiction, or that they will otherwise not be available when judgment is given.

The mere fact that the defendant is abroad is not by itself sufficient for this purpose (*Third Chandris Shipping Corporation* v *Unimarine SA*, CA, above; *Z Ltd* v *A* [1982] 1 All ER 556, CA). If the claimant has already obtained a freezing injunction but there is a real risk that the order will be frustrated by the defendant leaving the jurisdiction, the court has power to grant further relief by, for example, ordering the defendant not to leave the country and to surrender his passport (*Bayer AG* v *Winter* [1986] 1 All ER 733, CA; *Re Oriental Credit* [1988] 1 All ER 892).

The claimant who succeeds in obtaining a freezing injunction must normally give an *undertaking in damages* (see para 9.4.2.2.2 above) so as to protect the defendant if the claimant should fail in his claim or if the injunction turns out to have been wrongly granted (*Third Chandris Shipping Corporation* v *Unimarine SA*, above). But the decision whether to grant a freezing injunction rests on the balance of justice and convenience rather than on the financial standing of the claimant. An injunction should not, therefore, be refused merely because the claimant is unable to provide a satisfactory undertaking in damages (*Allen* v *Jambo Holdings Ltd* [1980] 2 All ER 502, CA, *per* Lord Denning MR at p. 505 and Templeman LJ at p. 506).

A claimant may also be required by the court to give an undertaking to pay any *expenses* reasonably incurred by an innocent third party, such as a banker, who is put to expense by the injunction in ascertaining whether any asset is within his possession or control (*Z Ltd* v *A* [1982] 1 All ER 556, CA, approving *Searose Ltd* v *Seatrain (UK) Ltd* [1981] 1 All ER 806).

Furthermore, an innocent third party affected by a freezing injunction who successfully applies to the court for a variation of the order is entitled to have his legal *costs* paid by the claimant, provided that they are not unreasonable in amount or unreasonably incurred. The burden of proving that the costs were reasonably incurred and that their amount is reasonable rests on the third party (*Project Development Co. Ltd SA* v *KMK Securities Ltd and others (Syndicate Bank intervening)* [1983] 1 All ER 465).

If a third party has notice of a freezing injunction which affects assets in his hands, it is a contempt of court for him knowingly to assist in the disposal of the assets to others whether or not the defendant himself has knowledge of the injunction (*Z Ltd* v

A, above). In *TDK Tape Distributor (UK) Ltd* v *Videochoice Ltd* [1985] 3 All ER 345, both the defendant and his solicitor were held to be guilty of contempt by dissipating the defendant's assets in breach of a freezing injunction even though the breach was merely 'negligent and inadvertent' rather than deliberate and intentional.

A disposal by the third party to the defendant himself is not a contempt of court unless it is known to be probable that the defendant will dissipate the assets, or otherwise dispose of them, in breach of the injunction (*Bank Mellat* v *Kazmi* [1989] 1 All ER 925, CA, *per* Nourse LJ at pp. 928–9).

According to the Court of Appeal in *Customs and Excise Commissioners* v *Barclays Bank plc* [2005] 1 WLR 2082, CA, a bank with notice of a freezing injunction owes a duty of care to the Customs and Excise Commissioners to prevent dissipation of money in the frozen accounts, and, on breach of that duty, the bank may be liable in damages for negligence. At the same time, the Court of Appeal indicated that the use of the law of contempt as a remedy for breach of the duty should be discouraged. The case was one in which the Commissioners had obtained freezing injunctions (which were subsequently frustrated by the bank) in support of claims for outstanding VAT.

9.4.2.6.3 *Vulnerable assets* If its purpose is to prevent the frustration of a judgment of an *English* court, a freezing injunction can be granted in respect of assets situated abroad, both *before* judgment (*Republic of Haiti* v *Duvalier* [1989] 1 All ER 456, CA; *Derby & Co. Ltd* v *Weldon (No. 1)* [1989] 1 All ER 469, CA; *Derby & Co. Ltd* v *Weldon (No. 2)* [1989] 1 All ER 1002, CA) and *after* judgment (*Babanaft International Co. SA* v *Bassatne* [1989] 1 All ER 433, CA).

However, in matrimonial proceedings for financial relief, the court should not grant a worldwide freezing injunction. It is unlikely that the divorce court would ever award a spouse the whole of the other spouse's assets. The protection of a freezing injunction which extends to all of those assets wherever situated is, therefore, not needed. In *Ghoth* v *Ghoth* [1992] 2 All ER 920, CA, the wife had been granted an interim maintenance order for £500 per month which she wished to protect. The husband had jewellery in England, $200,000 in a New York bank account, and other property in various countries. It was held that the freezing injunction granted to the wife would be limited to the jewellery in England and $50,000 in the New York account.

In the case of a *foreign* judgment which an English court is being asked to enforce, any injunction granted will normally be limited to assets within the jurisdiction of the English court. The court will only grant a worldwide freezing injunction in support of a foreign judgment in very exceptional circumstances. To do otherwise could cause confusion in, and resentment by, foreign nations (*Rosseel NV* v *Oriental Commercial and Shipping (UK) Ltd* [1990] 3 All ER 545, CA, where *Republic of Haiti* v *Duvalier* [1989] 1 All ER 456, CA, above, was distinguished as a special and unusual case).

To assist in making the correct order, the court has power to require the defendant to disclose the whereabouts of his foreign assets, both *before* judgment (*Republic of Haiti* v *Duvalier*, above) and *after* judgment (*Interpool Ltd* v *Galani* [1988] QB 738, CA; *Maclaine Watson & Co. Ltd* v *International Tin Council (No. 2)* [1988] 3 All ER 257, CA).

To protect the defendant against misuse of the information so obtained, the claimant will usually be required to give undertakings not to enforce the order in a foreign court, or to use in foreign proceedings any information disclosed by the defendant about his overseas assets without first obtaining the leave of the English court (*Derby & Co. Ltd* v *Weldon (No. 1)* [1989] 1 All ER 469, CA). Breach of these undertakings would constitute a contempt of court.

A freezing injunction should not be granted over foreign assets if there are sufficient assets within the jurisdiction (*Derby & Co. Ltd* v *Weldon (No. 2)* [1989] 1 All ER 1002, CA). By reason of its drastic and potentially oppressive nature, a pre-judgment injunction over foreign assets will usually only be granted in exceptional circumstances (*Derby & Co. Ltd* v *Weldon (No. 1)*, above).

Neither a pre-judgment nor a post-judgment freezing order granted in respect of overseas assets should be expressed to be unconditional (*Babanaft International Co. SA* v *Bassatne* [1989] 1 All ER 433, CA, *per* Kerr and Neill LJJ at pp. 446 and 450, respectively; *Ghoth* v *Ghoth* [1992] 2 All ER 920, CA, *per* Lord Donaldson MR at p. 922).

In particular, it should be qualified, so as to afford protection for third parties abroad who are indirectly affected by the order, by the inclusion of a proviso that, in so far as it purports to have extra-territorial effect, no person should be affected by it until it is declared enforceable, or recognised, or enforced, by an appropriate foreign court. This can be achieved by incorporating in the order the so-called 'Babanaft proviso'. This is a term in the injunction named after the *Babanaft* case, above, and refined in *Derby & Co. Ltd* v *Weldon (No. 2)*, above. The *Babanaft* proviso is now incorporated in the standard freezing injunction (*Practice Direction* to CPR, Part 25).

Most major banks, by virtue of their presence in this country, are subject to the jurisdiction of the English courts, and are, therefore, not protected by the usual *Babanaft* proviso. They may thus be placed in an impossible position if ordered to do something by an English court and the opposite by a foreign court.

In an attempt to deal with this problem, and to give additional protection to third parties with assets abroad, it was held by Clarke J at first instance in *Baltic Shipping Co.* v *Translink Shipping Ltd* [1995] 1 Lloyd's Rep 673 that a further proviso (afterwards known as the *Baltic* proviso) should be added to the usual *Babanaft* proviso contained in the standard freezing injunction. The *Baltic* proviso stipulates that, in respect of assets located outside England and Wales, nothing in the freezing injunction is to prevent a third party from complying with what it reasonably believes to be its obligations, contractual or otherwise, under the law of the country in which the assets are situated.

The *Baltic* case was approved by the Court of Appeal in *Bank of China* v *NBM LLC* [2002] 1 WLR 844, CA, in which it was held that the *Baltic* proviso should be included in a freezing injunction as a matter of course unless the facts of a particular case indicate that it is inappropriate. The *Baltic* proviso is now incorporated in the standard freezing injunction (*Practice Direction* to CPR, Part 25).

The court should take particular care before imposing an injunction on assets the restraint of which would bring the defendant's business to a standstill or inflict on him great loss that might not be fully compensated for by the claimant's undertaking in damages (*Rasu Maritima SA* v *Perusahaan* [1978] QB 644, CA). The court should not order the delivery up of the defendant's clothes, bedding, furnishings, tools of his trade, farm implements, livestock, or any machines or other goods which he uses for the purposes of a lawful business (*CBS United Kingdom* v *Lambert* [1982] 3 All ER 237, CA).

Where a bank, building society, or similar institution is involved in an application for a freezing injunction (whether as a defendant or as a third party), the court should give careful consideration to hearing the proceedings in private since public confidence in its finances is crucial to such an institution and the effects of publicity may harm not only the institution but also its innocent customers and depositors (*Polly Peck International* v *Nadir* (1991) *The Times*, 11 November, CA).

A freezing injunction must not be used as a means of exerting pressure on the defendant to settle the action (*Z Ltd* v *A* [1982] 1 All ER 556, CA; *PCW Ltd* v *Dixon* [1983] 2 All ER 158, para 9.4.2.6.4 below). Nor must it be used merely as an instrument of oppression, or for improving the position of the claimant as against other creditors of the defendant. To allow the freezing injunction to be so used would be to ignore its real object, which is to prevent the claimant from being cheated out of the proceeds of his action against the defendant should it be successful (*PCW Ltd* v *Dixon*, para 9.4.2.6.4 below).

9.4.2.6.4 *Variation and discharge of freezing injunctions* The court may afterwards *vary* the injunction so as to allow the restrained assets to be used, provided that the purpose for which they are to be used does not conflict with the policy underlying the freezing injunction, namely, to prevent the removal of assets from the jurisdiction or their dissipation within the jurisdiction.

In *Iraqi Ministry of Defence* v *Arcepey Shipping Co. SA* [1980] 1 All ER 480, the injunction was varied in a way which permitted the defendants to use the restrained assets to repay to a third party a debt incurred in the ordinary course of business within the jurisdiction, even though what remained of the assets would not be sufficient to satisfy the claimants' claim.

In another case, a freezing injunction was varied so as to allow the defendant to withdraw from his bank account the sum of £1,000 per week instead of £100 per week for his living expenses, to pay pressing bills amounting to £27,500, and to pay his solicitors £50,000 on account of their costs in the litigation. Lloyd J said that justice and convenience demanded that the defendant should be allowed the means to defend himself, pay his ordinary bills, and continue to live in the manner to which he had become accustomed.

There was no evidence that the defendant, who was a wealthy man with five children to educate, was dissipating his assets by living as he had always lived and paying bills such as he had always paid. The original figure of £100 per week for the living expenses of a wealthy man was unrealistically low and suggested that it had been set deliberately so as to put pressure on the defendant to settle the claimants' actions against him (*PCW Ltd* v *Dixon* [1983] 2 All ER 158, varied by consent on appeal at ibid., p. 697, CA).

In *Camdex International Ltd* v *Bank of Zambia (No. 2)* [1997] 1 All ER 728, CA, a freezing injunction was varied on appeal so as to exclude from its scope a large quantity of Zambian bank notes which had been printed in the United Kingdom and paid for by, but not yet delivered to, the defendants, who wished to transfer them to

Zambia for the purpose of allowing that country to repay debts to the World Bank and the International Monetary Fund. It was held that such a transfer would not be a dissipation of any assets available to the defendants' creditors since the bank notes had no value on the open market. Furthermore, not to exclude the bank notes from the scope of the injunction would amount to holding the defendants to ransom and would inflict great hardship on the Zambian people if Zambia was not able to honour its international debt obligations.

A variation of a freezing injunction will not be sanctioned where the purpose of the proposed variation is to facilitate repayment of a debt not incurred in the ordinary course of business and its effect would be to place assets out of the claimant's reach (*Atlas Maritime Co. SA* v *Avalon Maritime Ltd (No. 1)* [1991] 4 All ER 769, CA).

It is common for the court to specify in the order itself the maximum amount to be restrained, and to leave the defendant free to deal with the rest of his assets (*Z Ltd* v *A* [1982] 1 All ER 556, CA, *per* Lord Denning MR at p. 565). Where the order allows the defendant to spend a specified amount on 'ordinary living expenses', such expenses are limited to ordinary recurrent expenditure required to maintain the defendant in the style to which he is reasonably accustomed. They do not cover exceptional expenses, such as legal costs incurred by the defendant in employing Queen's Counsel to defend him on serious criminal charges (*TDK Tape Distributor (UK) Ltd* v *Videochoice Ltd* [1985] 3 All ER 345). Authority to incur exceptional expenses would have to be sought from the court in the form of a variation of the freezing injunction (ibid.).

To obtain a variation of the injunction it is not enough for the defendant merely to state that he owes money to someone. He must produce evidence to show that he has no other (unrestrained) assets available out of which the debt could be paid (*A* v *C (No. 2)* [1981] 2 All ER 126).

If the claimant has obtained a freezing injunction without notice to the defendant and without proper disclosure of material facts, the court will discharge the order where (a) the non-disclosure was not innocent (i.e., there was an intention to omit or withhold material information), and (b) the injunction would not have been granted in the first place if full disclosure had been made. If the non-disclosure was innocent, and the injunction could properly have been granted if there had been full disclosure, the court has a discretion to continue the order (*Behbehani* v *Salem* [1989] 2 All ER 143, CA, *per* Nourse LJ at p. 156; *Ali & Fahd Shobokshi Group Ltd* v *Moneim* [1989] 2 All ER 404). If new facts become known to the claimant, or if there is any material change in circumstances, after the grant of a freezing injunction without notice, the claimant is under a duty to inform the court of the new situation (*Commercial Bank of the Near East* v *A* [1989] 2 Lloyd's Rep 319).

9.5 **Rescission**

9.5.1 **Nature and effect of rescission**

Rescission is an equitable remedy and, therefore, discretionary. The right to rescind is the right to set aside a transaction (usually a contract) and be restored to one's former position (*restitutio in integrum*).

Rescission is not, strictly speaking, a judicial remedy because a party can rescind a contract without taking legal proceedings at all. A contract is rescinded if a party makes it clear that he refuses to be bound by it (*Abram Steamship Co. Ltd* v *Westville Shipping Co. Ltd* [1923] AC 773, HL, *per* Lord Atkinson at p. 781). He must usually give notice to the other party. If the other party cannot be communicated with (for example, where he has disappeared), the contract can be rescinded by the wronged party doing some overt act which is reasonable in the circumstances, such as asking the police to find his property, as in *Car & Universal Finance Co. Ltd* v *Caldwell* [1965] 1 QB 525, CA. This informal approach has not been followed in Scots law (see *MacLeod* v *Kerr* 1965 SC 253), and it was criticised by the Law Reform Committee, which recommended that until notice of rescission is communicated to the other party an innocent purchaser should acquire a good title to property (12th Report, *Transfer of Title to Chattels*, Cmnd 2958, 1966, para 16).

Although it is not essential to go to the court for rescission, it is desirable to do so for two reasons. First, a party who rescinds without good cause may be held to have repudiated the contract and be liable for breach. There may be a dispute over whether a party is or was entitled to rescind and this dispute may need to be resolved by the court. Secondly, the assistance of the court may be needed to obtain restitution of property which was handed over under the contract and which the other party now refuses to restore voluntarily.

In the absence of fraud, it is not possible to obtain damages as well as rescission. The reason is that the respective objects of rescission and damages are quite different. The object of rescission is to restore a party to the position he would have been in if the contract had not been made. The object of damages, on the other hand, is to place a party in the position he would have been in if the contract had been performed.

Sometimes, however, a party may be awarded a sum of money called an *indemnity* in addition to rescission where it would assist in restoring him more fully to his former position. An indemnity is an equitable remedy available to compensate a party for certain expenses incurred in performing obligations created by the contract. The limited scope of the remedy of indemnity, and the distinction between damages and an indemnity, is illustrated in the case of *Whittington* v *Seale-Hayne* (1900) 82 LT 49 in which a lease was rescinded and it was held that the claimant was also entitled to an indemnity in respect of certain items of expenditure.

It should be noted that, as a result of s. 2(1) of the Misrepresentation Act 1967, damages can now sometimes be awarded for the sort of misrepresentation that occurred in the *Whittington* case, where the defendant's innocent misrepresentation that his premises were in good sanitary condition led the claimant to take a lease of the premises in order to breed poultry. Thus, in *Walker* v *Boyle* [1982] 1 All ER 634 a purchaser of land was held entitled to rescind the contract and, by virtue of s. 2(1) of the 1967 Act, to an inquiry into damage suffered from loss of interest on his deposit. However, if the representor proves that he had reasonable ground to believe, and did believe up to the time the contract was made, that the facts represented were true, damages cannot be awarded and the representee would be able to claim only an indemnity.

9.5.2 Grounds for rescission

9.5.2.1 Fraudulent misrepresentation

A fraudulent misrepresentation is a false statement of fact which is made by the representor to the representee knowing it to be false, or without belief in its truth, or recklessly, without caring whether it is true or false, with the intention that it should be acted on and which is in fact acted on by the representee (*Derry* v *Peek* (1889) 14 App Cas 337, HL).

In these circumstances, the representee, on discovering the truth, can rescind the contract he was induced into making. In addition, he can claim damages for the tort of fraud or deceit if he has suffered loss as a result of relying on the misrepresentation.

9.5.2.2 Negligent misrepresentation

A negligent misrepresentation is a false statement of fact which satisfies the same criteria as a fraudulent misrepresentation except that it has not been made fraudulently but, instead, with no reasonable ground for belief in its truth. A contract entered into on the faith of such a misrepresentation may be rescinded. In addition, the representee may claim damages under s. 2(1) of the Misrepresentation Act 1967, as in *Watts* v *Spence* [1976] Ch 165 (para 9.3.2.9 above), if he has suffered loss. Negligent misrepresentation is also actionable at common law as a *tort* (*Hedley Byrne & Co. Ltd* v *Heller & Partners Ltd* [1964] AC 465, HL).

It has been held that, by reason of the wording of s. 2(1), the measure of damages for a misrepresentation giving rise to an action under the subsection is the measure of damages in tort for fraud or deceit and not that applicable where the tort is negligent misrepresentation at common law. This means that the representee is not limited (as he would be in a common law action for negligent misrepresentation) to claiming for loss which could be foreseen, but can also claim for that which was unforeseeable (*Royscott Trust Ltd* v *Rogerson* [1991] 3 All ER 294, CA).

9.5.2.3 Wholly innocent misrepresentation

A wholly innocent misrepresentation is a false statement of fact which is neither fraudulent nor negligent. It may have been made, for example, with an honest belief, and on reasonable grounds, that it is true. A contract entered into in reliance on such a misrepresentation may be rescinded. There is no *right* to claim damages, although the court has a *discretion* to award damages in *lieu* of rescission. On a claim for rescission, the court may, under s. 2(2) of the Misrepresentation Act 1967, declare the contract to be still subsisting and award damages in *lieu* of rescission if it would be equitable to do so.

Section 2(2) has been interpreted as meaning that, so long as the claimant once had a right to rescind, damages can be awarded under the subsection even though, by the time of the trial, rescission is no longer a viable remedy and would not be granted (*Thomas Witter Ltd* v *TBP Industries Ltd* [1996] 2 All ER 573).

9.5.2.4 Undue influence and unconscionable bargains

Contracts, gifts, and other transactions can be set aside by the victim of undue influence, which may be exerted *expressly*, as in *Williams* v *Bayley* (1866) LR 1 HL 200, HL (threat of prosecution of son for forgery), or may be *presumed* to have been exercised, as in *Re Craig* [1971] Ch 95 (domination of 84-year-old widower by his secretary/companion).

Transactions of an improvident nature made by poor and ignorant persons acting without independent advice can be set aside, unless the other party can show that the transaction is fair and reasonable (*Fry* v *Lane* (1888) 40 ChD 312, where two poor and ignorant men, without independent advice, sold property at considerably below the real value; the transactions were set aside by the court).

By statute, the High Court and the county courts have power to reopen extortionate credit bargains, including those entered into with moneylenders and financial institutions (Consumer Credit Act 1974, s. 137(1)). A credit bargain is 'extortionate' if the payments under it are 'grossly exorbitant', or if it 'otherwise grossly contravenes ordinary principles of fair dealing' (ibid., s. 138(1)). On reopening the transaction, the court is given a wide discretion to do justice between the parties (ibid., s. 137(1)). It may, *inter alia*, set aside the whole or part of any obligation imposed by the transaction on the debtor or a surety (ibid., s. 139(2)).

9.5.2.5 Non-disclosure in contracts *uberrimae fidei*

Contracts *uberrimae fidei* ('of the utmost good faith') are contracts in which there is a duty to make a full disclosure of material facts. By their nature they are contracts where one party alone has knowledge of facts which is denied to the other party (*Carter* v *Boehm* (1766) 3 Burr 1905 *per* Lord Mansfield at p. 1909). Examples are family arrangements (*Gordon* v *Gordon* (1821) 3 Swanst 400) and all types of insurance contract (*Lindenau* v *Desborough* (1828) 8 B & C 586 *per* Bayley J at p. 592).

Accordingly, a proposer for insurance must disclose all material facts to the insurer. Facts are 'material' if they would influence the judgment of a prudent insurer in deciding whether to accept the risk, and, if so, what the premium should be (*London Assurance* v *Mansel* (1879) 11 ChD 363). Failure to disclose all material facts will render the policy voidable at the option of the insurer, who may rescind it if he so wishes. (Note the warning given to the public about this by Fletcher Moulton LJ in *Joel* v *Law Union & Crown Insurance Co.* [1908] 2 KB 863, CA, at p. 885.)

9.5.2.6 Substantial misdescription of land

The purchaser under a contract for the sale of land may rescind if the vendor has substantially misdescribed the property; for example, where the land is described as registered freehold without disclosing that the title is only possessory and not absolute (*Re Brine and Davies's Contract* [1935] Ch 388; see also *Walker* v *Boyle* [1982] 1 All ER 634). This rule exists in order to ensure that the purchaser obtains what he bargained for, and that he is not compelled to take something else (*Knatchbull* v *Grueber* (1817) 3 Mer 124 *per* Lord Eldon LC at p. 146; *Flight* v *Booth* (1834) 1 Bing NC 370).

If the misdescription of the land is only slight and not material, so that the purchaser will obtain substantially, though not exactly, what he bargained for, the purchaser is not normally entitled to rescind but must be satisfied with financial compensation from the vendor (*M'Queen* v *Farquhar* (1805) 11 Ves Jr 467).

9.5.3 Loss of right of rescission

9.5.3.1 *Restitutio in integrum* impossible

A party may not be able to rescind a contract when the parties cannot be restored to their former positions; for example, where property has been consumed or destroyed and cannot, therefore, be handed back (*Clarke* v *Dickson* (1858) EB & E 148; *Vigers* v *Pike* (1842) 8 Cl & F 562). But *precise* restitution is not essential as long as *substantial* restitution is still possible so that equity can do what is just in the circumstances, including, if necessary, the ordering of financial adjustments between the parties (as in *Hulton* v *Hulton* [1917] 1 KB 813, CA, and *Cheese* v *Thomas* [1994] 1 All ER 35, CA).

In *Erlanger* v *New Sombrero Phosphate Co.* (1878) 3 App Cas 1218, HL, a company formed by E bought the lease of a phosphate mine for £55,000. E then formed another company and sold the lease to it for £110,000 without telling the shareholders of the second company the facts about the purchase, including the profit being made by the first company. This was in breach of the fiduciary duty owed by the promoter of a company to its shareholders. By the time the true facts were discovered, both the lease and the minerals were running out, although restitution was possible since the mine was not completely worked out. The House of Lords held that the sale of the lease to the second company would be rescinded on condition that the mine be handed back and an account rendered of the profits of working it.

In *O'Sullivan* v *Management Agency and Music Ltd* [1985] 3 All ER 351, CA, contracts entered into some years earlier by the 'pop' singer and composer, Gilbert O'Sullivan, with his managers, the defendants, were rescinded by the court (on the ground of undue influence) on terms which sought to do justice between the parties. The defendants were ordered to reconvey to the claimant the copyrights in his compositions, to deliver up to him the master tapes of his recordings, and to account for the profits due to him from the copyrights and recordings. However, the defendants were held to be entitled to an allowance representing reasonable remuneration (including a profit element) for work done by them in helping the claimant to enjoy a successful career.

9.5.3.2 By affirmation of the contract

The right to rescind will be lost if a party affirms the contract (whether by word or action) after learning the full facts. For instance, a party who, on discovering that he has been the victim of a fraudulent misrepresentation, takes a benefit under the contract will thereby lose his right to rescind it (*Clough* v *London & North Western Railway Co.* (1871) LR 7 Ex 26; *Urquhart* v *Macpherson* (1878) 3 App Cas 831, PC).

But a party must be in possession of *all* the facts before it can be concluded that what he has said or done amounts to an affirmation of the contract (*Central Railway of Venezuela* v *Kisch* (1867) LR 2 HL 99, HL).

Lapse of time between the date of the contract and a purported rescission is evidence of affirmation (*Clough* v *London & North Western Railway Co.* (1871) LR 7 Ex 26). In the case of a fraudulent misrepresentation, time begins to run only from the discovery of the truth (*Armstrong* v *Jackson* [1917] 2 KB 822). In the case of innocent misrepresentation (and probably negligent misrepresentation, too), time begins to run as soon as the contract is made. In *Leaf* v *International Galleries* [1950] 2 KB 86, CA, the claimant was induced to buy a picture on the faith of an innocent misrepresentation that it was a genuine Constable. Five years later he discovered that the picture was not by Constable and he claimed rescission of the contract. The Court of Appeal held that he had lost his right of rescission by lapse of time, even though he had started his action immediately on discovering the truth.

Affirmation bars only the right to rescind. A party who affirms a contract, though he cannot set it aside, can still claim damages for misrepresentation or for breach of contract.

9.5.3.3 By the intervention of innocent third-party rights for value

The right to rescind a contract will be lost if, before rescission, an innocent third party acquires for valuable consideration an interest in the subject-matter of the contract (*Clough* v *London & North Western Railway Co.* (1871) LR 7 Ex 26).

For example, X obtains P's goods under a contract induced by fraud. Before P discovers the fraud, X sells the goods to D, who is an innocent third party. It is too late for P to rescind his contract with X, and D may obtain a good title to the goods. If so, P loses his goods and must sue X, and not D, for damages in the tort of conversion—if X can be found and if he is worth suing.

9.6 Rectification

9.6.1 Nature of rectification

Rectification is an equitable remedy and, therefore, discretionary.

Sometimes, by reason of an omission or a mistake or fraud, a written instrument does not represent the true agreement of the parties. There may, for example, have been a mistake in recording the agreed rent in a lease (*Murray* v *Parker* (1854) 19 Beav 305; *Thomas Bates & Son Ltd* v *Wyndham's (Lingerie) Ltd* [1981] 1 All ER 1077, CA). Or, by an oversight, some terms previously agreed on orally may have been omitted from the later written instrument (as in *Joscelyne* v *Nissen* [1970] 2 QB 86, CA, para 9.6.3 below; *Craddock Bros Ltd* v *Hunt* [1923] 2 Ch 136, CA). A court of equity has power to rectify the instrument so that it does represent the true agreement of the parties.

The burden of proof, which rests on the party seeking rectification, is that applicable in all civil cases, namely the balance of probability. The claimant must prove, on the balance of probability, that the written instrument does not reflect the common intention of the parties (*Thomas Bates & Son Ltd* v *Wyndham's (Lingerie) Ltd* [1981] 1 All ER 1077, CA, *per* Buckley and Brightman LJJ at pp. 1085 and 1090, respectively).

The claimant does not have to satisfy the higher burden, demanded in criminal cases, of proving the matter beyond reasonable doubt (*Earl* v *Hector Whaling Ltd* [1961] 1 Lloyd's Rep 459, CA; *Joscelyne* v *Nissen* [1970] 2 QB 86, CA). Nevertheless, in some circumstances a high standard of evidence may be required, especially where sharp practice is alleged (*Thomas Bates & Son Ltd* v *Wyndham's (Lingerie) Ltd*, above, *per* Buckley LJ at p. 1086).

9.6.2 Instruments subject to rectification

Instruments that may be rectified include contracts, leases, conveyances, bills of exchange, transfers of shares, policies of marine insurance, marriage settlements, and wills.

The court had no power to rectify a will before the Administration of Justice Act 1982 became law (*Harter and Slater* v *Harter* (1873) LR 3 P & D 11). Section 20 of the 1982 Act conferred that power, thus implementing belatedly the recommendation of the Law Reform Committee (19th Report, *The Interpretation of Wills*, Cmnd 5301, 1973, paras 17–33). By s. 20, if the court is satisfied that a will fails to carry out the testator's intentions, as a result of a clerical error or of a failure to understand his instructions, it may order rectification of the will so as to carry out the testator's intentions. (For the meaning of 'clerical error', see *Wordingham* v *Royal Exchange Trust Co. Ltd* [1992] 3 All ER 204; *Re Segelman (deceased)* [1995] 3 All ER 676.)

9.6.3 Conditions for rectification

(a) There must be a prior concluded agreement between the parties, although it is not necessary for it to be a legally enforceable contract as long as there is an outward expression of agreement.

In *Joscelyne* v *Nissen* [1970] 2 QB 86, CA, the claimant applied for rectification of a written contract under which he had transferred his car-hire business to the defendant, who was his daughter. The parties shared a house, and, before the written contract was entered into, it had been agreed verbally that the defendant would pay certain household expenses, including the gas, electricity and coal bills, in respect of the claimant's part of the house. But this agreement was not contained in the written contract. The defendant paid the bills for a while but then refused to pay any more and denied any liability to do so. The Court of Appeal held that the written contract would be rectified so as to include the defendant's liability for the agreed expenses. The oral agreement between the parties was not in itself a legally binding contract. Nevertheless, it did represent the common intention of the parties, who had reached an outward expression of agreement in respect of the household expenses.

(b) This prior concluded agreement must have remained unchanged until it was put into writing in the instrument (*Marquis Townshend* v *Stangroom* (1801) 6 Ves Jr 328 *per* Lord Eldon LC at pp. 333–4). There can be no rectification if the parties deliberately changed the terms of their agreement or deliberately omitted a term from the written instrument.

(c) By mistake, the written instrument must fail to express the prior agreement of the parties. What is needed is a literal disparity between the language of the prior agreement and the later written instrument, as in *Joscelyne* v *Nissen* [1970] 2 QB 86, CA, above.

It is not enough to prove a *misunderstanding* of the language used where it is the same in both cases, so that, in fact, the written instrument has accurately recorded the prior agreement. In *Frederick E. Rose (London) Ltd* v *William H. Pim Jnr & Co. Ltd* [1953] 2 QB 450, CA, the claimants, who were London merchants, received an order for 'Moroccan horsebeans described here as *féveroles*'. They asked the defendants what '*féveroles*' were, and the defendants replied that they were just horsebeans and that they could obtain them for the claimants to resell. The parties orally agreed that the defendants would sell and the claimants would buy 'horsebeans'. The goods were also described as 'horsebeans' when the oral agreement was later embodied in a written contract.

It turned out that there were three types of Moroccan horsebeans: *fèves*, *féveroles*, and *fevettes*. The defendants supplied *fèves* and not *féveroles*. The claimants' customers claimed damages from the claimants, who now wished to claim damages from the defendants. The claimants, as a preliminary to claiming damages, first sought rectification of the written instrument so as to make it a contract for 'horsebeans described as *féveroles*'. The Court of Appeal held that the remedy of rectification was not available because both the prior oral agreement and the written contract were for 'horsebeans'. There had been a misunderstanding of the meaning of the word 'horsebeans', but there had been no mistake in recording the terms of the prior agreement. The parties all along had contracted to buy and sell 'horsebeans' and not '*féveroles*'.

Nor can mere *confusion* be used as the basis of a claim for rectification. Thus in *Cambro Contractors Ltd* v *John Kennelly Sales Ltd* (1994) *The Times*, 14 April, CA, rectification of a conveyance was refused where all that could be shown was that the parties and their solicitors had been confused about what land was included in the conveyance. This fell far short of establishing that the document contained a common mistake and failed to reflect the common intention of the parties.

9.6.4 Oral evidence

In a claim for rectification, oral evidence is admissible to prove that the intention of the parties was not properly expressed in the written instrument (*Marquis Townshend* v *Stangroom* (1801) 6 Ves Jr 328; *Craddock Bros Ltd* v *Hunt* [1923] 2 Ch 136, CA; *United States of America* v *Motor Trucks Ltd* [1924] AC 196, PC).

This is an exception to the parol evidence rule that evidence from witnesses will not be admitted to add to, vary, or contradict a written instrument. The exception is justified on the ground that rectification could not be an effective remedy without evidence of the prior oral agreement being admitted.

9.6.5 Rectification for unilateral mistake

Normally, a claim for rectification is based on a *common* mistake; that is, the written instrument fails to record the intention of *both* parties. However, rectification is also possible on the ground of *unilateral* mistake by the claimant alone on the conditions set out below (*Thomas Bates & Son Ltd v Wyndham's (Lingerie) Ltd* [1981] 1 All ER 1077, CA, applying *A. Roberts & Co. Ltd v Leicestershire County Council* [1961] Ch 555 and a dictum of Russell LJ in *Riverplate Properties Ltd v Paul* [1975] Ch 133, CA; see also *Paget v Marshall* (1884) 28 ChD 255).

(a) The claimant wrongly believed that the written instrument recorded the common intention of the parties.

(b) The defendant knew that the instrument did not record the common intention of the parties because of a mistake on the part of the claimant.

(c) The defendant failed to bring the mistake to the attention of the claimant.

(d) The claimant's mistake was favourable to the defendant (*Thomas Bates & Son Ltd v Wyndham's (Lingerie) Ltd*, above, *per* Buckley LJ at p. 1086) or, at least, was detrimental to the claimant (ibid., *per* Eveleigh LJ at p. 1090).

As long as the defendant's conduct is such as to make it inequitable for him to object to rectification, it is not necessary for the claimant to prove that the defendant was guilty of any sharp practice (ibid., *per* Buckley, Eveleigh, and Brightman LJJ at pp. 1086, 1090, and 1091, respectively).

9.6.6 Defences to a claim for rectification

Generally, rectification is subject to the same limitations as the remedy of rescission (para 9.5.3 above). Thus, rectification will not be ordered where the written instrument has been affirmed, or where a third party has acquired rights for value under the instrument and without notice of the mistake (*Bell v Cundall* (1750) Amb 101).

One important difference, however, is that impossibility of *restitutio in integrum* is not normally a bar to rectification (*Cook v Fearn* (1878) 48 LJ Ch 63, in which a marriage settlement was ordered to be rectified *after* the marriage had taken place).

9.6.7 Effect of an order of rectification

A new instrument need not be prepared. A copy of the court order can be endorsed on the existing instrument and that is enough (*White v White* (1872) LR 15 Eq 247). The order of rectification works retrospectively in that the instrument is to be read as if it had been originally drafted in its rectified form (*Craddock Bros Ltd v Hunt* [1923] 2 Ch 136, CA).

Rectification of a contract, and specific performance of the contract as rectified, can be decreed together in the same action (*Craddock Bros Ltd* v *Hunt*, above; *United States of America* v *Motor Trucks Ltd* [1924] AC 196, PC).

9.7 Declarations

A claimant may ask the High Court or a county court simply to declare the law on some disputed point. Alternatively, the court may make a declaration of its own motion even though not claimed by a party in the proceedings, as in *Cowan* v *Scargill* [1984] 2 All ER 750. Quite often a declaration can settle a dispute before the claimant's rights have been infringed.

The declaratory judgment is a relatively modern development. In general, the English courts had always denied any jurisdiction to make declarations without giving other relief (for example, damages or an injunction) at the same time. This attitude was possibly based on the unwillingness of the courts to try what they regarded as hypothetical cases.

In Scotland, the courts had jurisdiction to make declarations by reason of the action of *declarator*, which had existed for hundreds of years (*Russian Commercial & Industrial Bank* v *British Bank for Foreign Trade Ltd* [1921] AC 438, HL, *per* Lord Dunedin at p. 448). In England, it seems that the old Court of Chancery had no *inherent* power to issue declarations without at the same time granting consequential relief (*Guaranty Trust Co. of New York* v *Hannay & Co.* [1915] 2 KB 536, CA, *per* Bankes LJ at p. 568). However, from the middle of the nineteenth century the Court of Chancery was given a *statutory* power to make declarations without granting consequential relief (13 & 14 Vict, c. 35 (1850) (Special Case Act); Chancery Procedure Act 1852, s. 50).

But the major change was brought about in 1883 when, pursuant to s. 17 of the Supreme Court of Judicature Act 1875, new Rules of the Supreme Court (RSC) were made. By them, the power to make declarations of right was extended to all three divisions of the recently created High Court and could be exercised whether or not any other relief was claimed.

In *F* v *West Berkshire Health Authority* [1989] 2 All ER 545, HL, Lord Brandon (at p. 557) said it was wrong to suggest that the jurisdiction to make declarations arose under the Rules of the Supreme Court. (See also *Re S (hospital patient: court's jurisdiction)* [1995] 3 All ER 290, CA, *per* Sir Thomas Bingham MR at p. 296: the 'jurisdiction . . . to grant declaratory relief is not conferred, but is regulated, by Ord. 15, r. 16'.) But Lord Brandon's view that the power is instead part of the inherent jurisdiction of the High Court seems to conflict with Lord Goff's opinion in the same case (at p. 570), and is difficult to reconcile with earlier dicta in the House of Lords (see *Russian Commercial & Industrial Bank* v *British Bank for Foreign Trade Ltd* [1921] AC 438, HL, *per* Viscount Finlay at p. 444 and Lord Dunedin at p. 447; *Gouriet* v *Union of Post Office Workers* [1977] 3 All ER 70, HL, *per* Lord Edmund-Davies at p. 110).

The rule was formerly contained in RSC, Ord. 15, r. 16, which (substantially repeating the language of s. 50 of the Chancery Procedure Act 1852) stated that:

No action or other proceeding shall be open to objection on the ground that a merely declaratory judgment or order is sought thereby, and the court may make binding declarations of right whether or not any consequential relief is or could be claimed.

With effect from 26 March 2001, RSC, Ord. 15, r. 16 was replaced by CPR, r. 40.20, which states simply that:

The court may make binding declarations whether or not any other remedy is claimed.

Thus, any mention of 'rights' is now omitted. In addition, from April 1999 it has been possible for the court to grant *interim* declarations (see below).

The county courts have the same power to make declarations as the High Court (County Courts Act 1984, s. 38, as substituted by the Courts and Legal Services Act 1990), including the power to make declarations relating to land. Declarations, but not injunctions or specific performance, are available against the Crown in private law proceedings (Crown Proceedings Act 1947, s. 21).

In the field of private law, declarations are often sought for a determination of the correct interpretation of contracts, as in the case of *Eastham* v *Newcastle United Football Club Ltd* [1964] Ch 413, in which a declaration was granted that the former 'retain and transfer' system of the Football League, under which a player retained by his club could not obtain a transfer without the consent of that club, was in unlawful restraint of trade. The declaration was granted not only against the player's club (with whom he had a contract) but also against the Football League Ltd and the Football Association Ltd (with whom he did not).

In *Consorzio Veneziano di Armamento e Navigazione* v *Northumberland Shipbuilding Co. Ltd* (1919) 88 LJKB 1194, CA, Atkin LJ said (at p. 1201) that the action for a declaration was one of the most valuable contributions to the commercial life of the country. It provides a procedure whereby business people can go to court and have their disputes resolved relatively quickly. In the *Consorzio Veneziano* case, the Court of Appeal upheld a judge's decision to grant a declaration that certain contracts between the parties were still subsisting and had not been repudiated. Furthermore, it was made clear that the very act of applying for a declaration would not in itself usually amount to a repudiation of a contract (*per* Atkin LJ at p. 1200).

The declaration is a discretionary remedy, and the discretion is exercised with caution (*Russian Commercial & Industrial Bank* v *British Bank for Foreign Trade Ltd* [1921] 2 AC 438, HL). The court will not entertain hypothetical questions (ibid., *per* Lords Dunedin and Sumner at pp. 448 and 452, respectively). Despite the wide wording of the former RSC, Ord. 15, r. 16 (quoted above), it has been held that the power of the court will normally be limited to declaring contested legal rights, subsisting or future, of the parties to the case before it (*Guaranty Trust Co. of New York* v *Hannay & Co.* [1915] 2 KB 536, CA, *per* Pickford LJ at p. 562; *Gouriet* v *Union of Post Office Workers* [1977] 3 All ER 70, HL, *per* Lords Wilberforce and Diplock at pp. 85 and 100, respectively). There is, therefore, no jurisdiction to grant a declaration to a third party in connection with a contractual dispute between two other parties, even though the third party might be affected by the resolution of that dispute (*Meadows Indemnity Co. Ltd* v *Insurance Corporation of Ireland Ltd* [1989] 2 Lloyd's Rep 298, CA).

Where, however, the case concerns a serious justiciable issue of a moral or social nature which is not merely of academic interest, the court will be prepared to grant an appropriate declaration to a party with a genuine and legitimate interest in that issue (as opposed to a stranger or officious busybody) without insisting on the demonstration of a specific legal right vested in that party. To hold otherwise would be to 'confine the inherent jurisdiction of the court within an inappropriate straitjacket'.

Sir Thomas Bingham MR so said in *Re S (hospital patient: court's jurisdiction)* [1995] 3 All ER 290, CA, at p. 303. Here, the female companion of an elderly incapacitated patient was granted declaratory and injunctive relief against the patient's son and wife forbidding them to remove the patient from England to Norway. It was further said that, in any event, the companion had acquired a specific legal right which was liable to be infringed by the proposed action of the son and wife. This legal right had been obtained by assuming and discharging the responsibility for the patient's care and hospital treatment.

In view of the fact that there is no mention of 'rights' in CPR, r. 40.20, it remains to be seen whether cases like *Guaranty Trust Co. of New York v Hannay & Co.* [1915] 2 KB 536, CA, above, which confine the making of declarations to existing or future legal rights, are still good law. According to Lord Woolf CJ, *obiter*, in *Governor and Company of the Bank of Scotland v A Ltd* [2001] 1 WLR 751, CA, at [46], the power of the courts to make declarations can be used 'where there is a real dilemma which requires their intervention'. At the same time, if 'real' excludes 'hypothetical' it would seem to follow that cases like *Russian Commercial & Industrial Bank v British Bank for Foreign Trade Ltd* [1921] 2 AC 438, above, are unaffected by CPR, r. 40.20.

The declaration has become a valuable tool in determining personal status, although this sort of declaration can usually be granted only under the terms of a statute and not under CPR, r. 40.20. Thus, under what was s. 45 of the Matrimonial Causes Act 1973 an application for a declaration could be made in order to determine, *inter alia*, whether a person was legitimate, or whether a person had been validly married. In *Puttick v Attorney-General* [1980] Fam 1, the marriage of the petitioner (née Astrid Proll, the erstwhile terrorist) was held to be valid, but the judge, in his discretion, refused to make a declaration to that effect under s. 45 of the Matrimonial Causes Act 1973 because of the petitioner's fraud and other criminal activities. In another case, it was held that if s. 45 of the 1973 Act was available any declaration should be made under that provision rather than under the former RSC, Ord. 15, r. 16 (*Vervaeke v Smith* [1981] 1 All ER 55, CA; not argued on appeal to the House of Lords ([1982] 2 All ER 144), although the opinion of the Court of Appeal received the approval of Lord Simon, ibid., at p. 160).

In 1988, s. 45 of the Matrimonial Causes Act 1973 was repealed and replaced by part III of the Family Law Act 1986, which makes fresh provision for declarations of status and incorporates earlier case-law developments. Thus, for example, a person may apply to the High Court or a county court for a declaration that a marriage was at its inception a valid one (Family Law Act 1986, s. 55), or that he is the legitimate child of his parents, or that he has become or has not become a legitimated person (ibid., s. 56, as substituted by the Family Law Reform Act 1987).

It is also now possible to apply to the High Court, a county court or a magistrates' court for a declaration as to whether or not *a person named in the application* is or was the parent of another person named in the application (ibid., s. 55A, as inserted by the Child Support, Pensions and Social Security Act 2000, which also removed the former inability of the court to declare that a person is or was illegitimate).

If the truth of the proposition to be declared is proved to the satisfaction of the court, a declaration must be granted unless to do so would manifestly be contrary to public policy (Family Law Act 1986, s. 58(1), confirming *Puttick* v *Attorney-General*, above). No declaration available under the Family Law Act 1986 can be made by any court otherwise than under that Act (ibid., s. 58(4), confirming *Vervaeke* v *Smith*, above).

Under s. 58 of the Civil Partnership Act 2004, applications can be made to the High Court or a county court for a declaration that a civil partnership was valid at its inception, that it subsisted or did not subsist on a specified date, or that the validity of a dissolution, annulment, or legal separation obtained outside England and Wales in respect of a civil partnership is or is not entitled to recognition in England and Wales. If the truth of the proposition to be declared is proved to the satisfaction of the court, a declaration must be granted unless to do so would manifestly be contrary to public policy (ibid., s. 59). No declaration available under s. 58 can be made otherwise than under that section (ibid.). The court cannot make a declaration, either under s. 58 or otherwise, that a civil partnership was *void* at its inception.

As from April 1999, it has been possible for the courts to grant *interim* declarations at any time, including before proceedings are started (CPR, r. 25.1; on the usefulness of interim declarations, see *Governor and Company of the Bank of Scotland* v *A Ltd* [2001] 1 WLR 751, CA, *per* Lord Woolf CJ at [46]).

9.8 Private law remedies against public authorities

9.8.1 Introduction

Public bodies, such as central government departments, local government authorities, and statutory tribunals are obliged to keep within their powers ('*vires*'), to observe the principles of natural justice, and otherwise to act lawfully.

Ultra vires acts can take several forms, such as doing the wrong thing, acting in the wrong manner (for example, by using the wrong procedure or behaving negligently), or abusing a discretion.

An *ultra vires* act is *void*. One serious consequence of this rule is that money or property transferred under an *ultra vires* transaction may only be recoverable (if at all) with some difficulty. Thus, for example, since *ultra vires* transactions cannot be enforced by or against local authorities, a creditor who lends money to an authority for an *ultra vires* object will be unable to recover the amount of the debt by suing on the void contract of loan (*Hazell* v *Hammersmith and Fulham London Borough Council* [1991] 1 All ER 545, HL).

The principles of natural justice ensure that powers and duties are exercised in accordance with rules of 'fair play'. The principles are, first, *nemo judex in causa sua potest* ('no man can be a judge in his own cause'), sometimes called 'the rule against bias', and, secondly, *audi alteram partem* ('hear the other side'), which insists that both parties to a dispute should be heard and that the hearing should be a fair one.

The doctrine of *ultra vires* and the principles of natural justice are explained and illustrated further in Chapter 10.

In the main, the obligation of public authorities to act lawfully is enforced by means of the prerogative remedies, which are available only in *public law* proceedings for judicial review (see Chapter 10). However, some of the remedies outlined in the present chapter may be available in appropriate circumstances against public authorities in *private law* proceedings in order to prevent or redress wrongful conduct. These remedies are damages, injunctions, specific performance, and declarations.

9.8.2 Damages

A public authority which makes a contract of a type which it is within its powers to make may be sued in damages like a private individual if it commits a breach of that contract. If, on the other hand, an authority makes a contract which it has no power to make, the contract is *ultra vires* and void, and neither party can sue on it (*North West Leicestershire District Council* v *East Midlands Housing Association Ltd* [1981] 3 All ER 364, CA; *Rhyl Urban District Council* v *Rhyl Amusements Ltd* [1959] 1 WLR 465).

A public authority which commits a tort while acting *ultra vires* or in violation of the principles of natural justice may be sued in damages. Thus, public authorities have been successfully sued in trespass (*Cooper* v *Wandsworth Board of Works* (1863) 14 CB NS 180), false imprisonment (*Percy* v *Glasgow Corporation* [1922] AC 299, HL), and nuisance (*Managers of the Metropolitan Asylum District* v *Hill* (1881) 6 App Cas 193, HL, para 4.10.3.3).

9.8.3 Action for damages for breach of statutory duty

In some cases, an action for damages may be brought by a private individual against a public authority for breach of a duty imposed by statute. Whether such an action is available depends on the facts of each case and the interpretation of the particular statute (*Cutler* v *Wandsworth Stadium Ltd* [1949] AC 398, HL; *Lonrho Ltd* v *Shell Petroleum Co. Ltd (No. 2)* [1981] 2 All ER 456, HL; *Hague* v *Deputy Governer of Parkhurst Prison* [1991] 3 All ER 733, HL; *Stovin* v *Wise* [1996] 3 All ER 801, HL; *O'Rourke* v *Camden London Borough Council* [1997] 3 All ER 23, HL). The test is whether Parliament intended to confer on the claimant a private law cause of action for breach of statutory duty.

The statute may, for example, expressly allow an action for damages. On the other hand, the statute may be completely silent about it, as in *Reffell* v *Surrey County Council* [1964] 1 All ER 743, where the relevant legislation made no mention of any private law remedy. It was nevertheless held that an action for damages was available because the action was in respect of the kind of harm which the legislation was intended to prevent (injury by reason of insufficiently tough glass in school cloakroom door), the claimant schoolgirl belonged to the group which it was designed to protect, and the special remedy it provided (government minister's power to issue directions to education authorities or to apply to the court for a mandatory order) was not adequate for the protection of the claimant.

An action for damages for breach of statutory duty against a public authority is not available to a private individual if, on the construction of the relevant statute, there is no indication (express or implied) that Parliament intended to confer such a private law remedy. Thus, in *Hague* v *Deputy Governer of Parkhurst Prison* [1991] 3 All ER 733, HL, the House of Lords held that no action for damages lay against the prison authorities for breach of r. 43 (segregation from other prisoners) of the Prison Rules 1964 (made by the Home Secretary under the authority of the Prison Act 1952, and since replaced by the Prison Rules 1999). The House said that r. 43 was a purely preventive measure adopted, *inter alia*, for the purpose of conferring a necessary power to segregate disruptive prisoners and there was nothing in the legislation to show that Parliament intended to confer on a prisoner a private law remedy for breach of it.

Similarly, it has been held that Parliament did not intend to confer on a child a right of action against a local authority for breach of the statutory duties imposed by the child-minding legislation (*T (a minor)* v *Surrey County Council* [1994] 4 All ER 577), or by the special educational needs provisions of the Education Acts (*E (a minor)* v *Dorset County Council* [1995] 3 All ER 353, HL).

In subsequent cases, it has been held that no action for damages will lie at the suit of an individual in respect of a failure by a local authority under what was then part III of the Public Health Act 1936 to abate a statutory nuisance relating to the unhealthy state of its accommodation (*Issa* v *Hackney London Borough Council* [1997] 1 All ER 999, CA); a failure by a fire authority to provide an adequate water supply under what was s. 13 of the Fire Services Act 1947 (now s. 38 of the Fire and Rescue Services Act 2004) (*Capital and Counties plc* v *Hampshire County Council* [1997] 2 All ER 865, CA); or a failure by a health authority to provide after-care services for a discharged mental patient under s. 117 of the Mental Health Act 1983, as amended (*Clunis* v *Camden and Islington Health Authority* [1998] 3 All ER 180, CA). The remedy, if any, in cases of this nature is an application for judicial review in *public law* proceedings.

The statute may expressly provide for some other remedy, such as a complaint to a government minister or the imposition of a fine payable to the state. In this case, the court may decide that the specified remedy is exclusive of all other remedies so that an action for damages will not lie (*Pasmore* v *Oswaldtwistle Urban District Council* [1898] AC 387, HL; *London Borough of Southwark* v *Williams* [1971] 2 All ER 175, CA; *Clunis* v *Camden and Islington Health Authority*, CA, above).

However, an action for damages may still be possible where the specified remedy is inadequate for the protection of the person injured, as in *Reffell* v *Surrey County Council* [1964] 1 All ER 743, above, where it was said that an application by a government minister for a mandatory order against the education authority would have been useless in the circumstances (see also *Monk* v *Warbey* [1935] 1 KB 75, CA).

9.8.4 Injunctions

Injunctions are most commonly issued against public authorities to prevent the commission or continuance of nuisances. For example, in *Pride of Derby Angling Association Ltd* v *British Celanese Ltd and others* [1953] Ch 149, CA, the owners of fishing rights in the Trent and Derwent sued three separate bodies for polluting the rivers with municipal and industrial effluent. The defendants were British Celanese Ltd, an industrial company which was discharging chemicals and overheating the water; Derby Corporation, a local authority which was discharging sewage into the water; and the British Electricity Authority, a nationalised undertaking which was overheating the water in the rivers.

The decision of Harman J to grant injunctions against all three defendants was upheld by the Court of Appeal, although the operation of the injunction against Derby Corporation was suspended for 16 months to allow time for the Corporation to make alternative arrangements for sewage disposal.

9.8.5 Specific performance

A public authority may be sued for specific performance if it refuses to perform a contract which is both *intra vires* and specifically enforceable. Thus, in *Storer* v *Manchester City Council* [1974] 3 All ER 824, CA, the Court of Appeal ordered specific performance against a local housing authority of a contract to sell a council house to one of its tenants.

However, ordinary contractual principles apply in such cases, and, if there is no contract to enforce, specific performance will be refused. There must be a concluded contract, based on offer and acceptance, between the parties. In *Gibson* v *Manchester City Council* [1979] 1 All ER 972, HL, for example, the House of Lords held that the local housing authority had never made any offer to its tenant. There was, therefore, nothing for him to accept and no contract to enforce.

9.8.6 Declarations

As mentioned in para 9.7 above, an interested party can apply for a declaration of his rights under CPR, r. 40.20, in ordinary private law proceedings without at the same time claiming any other relief.

The declaration has proved in the past to be a valuable remedy in cases involving public authorities in all manner of situations. For example, in *Price* v *Sunderland Corporation* [1956] 3 All ER 153 a declaration was granted that the dismissal of teachers who refused to collect school dinner money was unlawful, while in *Sim* v *Rotherham Metropolitan Borough Council* [1986] 3 All ER 387 declarations were granted *in favour of* the employers to the effect that teachers were in breach of contract in refusing to 'cover' for absent colleagues, and that their employers were entitled to make deductions from salaries in such circumstances.

Other examples of the use of declarations in private law proceedings against public authorities are *Barnard* v *National Dock Labour Board* [1953] 2 QB 18, CA, in which the claimants were granted a declaration that their suspension from work was *ultra vires* and void, and *Ridge* v *Baldwin* [1964] AC 40, HL, where a chief constable dismissed from his post without being given a hearing was granted a declaration that his dismissal was null and void by reason of violation of the *audi alteram partem* principle of natural justice.

A declaration is available against the Crown in private law proceedings, whereas the remedies of injunction and specific performance are generally not (Crown Proceedings Act 1947, s. 21; and see *Dyson* v *Attorney-General* [1911] 1 KB 410, CA; *Congreve* v *Home Office* [1976] 1 QB 629, CA; *Factortame Ltd* v *Secretary of State for Transport (No. 1)* [1989] 2 All ER 692, HL; *Factortame Ltd* v *Secretary of State for Transport (No. 2)* [1991] 1 All ER 70, CJEC and HL; *M* v *Home Office* [1993] 3 All ER 537, HL, para 10.8.6.4).

In recent years, applications in private law proceedings for declarations under what is now CPR, r. 40.20, have been used as a means of testing the lawfulness of proposed surgical and medical procedures to be performed on adult mentally handicapped persons who are incapable of giving their own consent, and the courts have responded by making what are, in effect, *advisory* declarations. (See, for example, *T* v *T* [1988] Fam 62, where declarations were granted to the effect that it would not be unlawful to perform abortion and sterilisation operations on the defendant, a pregnant woman aged 19 who was epileptic and severely mentally handicapped.)

In *F* v *West Berkshire Health Authority* [1989] 2 All ER 545, HL, the House of Lords approved this practice. At the same time (Lord Griffiths dissenting on this point), their Lordships emphasised that a declaration is not essential in order to establish the lawfulness of such procedures since they may lawfully be performed at common law if they are in the best interests of the patient. According to their Lordships, the obtaining of a declaration is, however, *desirable* as a matter of good practice because a declaration of the court will establish whether the procedure is in the best interests of the patient and, therefore, lawful, and will protect the surgeon and others involved from subsequent criticism and legal action (see [1989] 2 All ER 545 *per* Lord Brandon at p. 552).

In determining whether the proposed procedure is in the best interests of the patient, the test is, applying *Bolam* v *Friern Hospital Management Committee* [1957] 2 All ER 118, whether it is accepted as appropriate treatment at the time by a reasonable body of medical opinion skilled in that particular form of treatment.

In what is a rapidly developing area of law, the use of the declaration has been extended, first, to cover the situation where the patient is not mentally handicapped but is otherwise refusing consent to treatment (as in *Re T (adult: refusal of medical treatment)* [1992] 4 All ER 649, CA, below), and, secondly, for the purpose of sanctioning the withdrawal of life-support facilities from a patient who is in a persistent vegetative state (as in *Airedale NHS Trust* v *Bland* [1993] 1 All ER 821, HL, below).

In *Re T (adult: refusal of medical treatment)*, CA, above, it was declared that it would not be unlawful to administer a blood transfusion to the patient (who was in her early 20s) following an emergency Caesarian section because her apparent decision to refuse treatment was vitiated by her medical condition and by reason of the undue influence of her mother, a Jehovah's Witness, who believed that the transfusion of blood was sinful. (See also *Re S (adult: refusal of medical treatment)* [1992] 4 All ER 671.)

Airedale NHS Trust v *Bland*, HL, above, concerned Tony Bland, a young man who had been in a persistent vegetative state for three and a half years after suffering catastrophic and irreversible brain damage due to a severe crushed chest injury sustained in the Hillsborough football ground disaster of April 1989. His parents and family now wanted him to be allowed to die peacefully and with dignity. It was decided that, in the best interests of the patient, declarations would be granted to the health authority responsible for Tony Bland's care to the effect that all life-sustaining treatment could lawfully be discontinued, and that there was no legal requirement thereafter to provide medical treatment, except for the sole purpose of allowing him to die peacefully with the greatest dignity and the least distress.

The House of Lords upheld these declarations but emphasised that the taking of positive steps to end a patient's life (euthanasia), such as by administering a drug to bring about death, remains unlawful.

Applying *F* v *West Berkshire Health Authority* [1989] 2 All ER 545, above, their Lordships said in the *Bland* case that an application for a declaration is the appropriate means of seeking the guidance of the court before withdrawing life-supporting treatment from a patient in a persistent vegetative state. It was said that, for the time being at least, doctors should as a matter of practice (designed to protect patients and doctors and to reassure patients' families and the public) continue to seek that guidance. The hope was expressed that, in time, applications for declarations could be limited to special cases as and when a body of experience and practice was built up in the Family Division of the High Court. (See *Airedale NHS Trust* v *Bland* [1993] 1 All ER 821, HL, *per* Lords Keith, Goff, and Lowry at pp. 862, 874, and 876, respectively.)

A standard form of declaration for use in cases of adult incapacity is set out in *Practice Note* (*Declaratory Proceedings: Medical and Welfare Decisions for Adults who lack Capacity*) [2002] 1 WLR 325. Applications must usually be made to, and will be heard by a judge of, the Family Division of the High Court (*Practice Direction* (*Declaratory Proceedings: Incapacitated Adults*), ibid.).

It is now more likely that public law, rather than private law, proceedings will be used to obtain a declaration against a public authority. As a result of procedural changes introduced in 1978 and the decision of the House of Lords in *O'Reilly* v *Mackman* [1982] 3 All ER 1124, HL, declaratory relief is not now normally available in private law proceedings under CPR, r. 40.20, against a public authority for infringement of a right protected by *public*, as opposed to *private*, law. Where it is desired to protect a *public law* right, the aggrieved citizen must normally proceed by way of an application for judicial review in public law proceedings under CPR, Part 54, rather than by way of a private law action.

If, however, the right to be protected is exclusively or substantially a matter of *private law* (as in *Price* v *Sunderland Corporation* [1956] 3 All ER 153, *Sim* v *Rotherham Metropolitan Borough Council* [1986] 3 All ER 387, and the medical cases cited above), then a private law action for a declaration remains available against a public authority notwithstanding the pre-eminence of public law proceedings for judicial review. Furthermore, *O'Reilly* v *Mackman* itself envisaged at least two exceptional cases involving *public law* rights where it may nevertheless be appropriate to use *private law* rather than judicial review proceedings to seek a declaration.

418   REMEDIES IN PRIVATE LAW

The first exception is where the alleged invalidity of a decision made by a public authority (i.e., the *public law* element in the case) is merely collateral to a claim for infringement of a right arising under *private law*. The second is where none of the parties objects to proceeding by way of a private law action instead of by means of an application for judicial review. It was said in *O'Reilly* v *Mackman* that other exceptions may be developed in the course of time on a case-by-case basis.

10

Remedies in public law

10.1 Introduction

It was noted in Chapter 9 that some of the remedies discussed there may be available in appropriate circumstances in *private law* proceedings against public authorities in order to prevent or redress wrongful conduct, including *ultra vires* acts and violations of the principles of natural justice (see para 9.8.1).

In addition, other remedies are available in *public law* proceedings for keeping public bodies, such as central government departments and local government authorities, within their powers, for ensuring that they observe the principles of natural justice, and for compelling them otherwise to act lawfully. These public law remedies form the basis of this chapter. They comprise the *non-prerogative* remedies of declarations (para 10.2 below) and relator actions (para 10.3 below), and the *prerogative* remedies of quashing, prohibiting, and mandatory orders (paras 10.5–10.7 below).

In 1982, in the important case of *O'Reilly* v *Mackman* [1982] 3 All ER 1124, HL, the House of Lords acknowledged that the distinction between *public* law and *private* law had assumed a new significance. This development has serious implications for the choice of remedies in cases involving public authorities, and means that private law and public law remedies are to be more rigidly segregated than before.

The effect of *O'Reilly* v *Mackman* on the availability of private law and public law remedies against public authorities is noted at greater length later in this chapter (paras 10.2.1 and 10.8.2.1 below) but, in brief, it may be said here that where it is sought to protect a right conferred by *public*, as opposed to *private*, law then the appropriate relief should be claimed in an application for judicial review in public law rather than in an ordinary action in private law.

The reasons for the procedural change ordered by the House of Lords in *O'Reilly* v *Mackman* are explained in para 10.8.2.1 below, together with the disadvantages of the change for the ordinary citizen and the advantages for public authorities. It is important, however, to appreciate from the outset that, on an application for judicial review, the applicant is not confined to the prerogative remedies but may claim in the same application appropriate *non-prerogative* relief in the form of a declaration, an injunction, or damages (see further, para 10.8.6 below). The old remedies are still available; only the procedure for obtaining them against public authorities has changed.

Prohibiting orders, injunctions, and declarations may be applied for *in advance of*, and quashing orders and damages may be claimed *after*, the commission of a wrongful act by a public authority.

10.2 Declarations

10.2.1 Nature and availability

A declaration from the court states the legal position of the parties but does not compel or forbid the taking of any action. Declarations are available against public authorities in both private law (para 9.8.6) and public law proceedings. However, as a result of *O'Reilly* v *Mackman* [1982] 3 All ER 1124, HL, declaratory relief cannot normally be claimed in private law proceedings for infringement of a right protected by *public*, as opposed to *private*, law.

In general, the correct procedure for protecting a *public law* right is for the aggrieved citizen to proceed by way of an application for judicial review under CPR, Part 54, rather than by way of a private law action for a declaration under CPR, r. 40.20. An attempt to protect a public law right by proceeding against a public authority in a private law action will in most cases be contrary to public policy, and, as such, an abuse of the process of the court which will result in the action being struck out (*O'Reilly* v *Mackman*, above, *per* Lord Diplock at p. 1134).

10.2.2 Scope

10.2.2.1 To challenge or confirm the validity of an administrative decision

In *Prescott* v *Birmingham Corporation* [1955] Ch 210, CA, the Court of Appeal granted to a ratepayer a declaration that the Corporation's scheme to provide free bus travel for old people resident in the city was *ultra vires* and void. It was said that the Corporation's scheme was an abuse of its statutory discretion to charge 'such fares as they may think fit'. The Corporation's statutory power to operate a transport service on commercial lines may have anticipated financial loss, but did not authorise the subsidisation of one class of the community at the expense of another merely out of benevolence or philanthropy.

In *Congreve* v *Home Office* [1976] 1 QB 629, CA, a private citizen obtained from the Court of Appeal a declaration that the Home Secretary's threat to revoke a television licence, bought at a fee of £12 before the expiry of an existing one and in anticipation of a rise in the fee to £18, was unlawful.

Both *Prescott* v *Birmingham Corporation* and *Congreve* v *Home Office*, above, were claims for a declaration in *private law* proceedings. If cases like them were to arise again, the decision in *O'Reilly* v *Mackman* [1982] 3 All ER 1124, HL, above, would insist that proceedings be started by way of an application for judicial review and not by means of a private law action for a declaration.

In appropriate circumstances, an application for a declaration can be used to challenge the validity of non-statutory government advice. Thus, in *Royal College of Nursing of the United Kingdom* v *DHSS* [1981] 1 All ER 545, HL (para 4.10.2.3), the House of Lords granted a declaration that advice on medically-induced abortions contained in a DHSS circular did not involve the performance of unlawful acts by nurses. In *Gillick* v *West Norfolk and Wisbech Area Health Authority* [1985] 3 All ER

402, HL (para 10.8.2.3 below), the House was prepared to hear a case in which advice in a DHSS circular on contraception was challenged, although in the end the House set aside the declaration made in the Court of Appeal to the effect that the advice was unlawful.

In the *Gillick* case, Lord Bridge, clearly perturbed by this new judicial incursion into the field of administrative discretion, gave a warning (at p. 427) that in such cases the court must be careful to confine itself to deciding only whether the proposition of law contained in the non-statutory advice is erroneous. In particular, the court must 'avoid . . . expressing *ex cathedra* opinions in areas of social and ethical controversy in which it has no claim to speak with authority or proffering answers to hypothetical questions of law which do not strictly arise for decision'.

In a case of great significance, *Council of Civil Service Unions* v *Minister for the Civil Service* [1984] 3 All ER 935, HL (the '*GCHQ* case'), the House of Lords held that judicial review is available to challenge the validity of an administrative decision where that decision is taken in the exercise of a delegated prerogative power. Although no declaration was granted in the GCHQ case, it was made clear that the minister's unilateral decision, taken under a delegated prerogative power, to withdraw trade union membership from civil servants employed at GCHQ had prima facie infringed the legitimate expectation of consultation enjoyed by those concerned. Only the pressing issue of national security prevented their Lordships from granting a declaration that the minister's decision was unlawful.

The issue of whether a power exercised *directly* under the prerogative is subject to judicial review was left open. However, the conventional wisdom, expressed in such cases as *Attorney-General* v *De Keyser's Royal Hotel Ltd* [1920] AC 508, HL, to the effect that the courts will inquire into the existence and extent of a claimed prerogative power, but not into the way a prerogative power is actually exercised in individual cases, was seriously questioned in the *GCHQ* case (see especially *per* Lords Scarman and Roskill at pp. 948 and 955–6, respectively).

It was later held by the Court of Appeal in *R* v *Secretary of State for Foreign and Commonwealth Affairs (ex parte Everett)* [1989] 1 All ER 655, CA, that judicial review can be used to question whether the Foreign Secretary, in the direct exercise of a prerogative power, has wrongly refused to issue a passport to a citizen. Although on the facts of the case no relief was granted, the Court of Appeal made it clear that the courts do have jurisdiction to inquire into the manner of exercise of a prerogative power, provided that the subject matter of the particular power is justiciable (see *per* O'Connor and Taylor LJJ at pp. 658 and 660, respectively).

It is thus now established that some aspects, at least, of the exercise of the royal prerogative are amenable to judicial review. Which aspects are reviewable will be for the courts to decide on a case-by-case basis. In *R* v *Secretary of State for the Home Department (ex parte Bentley)* [1993] 4 All ER 442, DC, it was decided that the exercise by the Home Secretary of the royal prerogative of mercy is reviewable by the courts. Although the court made no declaration or other formal order against the Home Secretary, it did suggest that he look again at the case of Derek Bentley with a view to deciding whether it would be right to grant him a partial posthumous pardon in recognition of the generally accepted view that he should not have been hanged in 1953. (For the background to this case, see para 7.5.8.)

Delivering the judgment of the court in the *Bentley* case, Watkins LJ said (at pp. 452–3) that it would be regrettable if the exercise of the prerogative of mercy were to be immune from legal challenge, regardless of the gravity of any legal errors made by the Home Secretary, given that this particular prerogative power is exercised by the minister on behalf of society as a whole and is an important feature of our criminal justice system. He pointed out that the *GCHQ* case had shown that the powers of the court cannot be ousted simply by invoking the word 'prerogative', and that the court can intervene when the nature and subject matter of a particular ministerial decision is amenable to the judicial process. It is amenable when the court is qualified to deal with the matter. It is not amenable when it involves such questions of policy that the court should not intervene because it is ill-equipped to do so.

Watkins LJ concluded (at p. 453) that there must be cases in which the exercise of the royal prerogative of mercy is reviewable. He gave the example of a situation where it was clear that a pardon had been refused to a person solely on the ground of sex, race, or religion. Here, he said, the court would be expected, and entitled, to interfere. Lord Roskill's statement in the *GCHQ* case ([1984] 3 All ER 935 at p. 956) that the exercise of the prerogative of mercy was a non-justiciable issue was described by Watkins LJ as a 'passing reference' and clearly *obiter*. (See further on justiciable issues, para 10.2.3 below.)

10.2.2.2 To challenge or confirm the validity of a judicial decision, whether of a statutory tribunal or of a voluntary tribunal

In *Barnard v National Dock Labour Board* [1953] 2 QB 18, CA, the claimants were granted a declaration that their suspension from work was *ultra vires* and void. The National Dock Labour Board had a duty under statutory regulations to delegate disciplinary functions to local boards. One such local board, the London Dock Labour Board, purported to sub-delegate these disciplinary functions, which included the power to suspend a worker, to the port manager. The port manager suspended the claimants from work and pay for refusing to obey instructions. They appealed to the statutory appeal tribunal, which dismissed their appeals. They then applied to the High Court for a declaration that their suspension was wrongful on the facts of the industrial dispute. When they later discovered that they had been suspended by the port manager and not by the local board, they claimed a further declaration that their suspension was *ultra vires* and void.

The Court of Appeal held that the local board had acted *ultra vires* because it had no power to delegate its disciplinary functions to the port manager. The power of suspension was a judicial, and not an administrative, function. Administrative functions can often be delegated, but judicial functions can only be delegated by a judicial tribunal if it is empowered to do so expressly or by necessary implication, as in *Local Government Board v Arlidge* [1915] AC 120, HL (para 10.5.3.2.2 below). There was nothing in the dock labour scheme which expressly authorised delegation of the power of suspension and authorisation could not be implied.

Barnard v National Dock Labour Board [1953] 2 QB 18, CA, was followed by the House of Lords in *Vine v National Dock Labour Board* [1957] AC 488, HL, although their Lordships were uncertain whether the function in question was judicial or administrative. Both of these cases were claims for a declaration in *private law* proceedings. However, since they involved questions of *private* law as well as questions of public law, cases similar to them could today still be commenced by way of a private law action under one of the exceptions recognised in *O'Reilly* v *Mackman* [1982] 3 All ER 1124, HL, above.

10.2.2.3 To challenge or confirm both the validity of delegated legislation and the legality of the exercise of delegated legislative powers

10.2.2.3.1 *Introduction* The declaration can be used as a remedy to strike down *ultra vires* delegated legislation and government action, although, as will be seen in para 10.2.2.3.2 below, it is not the only method available for challenging the *vires* of delegated legislation.

In *Agricultural, Horticultural & Forestry Industry Training Board* v *Aylesbury Mushrooms Ltd* [1972] 1 All ER 280, the Minister of Labour had purported to establish an industrial training board by means of an order made in 1966 under what was the Industrial Training Act 1964. He was under a statutory duty, before making any such order, to 'consult any organisation . . . appearing to him to be representative of substantial numbers of employers engaging in the activities concerned'. The minister did not consult the Mushroom Growers' Association, which represented about 85 per cent of all mushroom growers in England and Wales. Donaldson J granted a declaration that this omission rendered the minister's order invalid as against mushroom growers.

A particular regulation made under the authority of the Customs and Excise Management Act 1979 was declared *ultra vires* the Act in *R* v *Customs and Excise Commissioners (ex parte Hedges & Butler Ltd)* [1986] 2 All ER 164, DC, because it went further than was permitted by the Act.

In *R* v *Secretary of State for the Home Department (ex parte Simms)* [1999] 3 All ER 400, HL, it was held that the Home Secretary's policy (adopted in the exercise of his powers under the Prison Act 1952 and the Prison Rules 1964) of imposing an almost blanket ban on oral interviews between journalists and prison inmates was *ultra vires* and unlawful. The House of Lords granted a declaration to that effect. The Home Secretary's policy was unlawful because it effectively deprived a prisoner of what was described as 'a fundamental right' to seek to persuade a journalist, by means of oral interviews, to investigate the safety of the prisoner's conviction and to publicise his findings in an effort to gain access to justice for the prisoner (see *per* Lords Steyn, Hobhouse, and Millett at pp. 410, 422, and 424, respectively, and note that the Prison Rules 1964 have been replaced by the Prison Rules 1999 (SI 1999, No. 728)).

An effective blanket ban which prohibited prisoners from contacting the media by telephone for the purpose of commenting on matters of legitimate public interest relating to prisons and prisoners was declared unlawful in *R (Hirst)* v *Secretary of State for the Home Department* (2002) *The Times*, 10 April.

The House of Lords declared in *R (Daly)* v *Secretary of State for the Home Department* [2001] 2 WLR 1622, HL, that the Home Secretary's policy (set out in a Security Manual issued to prison governors) that prisoners be absent from their cells when their legally privileged correspondence was examined was both *ultra vires* the Prison Act 1952 and a violation of art. 8 (right to respect for correspondence) of the European Convention on Human Rights.

In *R (Mellor)* v *Secretary of State for the Home Department* [2001] 3 WLR 533, CA, the claimant, who was a mandatory life prisoner, sought judicial review of the Home Secretary's decision to reject his request for access to artificial insemination facilities in prison. There were no exceptional circumstances in the case of the claimant and his wife such as to satisfy the policy of the prison service on granting such access, and the claimant's request was rejected by the Home Secretary on the ground that there was no medical need for artificial insemination, and out of concern about the long-term stability of the marriage. The Court of Appeal refused a declaration that the decision was unlawful, irrational, or in violation of art. 12 (right to found a family) of the European Convention.

A decision of the Home Secretary was declared to be irrational in *R (Asif Javed)* v *Secretary of State for the Home Department* [2001] 3 WLR 323, CA. He had decided to designate Pakistan in the Asylum (Designated Countries of Destination and Designated Safe Third Countries) Order 1996 (SI 1996, No. 2671) as a country in respect of which there was 'in general no serious risk of persecution'. The Court of Appeal declared his decision to be irrational in view of the available evidence concerning the treatment of women and Ahmadis (a minority religious sect) in Pakistan (as described in the earlier House of Lords' case of *R* v *Immigration Appeal Tribunal (ex parte Shah)* [1999] 2 AC 629, HL).

10.2.2.3.2 *Alternative methods of challenging the validity of delegated legislation* It is possible to attack the validity of delegated legislation in other ways without seeking a declaration or otherwise resorting to judicial review. For example, it may be done by way of defence to a criminal charge in the magistrates' court or the Crown Court, in which event the trial court may be expected to decide the question of validity. Whether this course is available depends on the construction of the parent statute. If this indicates that judicial review is excluded, the criminal court can, in general, determine the question of validity.

This was decided by the House of Lords in *R* v *Wicks* [1997] 2 All ER 801, HL, doubting the correctness of *Bugg* v *DPP* [1993] 2 All ER 815, DC, in which it had been held that whether a criminal court had jurisdiction to investigate the validity of a by-law depended on the distinction between substantive invalidity (in which case the court could investigate) and procedural invalidity (in which case it could not).

Bugg v *DPP*, above, was eventually overruled by the House of Lords in *Boddington* v *British Transport Police* [1998] 2 All ER 203, HL. Here, the defendant, Mr Boddington, had been convicted by a stipendiary magistrate of an offence contrary to the British Railways Board's by-laws made under statutory authority, namely smoking a cigarette in a railway carriage where smoking was forbidden. His appeal against conviction by way of case stated to the Queen's Bench Divisional Court was dismissed, and he appealed to the House of Lords.

Their Lordships confirmed what was said in *R* v *Wicks*, above, and held that in such cases no distinction is to be made between substantive and procedural invalidity. It was further held that there was nothing in the parent statute or the by-laws to indicate that Parliament had intended to deprive the smoker of the opportunity to defend himself in a criminal court by alleging the invalidity of the by-law. However, on the facts, the appeal was dismissed because the relevant by-law, being wide enough to allow the imposition of a ban on smoking in all carriages, was not *ultra vires*.

As a result of *Boddington* [1998] 2 All ER 203, unless the relevant statutory provisions indicate the contrary, it is open to a defendant in criminal proceedings to defend himself by challenging the *procedural*, as well as the *substantive*, validity of the delegated legislation under which he stands charged.

Boddington, above, was considered in *Secretary of State for Defence* v *Percy* [1999] 1 All ER 732, where it was held that once a criminal court has decided that by-laws are invalid the maker of them must respect that decision and not put up, or keep up, notices of by-laws which are known to be invalid. It was further held, however, that members of the public are not entitled to go on to private land to remove such notices.

Another alternative way of challenging the validity of delegated legislation is on an appeal by way of case stated from the magistrates to the Queen's Bench Divisional Court, as in *Chester* v *Bateson* [1920] 1 KB 829, DC, in which reg. 2A(2) of the Defence of the Realm Regulations 1914 was held to be *ultra vires* the Defence of the Realm Consolidation Act 1914, or *Dunkley* v *Evans* [1981] 3 All ER 285, DC, where part of the West Coast Herring (Prohibition of Fishing) Order 1978 was held to be *ultra vires* the parent Act.

Or, again, the challenge may be made in an ordinary civil action between the parties, such as a claim for a sum of money, as in *Commissioners of Customs & Excise* v *Cure & Deeley Ltd* [1962] 1 QB 340, in which reg. 12 of the Purchase Tax Regulations 1945 was held to be *ultra vires* the Finance (No. 2) Act 1940.

10.2.2.3.3 *Severance* Instead of striking down a piece of delegated legislation totally, it may be possible in some cases to *sever* the bad part from the good part and leave the latter intact. Severance is not possible where the bad part is inextricably interconnected with the good part. Severance was achieved in the *Aylesbury Mushrooms* case [1972] 1 All ER 280 (para 10.2.2.3.1 above), where the minister's order was invalid as against mushroom growers but valid in relation to others affected by it, and in *Dunkley* v *Evans* [1981] 3 All ER 285, DC (para 10.2.2.3.2 above), where the order was invalid in relation to 0.8 per cent of the sea area covered by the order but valid in relation to the remaining 99.2 per cent.

In *DPP* v *Hutchinson* [1990] 2 All ER 836, HL, it was held that some of the Royal Air Force Greenham Common Byelaws 1985 were *ultra vires* the Military Lands Act 1892 and that their invalidity could not be cured by severance since that would result, in effect, in the enforcement of by-laws radically different from those in fact made. The appellants, who had been convicted of offences against the by-laws, were held to have been wrongly convicted and their appeals were allowed.

10.2.2.4 To clarify the rights of public employees

An example of this is provided by the case of *Ridge* v *Baldwin* [1964] AC 40, HL, which arose out of the dismissal of a Chief Constable without a hearing. He was granted a declaration that his dismissal was null and void. This meant that he had been lawfully employed all the time. He had also claimed damages and, in a compromise settlement, he received some back pay from the date of the wrongful dismissal to the date of the settlement and, from that date, a pension.

One important effect of *Ridge* v *Baldwin*, above, is that the applicability of natural justice no longer depends on any distinction between administrative, judicial, or quasi-judicial acts (see *Leech* v *Parkhurst Prison Deputy Governor* [1988] 1 All ER 485, HL, *per* Lord Oliver at p. 505). A person is entitled to a fair hearing if he has some right, interest, or legitimate expectation which it would be unfair to deprive him of without a hearing (see further, *Breen* v *Amalgamated Engineering Union* [1971] 2 QB 175, CA, *per* Lord Denning MR at p. 191).

Ridge v *Baldwin*, above, was applied by the House of Lords in *Chief Constable of the North Wales Police* v *Evans* [1982] 3 All ER 141, HL, a case in which the treatment of the respondent, a probationer constable, was described by Lord Hailsham LC (at p. 143) as 'little short of outrageous'.

The Chief Constable, acting on rumours about the respondent's private life, decided to dispense with the respondent's services. Without putting the rumours to him or allowing him to make any representations, the Chief Constable told the respondent that if he did not resign he would be discharged under the Police Regulations. The respondent resigned. He obtained leave to challenge the Chief Constable's actions by way of judicial review. The House of Lords held that the Chief Constable had acted unlawfully and in breach of his duty under the Police Regulations. He was wrong to assume that he had an absolute discretion under the regulations to dispense with the services of a probationer constable, and he had totally failed to observe the rules of natural justice.

The respondent was granted declarations to the effect that the Chief Constable had acted unlawfully and that the respondent, by reason of his unlawfully-induced resignation, was entitled to the same rights and remedies, not including reinstatement, as he would have had if the Chief Constable had unlawfully dispensed with his services under the regulations. A mandatory order to compel the reinstatement of the respondent was refused (see para 10.7.1 below).

10.2.2.5 To secure the recognition of a right, which is being denied, to engage in a particular occupation or activity

Examples of declarations being sought for this purpose are *Associated Provincial Picture Houses* v *Wednesbury Corporation* [1948] 1 KB 223, CA (unsuccessful claim for declaration that condition attached to licence for Sunday cinema performances was *ultra vires* and unreasonable), and *Pyx Granite Co.* v *Ministry of Housing & Local Government* [1960] AC 260, HL (successful claim for declaration that quarrying operations authorised by statute could be continued without obtaining further official permission).

10.2.3 Disadvantages of declarations

A declaratory judgment merely states the legal position of the parties. It does not order or prohibit the taking of any action. It cannot quash a decision made by a public authority within its jurisdiction (*Punton* v *Ministry of Pensions & National Insurance (No. 2)* [1964] 1 All ER 448, CA). Being merely declaratory, it cannot be directly enforced. Disobedience to it is not contempt of court (*Webster* v *Southwark London Borough Council* [1983] 2 WLR 217). It is, however, always assumed that a public authority, especially the Crown, will not flout the law once it has been declared by the court.

The remedy is discretionary and will be granted only in exceptional circumstances (*Maxwell* v *Department of Trade & Industry* [1974] 2 All ER 122, CA). A declaration must be appropriate in the circumstances (*Williams* v *Home Office (No. 2)* [1981] 1 All ER 1211 *per* Tudor Evans J at p. 1248).

Save in exceptional circumstances, the discretion should not be exercised so as to grant a declaration in a civil court to a defendant in criminal proceedings that the facts alleged by the prosecution do not in law prove the offence charged (*Imperial Tobacco Ltd* v *Attorney-General* [1980] 1 All ER 866, HL; *Attorney-General* v *Able* [1984] 1 All ER 277; *R* v *Director of Public Prosecutions (ex parte Camelot Group plc)* (1997) *The Independent*, 22 April, DC; *R (Rusbridger)* v *Attorney-General* [2003] 3 WLR 232, HL). Such a declaration would not be binding on the criminal court but might prejudice the criminal trial. Nor should a final declaration be made if there remains a dispute about the material facts of the case (*Commissioners of Inland Revenue* v *Rossminster Ltd* [1980] 1 All ER 80, HL).

The application for a declaration must raise a justiciable issue; i.e., an issue which is capable of determination by the application of legal principles. Questions of morality, as opposed to law, are non-justiciable in this sense. Thus, in *Cox* v *Green* [1966] 1 Ch 216, medical ethics were held not to be a justiciable issue.

The question whether a government decision or act was in fact necessitated by the requirements of national security is a non-justiciable issue (*Council of Civil Service Unions* v *Minister for the Civil Service* [1984] 3 All ER 935, HL), as is probably also the exercise of such prerogative powers as the making of treaties, the grant of honours, the dissolution of Parliament, and the appointment of ministers (ibid., *per* Lord Roskill at p. 956; *R* v *Secretary of State for Foreign and Commonwealth Affairs (ex parte Everett)* [1989] 1 All ER 655, CA, *per* O'Connor and Taylor LJJ at pp. 658 and 660, respectively; in *R* v *Secretary of State for Foreign and Commonwealth Affairs (ex parte Rees-Mogg)* [1994] 1 All ER 457, DC (para 4.7.1) the court inclined to the view that a question relating to the transfer or exercise of the prerogative power to conduct foreign security policy was a non-justiciable issue).

Questions of academic judgment, such as academic competence or what mark or class of degree should be awarded to a student, are beyond the experience of the courts and are likewise non-justiciable (*Clark* v *University of Lincolnshire and Humberside* [2000] 3 All ER 752, CA, *per* Sedley LJ at [12]).

There must be a real dispute between the parties; the action for a declaration cannot be used to seek an answer to hypothetical questions (*Russian Commercial & Industrial Bank* v *British Bank for Foreign Trade Ltd* [1921] AC 438, HL, *per* Lord Sumner at p. 452; *Blackburn* v *Attorney-General* [1971] 2 All ER 1380, CA; *R (Rusbridger)* v *Attorney-General* [2003] 3 WLR 232, HL).

The claimant must have *locus standi* ('a place of standing') in order to apply for a declaration in his own name against a public authority. *Locus standi* means, in effect, that the court will not grant the claimant a declaration unless he has a sufficient interest in the matter to which the application relates. It is a rule of procedure which is designed to discourage busybodies from meddling in other people's affairs and occupying the time of the courts. The question of *locus standi* is a fundamental one which affects the jurisdiction of the court. If the claimant has no *locus standi*, then the court has no jurisdiction to hear the case and it is not open to the parties to purport to confer the necessary jurisdiction by agreeing to waive the issue of *locus standi* (*R* v *Secretary of State for Social Services (ex parte Child Poverty Action Group)* [1989] 1 All ER 1047, CA).

A person has *locus standi* to take proceedings against a public authority if his legal rights are likely to be affected by the activities of the authority (as in *Ridge* v *Baldwin* [1964] AC 40, HL, para 10.2.2.4 above, and *Anisminic Ltd* v *Foreign Compensation Commission* [1969] 2 AC 147, HL, para 10.11.3 below). Some other special interest falling short of a legal right will probably also suffice.

10.3 **Relator actions**

If a claimant's legal rights are not affected, and he has no special interest above that of the general public, he has no *locus standi* to apply for a declaration in his own name. He must use a different procedure called the relator action, which involves bringing in the Attorney-General. This procedure is especially appropriate in the case of public wrongs because an ordinary citizen is not entitled to bring *private law* proceedings, either for a declaration or for an injunction, for the purpose of preventing or remedying a *public* wrong.

This was firmly established in *Gouriet* v *Union of Post Office Workers* [1977] 3 All ER 70, HL, in which the House of Lords held that the claimant (whom the Attorney-General had refused to assist) was not entitled to proceed with his application against the defendants for a declaration and an interim injunction to prevent the defendants' threatened boycott of mail to South Africa. In so holding, the House of Lords unanimously reversed a unanimous Court of Appeal.

It should be noted, however, that the *Gouriet* case does not affect the alternative remedy of judicial review. A private citizen who is unable to obtain the appropriate relief in private law, and who cannot enlist the help of the Attorney-General, may nevertheless seek to prevent or remedy a public wrong by means of an application for judicial review under s. 31 of the Supreme Court Act 1981 (*Commissioners of Inland Revenue* v *National Federation of Self-Employed & Small Businesses Ltd* [1981] 2 All ER 93, HL, *per* Lords Diplock, Scarman, and Roskill at pp. 103, 110, and 116–17, respectively). He may obtain a declaration under that section if he has a sufficient interest in the matter and the case is one in which a prerogative order could be granted.

A relator action is brought by the Attorney-General 'on the relation' (at the instance) of some other person. The Attorney-General, as an officer of the Crown, represents the public interest. The relator action derives from a special power exercised by the Crown to ensure that public authorities keep within their powers. The Attorney-General can bring proceedings *ex officio* on his own initiative, but they are more often brought at the instance of a 'relator'. In either case, the decision whether to proceed is within the Attorney-General's unfettered discretion. Thus, the refusal of the Attorney-General to give his consent to a relator action is not subject to review by the courts (*London County Council* v *Attorney-General* [1902] AC 165, HL; *Gouriet* v *Union of Post Office Workers* [1977] 3 All ER 70, HL).

The conduct of relator proceedings is under the control of the Attorney-General (*Gouriet's* case, above, *per* Lord Wilberforce, Viscount Dilhorne, Lord Edmund-Davies, and Lord Fraser at pp. 80, 94, 106, and 117, respectively), but the costs must be paid by the relator.

The relator action is a potentially valuable procedure. The ordinary citizen, who may not have a sufficient interest to enable him to sue for a declaration or an injunction in his own name, may yet see justice done by calling in the Attorney-General, although it seems that he will not agree to a relator action where the defendant is a central government department or a government minister. An example of a relator action is provided by the case of *Attorney-General (on the relation of Yapp)* v *Fulham Corporation* [1921] 1 Ch 440. Here, the Corporation had a statutory power to set up public wash-houses where people could wash their own clothes. The Corporation set up and ran a municipal laundry where people's clothes were washed by employees of the Corporation. In an action brought by the Attorney-General on the relation of a ratepayer, Sargant J held that the Corporation was acting *ultra vires*. He granted a declaration to that effect and issued an injunction restraining the Corporation, its officers, servants, and agents, from acting in contravention of the declaration.

The development of judicial review has led to a decline in the use of relator actions, which are now rare.

10.4 The prerogative remedies and judicial review

10.4.1 Introduction

The three prerogative remedies available against public authorities on an application for judicial review are mandatory orders, prohibiting orders, and quashing orders. They were formerly known, and are sometimes still referred to as, mandamus, prohibition, and certiorari. Now, all references in any primary or secondary legislation to orders of mandamus, prohibition, and certiorari are to be read instead as references to mandatory, prohibiting, and quashing orders, respectively (Supreme Court Act 1981, s. 29, as amended by the Civil Procedure (Modification of Supreme Court Act 1981) Order 2004, SI 2004, No. 1033). The remedies were called 'prerogative writs' until their title was changed to 'prerogative orders' in 1938. Originally, they were writs brought by the King against his own officials to ensure that they exercised their functions properly and did not abuse their powers.

In modern times, the prerogative orders are available to both private citizens and public authorities. Moreover, *non-prerogative* remedies in the form of a declaration, an injunction, or damages can be claimed and granted on an application for judicial review. Prohibiting orders, injunctions, and declarations may be applied for *in advance of*, and quashing orders and damages may be claimed *after*, the commission of a wrongful act by a public authority.

Although set within the framework of *public* as distinct from *private* law, judicial review proceedings are essentially *civil* as opposed to *criminal* in character. The Crown is usually the nominal claimant but it is generally a private citizen who benefits from the proceedings if they are successful. It has accordingly been held that judicial review proceedings are not 'brought by or at the instigation of a public authority', within the meaning of s. 22(4) of the Human Rights Act 1998, so that a claimant is not entitled retrospectively to claim damages against a public authority under s. 7(1)(b) of the 1998 Act for acts committed before the Act came into force (*R (Ben-Abdelaziz)* v *Haringey London Borough Council* [2001] 1 WLR 1485, CA).

10.4.2 Grounds for judicial review

Historically, the usual ground for judicial review was based on the *ultra vires* doctrine (see para 9.8.1). This embraced such unlawful activity as doing the wrong thing, using the wrong procedure, and abusing a discretion.

Perhaps the best known attempt to classify the circumstances in which the courts will intervene to prevent, or remedy, abuses of power by public authorities was in *Associated Provincial Picture Houses Ltd* v *Wednesbury Corporation* [1948] 1 KB 223, CA, in which Lord Greene MR laid down the so-called '*Wednesbury* principles' of unreasonableness or perversity.

One of the most important modern attempts at classifying the grounds for judicial review, though not meant to be exhaustive, is that contained in Lord Diplock's opinion in *Council of Civil Service Unions* v *Minister for the Civil Service* (the '*GCHQ* case') [1984] 3 All ER 935, HL, with which the other Law Lords expressly agreed. Lord Diplock said (at p. 950) that the grounds on which administrative action is subject to control by judicial review can be classified under three heads:

(a) *illegality* (covering substantive *ultra vires*), where the public authority has made an error of law, such as purporting to exercise a power it does not possess;

(b) *irrationality* (in the *Wednesbury* sense of 'perversity'), where the authority has acted so unreasonably that no reasonable authority would have made the decision;

(c) *procedural impropriety* (covering procedural *ultra vires*, violation of the principles of natural justice, and legitimate expectation), where the authority has failed in its duty to act fairly.

This tripartite classification goes further than the *ultra vires* doctrine.

10.4.3 Proportionality

The three grounds listed above cover the matters dealt with in para 10.2.2 above, and paras 10.5.3, 10.6.3, and 10.7.3 below. Lord Diplock envisaged that further grounds might be added on a case-by-case basis. In particular, he suggested ([1984] 3 All ER 935 at p. 950) that English law might possibly adopt the principle of 'proportionality' at some time in the future.

This principle, which is recognised by the administrative law of several Member States of the European Union, dictates that administrative action can be struck down (even though not perverse or absurd) if it is *disproportionate to the benefit it seeks to obtain or the mischief it seeks to avoid*. It is intended to cover the situation where a sledgehammer is taken to crack a nut when there is an efficient pair of nutcrackers readily available (*R* v *Secretary of State for the Home Department (ex parte Brind)* (1989) *The Times*, 30 May, DC, *per* Watkins LJ; decision affirmed by the Court of Appeal at [1990] 1 All ER 469; decision of the Court of Appeal affirmed by the House of Lords, *sub nom. Brind* v *Secretary of State for the Home Department* [1991] 1 All ER 720).

These words of Watkins LJ were spoken in a case in which several journalists challenged the legality of the restrictions imposed by the Home Secretary on the broadcasting of interviews with members and representatives of some named Northern Ireland extremist groups. One of the grounds of challenge was that the restrictions were a disproportionate response to the mischief they sought to control, namely the excessive amount of publicity given to such persons and groups.

In upholding the legality of the Home Secretary's actions, the House of Lords made it clear that, while the principle of proportionality as understood elsewhere in Europe is not recognised by English law as a separate ground for reviewing administrative action, it is a relevant factor in a challenge based on *Wednesbury* unreasonableness or perversity. In other words, whereas administrative action which is not in itself irrational will not be struck down by the court merely because it is 'out of proportion' to the occasion, it may be struck down where that lack of proportion indicates irrationality (see, e.g., [1991] 1 All ER 720 *per* Lord Ackner at p. 735).

The principle of proportionality as a separate ground of challenge to administrative decision-making was rejected by a majority of their Lordships on the facts of *Brind* v *Secretary of State for the Home Department* for the reason that it would require the judges to consider the merits of the decision under review, and, if necessary, to substitute their own decision (although, presumably, Lord Templeman would have gone this far: see [1991] 1 All ER 720, 726).

It was suggested by two of the Law Lords that rejection of the proportionality principle in the present case did not close the door to its possible development in future cases, as envisaged by Lord Diplock in *Council of Civil Service Unions* v *Minister for the Civil Service* [1984] 3 All ER 935, HL (see *Brind* [1991] 1 All ER 720 *per* Lords Bridge and Roskill at pp. 724 and 725, respectively). Another considered that the recognition of the proportionality principle was surrounded by disadvantages which made it an impractical proposition (ibid., *per* Lord Lowry at p. 739).

That the proportionality doctrine is still not part of English domestic law, and is, therefore inapplicable in cases not involving European Community law or human rights issues, is clear from the decision of the Court of Appeal in *R (Association of British Civilian Internees: Far East Region)* v *Secretary of State for Defence* [2003] 3 WLR 80, CA (para 10.4.5 below).

10.4.4 Irrationality

The so-called '*Wednesbury* principles' of unreasonableness or perversity originated in the judgment of Lord Greene MR in *Associated Provincial Picture Houses Ltd* v *Wednesbury Corporation* [1948] 1 KB 223, CA. These principles are concerned with scrutinising, *inter alia*, what Lord Diplock later described as 'a decision which is so outrageous in its defiance of logic or of accepted moral standards that no sensible person . . . could have arrived at it' (*Council of Civil Service Unions* v *Minister for the Civil Service* [1984] 3 All ER 935 at p. 951).

A successful challenge to an administrative decision on the irrationality ground involves crossing, in every case, a high threshold. It is necessary to demonstrate that the decision is unreasonable in that it is 'beyond the range of responses open to a reasonable decision-maker' (*R v Ministry of Defence (ex parte Smith)* [1996] 1 All ER 257, CA, *per* Sir Thomas Bingham MR and Thorpe LJ at pp. 263 and 272, respectively, quoting from the submissions of counsel for the applicants). The threshold is lower where the decision affects human rights, especially the right to life. It is higher where the decision is of a 'policy-laden, esoteric or security-based nature' and remote from the ordinary experience of judges (ibid., *per* Sir Thomas Bingham MR at p. 264).

The irrationality threshold was held to have been crossed in *R (Asif Javed) v Secretary of State for the Home Department* [2001] 3 WLR 323, CA. Here, the Court of Appeal held that the Home Secretary had acted irrationally in deciding, against the weight of the available evidence, to designate Pakistan as a country in respect of which there was, for political asylum purposes, 'in general no serious risk of persecution'. (See further on this case, para 10.2.2.3.1 above.)

The decision of the Secretary of State for Health that the inquiry into the case of Harold Frederick Shipman (the general practitioner convicted in January 2000 of murdering 15 of his patients) would be held in private was held to be irrational in *R (Wagstaff) v Secretary of State for Health* [2001] 1 WLR 292, DC. The Secretary of State had not adequately consulted the claimant families and friends of the victims, or properly considered their wishes, before making his decision. In addition, he had failed adequately to take into account material considerations in the form of the persuasive reasons for holding the inquiry in public, which far outweighed those in favour of holding it in private. The matter was remitted to the Secretary of State for reconsideration, and it was eventually decided that the inquiry would be held in public under the chairmanship of a High Court judge. (See further on this case, para 10.4.5 below.)

The decision whether there should be an inquiry on a matter of public concern is essentially a political decision for a government minister, as is the decision whether it should be held in public. It is his responsibility to weigh up the rival considerations and determine whether the balance of advantage lies in having a public or a private inquiry. The courts are reluctant to interfere with this political balancing exercise.

The facts of the *Wagstaff* case, above, were highly exceptional and the case is probably the first one in which a minister's decision to hold an inquiry in private has been held to be unlawful. There certainly appears to be no presumption that all such inquiries will be held in public. It is, therefore, not surprising that *Wagstaff* was distinguished in *R (Persey) v Secretary of State for the Environment, Food and Rural Affairs* [2002] 3 WLR 704, DC (where it was held not irrational to order a private, as opposed to a public, inquiry into an outbreak of foot and mouth disease), and in *R (Howard) v Secretary of State for Health (Note)* [2002] 3 WLR 738 (where it was held not irrational to order a private inquiry into how the National Health Service handled allegations about the conduct of a medical practitioner).

In *R (H) v Ashworth Special Hospital Authority* [2003] 1 WLR 127, CA, a mental health review tribunal was held to have acted irrationally when it decided to discharge a patient with a history of serious violence, and in respect of whom previous attempts at release into the community had been unsuccessful, without there being any after-care facilities available to him. The tribunal's decision was described as 'one which no reasonable tribunal could properly have made' ([2003] 1 WLR 127, CA, *per* Dyson LJ at [68]).

In *R (Association of British Civilian Internees: Far East Region) v Secretary of State for Defence* [2003] 3 WLR 80, CA, the minister's decision to limit the potential beneficiaries of an *ex gratia* compensation scheme, established to compensate 'British' civilians who had been interned by the Japanese during the Second World War, to those born in the United Kingdom, or those with at least one parent or grandparent born in the United Kingdom, was held not to be irrational. The birth criteria adopted by the minister were judged to be a rational method of achieving the scheme's objective of paying to British civilian internees the debt of honour owed by the United Kingdom, even though their effect was to exclude some civilians who could not satisfy those criteria but who were British subjects by the time of their internment. (See further on this case, para 10.4.5 below.)

10.4.5 Legitimate expectation

The doctrine of legitimate expectation as a ground for judicial review has developed in relation to natural justice and reasonableness in situations where a citizen with no private law rights can legitimately expect in public law to be treated fairly by a public body (see further, para 10.8.4.3 below). The doctrine was explained in the following words by Lord Fraser in *Council of Civil Service Unions v Minister for the Civil Service* [1984] 3 All ER 935, HL (the '*GCHQ* case'), at pp. 943–4:

But even where a person claiming some benefit or privilege has no legal right to it, as a matter of private law, he may have a legitimate expectation of receiving the benefit or privilege, and, if so, the courts will protect his expectation by judicial review as a matter of public law . . . Legitimate . . . expectation may arise either from an express promise given on behalf of a public authority or from the existence of a regular practice which the claimant can reasonably expect to continue.

A legitimate expectation can thus arise in two ways: from an express promise, as in *R v Liverpool Corporation (ex parte Liverpool Taxi Fleet Operators' Association)* [1972] 2 QB 299, CA, and *Attorney-General of Hong Kong v Ng Yuen Shiu* [1983] 2 All ER 346, PC, or from a regular practice, as in the *GCHQ* case itself.

In the *Liverpool* case, above, the Corporation promised to consult the Taxi Fleet Operators' Association before deciding to grant new taxi licences. The Corporation later decided to increase the number of licences without consultation. The Court of Appeal quashed the Corporation's decision because it would have been unfair to allow it to go back on its promise. In *Attorney-General of Hong Kong v Ng Yuen Shiu*, above, a senior immigration officer announced a government policy that illegal immigrants would be interviewed, and each case treated on its merits, before being sent back home. It was held that this announcement gave rise to a legitimate expectation that an illegal immigrant would be allowed to make representations before the authorities decided on repatriation.

The doctrine of legitimate expectation was central to the *GCHQ* case [1984] 3 All ER 935, HL. Although the House of Lords held on the facts that the minister had acted lawfully in withdrawing the right to trade union membership from civil servants employed at Government Communications Headquarters (GCHQ) without consulting them, it was made abundantly clear that the decision might have been different if the case had not involved national security at all, or if the minister had failed to produce any evidence (beyond a bald assertion) that it did. The civil servants at GCHQ had no *legal right* to prior consultation. Nevertheless, their Lordships were unanimous in the view that the *regular practice* of consultation between management and the unions about changes in terms and conditions of employment had given rise to a *legitimate expectation* of consultation.

If a public body gives notice of a change in policy (as in *R v Secretary of State for the Home Department (ex parte Hargreaves)* [1997] 1 All ER 397, CA, below) this destroys any expectations based on the earlier policy. If this were not the case, the government, for example, would never be able to change its policies. However, the public body may be compelled to honour the old policy if the change is irrational, or if a promise has been made to the particular claimant, as in *R v North and East Devon Health Authority (ex parte Coughlan)* [2000] 3 All ER 850, CA, below.

In the *Hargreaves* case, above, three prison inmates applied for judicial review of the Home Secretary's decision to implement a new home leave scheme under which eligibility to apply for leave arose after serving *one half* of a sentence. When the applicants began their sentences, eligibility arose after serving *one third* of a sentence and they were given a notice confirming this and explaining that home leave was a privilege. They also signed an 'inmate compact', under which the prison promised to consider them for home leave when they became eligible. The applicants' challenge to the Home Secretary's decision was based, *inter alia*, on the doctrine of legitimate expectation.

It was held that the applicants had no legitimate expectation of being considered eligible for leave after serving one third of their sentences since, otherwise, the Home Secretary's unfettered statutory discretion to change policy would be restricted. Notwithstanding the notice and compact, the most they could legitimately expect was that their cases would be examined individually in the light of whatever policy the Home Secretary lawfully adopted for the time being.

A public body can also depart from a policy, and defeat a legitimate expectation, if the departure is justified by an 'overriding public interest'. Whether such a public interest exists is for the court, and not the public body itself, to decide (*R v North and East Devon Health Authority (ex parte Coughlan)* [2000] 3 All ER 850, CA, below).

Most of the case law on legitimate expectation has been concerned with a 'procedural' expectation; i.e., an opportunity to make representations before a decision is made, as in *Attorney-General of Hong Kong v Ng Yuen Shiu* [1983] 2 All ER 346, PC, above (legitimate expectation of an interview), and the *GCHQ* case [1984] 3 All ER 935, HL, above (legitimate expectation of consultation).

However, since the doctrine of legitimate expectation imposes a general duty to act fairly, it is not confined to cases where the expectation is to be consulted or to be given an opportunity to make representations before a decision is made (*R v Secretary of State for the Home Department (ex parte Ruddock)* [1987] 2 All ER 518 *per* Taylor J at p. 531). The law recognises also a legitimate expectation arising in respect of a 'substantive' benefit, as in *R v North and East Devon Health Authority (ex parte Coughlan)* [2000] 3 All ER 850, CA.

In *Coughlan*, a patient with severe physical disabilities had been assured that she had a home for as long as she wished in the NHS unit in which she lived. It was later decided to close the unit for practical and financial reasons. It was held that since the assurance had induced a legitimate expectation of a substantive benefit, the authority would not be allowed to go back on it since to do so would be unfair and an abuse of power. (See also *R* v *Secretary of State for the Home Department (ex parte Khan)* [1984] 1 WLR 1337, CA.)

It seems that the legitimate *substantive* expectation doctrine will be more readily applied where the decision under review affects only a small group of people (*Coughlan* [2000] 3 All ER 850, CA, *per* Lord Woolf MR at p. 872). This provides a further explanation of the difference in outcome between the *Hargreaves* case [1997] 1 All ER 397, CA, above (where the Home Secretary's decision affected a large number of prison inmates), and the *Coughlan* case, above (where the health authority's decision affected only eight individuals).

The public body must have made a definite promise, assurance, or representation which is clear and unambiguous. Whether a legitimate expectation exists is a question of fact in each case. No legitimate expectation was held to have arisen in *R (Wagstaff)* v *Secretary of State for Health* [2001] 1 WLR 292, DC. Here it was decided that since the Secretary of State had given no clear and unambiguous representation that the inquiry into the case of Harold Shipman would be held in public, and since there was no clearly established practice of holding such inquiries in public, the claimants (who were the families and friends of Shipman's victims) had failed to establish a legitimate expectation of a public inquiry. (The challenge to the Secretary of State's decision succeeded on another ground, that of irrationality: see para 10.4.4 above.)

Similarly, in *R (Association of British Civilian Internees: Far East Region)* v *Secretary of State for Defence* [2003] 3 WLR 80, CA (para 10.4.4 above), it was held that the claimants had no legitimate expectation that they would be covered by a compensation scheme since the original statement announcing the scheme did not contain a clear and unequivocal representation to that effect.

It seems that *detrimental reliance* is normally an ingredient of the doctrine of legitimate expectation. Thus, in order to enforce a legitimate expectation the claimant will usually be required to show that he has suffered hardship (not necessarily financial loss) in reliance on the public body's promise, assurance, or representation.

However, it would appear not to be necessary to show detrimental reliance in a case where the public body is attempting to deviate from an established policy in relation to particular persons. Here, the detrimental reliance requirement may be overridden by the fairness requirement that consistency of treatment should be observed. A legitimate expectation was held to have arisen in *R (Bibi)* v *Newham London Borough Council* [2002] 1 WLR 237, CA, where a local authority, in the mistaken belief that it was under a duty to do so, promised to provide the claimant refugees and their families with legally secure accommodation within 18 months. The claimants had not altered their position on the faith of the authority's promise since for many years they had been occupying temporary accommodation provided by the authority. The authority argued that, in the absence of bad faith, a substantive legitimate expectation can only arise where there has been detrimental reliance.

The argument was rejected, it being held that it is not always necessary to show detrimental reliance in order to rely on either a substantive or a procedural legitimate expectation. A declaration was granted that the authority was under a duty to consider the claimants' request for suitable housing on the basis that they had a legitimate expectation that they would be provided with suitable secure accommodation.

Although there are similarities between the doctrine of legitimate expectation in *public* law and the doctrine of estoppel in *private* law, the courts have warned that the two doctrines should be kept separate. There are good reasons for this. Remedies available in public law, unlike those available in private law, have to take into account the wider interests of the general public (see *R (Reprotech Ltd)* v *East Sussex County Council* [2003] 1 WLR 348, HL, *per* Lord Hoffmann at [34]). In addition, legitimate expectation and abuse of power are now so sufficiently well developed as public law concepts that there is no need to import the private law estoppel doctrine into public law.

As Lord Hoffmann pointed out in the *Reprotech* case, above, at [35]: 'public law has already absorbed whatever is useful from the moral values which underlie the private law concept of estoppel and the time has come for it to stand upon its own two feet'. (See also *South Bucks District Council* v *Flanaghan* [2002] 1 WLR 2601, CA, *per* Keene LJ at [16]; *R (Bloggs 61)* v *Secretary of State for the Home Department* [2003] 1 WLR 2724, CA, *per* Auld LJ at [39]–[40]; *R (Munjaz)* v *Mersey Care NHS Trust* [2003] 3 WLR 1505, CA, *per* Hale LJ, delivering the judgment of the court, at [79].)

10.4.6 Overlapping of grounds for judicial review

The grounds contained within the tripartite classification suggested by Lord Diplock in *Council of Civil Service Unions* v *Minister for the Civil Service* [1984] 3 All ER 935, HL (para 10.4.2 above), are not mutually exclusive. His three grounds overlap. It is possible, therefore, that the same facts may fall under more than one head. As Lord Irvine LC explained in *Boddington* v *British Transport Police* [1998] 2 All ER 203, HL, at p. 208:

Categorisation of types of challenge assists in an orderly exposition of the principles underlying our developing public law. But these are not watertight compartments because the various grounds for judicial review run together. The exercise of a power for an improper purpose may involve taking irrelevant considerations into account, or ignoring relevant considerations; and either may lead to an irrational result. The failure to grant a person affected by a decision a hearing, in breach of principles of procedural fairness, may result in a failure to take into account relevant considerations.

The overlapping nature of the grounds for judicial review may be illustrated by reference to *Wheeler* v *Leicester City Council* [1985] 2 All ER 1106, HL, in which an administrative decision was condemned on a number of grounds. (See also *R* v *Lewisham London Borough Council (ex parte Shell UK Ltd)* [1988] 1 All ER 938, DC, in which the *Wheeler* case was applied.)

In *Wheeler*, the council passed a resolution to ban the Leicester rugby club from using a recreation ground (on which it trained and played its club matches) for 12 months. The ban was imposed because the club, while opposing apartheid in South Africa, had refused the council's request to condemn the tour of an English rugby team to that country and to put pressure on three Leicester players not to join the tour.

The council's resolution was quashed by a quashing order, the House of Lords holding that the ban was unreasonable, unfair, and a procedural impropriety in that the council had used illegitimate pressure in an attempt to coerce the club into accepting the council's own policy on the council's own terms. It was further held that the council had misused its power by punishing the club when it had done no wrong, as in *Congreve* v *Home Office* [1976] 1 QB 629, CA (para 10.2.2.1 above).

Lord Diplock's illegality ground covers a number of different issues which are often treated separately, such as excess of jurisdiction, misuse of power, error of law, failing to perform a statutory duty, fettering a discretion, taking into account irrelevant considerations and leaving out of account relevant considerations, and unauthorised delegation of the exercise of a discretion.

Misuse of power, and taking into account irrelevant considerations, were among the grounds on which a local authority's ban on deer hunting on certain of its land was challenged in *R* v *Somerset County Council (ex parte Fewings)* [1995] 3 All ER 20, CA. The majority of the councillors who voted for the ban did so on the ethical ground of animal cruelty. The claimants, who were regular hunters on the land, argued that the ban was unlawful because it had been made on purely moral grounds and ignored the purpose of the local authority's statutory power to acquire land for the 'benefit, improvement or development of their area'. The Court of Appeal, by a majority, held that the authority had failed to exercise its discretion in accordance with the stated statutory purpose. The ban was quashed by a quashing order. There was disagreement over the morality argument. Two of the three judges in the Court of Appeal thought that the cruelty argument is not necessarily irrelevant to a local authority's consideration of what is for the benefit of the area.

R v *Secretary of State for the Home Department (ex parte Venables)* [1997] 3 All ER 97, HL, was an application for judicial review by the two boys who were convicted at the age of ten of the murder of a two-year-old child, James Bulger. They were both subjected to the mandatory sentence of detention during Her Majesty's pleasure. The trial judge recommended that the 'penal' (or 'tariff') element in their sentence should be eight years. Later, the Lord Chief Justice recommended an increase to ten years and the Home Secretary decided that it should be increased to 15 years. The Home Secretary's decision was quashed by a quashing order because (a) he had failed to take into account the possibility of early release based on the progress and development of the applicants during their detention, and (b) he had taken into account irrelevant considerations, namely the public protests about the level of the tariff to be fixed in the case of the applicants. (See also *Pierson* v *Secretary of State for the Home Department* [1997] 3 All ER 577, HL, and *R* v *Secretary of State for the Home Department (ex parte Hindley)* [2000] 2 All ER 385, HL.)

There is overlap between irrationality and the emerging proportionality principle. However, it is clear from *R (Association of British Civilian Internees: Far East Region)* v *Secretary of State for Defence* [2003] 3 WLR 80, CA (para 10.4.4 above), applying *Brind* v *Secretary of State for the Home Department* [1991] 1 All ER 720, HL (para 10.4.3 above), that proportionality has not replaced the test of irrationality in cases which do not raise questions of human rights or European Community law.

Quite often, but not always, the result will be the same whether irrationality or proportionality is applied as the test. But it does not follow that the tests are the same. Under the proportionality doctrine, for example, the depth of the court's review is greater since the court may be required to 'assess the balance which the decision maker has struck, not merely whether it is within the range of rational or reasonable decisions' (*R (Daly)* v *Secretary of State for the Home Department* [2001] 2 WLR 1622, HL, *per* Lord Steyn at [27]; the *British Civilian Internees* case, CA, above, *per* Dyson LJ, delivering the judgment of the court, at [33]). In other words, under the proportionality approach the court may well have to consider the *merits* of the decision under review, and, if necessary, to substitute its own decision for that of the decision-maker. This is a much more fundamental approach than simply examining the *rationality* of a decision under the *Wednesbury* principles.

In the *British Civilian Internees* case, above, the Court of Appeal expressed the opinion that since there is now 'little daylight' between irrationality and proportionality it is difficult to find any justification for retaining irrationality as a separate ground for judicial review. However, in view of the fact that the decision of the House of Lords in *Brind* v *Secretary of State for the Home Department* [1991] 1 All ER 720, HL (para 10.4.3 above), is still binding on the Court of Appeal, the abolition of irrationality is a matter for the House of Lords, not the Court of Appeal (*British Civilian Internees* case [2003] 3 WLR 80, CA, *per* Dyson LJ at [34]–[37]).

The legitimate expectation doctrine and natural justice overlap. Both may result in entitlement to a hearing, but they have different bases. Natural justice demands a hearing because of the importance of the right or interest involved. This right or interest is regarded as so important that it deserves protection (see, e.g., *Ridge* v *Baldwin* [1964] AC 40, HL, para 10.2.2.4, above, where the Chief Constable's summary dismissal affected his livelihood and reputation). Legitimate expectation is not based on rights or interests, but on the fact that the decision-making public body has encouraged the expectation.

The overlap between legitimate expectation and natural justice is illustrated in *R* v *Great Yarmouth Borough Council (ex parte Botton Brothers)* (1988) 56 P & CR 99. Here, the council decided to grant planning permission for a new amusement arcade to rival the one owned by the claimants, who argued that they should have been allowed to make representations before the council made its decision. It was held that the claimants, although they had no legitimate expectation of a hearing because the council had made no promise and there was no past practice of granting hearings in such circumstances, should have been allowed to make representations under the fair hearing rule of natural justice since the council was contemplating a decision which could seriously affect the livelihoods of the claimants.

10.5 Quashing orders (formerly certiorari)

10.5.1 Nature

A quashing order is a remedy which is used to bring before the High Court the decision of some inferior court, tribunal, or other public authority so that its legality can be examined. The decision will be quashed if it is found to be invalid. Disobedience to a quashing order, as by refusing to submit the record of a case to the High Court for review, is punishable as a contempt of court.

10.5.2 **Availability**

10.5.2.1 Introduction

Quashing orders have been sought successfully to quash the decisions of, *inter alia*, the Crown Court, a county court, a magistrates' court, a coroner's court, the board of visitors of a prison, the Medical Appeal Tribunal, the Criminal Injuries Compensation Board, the Advertising Standards Authority, a mental health review tribunal, the registrar of companies, a local election court, a local valuation court, a local authority, an immigration officer, and of a government minister made after a public inquiry.

An applicant for a quashing order must have sufficient interest in the matter to which the application relates (para 10.8.4 below).

10.5.2.2 Quashing orders and bodies exercising a public function

Until recently, a quashing order was not available to question the decisions of voluntary (i.e., non-statutory) domestic tribunals, whose powers generally exist within the field of *private* law and, therefore, outside the reach of prerogative remedies (*Ex parte Fry* [1954] 2 All ER 118, DC; *Buckoke* v *Greater London Council* [1971] 1 Ch 655, CA).

By an oversight, the Divisional Court was prepared to (but did not) grant prerogative relief in *R* v *Aston University Senate (ex parte Roffey)* [1969] 2 QB 538, DC (para 3.4.2). This preparedness to invoke the prerogative jurisdiction against a voluntary domestic tribunal was criticised by the Court of Appeal in *Herring* v *Templeman* [1973] 3 All ER 569, CA.

However, now that it is clear that the jurisdiction to review judicially the decisions of inferior bodies depends not on any distinction between their statutory or non-statutory method of creation but on whether a particular body is exercising a sufficiently public function, there can be no doubt that the universities have a *public* character which is sufficient to bring their decisions within the High Court's supervisory jurisdiction in appropriate circumstances (*Page* v *Hull University Visitor* [1993] 1 All ER 97, HL).

It should be noted that although the prerogative remedies are not available against the members of a voluntary tribunal in respect of the exercise of a purely *private* function, the private law remedies of declaration and injunction *are* available.

A quashing order does not lie directly against the Crown, although, in practice, this limitation is not a serious obstacle since it *is* available against individual ministers of the Crown. It is also available against the Ombudsman (the Parliamentary Commissioner for Administration), a creature of statute (*R* v *Parliamentary Commissioner for Administration (ex parte Dyer)* [1994] 1 All ER 375, DC, where, however, relief was refused on the facts). It is not, however, available against the Parliamentary Commissioner for Standards (*R* v *Parliamentary Commissioner for Standards (ex parte Al Fayed)* [1998] 1 All ER 93, CA; see further, para 10.8.2.4 below).

A quashing order probably cannot be used as a means of challenging delegated legislation. This is because a quashing order applies to acts of a *judicial* nature, as opposed to legislative or administrative acts (*R* v *Electricity Commissioners (ex parte London Electricity Joint Committee Co. (1920) Ltd)* [1924] 1 KB 171, CA; *R* v *Legislative Committee of the Church Assembly (ex parte Haynes-Smith)* [1928] 1 KB 411, DC; for legitimate methods of challenging delegated legislation, see para 10.2.2.3 above).

10.5.2.3 Quashing orders and the Crown Court

10.5.2.3.1 *Introduction* Whether a quashing order is available to question the decisions of the ordinary courts of law depends on the status of the individual court. The decisions of superior courts of record, which include the House of Lords, the Court of Appeal, and the High Court itself, are not subject to judicial review. The Employment Appeal Tribunal is also a superior court of record, and, as such, is not amenable to the prerogative jurisdiction of the High Court (Employment Tribunals Act 1996, s. 20). On the other hand, decisions of the county courts and magistrates' courts, which are inferior courts, *can* be challenged by means of judicial review.

The Crown Court is in an anomalous position as regards judicial review. It is a superior court of record (Supreme Court Act 1981, s. 45(1)) and yet, in some circumstances, its decisions are subject to judicial review.

10.5.2.3.2 *Supreme Court Act 1981, s. 29(3)* As amended by the Civil Procedure (Modification of Supreme Court Act 1981) Order 2004 (SI 2004, No. 1033), s. 29(3) provides as follows:

In relation to the jurisdiction of the Crown Court, other than its jurisdiction in matters relating to trial on indictment, the High Court shall have all such jurisdiction to make mandatory, prohibiting or quashing orders as the High Court possesses in relation to the jurisdiction of an inferior court.

10.5.2.3.3 *'Matters relating to trial on indictment'* The object of s. 29(3) in exempting from challenge matters relating to trial on indictment is to prevent the delay that might otherwise occur in criminal trials if an application for judicial review were available at an intermediate stage (*Smalley* v *Crown Court at Warwick* [1985] 1 All ER 769, HL, *per* Lord Bridge at p. 779; *DPP* v *Crown Court at Manchester and Ashton* [1993] 2 All ER 663, HL, *per* Lord Slynn at p. 669; *DPP* v *Crown Court at Manchester and Huckfield* [1993] 4 All ER 928, HL, *per* Lord Browne-Wilkinson at p. 933).

No decision of the Crown Court on a 'matter relating to trial on indictment' can be challenged in the High Court by means of an application for a quashing order. Thus, if a judge in the Crown Court at a trial on indictment should impose an *ultra vires* penalty, or act in violation of natural justice, or err in law, his action cannot be quashed by a quashing order, and the aggrieved party's only remedy will be an appeal to the Court of Appeal, criminal division, either against sentence, or against conviction on the ground that the conviction is unsafe, or against both.

The precise scope of the statutory phrase, 'matters relating to trial on indictment', is unclear. Lord Bridge suggested two 'helpful pointers'. The first, mentioned in *Smalley* v *Crown Court at Warwick*, above (at p. 780, the other Law Lords agreeing with him), is to ask whether the decision in question is a 'decision affecting the conduct of a trial on indictment'. The second 'helpful pointer', suggested in *Sampson* v *Crown Court at Croydon* [1987] 1 All ER 609, HL (at p. 613), is to ask whether the decision is an 'integral part of the trial process'.

Lord Bridge's 'helpful pointers' have not been entirely successful in unlocking the meaning of the words, 'matters relating to trial on indictment'. Both in *DPP* v *Crown Court at Manchester and Ashton* [1993] 2 All ER 663, HL, and in *DPP* v *Crown Court at Manchester and Huckfield* [1993] 4 All ER 928, HL, the House of Lords indicated that much difficulty had been caused in previous cases because the Divisional Court had treated the 'pointers' as if they constituted a statutory definition or test.

Notwithstanding this criticism, an examination of the authorities by Lord Browne-Wilkinson in *DPP* v *Crown Court at Manchester and Huckfield*, above, led him (at p. 934) to the conclusion (while eschewing any attempt to lay down a comprehensive definition of the meaning of the statutory phrase) that there may be a *third* 'helpful pointer', namely, whether the decision arose in the issue (including the costs thereof) between the Crown and the defendant as formulated by the indictment. If it did, then it is, according to Lord Browne-Wilkinson, probably excluded from review by s. 29(3) since review could delay the defendant's trial. If it did not, then the decision might well not be excluded from review by s. 29(3) since it is collateral to the indictment and reviewing it would not delay the trial.

The following decisions of the Crown Court have been held to constitute 'matters relating to trial on indictment', and, as such, to be excluded from judicial review: an order refusing costs to an acquitted defendant (*Ex parte Meredith* [1973] 2 All ER 234, DC); a decision refusing to grant legal aid to defendants for their trial on indictment (*R* v *Crown Court at Chichester (ex parte Abodundrin)* (1984) 79 Cr App R 293, DC); an order refusing the defendant his pre-trial costs (*R* v *Central Criminal Court (ex parte Spens)* [1993] COD 194); a decision at the pre-trial stage not to disclose sensitive material to the defence (*R* v *Crown Court at Southwark (ex parte Johnson)* [1992] COD 364, DC); an order made at the trial for disclosure to the defence of inadmissible statements made during an informal complaints procedure (*R* v *Crown Court at Chelmsford (ex parte Chief Constable of the Essex Police)* [1994] 1 All ER 325, DC); a decision about the starting date of the trial (*R* v *Southwark Crown Court (ex parte Ward)* (1994) *The Times*, 19 August, DC); an order that an indictment should lie on the file and not be proceeded with without leave of the court (*R* v *Central Criminal Court (ex parte Raymond)* [1986] 2 All ER 379, DC—approved by the House of Lords in *DPP* v *Crown Court at Manchester and Ashton* [1993] 2 All ER 663, HL); a decision to quash an indictment for lack of jurisdiction (*DPP* v *Crown Court at Manchester and Huckfield* [1993] 4 All ER 928, HL).

The prohibition imposed by s. 29(3) of the Supreme Court Act 1981 against reviewing Crown Court decisions which relate to trial on indictment cannot be circumvented by using instead the general power of judicial review conferred by s. 31 of the same Act. This is because the Crown Court is a 'superior court', and the High Court has no jurisdiction to review the decisions of a superior court except where expressly granted by s. 29(3) (*R* v *Crown Court at Chelmsford (ex parte Chief Constable of the Essex Police)* [1994] 1 All ER 325, DC; see also *R* v *Director of Public Prosecutions (ex parte Kebeline)* [1999] 4 All ER 801, HL, para 10.5.2.3.5 below).

10.5.2.3.4 *Matters not relating to trial on indictment* Section 29(3) of the Supreme Court Act 1981 clearly implies that decisions of the Crown Court on other aspects of its jurisdiction, not concerned with trials on indictment, shall be capable of challenge by means of a quashing order.

In *Smalley v Crown Court at Warwick* [1985] 1 All ER 769, HL, for example, an order which caused the forfeiture of a recognisance given by a surety for a defendant who failed to surrender to his bail at the Crown Court was held to be reviewable. The House of Lords said that the order was not a 'matter relating to trial on indictment' within s. 29(3) since it did not affect the conduct of the trial in any way.

Smalley's case was applied in somewhat unusual circumstances in *R v Crown Court at Maidstone (ex parte Gill)* [1987] 1 All ER 129, DC, in which the applicant sought a quashing order to quash a forfeiture order made by the Crown Court in respect of two motor cars. The applicant had lent the cars to his son. Without the applicant's knowledge, the son used the cars for two separate journeys to supply prohibited drugs. Later the son pleaded guilty to a charge of supplying heroin and was sentenced to a term of imprisonment; a second charge was not proceeded with.

The trial judge exercised a statutory power to make an order forfeiting the applicant's two cars. In judicial review proceedings in the High Court, it was held that the forfeiture order was not a 'matter relating to trial on indictment' within s. 29(3) of the Supreme Court Act 1981 because it did not affect the conduct of the trial of the applicant's son in any way. It followed that the High Court was not precluded by s. 29(3) from considering the application to quash the forfeiture order. Having considered the merits, the court quashed the order on the ground that one of the cars had not, in fact, been used in the offence to which the son had pleaded guilty, and that, although the other car had been so used, the applicant had had no reason to suspect that it would be used to transport prohibited drugs.

If it had been held in *Gill* that the forfeiture order was a 'matter relating to trial on indictment', the implications would have been very serious. The applicant's remedy by way of judicial review would have been barred by s. 29(3), and, indeed, he would have had no remedy at all against the imposition of an unlawful order by the Crown Court. There could be no appeal to the Court of Appeal, criminal division, because the applicant had been neither 'convicted' nor 'sentenced' (see [1987] 1 All ER 129, DC, *per* Lord Lane CJ at p. 131).

10.5.2.3.5 *Decisions to stay criminal proceedings as an abuse of the process of the court* An application to stay criminal proceedings on the ground that they are an abuse of the process of the court is a 'matter relating to trial on indictment' within s. 29(3). It follows that the decision of the trial judge at the Crown Court on such an application is not subject to judicial review.

This was decided by the House of Lords in *DPP v Crown Court at Manchester and Ashton* [1993] 2 All ER 663, HL, reversing the Queen's Bench Divisional Court. Two earlier decisions to the contrary given by the Divisional Court—*R v Central Criminal Court (ex parte Randle)* (1990) [1992] 1 All ER 370, DC, and *R v Crown Court at Norwich (ex parte Belsham)* [1992] 1 All ER 394, DC—were overruled. In addition, the correctness of two other decisions which had been influenced by the *Randle* and *Belsham* cases—*R v Crown Court at Southwark (ex parte Customs and*

Excise Commissioners) [1993] 1 WLR 764, DC, and *R* v *Central Criminal Court (ex parte Spens)* [1993] COD 194, DC—was doubted. *R* v *Central Criminal Court (ex parte Raymond)* [1986] 2 All ER 379, DC (para 10.5.2.3.3 above), was approved.

In *R* v *Central Criminal Court (ex parte Director of Serious Fraud Office)* [1993] 2 All ER 399, DC, the Divisional Court held that a decision taken before a trial begins to dismiss fraud charges under what was s. 6 of the Criminal Justice Act 1967, because the evidence is not sufficient to secure a conviction, is not a 'matter relating to trial on indictment', and is, therefore, reviewable in the High Court. However, the status of this case was thrown into doubt following the decisions of the House of Lords in *DPP* v *Crown Court at Manchester and Ashton* [1993] 2 All ER 663, HL, and *DPP* v *Crown Court at Manchester and Huckfield* [1993] 4 All ER 928, HL.

According to the House of Lords in *Ashton*, the *Serious Fraud Office* decision was not based on the discredited cases of *Randle* and *Belsham* (above), which were merely 'referred to' therein, and, since the *Serious Fraud Office* case was not concerned with the same issues as *Randle, Belsham*, and *Ashton*, it was 'undesirable' to say anything about it (see *per* Lord Slynn at [1993] 2 All ER 663, 670).

According to the House of Lords in *Huckfield*, however, 'much weight' was attached to *Randle* and *Belsham* in the *Serious Fraud Office* case, although again their Lordships were not prepared to express a view on the correctness of the latter decision but pointed out that it was concerned with a different issue and one which may raise special considerations (see *per* Lord Browne-Wilkinson at [1993] 4 All ER 928, 934).

Further doubt was cast on the correctness of the *Serious Fraud Office* case in *R (Snelgrove)* v *Crown Court at Woolwich* [2005] 1 WLR 3223, DC, which was concerned with the Crime and Disorder Act 1998, sch. 3, para 2. As amended by the Criminal Justice Act 2003, this paragraph (similarly worded to what was s. 6 of the Criminal Justice Act 1967) gives a judge the power to dismiss charges against a defendant before the start of the trial if it appears that the evidence would not be sufficient for the defendant to be properly convicted. The Divisional Court held that a judge's decision not to dismiss the charges is a 'matter relating to trial on indictment' and, as such, not reviewable in the High Court.

In coming to this conclusion, the Divisional Court declined to follow its earlier decision in the *Serious Fraud Office* case. Two reasons were given. First, the latter decision was said to be 'already open to question' when it was given since it appeared to be at variance with the 'helpful pointers' (para 10.5.2.3.3, above) articulated by the House of Lords in *Smalley* v *Crown Court at Warwick* [1985] 1 All ER 769, HL, and *Sampson* v *Crown Court at Croydon* [1987] 1 All ER 609, HL. Secondly, the *Serious Fraud Office* case could not, in any event, survive the subsequent House of Lords' decisions in *Director of Public Prosecutions* v *Crown Court at Manchester and Ashton* [1993] 2 All ER 663, HL, and *Director of Public Prosecutions* v *Crown Court at Manchester and Huckfield* [1993] 4 All ER 928, HL.

Delay is the commonest ground for applying to stay criminal proceedings which are alleged to be an abuse of the process of the court.

According to the Court of Appeal when issuing guidance to trial judges and lawyers in *Attorney-General's Reference (No. 1 of 1990)* [1992] 3 All ER 169, CA, proceedings should rarely be stayed on the ground of delay in bringing the matter to trial even where the delay is unjustifiable. They should rarely be stayed in the absence of any fault on the part of the prosecuting authorities, and should never be stayed where the delay is due to the complexity of the case or to the actions of the defendant himself. When the judge is contemplating the imposition of a stay in an appropriate case, the correct test to apply is whether the defendant has shown on the balance of probabilities that the delay would cause him serious prejudice in that he could no longer receive a fair trial.

Delivering the judgment of the court in *Attorney-General's Reference (No. 1 of 1990)*, above, Lord Lane CJ (at pp. 176–7) expressed concern about the public suspicion and mistrust that would arise if stays became a matter of routine, and he hoped that the court's decision would lead to a substantial reduction in the number of applications for stays on the ground of delay.

The case of *R v Bow Street Metropolitan Stipendiary Magistrate (ex parte DPP)* [1992] COD 267, DC, was concerned with the decision of a stipendiary magistrate that proceedings against three police officers for conspiracy to pervert the course of justice in the 'Guildford Four' case 17 years earlier were an abuse of the process of the court and should be stayed. This was not a case governed by s. 29(3) of the Supreme Court Act 1981 since it did not involve the Crown Court; the magistrate's decision was, therefore, reviewable.

On judicial review, the decision was quashed because, in spite of the long delay and the adverse publicity the case had received, the Divisional Court was convinced that a fair trial was still possible. The three officers were acquitted by a jury at their subsequent trial at the Central Criminal Court. (See *The Times*, 20 May 1993; see also *R v Croydon Justices (ex parte Dean)* [1993] 3 All ER 129, DC, where on the facts it was held to be an abuse of process for the Crown Prosecution Service to prosecute a defendant who had been told by the police that he would not be prosecuted, and the decision of the magistrates to commit the defendant for trial at the Crown Court was quashed by a quashing order.)

When three former police officers involved in the 'Birmingham Six' investigation were prosecuted for conspiracy to pervert the course of justice and perjury, the case was withdrawn from the jury by the trial judge at the Crown Court on the ground that to proceed with the trial would amount to an abuse of process. The judge, Garland J, was of the opinion that the defendants would not receive a fair trial due to prejudicial publicity and due to the risk that the jury would be confused by the issues involved in the case. Since the decision was on a 'matter relating to trial on indictment' within s. 29(3), it was not, in the light of *DPP v Crown Court at Manchester and Ashton* [1993] 2 All ER 663, HL, reviewable. Garland J's decision caused considerable surprise and consternation; it raised the issues of public suspicion and mistrust of the criminal justice system about which Lord Lane CJ had spoken in *Attorney-General's Reference (No. 1 of 1990)*, above.

The prohibition against reviewing the trial judge's decision to stay proceedings as an abuse of process cannot be circumvented by instead challenging the prosecutor's decision to prosecute. Thus, in *R v Director of Public Prosecutions (ex parte Kebeline)* [1999] 4 All ER 801, HL, it was held that a decision of the Director of Public Prosecutions to consent to a prosecution is not reviewable in the absence of dishonesty, bad faith, or exceptional circumstances.

10.5.3 Scope

10.5.3.1 Where the decision is *ultra vires*

Decisions were quashed by quashing orders on this ground in *R* v *London Borough of Hillingdon (ex parte Royco Homes Ltd)* [1974] QB 720, DC, and *Westminster City Council* v *Greater London Council* [1986] 2 All ER 278, HL.

In the *Royco Homes* case, a planning permission was quashed where the planning authority had attached conditions to the permission which were so unreasonable as to be *ultra vires* the planning legislation. The planning authority was held to have acted *ultra vires* in spite of the fact that there was a statutory power to impose 'such conditions as they think fit'. In the *Westminster* case, decisions of the GLC (taken without consulting the London boroughs and to be implemented after its abolition on 1 April 1986) to provide grants totalling £76m for the Inner London Education Authority, an arts centre, and 900 voluntary organisations were quashed on the ground that they were *ultra vires* the relevant legislation. (See also *Wheeler* v *Leicester City Council* [1985] 2 All ER 1106, HL, para 10.4.6 above.)

10.5.3.2 Where the decision was reached in violation of the principles of natural justice

10.5.3.2.1 *The rule against bias* This principle is expressed in the maxim *nemo judex in causa sua potest* ('no man can be a judge in his own cause'). A useful illustration is provided by *R* v *Sunderland Justices* [1901] 2 KB 357, CA. Here, the local authority in Sunderland had bought a public house in order to demolish it so that a road could be widened. The local authority then made an agreement with a brewery under which the local authority would surrender the licence for the public house it had bought and would not object to the brewery's application for a licence relating to new premises nearby, provided the brewery paid the authority £10,000. The licensing committee of justices who heard the brewery's application included the chairman of the authority's highways committee, who had organised the agreement between the brewery and the local authority, and four other Sunderland councillors. The Court of Appeal ruled that the grant of the licence to the brewery would be set aside because of the committee's bias.

The rule against bias was held to have been violated in the unusual case of *R* v *Barnsley Metropolitan Borough Council (ex parte Hook)* [1976] 3 All ER 452, CA. H was a trader with a stall in a Barnsley market. One evening, after the market and the public lavatories had closed, he satisfied an urgent need to relieve himself by urinating in a side street. His behaviour came to the attention of the local council, which decided to ban him for life from trading in the market. Although H was given an opportunity to put his case before the council meeting at which this decision was taken, the market manager, who had brought the complaint before the council in the first place, was present throughout that meeting and in a position to give the council his view of the evidence.

H applied for a quashing order to quash the council's decision. The Court of Appeal was unanimous in holding that the order would be granted. The rules of natural justice applied to the situation because the council was determining questions affecting the rights of a citizen. The rules had been violated as the council had heard the market manager's evidence in the absence of H or his representatives, and the market manager, who was in effect a prosecutor, had been present throughout the deliberations leading to the decision to ban H from the market.

Two of the three judges in the appeal found an additional reason in favour of H. They said that the court could also interfere by means of a quashing order to quash the council's decision on the ground that the punishment inflicted on H was excessive and out of proportion to the occasion ([1976] 3 All ER 452, CA, *per* Lord Denning MR at p. 457 and Sir John Pennycuick at p. 461).

One of the most serious aspects of bias is the existence of a *direct pecuniary interest* in the outcome of a case. Thus, in the well-known case of *Dimes* v *Grand Junction Canal Co.* (1852) 3 HL Cas 759, HL, a decision adverse to the defendant, given by no less a judicial figure than the Lord Chancellor himself, was set aside on it appearing that his Lordship was the owner of several thousand pounds' worth of shares in the claimant company. There was no evidence that the Lord Chancellor had been influenced in fact by his pecuniary interest in the company, but, where a judge has a direct pecuniary interest in the outcome of a case, bias is conclusively presumed, and the maintenance of public confidence in the administration of justice demands that his decision should not be allowed to stand.

There will not be an automatic breach of the rule against bias where the judge's interest in the case, whether pecuniary or otherwise, is merely *indirect* or *remote*. In this situation, bias is not presumed: a real danger or possibility of bias would need to be established in accordance with the test endorsed in *R* v *Gough* [1993] 2 All ER 724, HL (see below). In *R* v *Mulvihill* [1990] 1 All ER 436, CA, for example, the defendant's sole ground of appeal against his conviction for conspiracy to rob was that the trial judge at the Crown Court had failed to disclose that he owned 1,650 shares in the National Westminster Bank plc, one of whose branches was involved in the charge against the defendant. The Court of Appeal had no difficulty in dismissing the appeal. It was said that, save possibly in very exceptional circumstances, a judge in a criminal trial is not bound to declare that he has a *remote* interest, such as a shareholding, in a company whose premises are alleged to be the scene of the crime (ibid., *per* Brooke J at p. 441).

The allegation of bias must be more than a flimsy pretext to have an inconvenient decision set aside (see, e.g., *R* v *Mulvihill*, above). The proper test is whether the facts disclose that there was a *real danger or possibility of bias*. This was finally established by the House of Lords in *R* v *Gough* [1993] 2 All ER 724, HL, which arose out of an appeal in a criminal case in which the appellant had been convicted of conspiracy to rob. One member of the jury was the next door neighbour of the appellant's brother, who had originally been charged with robbery but did not stand trial. The Court of Appeal, applying the real danger of bias test and holding that, in the circumstances, there was no danger that the appellant might not have had a fair trial, dismissed the appeal. The House of Lords affirmed this decision.

In *R* v *Gough*, above, the House of Lords subjected the previous case law to a detailed examination. This analysis disclosed two strands of authority on the correct test to be applied to alleged bias.

First, some cases had laid down that the test was whether there was a 'real danger' (sometimes expressed as a 'real likelihood') of bias (see, e.g., *R* v *Sunderland Justices* [1901] 2 KB 357, CA, above; *R* v *Camborne Justices (ex parte Pearce)* [1955] 1 QB 41, DC, below).

Secondly, other cases had held that the test was whether a reasonable person might reasonably suspect bias (sometimes expressed as the 'reasonable suspicion' or 'appearance of bias' test) (see, e.g., *R* v *Sussex Justices (ex parte McCarthy)* [1924] 1 KB 256, DC, below).

The first test is the more rigid of the two. With the second test, there is a risk that it can be used to overturn decisions on the flimsiest pretext of bias (see *R* v *Camborne Justices (ex parte Pearce)* [1955] 1 QB 41, DC, below, *per* Slade J at pp. 51–2; *R* v *Gough* [1993] 2 All ER 724, HL, *per* Lords Goff and Woolf at pp. 732 and 740, respectively).

In *R* v *Gough*, the House of Lords unanimously supported the first test and preferred to state it in terms of *real danger*, rather than real likelihood, in order to emphasise that the test is concerned with the *possibility* and not the probability of bias (see *per* Lord Goff at p. 737).

Some of the previous case law had suggested that, as regards legal proceedings, the test varied according to whether the person against whom bias was alleged was a justice of the peace or other judicial figure, or whether he was a member of a jury. Their Lordships in *R* v *Gough* further held that the 'real danger of bias' test applies in like manner (with one exception, relating to justices' clerks, noted below) in all cases of alleged bias on the part of justices of the peace, members of other inferior courts, and jurors.

In *R* v *Gough*, the House affirmed the following principles. First, evidence of *actual* bias is not required (see the explanation of this given at [1993] 2 All ER 724 *per* Lords Goff and Woolf at pp. 728 and 740, respectively).

Secondly, explaining and approving *Dimes* v *Grand Junction Canal Co.* (1852) 3 HL Cas 759, HL, above, in the exceptional case of a person acting in a judicial capacity who has a *direct* pecuniary or proprietary interest in the outcome of the proceedings before him, bias is assumed and his decision will be set aside without further inquiry. (See further, *R* v *Bow Street Metropolitan Stipendiary Magistrate (ex parte Pinochet Ugarte) (No. 2)* [1999] 1 All ER 577, HL, below.)

Thirdly, expressly approving *R* v *Camborne Justices (ex parte Pearce)* [1955] 1 QB 41, DC, below, in all other cases the test to be applied is whether in the light of all the circumstances there was a 'real danger or possibility of bias' in the sense that the person against whom bias is alleged might have unfairly disfavoured the claim of a party to the proceedings.

Fourthly, in cases involving justices' clerks it must be shown that not only was there a real danger or possibility of bias on the part of the clerk but also that his participation in the decision-making process created a real danger that his bias would prejudice the justices against a party to the proceedings. This depends on whether the clerk was requested to, and did, give advice to the justices. If he was so requested, the court of review may be prepared to infer from the facts that there was a real danger of the clerk's bias adversely infecting the justices' views. If he was not so requested, any bias on his part would not normally influence the outcome of the proceedings.

As mentioned above, the decision in *R* v *Camborne Justices (ex parte Pearce)* [1955] 1 QB 41, DC, was expressly approved by the House of Lords in *R* v *Gough*. In the

Camborne Justices case, a county council had prosecuted a trader under what was the Food and Drugs Act 1938. The trader complained later that the justices' clerk, who had been invited to advise the magistrates in their private room, was a member of the county council, and that there was, therefore, a 'reasonable suspicion' of bias on his part. The Divisional Court refused to interfere because the justices' clerk, although a county councillor, was not a member of the council's health committee (which had initiated the prosecution) and in the circumstances there was no 'real danger' of bias on his part. It was made clear that a mere suspicion of bias is not a sufficient basis on which to challenge the validity of a decision.

The status of some of the previous cases on bias is in doubt following *R v Gough* [1993] 2 All ER 724, HL. Insofar as they applied the 'reasonable suspicion'/'appearance of bias' test they are wrongly decided. If cases similar to these were to arise again, they would, in the light of *Gough*, need to be decided in accordance with what is now the correct test, the 'real danger or possibility of bias' test. In some of the earlier cases, however, the conclusion on the facts may have been the same if the 'real danger or possibility of bias' test had been applied.

This cannot be said of one of the most well known of them, *R v Sussex Justices (ex parte McCarthy)* [1924] 1 KB 256, DC. Here, the applicant's conviction for dangerous driving was quashed on it appearing that the clerk to the justices was a member of a firm of solicitors acting for the claimant in a civil action against the applicant for damage sustained in the collision. This was in spite of the fact that, although he retired with them when they went to consider their decision, the clerk was not called on to advise the magistrates and did not refer to the case. Applying the correct test, and what was said in *Gough* about its application to justices' clerks (see above), it cannot be concluded that there was a real danger or possibility that the clerk on the facts of the *Sussex Justices* case could have influenced the outcome of the proceedings.

It was in the *Sussex Justices* case that Lord Hewart CJ (at p. 259) uttered his famous dictum that 'it is not merely of some importance but is of fundamental importance that justice should not only be done, but should manifestly and undoubtedly be seen to be done'. According to their Lordships in *R v Gough* [1993] 2 All ER 724, HL, repeating what was said earlier in *R v Camborne Justices (ex parte Pearce)* [1955] 1 QB 41, DC, *per* Slade J at pp. 51–2, Lord Hewart CJ's judgment in the *Sussex Justices* case was the cause of much misunderstanding, and the continued citation of his dictum was in danger of giving the erroneous impression that 'it is more important that justice should appear to be done than that it should in fact be done' (*R v Gough*, above, *per* Lord Woolf at p. 740).

The 'real danger or possibility of bias' test propounded in *R v Gough*, above, was applied in *R v Inner West London Coroner (ex parte Dallaglio)* [1994] 4 All ER 139, CA, where decisions of the coroner inquiring into the disaster which befell the passenger launch *Marchioness* on the Thames in August 1989 were quashed by a quashing order after he had described some relatives and survivors as 'mentally unwell' and referred to one bereaved mother as 'unhinged'. The coroner's language was held to have given rise to a real possibility that he had unconsciously allowed himself to become biased against the relatives.

The test was also applied in *R v Wilson* (1995) *The Times*, 24 February, CA, in which it was held that a real danger or possibility of bias had been caused by the presence on a jury of the wife of a prison officer who worked with the defendants while they were in prison on remand. The defendants' convictions were quashed and a retrial ordered. (See also *In Re Medicaments and Related Classes of Goods (No. 2)* [2001] 1 WLR 700, CA, in which it was held that, in the circumstances, a lay member of the Restrictive Practices Court should have excused herself from hearing a particular application.)

In the *Medicaments* case, above, the Court of Appeal suggested a 'modest adjustment' to the test in *R* v *Gough*, above, with a view to, first, making it clear that the test of bias is based objectively on public perception rather than on a judge's subjective view of the facts, and, secondly, bringing the test into line with the case law of most other common law countries, of Scotland, and of the European Court of Human Rights (see [2001] 1 WLR 700 *per* Lord Phillips MR at pp. 726–7). This adjustment was approved by the House of Lords in *Porter* v *Magill* [2002] 2 AC 357, HL, with a slight modification which prefers 'a real possibility' to 'a real danger' of bias. As stated by Lord Hope in *Porter* v *Magill* (at [103]), the test is now 'whether the fair-minded and informed observer, having considered the facts, would conclude that there was a real possibility that the tribunal was biased'.

Before the decision of the House of Lords in *R* v *Bow Street Metropolitan Stipendiary Magistrate (ex parte Pinochet Ugarte) (No. 2)* [1999] 1 All ER 577, HL, it had been assumed (wrongly, as it turned out) that the automatic disqualification rule (i.e., that bias is conclusively presumed) applies only to a judge who has a pecuniary or proprietary interest in the outcome of a case, as in *Dimes* v *Grand Junction Canal Co.* (1852) 3 HL Cas 759, HL, above. In *Pinochet (No. 2)*, the House of Lords extended the automatic disqualification rule to a judge with a *non-financial* interest whose decision may promote a cause in which he is involved together with one of the parties to the case. Lord Browne-Wilkinson said ([1999] 1 All ER 577 at p. 588) that there is 'no good reason' for limiting the automatic disqualification rule to cases of pecuniary or proprietary interest. He continued:

The rationale of the whole rule is that a man cannot be a judge in his own cause . . . [I]f, as in the present case, the matter at issue does not relate to money or economic advantage but is concerned with the promotion of the cause, the rationale disqualifying a judge applies just as much if the judge's decision will lead to the promotion of a cause in which the judge is involved together with one of the parties.

(See also *per* Lords Goff, Hope, and Hutton at pp. 591–2, 595–6, and 597, respectively.)

Pinochet (No. 2) arose out of the attempt by the former head of state of Chile, Senator Pinochet, to avoid being extradited to Spain to face trial for crimes against humanity. In *R* v *Bow Street Metropolitan Stipendiary Magistrate (ex parte Pinochet Ugarte) (Amnesty International intervening)* [1998] 4 All ER 897, HL, proceedings in which Amnesty International was granted leave to take part, it had been held by a majority of three to two that the Senator could not claim immunity from the criminal processes, including extradition, of the United Kingdom in respect of acts of torture and hostage-taking since such acts were not a function of a head of state.

A few days after judgment was given, it was discovered that one of the Law Lords in the majority, Lord Hoffmann, was a director and chairperson of Amnesty International Charity Ltd, which had been formed to carry out Amnesty International's charitable purposes. Senator Pinochet petitioned the House of Lords to set aside its judgment.

In R v Bow Street Metropolitan Stipendiary Magistrate (ex parte Pinochet Ugarte) (No. 2) [1999] 1 All ER 577, HL, it was held unanimously by five Law Lords who had not been involved in the earlier appeal proceedings that Lord Hoffmann was automatically disqualified from hearing that appeal since, in effect, his failure to disclose his connection with Amnesty International meant that he was acting as a judge in his own cause. Accordingly, the judgment in R v Bow Street Metropolitan Stipendiary Magistrate (ex parte Pinochet Ugarte) (Amnesty International intervening) [1998] 4 All ER 897, HL, above, was set aside and the matter was referred to another appellate committee of the House for rehearing. (The rehearing, at which the Senator was again unsuccessful, took place before seven Law Lords, and is reported as R v Bow Street Metropolitan Stipendiary Magistrate (ex parte Pinochet Ugarte) (Amnesty International intervening) (No. 3) [1999] 2 All ER 97.)

In Pinochet (No. 2) [1999] 1 All ER 577, above, the House of Lords emphasised that the facts of the case were exceptional and that a judge is not expected to stand down, or disclose his position to the parties, in every case involving a charity with whose work he is connected. He need do this only where he has an active role as trustee or director of a charity which has a close connection with a party to the litigation (see, e.g., per Lord Browne-Wilkinson at [1999] 1 All ER 577, 589).

The Pinochet (No. 2) judgment encouraged the making of allegations of judicial bias in a number of subsequent cases in which applications were made to the Court of Appeal for permission to appeal. Although these cases were not so much concerned with the automatic disqualification rule as with the real danger or possibility of bias test laid down in R v Gough [1993] 2 All ER 724, HL, above, their potential impact on public confidence in the legal process led to them being dealt with by a particularly strong Court of Appeal, comprising the Lord Chief Justice, the Master of the Rolls, and the Vice-Chancellor. They are reported at [2000] 1 All ER 65, beginning with Locabail (UK) Ltd v Bayfield Properties Ltd.

The 28-page judgment in these cases emphasises, first, the fundamental nature of a litigant's right to a fair trial by an impartial tribunal, and, secondly, that a judge would be as wrong to accede to a tenuous or frivolous objection to his partiality as he would to ignore one of substance. Thus, it seems that no objection can ever be soundly based on the judge's religion, ethnic or national origin, gender, age, class, means, or sexual orientation, and only rarely can an objection be soundly based on the judge's educational, service or employment background, previous political associations, membership of social, sporting or charitable bodies, Masonic associations, previous judicial decisions, or extra-judicial utterances (see [2000] 1 All ER 65 at p. 77, and the later case of Seer Technologies Ltd v Abbas (2000) The Times, 16 March, in which an application by an Arab defendant that an inquiry as to damages should not be heard by a Jewish judge was dismissed).

On the other hand, a real possibility of bias may well be thought to arise, for example, out of personal friendship or animosity between the judge and any member of the public involved in the case, or if for any other reason there are real grounds for suspecting that the judge's inability to ignore extraneous considerations and prejudices will prevent him from dealing objectively with the case before him. In these instances, the judge, although not automatically disqualified, will be expected to excuse himself from trying the case where there is real doubt over his impartiality (see [2000] 1 All ER 65 at pp. 77–8). Failure to do so will lead, on appeal, to the setting aside of his decision if a real possibility of bias is established.

In the event, permission to appeal was granted in only one of the five cases dealt with in the judgment in *Locabail (UK) Ltd* v *Bayfield Properties Ltd* [2000] 1 All ER 65, CA. This was a case in which the Court of Appeal concluded (reluctantly) that there was a real possibility of bias on the part of a recorder who had decided a personal injury case in the county court in favour of the claimant. The recorder had views which were pro-claimant and anti-insurer, and had frequently expressed those views in legal publications. Taking a broad, commonsense approach, the Court of Appeal held that the possibility could not be excluded that the recorder might have unconsciously leant in favour of the claimant and against the defendant (who was supported by an insurance company) in deciding the factual issues between them (see [2000] 1 All ER 65 at pp. 92–3). The defendant's appeal was allowed and a retrial ordered.

It has been held that the automatic disqualification rule applies to a member of a disciplinary body who has played, or is deemed to have played, an active part in the earlier prosecution of complaints against an individual. Thus, in *In Re P (A Barrister)* [2005] 1 WLR 3019, the Visitors to the Inns of Court, applying *Pinochet (No. 2)*, above, held that a lay representative chosen to sit on the panel of Visitors to hear the appeal of a barrister found guilty of misconduct by a disciplinary body was automatically disqualified since she was also a member of the Professional Conduct and Complaints Committee of the Bar Council, the body responsible for deciding whether to proceed with complaints of misconduct against barristers. Even though she had not been a party to the decision to proceed against the barrister, it was decided that she would be a judge in her own cause if she was allowed to hear the appeal as a Visitor since each member of the Professional Conduct and Complaints Committee was held to have a common interest with the Committee as a whole in securing the conviction of accused barristers and in defeating any appeals.

On the other hand, the automatic disqualification rule does not apply to a member of a disciplinary body who has taken no active part in the prosecution of the complaints. *Pinochet (No. 2)* [1999] 1 All ER 577, HL, was distinguished on the facts by the Judicial Committee of the Privy Council in *Meerabux* v *Attorney-General of Belize* [2005] 2 WLR 1307, PC. Here, the chairman of a disciplinary body investigating complaints of misbehaviour in office against a former judge was held not to be *automatically* disqualified simply because he was also a member of the local Bar Association since he had had no role in the earlier prosecution of the complaints. Nor, applying *Porter* v *Magill* [2002] 2 AC 357, HL, was he *apparently* biased by reason of membership of the Bar Association.

It was held by the Court of Appeal in *R* v *Abdroikov* [2005] 1 WLR 3538, CA, that the presence on a jury of an eligible person who happens to be connected with the administration of justice (a serving police officer and a solicitor employed by the Crown Prosecution Service) does not by itself give rise to a real possibility of bias. It might be otherwise if such a juror fails to bring to the attention of the trial judge any special knowledge he has about individuals involved in the case or about the facts of the case.

It is extremely unwise for a judge to refer in public outside court to a case he is currently trying since there is a danger that he may appear to be biased against one side or the other. If his public comments are reported in the media and he disagrees with the contents of the report, he must subsequently make clear in court his disagreement, and what are the inaccuracies in the report, so as to remove any appearance of bias. This was decided in *R v Batth* [1990] Crim LR 727, CA, where the Recorder of London was reported in a newspaper as having referred in a public address to 'murderous Sikhs' involved in a case he was trying at the time. His explanation in court that the report was 'as usual inaccurate' was held not to be sufficient to dispel the appearance of bias raised on the facts, and his reported comments were described as 'deplorable'.

10.5.3.2.2 *The duty to act fairly* There have been numerous cases involving a breach of the *audi alteram partem* rule of natural justice. This rule imposes a duty to act fairly and demands that a person should be told of the nature of the charges against him, that he should be allowed adequate time to prepare his defence, and that he should be given a fair hearing.

The requirement of fairness does not, however, demand that during the investigative process a person should be given particulars of the criminal conduct of which he is suspected by the police or other appropriate authorities. Thus, in *R v Serious Fraud Office (ex parte Nadir)* (1990) *The Independent*, 16 October, Asir Nadir, then chairman and chief executive of Polly Peck International plc, was refused permission to apply for judicial review to compel the Serious Fraud Office to provide him with brief details of the transactions which were suspected of involving criminality. (Mr Nadir later absconded while on bail and flew to the Turkish Republic of Northern Cyprus, with which the United Kingdom has no extradition treaty, in May 1993.)

Similarly, the public and the media have no right to be told the names of persons who are charged with, or under investigation for, criminal activity: *R v Secretary of State for the Home Department (ex parte Westminster Press)* [1992] COD 303, DC, in which an application for judicial review of the legality of a police circular advising use of the formula, 'a person has been charged', was dismissed.

A person is entitled to a fair hearing before a decision adversely affecting his interests is made by a public authority if, in the absence of any other more specific provision, he has a legitimate or reasonable expectation of being accorded such a hearing. The expectation may, for example, be based on some statement or undertaking by, or on behalf of, the public authority (*Attorney-General of Hong Kong v Ng Yuen Shiu* [1983] 2 All ER 346, PC; *Council of Civil Service Unions v Minister for the Civil Service* (the 'GCHQ case') [1984] 3 All ER 935, HL), or it may arise by implication from the seriousness of the circumstances.

In *R v Norfolk County Council (ex parte M)* [1989] 2 All ER 359, it was held that a person is entitled to a fair hearing before having his name entered on a child abuse register since the decision to register his name is not merely an internal administrative procedure of the local authority but affects substantially the rights of the alleged abuser. *R v Norfolk County Council (ex parte M)*, above, was approved by the Court of Appeal in *R v Harrow London Borough Council (ex parte D)* [1990] 3 All ER 12, CA, although, on the facts, the court refused to quash the local authority's decision to place the applicant's name on the register as an abuser since she had been given an opportunity to give her explanation of how her children had come to be injured and had done so.

Even the ordinary courts of law are capable, on occasions, of acting in breach of the fair hearing rule. In *Ex parte Bowgin* (1965) 109 SJ 76, DC, a quashing order was granted to quash the decision of a magistrates' court where the conviction of a person for careless driving had been announced before he had been given the opportunity to present his defence in full. In *R v Marylebone Magistrates' Court (ex parte Perry)* [1992] Crim LR 514, DC, and *R v Marylebone Magistrates' Court (ex parte Joseph)* (1993) *The Times*, 7 May, DC, convictions were quashed on the ground that the defendants had not had a fair hearing where a stipendiary magistrate had failed to give their cases his undivided attention during the taking of evidence. In *Perry* he had signed warrants relating to other cases, while in *Joseph* he had read a newspaper law report, instead of devoting his full attention to the evidence in the proceedings before him.

In *R v Birmingham Justices (ex parte Lamb)* [1983] 3 All ER 23, DC, it was the *prosecutor* who sought judicial review after the magistrates had dismissed informations against the accused without a hearing! It was held that the magistrates had abused their powers, but the court, in the exercise of its discretion, declined to grant quashing and mandatory orders because to have done so would have been unfair to the accused in the circumstances. (See also *R v Dorchester Justices (ex parte Director of Public Prosecutions)* [1990] RTR 369, DC, *R v Watford Justices (ex parte Director of Public Prosecutions)* [1990] RTR 374, DC, and *R v Milton Keynes Justices (ex parte Director of Public Prosecutions)* [1991] COD 498, DC, in all of which the magistrates were held to have wrongly dismissed informations without hearing any evidence; *R v Swansea Justices (ex parte Director of Public Prosecutions)* (1990) *The Times*, 30 March, DC, and *R v Sutton Justices (ex parte Director of Public Prosecutions)* [1992] 2 All ER 129, DC, where it was held that the magistrates had acted too hastily in dismissing informations—in the *Swansea* case simply because the prosecution witnesses had not appeared and the magistrates were not prepared to wait half an hour for them to arrive, and in the *Sutton* case because counsel for the prosecution was late and the magistrates were not prepared to wait another ten minutes for his arrival.)

Where the magistrates act unreasonably in dismissing an information (as in the *Sutton Justices* case, above), in consequence of which the defendant is acquitted, the acquittal is a nullity. Notwithstanding the fact that it appears to involve a contradiction in terms, the acquittal can be quashed by a quashing order even though it is already null and void (*R v Hendon Justices (ex parte Director of Public Prosecutions)* [1993] 1 All ER 411, DC, not following the *Sutton Justices* case on this point). Where the prosecution still wishes to proceed on the information, a mandatory order directing the magistrates to hear the information will usually be a more appropriate remedy than a quashing order (the *Hendon Justices* case, above, where the DPP sought a quashing order but was granted a mandatory order instead).

The failure to provide a fair hearing (or any hearing at all) will usually be the fault of the court, tribunal, or other public body itself. Although unfairness to the defendant in criminal proceedings resulting from a failure by the prosecution to disclose the existence of witnesses who could have given evidence favourable to the defence may lead to the defendant's conviction being quashed on *appeal*, it is doubtful whether this unfairness is reviewable by means of a *quashing order* (*Al-Mehdawi* v *Secretary of State for the Home Department* [1989] 3 All ER 843, HL, *per* Lord Bridge at p. 848, explaining and distinguishing *R* v *Leyland Magistrates (ex parte Hawthorn)* [1979] QB 283, DC; see also *R* v *West Sussex Quarter Sessions (ex parte Albert & Maud Johnson Trust Ltd)* [1974] QB 24, CA, approved by the House of Lords in the *Al-Mehdawi* case).

However, the unfairness may be reviewable by means of a quashing order where the defendant has no right of appeal and he was convicted on the basis of flawed evidence unwittingly relied on by the prosecution (*R* v *Bolton Justices (ex parte Scally)* [1991] 2 All ER 619, DC, in which the applicant's conviction for driving while the proportion of alcohol in her blood exceeded the prescribed limit was quashed by a quashing order on the ground that, after the conviction, it was discovered that the cleansing swabs used at the time the blood specimens were taken had been impregnated with alcohol in the same concentration as in beer and that this could have led to inaccurately high results when the specimens were subjected to laboratory analysis).

A failure to obtain a hearing (whether the case raises issues of *public* or *private* law) is not reviewable where it is due to the fault of the aggrieved party's own advisers rather than that of the court or tribunal. This is because the fault is outside the knowledge and control of the court or tribunal, and does not, therefore, of itself involve any procedural impropriety or violation of natural justice on the part of the court or tribunal.

Thus, in a case where an immigration adjudicator had given an immigrant the opportunity of presenting his case against deportation but the opportunity was lost through the fault of the immigrant's solicitor, it was held that there had been no procedural impropriety or denial of natural justice, and that, accordingly, the adjudicator's decision given in the absence of the immigrant was not reviewable (*Al-Mehdawi* v *Secretary of State for the Home Department* [1989] 3 All ER 843, HL, reversing the decision of the Court of Appeal in the case and also overruling *R* v *Diggines (ex parte Rahmani)* [1985] 1 All ER 1073, CA). This situation is akin to those wherein, owing to a solicitor's negligence, judgment is given in default of appearance, an action is dismissed for want of prosecution, or a claim becomes statute-barred (*Al-Mehdawi* v *Secretary of State for the Home Department*, above, *per* Lord Bridge at p. 849). In these circumstances, the aggrieved person's remedy is to sue his solicitor for the tort of negligence.

In *R* v *Medical Appeal Tribunal (Midland Region) (ex parte Carrarini)* [1966] 1 WLR 883, DC, an applicant for a disability pension who wished to introduce in evidence a consultant's report favourable to himself was refused an adjournment. The tribunal decided the case against the applicant solely on the basis of the report of its own consultant. A quashing order was granted to quash the tribunal's decision because the tribunal had ignored the requirements of natural justice by refusing to let in expert evidence to controvert its own expert evidence.

There have been a number of cases where decisions have been quashed because the court of first instance had refused an adjournment and proceeded in the absence of one of the parties. Examples are *R* v *Thames Magistrates' Court (ex parte Polemis)* [1974] 2 All ER 1219, DC (magistrates' court refusing an adjournment); *Priddle* v *Fisher & Sons* [1968] 1 WLR 1478, DC (employment tribunal refusing an adjournment).

In *Ostreicher v Secretary of State for the Environment* [1978] 3 All ER 82, CA, the Court of Appeal held that in deciding whether an adjournment should have been granted, a distinction must be drawn between judicial proceedings before a court on the one hand and an administrative inquiry on the other. The court refused to quash an inspector's decision following a public inquiry (into a compulsory purchase order) which the Secretary of State had refused to adjourn.

In *R v Hull Prison Board of Visitors (ex parte St Germain and others)* [1979] 1 All ER 701, CA, which arose out of the riot at Hull prison in 1976, the Court of Appeal decided as a preliminary matter that, in principle, a quashing order is available to review the disciplinary decisions of a prison board of visitors. This decision was approved by the House of Lords in *O'Reilly v Mackman* [1982] 3 All ER 1124, HL. In *R v Hull Prison Board of Visitors (ex parte St Germain and others) (No. 2)* [1979] 3 All ER 545, DC, the Queen's Bench Divisional Court actually granted a quashing order to quash some decisions of a prison board of visitors. One decision was set aside because the visitors had improperly refused a prisoner permission to call witnesses, and the others because prisoners were not given an opportunity to dispute the hearsay evidence of prison officers. Although not bound by the strict rules of evidence applicable in criminal cases in the courts, the board of visitors had an overriding duty to provide the accused with a fair hearing.

In *R v Deputy Governor of Camphill Prison (ex parte King)* [1984] 3 All ER 897, CA, the Court of Appeal had held that a quashing order was not available to review disciplinary decisions of the *prison governor* on the ground that, unlike a board of visitors, he was not an impartial and independent body but was instead a manager appointed by, and answerable to, the Home Secretary. That decision was, however, unanimously overruled by the House of Lords in the similar case of *Leech v Parkhurst Prison Deputy Governor* [1988] 1 All ER 485, HL. Their Lordships took the view that a prison governor, when exercising his statutory power to find an offence proved and to make an award against a prisoner in disciplinary proceedings, is acting no less independently of the Home Secretary than a board of visitors (see *per* Lords Bridge and Oliver at pp. 497–8 and 509, respectively).

The fact that a governor can exercise a statutory power which affects the rights or legitimate expectations of citizens makes it imperative that the power be exercised in accordance with the principles of natural justice, and that any allegation of abuse be subject to judicial review by the courts of law. Accordingly, in *Leech* the applicant's conviction for a prison disciplinary offence (being in possession of a device adapted for the smoking of a controlled drug), in respect of which he had been awarded 28 days' loss of remission, was quashed by a quashing order where he had been found guilty by the deputy governor before being given any opportunity to state his defence or cross-examine the reporting prison officer.

The right to a fair hearing implicit in the rule, *audi alteram partem*, does not demand that in every case a person should be heard *orally*. The making of *written representations* may satisfy the rule (*Board of Education v Rice* [1911] AC 179, HL, *per* Lord Loreburn LC at p. 182; *Local Government Board v Arlidge* [1915] AC 120, HL; *Lloyd v McMahon* [1987] AC 625, HL; *R v Army Board (ex parte Anderson)* [1991] 3 All ER 375, DC, *per* Taylor LJ at p. 387).

Nor does the requirement of a fair hearing necessarily demand *legal representation* by counsel or a solicitor before a tribunal or other hearing, even though it is available before the ordinary courts of law (*Pett* v *Greyhound Racing Association* [1969] 1 QB 125, CA; *Fraser* v *Mudge* [1975] 3 All ER 78, CA; *Maynard* v *Osmond* [1977] 1 All ER 64, CA). It has been said that in hearings before non-statutory domestic tribunals, justice can often be done better by a good layman than by a bad lawyer (*Enderby Town Football Club Ltd* v *Football Association Ltd* [1971] Ch 591, CA, *per* Lord Denning MR at p. 605). Whether legal representation should be permitted as a matter of discretion will depend on the facts of each individual case. Relevant factors which may tip the scales in favour of permitting it are the gravity of the charge against a person or of the consequences to him should a decision go against him (*Pett* v *Greyhound Racing Association*, above, and see *Ezeh* v *United Kingdom* (2003) 15 BHRC 145, ECHR).

A person is always free to represent himself at proceedings before a court or tribunal as a 'litigant in person'. In addition, of respectable antiquity is the right (recognised in *Collier* v *Hicks* (1831) 2 B & Ad 663) in civil proceedings to have the assistance of a 'McKenzie friend' (named after the case of *McKenzie* v *McKenzie* [1971] P 33, CA).

A 'McKenzie friend' is a member of the public who accompanies a litigant in person in order to assist him by quietly making suggestions, quietly giving advice, or by taking notes. He does not have rights of audience, and is, therefore, not entitled to address the court. A decision to deprive a litigant in person of his right to a McKenzie friend can be challenged in judicial review proceedings.

The right to a McKenzie friend is not an absolute one since the court may require the friend to leave if it becomes clear during the proceedings that his assistance is harmful to the efficient administration of justice, such as where the litigant in person is encouraged to waste time, or is advised to ask immaterial questions or to raise irrelevant issues.

In *R* v *Leicester City Justices (ex parte Barrow)* [1991] 3 All ER 935, CA, a magistrates' court refused to allow the applicants to be assisted by a member of an anti-poll tax organisation, acting as a McKenzie friend, in proceedings brought by a local authority for liability orders under the community charge legislation. The liability orders were made and the applicants sought judicial review. The Court of Appeal held that since there was no evidence that the applicants, or their chosen McKenzie friend, had any intention of disrupting or abusing the process of the court, the magistrates were wrong to have denied the applicants the help of a friend. In the circumstances, that denial had created the appearance of unfairness and the liability orders made against the applicants were quashed by a quashing order. (The *Barrow* case was followed in *R* v *Highbury Corner Magistrates' Court (ex parte Watkins)* (1992) *The Times*, 22 October, DC, and *R* v *Wolverhampton Stipendiary Magistrate (ex parte Mould)* (1992) *The Times*, 16 November, DC.)

The Court of Appeal's plea in the *Barrow* case that the term 'McKenzie friend' no longer be used because, *inter alia*, it wrongly implies the status of an unqualified legal assistant known to the law (see [1991] 3 All ER 935, CA, *per* Lord Donaldson MR at p. 947 and Staughton LJ at p. 948) has not been heeded (see *R* v *Bow County Court (ex parte Pelling)* [1999] 4 All ER 751, CA, *per* Lord Woolf MR at p. 757, and *In Re O (Children) (Hearing in Private: Assistance)* [2005] 3 WLR 1191, CA, *per* Wall LJ at [125]).

In *R* v *Bow County Court* (*ex parte Pelling*), above, the Court of Appeal said, *obiter*, that a person who earns his living by appearing regularly as a McKenzie friend must exercise considerable restraint in order to avoid abusing his true role (as an assistant to the litigant in person) by indirectly running the case instead (see *per* Lord Woolf MR at p. 758; see also *Paragon Finance plc* v *Noueiri* (*Practice Note*) [2001] 1 WLR 2357, CA).

Guidance on the use of McKenzie friends in family proceedings was issued by the Office of the President of the Family Division in 2005 (see [2005] Fam Law 405). This guidance was amplified by the Court of Appeal in *In Re O* (*Children*) (*Hearing in Private: Assistance*) [2005] 3 WLR 1191, CA, where it was made clear that the purpose of allowing the assistance of a McKenzie friend is to further the interests of justice 'by achieving a level playing field and ensuring a fair hearing' ([2005] 3 WLR 1191 *per* Wall LJ at [128]). It was held that the European Convention on Human Rights, art. 6 (right to a fair trial), is automatically invoked in every application to the court for the assistance of a McKenzie friend, that in family proceedings (even where they involve children and are being heard in private) the presumption in favour of allowing such assistance is a strong one, and that assistance should not be refused without good reason. If the judge thinks there is a good reason for refusal, he should identify it and explain it fully.

The judges at first instance in two of the three cases reported as *In Re O*, above, were held to have effectively deprived the appellants of a fair hearing by denying them the assistance of a McKenzie friend. They were severely criticised by the Court of Appeal for having done so (see *per* Wall LJ at [23], [36], and [51]–[56]).

The right to a fair hearing does not entitle a citizen to have his case decided by the same person who heard the evidence initially. This limitation is particularly relevant in the case of statutory inquiries where the evidence is heard by an inspector appointed by the minister, but it is the minister himself who makes the final decision. In *Local Government Board* v *Arlidge* [1915] AC 120, HL, the Hampstead Borough Council made a closing order against a house as being unfit for human habitation. The lessee of the house, A, appealed to the Local Government Board and a public inquiry, at which A presented his case, was held by one of the Board's housing inspectors. The Board confirmed the closing order and A applied for a quashing order to quash its decision on the following grounds:

(a) that the appeal had not been decided properly by the Board because the decision did not disclose the identity of the officer of the Board by whom the appeal was decided;

(b) that he had not been allowed to be heard orally before the Board;

(c) that he had not been allowed to see the inspector's report.

The House of Lords held that a quashing order would be refused. The Board was bound to observe the principles of natural justice but had done so. The Board had followed its customary procedure. It was entitled to act through delegates without revealing their identities. A was not entitled to an oral hearing before the Board, and the Board was not obliged to disclose the inspector's report to A.

Their Lordships made it clear in the *Arlidge* case, above, that the standards expected of public authorities when deciding questions affecting the rights and property of the individual are not as high as those expected of the ordinary courts of law. Accordingly, in *Bushell* v *Secretary of State for the Environment* [1980] 2 All ER 608, HL, the House of Lords held that the refusal of an inspector at a public inquiry to allow cross-examination of the Secretary of State's witnesses on traffic prediction methods did not amount to a denial of natural justice. In an ordinary court of law, such an omission would not be countenanced.

At the same time, their Lordships in the *Arlidge* case emphasised that public authorities must act judicially, at least to the extent of dealing with such questions impartially and of giving the parties the opportunity of adequately presenting their cases—unless, it may now be added, a party could have had no legitimate expectation of being heard in view of his own behaviour (*Cinnamond* v *British Airports Authority* [1980] 2 All ER 368, CA), or unless the requirement of a fair hearing has been waived by implication in an emergency which has led to the *provisional* suspension of a licence or permit in order to avoid the possible loss of many lives (*R* v *Secretary of State for Transport (ex parte Pegasus Holidays (London) Ltd)* [1989] 2 All ER 481—Romanian organisation's permit to operate charter flights provisionally suspended pending an inquiry into the competence of their pilots).

10.5.3.2.3 *The duty to act fairly: giving reasons for decisions* It is not, and never has been, a general requirement of natural justice that reasons for decisions should be given (*R* v *Gaming Board for Great Britain (ex parte Benaim and Khaida)* [1970] 2 QB 417, CA, *per* Lord Denning MR at p. 431; *R* v *Immigration Appeal Tribunal (ex parte Khan (Mahmud))* [1983] 2 All ER 420, CA, *per* Lord Lane CJ at p. 423; *Doody* v *Secretary of State for the Home Department* [1993] 3 All ER 92, HL, *per* Lord Mustill at p. 110).

In appropriate circumstances, however, a duty to give reasons may be *implied* (*R* v *Civil Service Appeal Board (ex parte Cunningham)* [1991] 4 All ER 310, CA, below; *Doody* v *Secretary of State for the Home Department*, above, *per* Lord Mustill at p. 110).

The fact that natural justice does not generally demand that reasons be given means, first, that a quashing order will not lie to quash a decision on the ground only that no reasons for it were given, and secondly, that a mandatory order is not available to compel the giving of reasons (*R* v *Gaming Board for Great Britain (ex parte Benaim and Khaida)*, CA, above, where a mandatory order was refused because the Gaming Board had given the applicants all the information which had led to doubts about their suitability for a gaming licence and had given them full opportunity to deal with it, and a quashing order was refused because the Board was not bound to give reasons for its decisions).

In *Payne* v *Lord Harris of Greenwich* [1981] 2 All ER 842, CA, it was held that the Parole Board was not under a duty to give reasons for not recommending the release of a person from prison. This was held to be so in the case of a prisoner serving a life sentence whether that life sentence was a mandatory one (as in *Payne* v *Lord Harris of Greenwich*), or a discretionary one (as in *R* v *Secretary of State for the Home Department (ex parte Gunnell)* (1984) *The Times*, 7 November, CA, and *R* v *Parole Board (ex parte Bradley)* [1990] 3 All ER 828, DC).

In *R* v *Parole Board (ex parte Wilson)* [1992] 2 All ER 576, CA, in which the

76 year-old applicant had already served 20 years of a discretionary life sentence imposed for buggery, the Court of Appeal declined to follow its two earlier decisions in *Payne* and *Gunnell,* above, and held that a prisoner serving a discretionary life sentence who applies for release on licence is entitled to see any material to be placed before the Parole Board which suggested that he was still a danger to the public. It was said that fairness demands disclosure of the material because, without knowing what is being said against him, the applicant will be unable to make effective representations on the issue of dangerousness.

Payne was rejected as no longer reflecting established views on prisoners' rights and on the ground that, although it was a decision of the Court of Appeal, it was distinguishable (and, accordingly, not binding in the circumstances) since it concerned a *mandatory* life sentence. *Gunnell,* while not distinguishable, was rejected for the reason that, applying *Williams* v *Fawcett* [1985] 1 All ER 787, CA (para 5.3.2.3.1), in a case involving the liberty of the subject the Court of Appeal is not bound by an earlier decision of its own if following it might produce injustice.

Payne was eventually overruled by the House of Lords in *Doody* v *Secretary of State for the Home Department* [1993] 3 All ER 92, HL. This case was decided at a time when the penal element of a life sentence was ultimately determined by the Home Secretary. It was held that in view of the fact that a discretionary life prisoner enjoyed certain rights as a result of *R* v *Parole Board (ex parte Wilson)* [1992] 2 All ER 576, CA, with which the House was in agreement, it was fair that a *mandatory* life prisoner should be entitled to certain rights in respect of the Home Secretary's decision about the penal element in his sentence.

The reasoning in *Doody* has since been extended to cover category A prisoners. These are prisoners who are high security risks, whose escape would present great danger, and for whom release on licence by the Parole Board is virtually inconceivable. It follows that any decision to keep a prisoner in category A has serious implications for his liberty. A prisoner who aspires to release on licence will need to obtain classification in a lower category (see *R* v *Secretary of State for the Home Department (ex parte Duggan)* [1994] 3 All ER 277, DC).

Not even the ordinary courts of law have a general duty, either by statute or at common law, to give reasons for their decisions, although the superior courts invariably do so in practice as a means of convincing the parties that justice has been done. Reasons for a decision may also provide grounds for a possible appeal to a higher court. (See *per* Rose LJ in *R* v *Snaresbrook Crown Court (ex parte Lea)* (1994) *The Times,* 5 April, DC; *Flannery* v *Halifax Estate Agencies Ltd* [2000] 1 All ER 373, CA; *Douglas* v *Hello! Ltd* [2001] 2 WLR 992, CA, *per* Brooke LJ at [8].)

The present trend is towards *implying* a requirement that courts should give reasons (see, e.g., *R* v *Warwick Crown Court (ex parte Patel)* [1992] COD 143; *R* v *Harrow Crown Court (ex parte Dave)* [1994] 1 All ER 315, DC; *R* v *Snaresbrook Crown Court (ex parte Input Management Ltd)* (1999) *The Times,* 29 April, DC; *Flannery* v *Halifax Estate Agencies Ltd,* CA, above (county court); *Coleman* v *Dunlop Ltd* [1998] PIQR P398, CA (county court)). Guidance on the giving of reasons was provided to judges sitting at first instance by the Court of Appeal in *English* v *Emery Reimbold & Strick Ltd (Practice Note)* [2002] 1 WLR 2409, CA.

By s. 10 of the Tribunals and Inquiries Act 1992, reasons must be given, if requested, for the decisions of tribunals listed in the Act and of ministers holding statutory inquiries. The reasons given must be proper, intelligible, and adequate (*Re Poyser and Mills's Arbitration* [1964] 2 QB 467 *per* Megaw J at p. 478; this description was approved by the House of Lords in *Westminster City Council* v *Great Portland Estates plc* [1984] 3 All ER 744, HL, and in *Save Britain's Heritage* v *Secretary of State for the Environment* [1991] 2 All ER 10, HL; and see *Mountview Court Properties Ltd* v *Devlin* (1970) 114 SJ 474, DC, in which a case was sent back to a rent assessment committee with a direction to state its reasons fully and adequately).

The tribunal's reasons may be given in writing or orally. The reasons, even if given orally, are taken to form part of the decision and accordingly are incorporated in the record of the proceedings (Tribunals and Inquiries Act 1992, s. 10(6)). The result of this rule is that the tribunal's decision can be challenged by a quashing order on the ground of error of law on the face of the record. This procedure is explained more fully in para 10.5.3.3 below.

A mental health review tribunal must give *written* reasons for its decisions (Mental Health Review Tribunal Rules 1983, SI 1983, No. 942, r. 23(2)). The reasons must be proper and adequate in the sense of enabling the patient to know whether the tribunal has made any error of law in reaching its decision (*Bone* v *Mental Health Review Tribunal* [1985] 3 All ER 330; *R* v *Mental Health Review Tribunal (ex parte Clatworthy)* [1985] 3 All ER 699). The tribunal should indicate, with reasons, which expert evidence (if any) has been accepted and which has been rejected (*R (H)* v *Ashworth Special Hospital Authority* [2003] 1 WLR 127, CA).

The Criminal Injuries Compensation Authority (formerly the Criminal Injuries Compensation Board) is bound to give reasons for a decision to refuse or reduce an award of compensation under the Criminal Injuries Compensation Scheme. In *R* v *Criminal Injuries Compensation Board (ex parte Moore)* [1999] 2 All ER 90, Sedley J commended it as good practice that the reasons should be given in writing. However, he refused to elevate this practice to the status of a principle of public law since to do so would come too close to usurping the Parliamentary function of legislating.

With regard to tribunals and other public authorities not covered by any statutory provisions, recent case law reveals a clearly discernible movement in favour of the giving of at least outline reasons for decisions. As Lord Clyde put it in *Stefan* v *General Medical Council* [1999] 1 WLR 1293, PC, at p. 1300:

The trend of the law . . . towards an increased recognition of the duty upon decision-makers of many kinds to give reasons . . . is consistent with current developments towards an increased openness in matters of government and administration. But the trend is proceeding on a case by case basis . . . and has not lost sight of the established position of the common law that there is no general duty, universally imposed on all decision-makers . . . [T]he law does not at present recognise a general duty to give reasons for administrative decisions. But it is well established that there are exceptions where the giving of reasons will be required as a matter of fairness and openness . . . There is certainly a strong argument for the view that what were once seen as exceptions to a rule may now be becoming examples of the norm, and the cases where reasons are not required may be taking on the appearance of exceptions.

In *R* v *Civil Service Appeal Board (ex parte Bruce)* [1989] 2 All ER 907, CA, the Court of Appeal said, *obiter*, that in order to create a greater sense of fairness in its proceedings it was desirable as a matter of policy and discretion for the Civil Service Appeal Board to give a brief statement of its reasons. In *R* v *Civil Service Appeal Board (ex parte Cunningham)* [1991] 4 All ER 310, CA, the Court of Appeal went further and held that, in certain circumstances, the Board is bound by law to give reasons when deciding whether a dismissal is fair or unfair, and, if unfair, when assessing the appropriate financial compensation. The approach adopted in this case was approved by the House of Lords in *Doody* v *Secretary of State for the Home Department* [1993] 3 All ER 92, HL, above.

In deciding whether a particular tribunal should give reasons for its decisions when it is not obliged to do so by any statutory provisions, regard must be had to the nature and function of the tribunal and to its decisions.

It has been suggested, for instance, that a distinction must be made between those decisions which obviously demand the giving of reasons (as where personal liberty is involved) and those decisions where the giving of reasons would be completely inappropriate (*R* v *Higher Education Funding Council (ex parte Institute of Dental Surgery)* [1994] 1 All ER 651, DC, where it was held that the HEFC was not compelled to give reasons for its decision to lower the research rating of a London University dental college following a research assessment exercise, but note that the decision on the actual facts of this case was later exposed to some criticism in *R (Wooder)* v *Feggetter* [2002] 3 WLR 591, CA, below; see also *R* v *English Nursing Board (ex parte Roberts)* [1994] COD 223, DC, in which the investigating committee of the English Nursing Board was held not to be obliged to give reasons for a decision to take no action on a complaint from a member of the public, and *R* v *University College London (ex parte Idriss)* [1999] Ed CR 462, where it was held that there was no requirement for a university to give reasons for refusing to admit an applicant to a course).

In *R (Wooder)* v *Feggetter* [2002] 3 WLR 591, CA, in which the claimant's personal liberty was at stake, it was held that reasons must be given for a decision made under s. 58 of the Mental Health Act 1983 to administer medical treatment to a competent non-consenting adult.

It was held in *R* v *Director of Public Prosecutions (ex parte Manning)* [2000] 3 WLR 463, DC, that, although the Director of Public Prosecutions is under no duty to give reasons for his refusal to prosecute, he should normally do so where his decision concerns a death in custody in respect of which a coroner's jury has already returned a verdict of unlawful killing implicating a clearly identified, though unnamed, person whose whereabouts are known. The Director's refusal to prosecute in the particular case was quashed.

A duty to give reasons may be implied even where to do so runs counter to the express words of a statute. In *R* v *Secretary of State for the Home Department (ex parte Fayed)* [1997] 1 All ER 228, CA, the Fayed brothers' applications for naturalisation under the British Nationality Act 1981 had been refused by the Home Secretary. The brothers challenged his decision on the grounds, first, that they had not been given an opportunity to make representations, and, secondly, that the Home Secretary had given no reasons for his decision. At first instance, both grounds were rejected for the same reason: they were incompatible with the unequivocal language of what was then s. 44(2) of the 1981 Act that the 'Secretary of State shall not be required to assign any reason for the grant or refusal of any application under this Act the decision on which is at his discretion'.

On appeal, however, the Court of Appeal found in favour of the Fayeds. It was held that, notwithstanding the wording of s. 44(2), the Home Secretary must act fairly in exercising his discretion to grant or refuse citizenship, and fairness demands that the applicant should be given sufficient information about the Home Secretary's concerns as to enable the applicant to make what representations he can before the final citizenship decision is taken. The Home Secretary's decisions to reject the Fayeds' applications for citizenship were quashed. (Section 44(2) of the British Nationality Act 1981 was repealed by the Nationality, Immigration and Asylum Act 2002.)

10.5.3.3 Where there is an error of law

The jurisdiction of the High Court to quash a decision on this ground is quite independent of the two grounds already considered, namely, the doctrine of *ultra vires* and the principles of natural justice. A decision which is *ultra vires,* or made in violation of the principles of natural justice, is void, whereas a decision containing an error of law is merely voidable; i.e., it is valid until avoided by a quashing order from the court.

Again, while there are several remedies available for challenging *ultra vires* actions (damages, injunctions, declarations, quashing orders, prohibiting orders, habeas corpus), the only remedy usually available for challenging error of law is the quashing order. A declaration, for instance, is probably not available as a remedy for error of law (this question was raised, but not answered, in *Re Tillmire Common* [1982] 2 All ER 615).

Originally, the courts would only interfere on the ground of error of law where the error appeared *on the face of the record.* Thus, they would not interfere with the decision of an inferior tribunal if it had acted within its jurisdiction, observed the principles of natural justice, and the proceedings were regular on the face of them. However, a tribunal which acted within its jurisdiction and in accordance with natural justice may nevertheless have had its decision quashed if the proceedings were *not regular on the face of them;* that is, where the tribunal had erred in law and the error appeared on the face of the record of the proceedings.

The power of the courts to interfere on the ground of an error of law on the face of the record was reaffirmed for modern times by the Court of Appeal in *R v Northumberland Compensation Appeal Tribunal (ex parte Shaw)* [1952] 1 KB 338, CA, a case in which the judgment of Denning LJ gives a useful account of the history of this piece of jurisdiction.

S, the clerk to a hospital board, lost his job on the introduction of the National Health Service. He was entitled by statute to compensation, but there was some doubt about how it should be calculated. S claimed that his earlier service in local government should be taken into account as well as his service with the hospital board. The relevant statutory regulations required both to be taken into account, but the Northumberland Compensation Appeal Tribunal decided that only S's service with the hospital board should count. This error of law did not actually appear on the face of the record because the record was incomplete in that it did not mention S's correct submission concerning the periods of employment that should be taken into account.

The Tribunal, however, admitted the error and the Court of Appeal quashed its decision. Denning LJ said (at p. 354):

We have here a simple case of error of law by a tribunal, an error which they frankly acknowledge. It is an error which deprives Mr Shaw of the compensation to which he is by law entitled. So long as the erroneous decision stands, the compensating authority dare not pay Mr Shaw the money to which he is entitled lest the auditor should surcharge them. It would be quite intolerable if in such case there were no means of correcting the error . . . I am clearly of opinion that an error admitted openly in the face of the court can be corrected by certiorari as well as an error that appears on the face of the record. The decision must be quashed, and the tribunal will then be able to hear the case again and give the correct decision.

In *R v Medical Appeal Tribunal (ex parte Gilmore)* [1957] 1 QB 574, CA, statutory regulations provided, in effect, that if a one-eyed man lost the sight of his good eye in an industrial accident his degree of disablement should be assessed at 100 per cent. The Medical Appeal Tribunal erred in law in assessing G's disablement at only 20 per cent. G applied for a quashing order to quash the Tribunal's decision on the ground of error of law on the face of the record. His case was complicated by three factors:

(a) the usual time-limit for applying for a quashing order had expired;

(b) the error did not appear on the face of the record because the record made no mention of either eye;

(c) it was provided by statute that the decision of the Tribunal 'shall be final'.

As to (a), the Court of Appeal exercised its discretion to extend the time-limit since G had not been guilty of any delay in claiming relief. As to (b), the Tribunal's written decision had quoted from a specialist's report, which dealt with the injuries to G's eyes. The Tribunal had thereby made the full report part of the record. Taken together, the decision and the specialist's report gave a full record, on the face of which was an error of law, for it was manifest that the statutory regulations had been misconstrued. As to (c), it was held that, although the words 'shall be final' successfully excluded an *appeal*, they were not explicit enough to preclude judicial review, by means of an application for a quashing order, for error of law on the face of the record (see further, para 10.9.1 below).

The Tribunal's decision was quashed and the value of quashing orders as a remedy was demonstrated once more. Since it was expressly provided that G had no right of appeal against the tribunal's decision, it was only by recourse to a quashing order that he was able to obtain the industrial injuries benefit to which he was lawfully entitled.

The 'record' of proceedings includes the document which initiates the proceedings, the pleadings (if any), the decision of the tribunal (*R v Northumberland Compensation Appeal Tribunal (ex parte Shaw)* [1952] 1 KB 338, CA, *per* Denning LJ at p. 352), and all those documents which appear from the formal order to have formed the basis of the decision (*Baldwin & Francis Ltd v Patents Appeal Tribunal* [1959] AC 663, HL, *per* Lord Denning at p. 690), such as the applicant's submission in *R v Northumberland Compensation Appeal Tribunal (ex parte Shaw)*, above, and the specialist's report in *R v Medical Appeal Tribunal (ex parte Gilmore)*, above.

In *R* v *Southampton Justices (ex parte Green)* [1976] QB 11, CA, affidavits sworn by the chairman of, and the clerk to, the justices were held by the Court of Appeal to be part of the record. The affidavits, which gave reasons for a decision, contained errors of law and the justices' decision was quashed.

The 'record' also extends to the reasons given in an oral judgment and set out in the official transcript of the proceedings (*R* v *Crown Court at Knightsbridge (ex parte International Sporting Club (London) Ltd)* [1981] 3 All ER 417, DC).

The record, however, does not include the evidence given before the tribunal, unless the tribunal chooses to incorporate it as part of the record (*R* v *Northumberland Compensation Appeal Tribunal (ex parte Shaw)*, CA, above, *per* Denning LJ at p. 352; *Re Allen and Matthews's arbitration* [1971] 2 QB 518).

In more recent times, it has become clear that the jurisdiction of the High Court to interfere with the decision of an inferior court or other body on the ground of error of law is not limited to cases where the error appears on the face of the record of the proceedings. It extends to *any* error of law—whether or not it appears on the face of the record.

In *Page* v *Hull University Visitor* [1993] 1 All ER 97, HL, the House of Lords, although divided on some issues, was unanimous in the view that the distinction between errors of law on the face of the record and other errors of law had been swept away by their Lordships' decision in *Anisminic Ltd* v *Foreign Compensation Commission* [1969] 2 AC 147, HL (para 10.11.3 below), as explained by Lord Diplock in *Re Racal Communications Ltd* [1980] 2 All ER 634, HL (para 10.11.2 below), and by the whole House in *O'Reilly* v *Mackman* [1982] 1 All ER 1124, HL. In *Page*, above, the House declared that the distinction was no longer meaningful following the extension of the *ultra vires* doctrine achieved in the *Anisminic* case.

As a result of *Page*, the position now is that any error of law made by an inferior court or other body—whether acting within, or outside, its jurisdiction—is in general capable of correction by means of a quashing order. However, according to the decision of the majority there is an exception in the case of the visitor to a university (or other educational or ecclesiastical foundation) (see, e.g., [1993] 1 All ER 97, HL, *per* Lord Browne-Wilkinson at p. 109, and see also *R* v *Visitors to the Inns of Court (ex parte Calder)* [1993] 2 All ER 876, CA, para 10.8.2.4 below, in which *Page* was applied).

It was held in *Page* that since the visitor has exclusive jurisdiction to decide disputes arising under the domestic law of the university, judicial review is not available to correct the visitor's decision, whether right or wrong, and whether on a question of fact or of law, so long as that decision was made within his jurisdiction and in accordance with the principles of natural justice. A review of the authorities convinced the majority in *Page* that this had been the position for 300 years. It followed that the High Court had no jurisdiction to review the Hull University Visitor's interpretation of the University statutes at the instigation of Mr Page, whose employment as a lecturer had been terminated (lawfully according to the visitor) on the ground of redundancy. It was further held (unanimously) that, in any event, the visitor's decision contained no error of law.

The reasons given by the majority in *Page* were, first, that the visitor is not called on to apply the common law familiar to the judiciary but, instead, a peculiar or domestic law of which he is appointed the sole judge, and, secondly, that the perceived advantages of the visitorial jurisdiction—swiftness, cheapness, and finality—would be lost if the visitor's decisions could be challenged by judicial review.

After *Page*, it is possible to challenge the visitor's decision by judicial review and have it quashed by a quashing order in three cases. First, if the visitor has acted outside his jurisdiction; i.e., where he had no power to adjudicate on the dispute in the first place. Secondly, if he has abused his powers by acting in a way which is incompatible with his judicial role. However, he is not to be regarded as having abused his powers simply by making a decision which a superior court might regard as containing a mistake of law (see [1993] 1 All ER 97 *per* Lord Griffiths at p. 100). Thirdly, if he has violated the principles of natural justice.

10.6 Prohibiting orders (formerly prohibition)

10.6.1 Nature

A prohibiting order is a remedy which is used to prevent inferior courts, tribunals, and other public authorities from exceeding their jurisdiction and to prevent them from violating the principles of natural justice. It may be issued, for example, to prohibit a tribunal from exceeding, or continuing to exceed, its jurisdiction by hearing a case which it is outside its powers to hear.

A prohibiting order is in nature similar to a quashing order, with the difference that the former is concerned with the future while the latter is concerned with the past. Although a prohibiting order and a quashing order may be sought separately, the two are often applied for together in the same proceedings. Thus, a decision may be set aside by a quashing order, and further unlawful action, such as an attempt to implement the quashed decision, forbidden by a prohibiting order. Like a quashing order, a prohibiting order is a discretionary remedy. Disobedience to it is punishable as a contempt of court.

10.6.2 Availability

Prohibiting and quashing orders are available against the same bodies. Prohibiting orders have been sought successfully against an ecclesiastical court, a county court, a magistrates' court, Income Tax Commissioners, rent tribunals, licensing authorities, police authorities, and government ministers.

An applicant for a prohibiting order must have a sufficient interest in the matter to which the application relates (para 10.8.4 below).

In one case, a prohibiting order is available where a quashing order is not. A quashing order will not lie to challenge proceedings in *ecclesiastical courts* because ecclesiastical law is substantially different from the general law applied in the High Court. A prohibiting order, however, *will* lie against ecclesiastical courts to restrain excesses of jurisdiction and violations of natural justice (*R v North (ex parte Oakey)* [1927] 1 KB 491, CA).

A prohibiting order is not available against voluntary (i.e., non-statutory) domestic tribunals. Nor does it lie directly against the Crown since it emanates from the Crown, but it *is* available against individual ministers of the Crown. A prohibiting order will not lie against a deliberative body, such as the Legislative Committee of the Church Assembly, because it does not have the power to decide questions affecting the rights of subjects (*R* v *Legislative Committee of the Church Assembly (ex parte Haynes-Smith)* [1928] 1 KB 411, DC).

10.6.3 Scope

A prohibiting order will lie on two of the same grounds as a quashing order.

(a) To restrain an *ultra vires* act

This is illustrated by the case of *R* v *Electricity Commissioners (ex parte London Electricity Joint Committee Co (1920) Ltd)* [1924] 1 KB 171, CA. Here, the Electricity Commissioners had a statutory duty to make schemes for the supply of electricity. This involved the creation of electricity districts. For the London area, the Commissioners proposed a scheme which was, in effect, for two electricity districts when the statute allowed only one. The Court of Appeal held that the scheme was *ultra vires* and issued a prohibiting order to prevent the Commissioners from going any further with it. Similarly, in *R* v *Minister of Health (ex parte Davis)* [1929] 1 KB 619, CA, the Court of Appeal granted a prohibiting order to prevent confirmation by the minister of an *ultra vires* housing scheme.

A prohibiting order may be used to prevent a local authority from acting unlawfully. In *R* v *Greater London Council (ex parte Blackburn)* [1976] 3 All ER 184, CA, it was held by the Court of Appeal that the rules made by the GLC for the censorship of films were *ultra vires*. The GLC was applying the statutory test of 'tendency to deprave or corrupt' under the Obscene Publications Act 1959, which did not apply to films, instead of applying the common law test of indecency. In consequence, films were being shown contrary to law. The court was minded to grant a prohibiting order but did not do so, preferring instead to give the GLC an opportunity either to change its rules so as to incorporate the proper test or to stop censoring films for adults altogether.

(b) To restrain a violation of the principles of natural justice

In *R* v *Kent Police Authority (ex parte Godden)* [1971] 2 QB 662, CA, the Police Authority took steps compulsorily to retire G, a chief inspector of police, on medical grounds. The Authority proposed to have G examined by the chief medical officer of the force. He was the same doctor who two years earlier had examined G for some other reason and had found him to be suffering from a paranoid mental illness. G himself went to a consultant psychiatrist who reported that G was mentally completely normal. The Police Authority refused to make available to G's doctors all the information which was available to its chief medical officer. G applied for prohibiting and mandatory orders.

The Court of Appeal granted both remedies. A prohibiting order was issued against the chief medical officer of the force to prevent him from deciding whether G was permanently disabled. The chief medical officer had already formed an opinion adverse to G and had thus committed himself to a view of G's mental condition which would preclude any appearance of impartiality at a second examination. A mandatory order was issued against the Police Authority compelling it to supply to G's doctors all the information available to its own medical examiner. It should be noted that justice did not demand that G himself should be allowed to see all the material ([1971] 2 QB 662, CA, *per* Lord Denning MR at p. 670).

In *R* v *Liverpool Corporation (ex parte Liverpool Taxi Fleet Operators' Association)* [1972] 2 QB 299, CA, the Court of Appeal issued a prohibiting order against a taxicab licensing authority. The authority was prohibited from acting on a resolution to increase the number of taxicab licences without first considering representations from interested parties.

A prohibiting order was issued in *R* v *Telford Justices (ex parte Badhan)* [1991] 2 All ER 854, DC, in order to stop committal proceedings in respect of a rape allegedly committed in 1973 or 1974 and which had not even been reported until 14 or 15 years after the event. It was held that, by reason of delay, the committal proceedings would be an abuse of the process of the court and the defendant would be so prejudiced in the preparation and conduct of his defence as to be unlikely to receive a fair trial in any subsequent proceedings.

10.7 Mandatory orders (formerly mandamus)

10.7.1 Nature

A mandatory order directs that a public legal duty be performed. It is regarded as a residuary remedy with the result that it will not usually be granted where there is available some alternative remedy for enforcing a particular public legal duty (para 10.7.4 below).

A mandatory order appears to be similar in nature to a mandatory injunction. The difference between the two is that an injunction is an *equitable* remedy which is normally available only in *private* law, while a mandatory order is a *common law* remedy, based on royal authority, which is available only in *public* law (*Glossop* v *Heston & Isleworth Local Board* (1879) 12 ChD 102, CA).

Although a mandatory order may be sought separately, it is often applied for in the same proceedings together with a quashing order. Thus, the decision of a tribunal which is *ultra vires* may be quashed by a quashing order and the tribunal compelled to hold a proper rehearing by means of a mandatory order. However, if a mandatory order alone is granted it is to be implied that the tribunal's decision was a nullity without its being quashed by a quashing order (see, e.g., *R* v *Hendon Justices (ex parte Director of Public Prosecutions)* [1993] 1 All ER 411, DC).

A mandatory order, like a quashing order and a prohibiting order, is a discretionary remedy. In the exercise of its discretion, the court may refuse to grant it where the applicant has delayed unduly in seeking relief, or where the defendant public body has tried but failed, through circumstances beyond its control, to perform its duty, or where it is unnecessary (*R* v *Commissioner of Police of the Metropolis (ex parte Blackburn) (No. 3)* [1973] QB 241, CA; *R* v *Bristol Corporation (ex parte Hendy)*

[1974] 1 All ER 1047, CA, *per* Scarman LJ at p. 1051; *R v Northumberland Compensation Appeal Tribunal (ex parte Shaw)* [1952] 1 KB 338, CA, *per* Morris LJ at p. 357).

A mandatory order will not be granted to compel the performance of something which is impossible. Thus, in *R v London & North Western Railway Co.* (1851) 16 QB 864, the expiry of compulsory purchase powers was held to have barred the grant of a mandatory order to command the purchase of land necessary for constructing a branch railway line.

A mandatory order may be refused if it is impractical to order it, even though it may be the only satisfactory remedy for the applicant in the particular circumstances of his case. In *Chief Constable of the North Wales Police v Evans* [1982] 3 All ER 141, HL (para 10.2.2.4 above), the House of Lords declined to grant a mandatory order compelling the Chief Constable to reinstate a dismissed probationer constable on the ground, *inter alia*, that to do so would be tantamount to a usurpation by the courts of the powers of the Chief Constable. Instead, a declaration was granted that the Chief Constable had acted unlawfully. If, however, the making of a declaration would, on the facts of the case, amount to a usurpation of a public authority's function, a mandatory order should be preferred (*Shah v Barnet London Borough Council* [1983] 1 All ER 226, HL).

Disobedience to a mandatory order is punishable as a contempt of court. In *R v Poplar Borough Council (ex parte London County Council) (No. 1)* [1922] 1 KB 72, CA, the Poplar Borough Council refused to pay a sum of money in respect of rates to the London County Council as required by statute. The Court of Appeal issued a mandatory order against the Borough Council commanding it to pay over the money, and, if necessary, to levy a rate for the purpose. When the Borough Council refused to obey the mandatory order, the County Council, in *R v Poplar Borough Council (ex parte London County Council) (No. 2)* [1922] 1 KB 95, CA, initiated contempt proceedings and some of the disobedient borough councillors were imprisoned.

10.7.2 Availability

A mandatory order is available to the Crown, to the private citizen, and to one public authority against another, as in the *Poplar* cases noted in para 10.7.1 above. Mandatory orders have been issued against the Crown Court, a county court judge, a magistrates' court, Income Tax Commissioners, a housing appeal tribunal, licensing authorities, planning authorities, other local authorities, the registrar of companies, and against government ministers.

The applicant for a mandatory order must have a sufficient interest in the matter to which the application relates (para 10.8.4 below).

A mandatory order does not lie against the Crown since it emanates from the Crown. Moreover, unlike quashing and prohibiting orders, a mandatory order was not always available against servants of the Crown, such as individual ministers.

A mandatory order was held to lie (provided that the applicant had a sufficient interest) against a Crown servant to compel him to perform a statutory duty which was owed to a member of the public as well as to the Crown (see *M v Home Office* [1993] 3 All ER 537, HL, *per* Lord Woolf at p. 558). Thus, in *R v Commissioners for Special Purposes of the Income Tax* (1888) 21 QBD 313, CA, the Court of Appeal issued a mandatory order against the Special Commissioners of Income Tax ordering them to make orders for the repayment of some income tax which a taxpayer had overpaid. It had been argued that a mandatory order would not lie against the Special Commissioners. The court held, however, that it would lie against them since they owed a statutory duty to the public, as well as to the Crown, to make orders for repayment of overpaid income tax.

Historically, a mandatory order did not lie against a Crown servant to compel him to perform, as an agent, a duty which was owed only to the Crown (*R v Lords Commissioners of the Treasury* (1872) LR 7 QB 387, DC; *R v Secretary of State for War* [1891] 2 QB 326, CA). This was because a third party has no power to compel an agent to perform a duty which is owed only to the agent's principal. However, according to the House of Lords in *M v Home Office* [1993] 3 All ER 537, HL, the common law position has changed and the current rule is that a mandatory order will lie against a Crown servant, such as a minister, without having to show that he was placed under some sort of statutory duty (see [1993] 3 All ER 537 *per* Lord Woolf at p. 560). (For the facts of *M v Home Office*, see para 8.3, and see also para 10.8.6.4 below.)

10.7.3 Scope

A mandatory order is used primarily to enforce the performance of a *public duty*, whether statutory or otherwise. But it may also be used to ensure that a *discretion* conferred by statute is exercised properly according to law. It may be granted, for example, where a minister is exercising his discretion in such a way as to frustrate the underlying policy of the statute (*Padfield v Minister of Agriculture, Fisheries & Food* [1968] AC 997, HL, below), or where irrelevant factors are taken into consideration (*R v Port of London Authority (ex parte Kynoch Ltd)* [1919] 1 KB 176, CA, in which, however, a mandatory order was refused on the facts).

Courts and tribunals are under a public duty to hear and determine all cases within their jurisdiction provided that those cases have been initiated according to the correct procedure. In *R v Huddersfield County Court Judge (ex parte Beaumont Ashton Ltd)* (1967) *The Times*, 24 October, DC, a county court judge refused to hear an application for the renewal of the tenancy of a shop on the ground that he did not have jurisdiction to do so. The Divisional Court held that he did have jurisdiction, and granted a mandatory order commanding him to hear and determine the application.

While a mandatory order may be issued to order a judge to hear and determine a case, it will not be issued so as to control the judge in his conduct of the case, or to instruct him what evidence to admit or reject (*R v Sir Robert Carden* (1879) 5 QBD 1, DC, *per* Cockburn CJ at p. 5). In *R v Wells Street Stipendiary Magistrate (ex parte Seillon)* [1978] 3 All ER 257, DC, the Divisional Court refused to interfere by means of a mandatory order where a stipendiary magistrate refused to allow a particular line of cross-examination by defence counsel during committal proceedings.

A mandatory order is available to command the performance of the public duties owed by local authorities (see, e.g., *R v Bedwellty Urban District Council (ex parte Price)* [1934] 1 KB 333, DC, where a local authority was compelled by a mandatory order to produce its accounts for the inspection of the agent of a ratepayer).

A mandatory order has also been used to ensure that a local authority abides by its own standing orders (*R v Hereford Corporation (ex parte Harrower and others)* [1970] 1 WLR 1424, DC).

A mandatory order may issue against a licensing authority to compel the proper performance of its public duties. In *R v Weymouth Borough Council (ex parte Teletax (Weymouth) Ltd)* [1947] KB 583, DC, a taxicab licensing authority misunderstood its statutory licensing powers and refused to transfer cab licences to a new owner of five cabs which had existing licences. The court held that the authority's power was to license cabs and not their owners or drivers. Accordingly, if a licensed cab was sold, the buyer was entitled to use the cab until the licence expired. A mandatory order was granted commanding the authority to register the transfer of the licences.

A mandatory order may lie against a police authority to compel the enforcement of the criminal law. In *R v Commissioner of Police of the Metropolis (ex parte Blackburn)* [1968] 2 QB 118, CA, the Court of Appeal was prepared to issue a mandatory order against the police but did not do so. The remedy had become unnecessary because the Commissioner withdrew his instructions not to enforce the gaming laws in London. The Court of Appeal was of the clear opinion that a mandatory order would be available against the police in appropriate circumstances.

This view was repeated in *R v Commissioner of Police of the Metropolis (ex parte Blackburn) (No. 3)* [1973] QB 241, CA, in which the same applicant sought a mandatory order to compel the Commissioner to enforce the laws against pornography. The Court of Appeal held that it would only interfere in the extreme case where a police authority was not carrying out its duty of enforcing the law. It would not interfere with the discretion which the authority had in performing the duty of law enforcement as long as that duty was being carried out. On the facts, a mandatory order was refused because the Commissioner was doing what he could to enforce the law under difficult conditions and no more could reasonably be expected of him.

A mandatory order may issue to order a public authority to exercise a *discretion* one way or the other where it has failed to do so, or to exercise it in a lawful manner (as by taking into account only relevant considerations and excluding irrelevant ones), or to give reasons for its decision to exercise the discretion in a particular way. The justification for the court's power to interfere in these circumstances is that a complete failure to exercise a discretion which is required by law to be exercised is unlawful, and that an improper or capricious exercise of a discretion is tantamount to a failure to exercise it at all (*Commissioners of Inland Revenue v National Federation of Self-Employed & Small Businesses Ltd* [1981] 2 All ER 93, HL, *per* Lord Scarman at p. 111).

In dealing with a public authority's failure to exercise a discretion, the court will only compel the authority to act; it will not substitute its own decision for that of the authority. In *Padfield v Minister of Agriculture, Fisheries & Food* [1968] AC 997, HL, it was held by the House of Lords that the minister had wrongfully refused to exercise a discretion to refer a complaint about the operation of a milk marketing scheme to a committee of investigation as required by statute. The case was sent back to the Divisional Court of the Queen's Bench Division for a mandatory order to be issued commanding the minister to consider the complaint according to law.

The minister in the *Padfield* case had a statutory discretion to refer a complaint for consideration and report 'if the minister in any case so directs'. The House of Lords emphatically denied that those words gave the minister a completely unfettered discretion to refer a complaint or not. Their Lordships said that the minister's discretion must be exercised in accordance with the intention of the statute which conferred it. The obvious intention of Parliament was that an independent committee, established for the purpose, should investigate complaints about the operation of the milk marketing scheme and that its reports should be available to Parliament. It was not a proper exercise of his discretion for the minister to take no action at all on a legitimate complaint, or to ignore the merits of the complaint and take into account irrelevant considerations (such as political embarrassment to himself if he later felt obliged to implement the committee's recommendations).

All the minister's reasons for refusing to order an inquiry into the complaint were held to be bad in law. It was argued on behalf of the minister that he was not bound to give reasons for refusing to refer a complaint, and that if he had given no reasons his decision could not have been questioned. But the House of Lords made it absolutely clear that his decision could be challenged in those circumstances. If the minister failed, or refused, to give any reason for his decision, the court may infer that he had no good reason to give and may issue a prerogative order against him. 'Reason' means something more than a mere conclusion. If, however, an intelligible and salient reason is given, it is immaterial that matters of detail are not included (*Elliott v London Borough of Southwark* [1976] 2 All ER 781, CA). In *Padfield*, the House of Lords considered that the minister's silence would amount to an attempt to frustrate or thwart the underlying policy of the statute and, as such, would be an abuse of the discretion conferred on him by the statute.

10.7.4 Mandatory orders and alternative remedies

A mandatory order is a discretionary remedy and will not normally be granted where there is some adequate alternative remedy or procedure available for enforcing a particular public legal duty. The alternative remedy or procedure may be provided by a particular statute, or it may be available in the form of an action in tort.

In *Pasmore v Oswaldtwistle Urban District Council* [1898] AC 387, HL, a local authority had a statutory duty to provide such sewers as were necessary for draining their district. In default, the Act provided that a complaint could be made to the Local Government Board, which had power to order performance of the duty within a fixed time and to enforce its order by an application for a mandatory order. The owner of a paper mill applied to the court for a mandatory order to command the local authority to provide sewers adequate to carry effluent from the mill. The House of Lords held that the statute, by implication, had excluded an application for a mandatory order by a private person. The proper remedy, as provided by the statute, was to complain to the Local Government Board.

Before deciding whether a mandatory order is excluded by the availability of an alternative remedy, the court will consider carefully whether that other remedy is adequate in terms of convenience and effectiveness (*R v Poplar Borough Council (ex parte London County Council) (No. 1)* [1922] 1 KB 72, CA, *per* Scrutton LJ at p. 94; *R v Paddington Valuation Officer (ex parte Peachey Property Corporation Ltd) (No. 2)* [1966] 1 QB 380, CA, *per* Lord Denning MR at p. 400).

The alternative remedy was considered to be inadequate in three cases already discussed elsewhere. First, in *R v Poplar Borough Council (ex parte London County Council) (No. 1)* [1922] 1 KB 72, CA (para 10.7.1 above), a mandatory order was granted despite the existence of a specific statutory remedy. That remedy—the levying of distress on the goods of the Council and/or those of its members—was held to be utterly inadequate for ensuring payment of the sum due to the County Council.

Secondly, in *R v Bedwellty Urban District Council (ex parte Price)* [1934] 1 KB 333, DC (para 10.7.3 above), a mandatory order was granted even though the statute provided another remedy, namely, a criminal penalty not exceeding £5 to be imposed on the appropriate officer of the local authority for refusing to allow inspection of the accounts. It was held that such a remedy could lead the applicant no nearer to inspection of the accounts and was not, therefore, as convenient and effective as a mandatory order.

Thirdly, in *R v Commissioner of Police of the Metropolis (ex parte Blackburn)* [1968] 2 QB 118, CA (para 10.7.3 above), it was argued that Mr Blackburn had an equally effective and convenient remedy in that he could himself seek to enforce the gaming laws by starting private prosecutions on his own. A mandatory order was refused on the facts, but the Court of Appeal made it clear that they would not have regarded the alternative remedy as adequate. Salmon LJ (at p. 145) described it as 'fantastically unrealistic', and Edmund-Davies LJ (at p. 149) said that 'only the most sardonic could regard the launching of a private prosecution . . . as being equally convenient, beneficial and appropriate' as a mandatory order.

10.8 Application for judicial review: Civil Procedure Rules, Part 54, and Supreme Court Act 1981, section 31

10.8.1 Introduction

Although the former complicated procedure for obtaining the prerogative remedies was replaced in 1933 with a simpler, common procedure available under rules of court, other, more serious, deficiencies remained.

In particular, it was not possible to claim an ordinary remedy, like damages or an injunction, in the same proceedings in addition to, or instead of, a prerogative remedy. And the *locus standi* required to apply for prerogative relief was not quite the same for each remedy, although there were signs that the differences were diminishing.

In 1976, the Law Commission recommended a new procedure to be called 'application for judicial review' (Law Com. No. 73, *Remedies in Administrative Law*, Cmnd 6407, 1976). By 1978, most of the Law Commission's recommendations had been implemented, not by statute but by changes in the Rules of the Supreme Court. The new RSC, Ord. 53, came into force on 11 January 1978. Some of the more important provisions of Ord. 53 were put into statutory form, with effect from 1 January 1982, by s. 31 of the Supreme Court Act 1981. Order 53 was replaced by Part 54 of the Civil Procedure Rules in October 2000.

It should be noted that, unless the circumstances are exceptional, an application for judicial review should not normally be made by the applicant, or entertained by the court, until *after* the objectionable decision has been reached or the impugned hearing concluded. To seek relief from the High Court too soon will in many cases turn out to be an unnecessary waste of time and money. Furthermore, for the High Court to entertain challenges to decisions made at an interim stage of proceedings in inferior courts and tribunals could cause havoc there (see, e.g., *R v Association of Futures Brokers & Dealers Ltd (ex parte Mordens Ltd)* [1991] COD 40, where an application for judicial review of two interlocutory decisions of a commissioner appointed to hear the applicants' appeal against the refusal of the Association to admit them to membership was dismissed.)

10.8.2 Availability

10.8.2.1 Introduction: *O'Reilly* v *Mackman*

The procedure by way of application for judicial review *must* be used in order to obtain mandatory, prohibiting, or quashing orders (CPR, r. 54.2; Supreme Court Act 1981, s. 31(1), as amended). In addition, the procedure *may* be used to obtain a declaration or an injunction (CPR, r. 54.3; Supreme Court Act 1981, s. 31(2)), although these remedies continue to exist as private law remedies also.

It was at first thought that a declaration in a matter of public law could be sought, according to the circumstances, *either* on an application for judicial review *or* in a private law action under CPR, r. 40.20. This was the intention of the Law Commission expressed in its report published in 1976, para 10.8.1 above.

However, in *O'Reilly* v *Mackman* [1982] 3 All ER 1124, HL, the House of Lords held that a person seeking to challenge the decision of a public authority on the ground that it infringes a right protected by *public* law must, as a general rule, proceed by way of an application for judicial review rather than by way of a private law action for a declaration. Failure to use the correct procedure will result in the action being struck out as an abuse of the process of the court.

The leading speech in *O'Reilly* v *Mackman* was prepared by Lord Diplock, with whom the other Law Lords expressly agreed. It is clear from Lord Diplock's opinion that the major grounds for the decision were, first, the improved usefulness of (the former) RSC, Ord. 53 since the changes introduced in 1978, and, secondly, the desire to protect public authorities from 'groundless, unmeritorious or tardy harassment' ([1982] 3 All ER 1124 at p. 1133).

There can be no doubt that judicial review possesses safeguards against its abuse as a remedy which are not available in private law actions. These safeguards are the requirement of permission of a High Court judge to apply for judicial review, the limited form of discovery of documents, the fact that cross-examination will only be allowed when justice demands it, the three-month time-limit within which the application must be made, and the fact that any relief sought is granted only at the discretion of the court. (See, e.g., *R v Secretary of State for Social Services (ex parte Association of Metropolitan Authorities)* [1986] 1 All ER 164; *R v Monopolies and Mergers Commission (ex parte Argyll Group plc)* [1986] 2 All ER 257, CA; *R v Secretary of State for the Environment (ex parte Walters)* (1997) *The Times*, 2 September, CA, in which it was held on appeal that the judge, after taking into account all the circumstances, had correctly exercised his discretion in refusing relief to the applicant despite the judge's finding that the relevant statutory consultation process had not been properly followed.)

While these safeguards may undoubtedly prevent abuse of the legal process and the unreasonable harassment of public authorities, they may also be seen as weapons with which to oppress the ordinary citizen seeking to remedy the excesses of the administration. Therein lies the danger to justice of the House of Lords' decision that an application for judicial review will in most cases exclude a private law action for a declaration under CPR, r. 40.20.

10.8.2.2 Criticism of *O'Reilly* v *Mackman*

In *O'Reilly* v *Mackman* [1982] 3 All ER 1124, HL, it may well be that the disadvantages to the citizen of the procedure under (the former) Ord. 53 were played down and that his interests were placed a poor second best behind the desire to protect the administration. The advantages to the citizen (and corresponding disadvantages to public authorities) flowing from a private law action are real and well recognised. He does not require permission of anyone to commence his action; there is thus no judicial 'filter' at this stage. There are no special restrictions on the discovery of documents or on cross-examination. The limitation period is usually six years. The remedy of damages is not discretionary.

In *Doyle* v *Northumbria Probation Committee* [1991] 4 All ER 294, Henry J (at p. 300) referred to the difficulties caused by *O'Reilly* v *Mackman*. The need to choose the correct procedure in litigation where public law and private law mix creates what he described as a 'formidable extra hurdle' for claimants, and gives rise to the potential for expensive appeals.

That the rule in *O'Reilly* v *Mackman* involves a risk of creating procedural over-technicality was acknowledged by the House of Lords itself in *Roy* v *Kensington and Chelsea and Westminster Family Practitioner Committee* [1992] 1 All ER 705, HL. Lord Lowry said (at pp. 729–30) that the rule in *O'Reilly* v *Mackman* is subject to exceptions based on the nature of the claim and on the undesirability of erecting procedural barriers, and that unless the procedure adopted in a case is ill-suited to dispose of the issues raised in it there is much to be said for allowing the case to be heard rather than for the court to entertain a debate concerning the form of the proceedings.

In *Roy*, above, it was held that the proceedings had been properly brought by private law action rather than by judicial review: the proceedings were dominated by the claimant's private law rights; his claim might involve disputed issues of fact; the order sought (the payment of money due) could not be granted in judicial review proceedings; and it was said that, when individual private rights are claimed, there should not be a need for permission or a special time-limit and the relief should not be discretionary (see [1992] 1 All ER 705 *per* Lord Lowry at p. 729).

In a subsequent case, Lord Lowry, referring to his criticism of *O'Reilly* v *Mackman* in *Roy*, above, expressed the hope that *O'Reilly* v *Mackman* would one day be the subject of further consideration by their Lordships (*Equal Opportunities Commission* v *Secretary of State for Employment* [1994] 1 All ER 910, HL, at p. 926). A superficial re-examination was later conducted in *Mercury Communications Ltd* v *Director General of Telecommunications* [1996] 1 All ER 575, HL (para 10.8.2.3 below).

10.8.2.3 Protecting legal rights in a private law claim

It was recognised in *O'Reilly* v *Mackman* [1982] 3 All ER 1124, HL, that neither (the former) Ord. 53 nor s. 31 of the Supreme Court Act 1981 makes judicial review the exclusive procedure for obtaining relief for the infringement of rights protected by public law. Lord Diplock himself (at p. 1134) identified two exceptions where a private law action may still be used, but preferred to leave other exceptions to be developed on a case-to-case basis. The first exception mentioned by Lord Diplock is where the invalidity of the impugned decision arises as a collateral issue in a claim for infringement of a right arising under *private* law. The second exception is where none of the parties objects to proceeding by way of a private law action instead of by way of an application for judicial review.

The courts have had to consider the implications of *O'Reilly* v *Mackman* on a number of subsequent occasions. In *Davy* v *Spelthorne Borough Council* [1983] 3 All ER 278, HL, the House of Lords upheld the claimant's right to bring a private law action for damages in negligence against a local planning authority. His action was concerned with infringement of his *private law* rights and did not raise any issue of *public law*. He was not, therefore, obliged to use judicial review proceedings.

In *Wandsworth London Borough Council* v *Winder* [1984] 3 All ER 976, HL, the local authority brought a claim against the defendant for arrears of rent and possession of his flat on the ground that he had not paid the rent lawfully due. In his defence, the defendant denied that he was in arrears because the local authority had exceeded its statutory powers in increasing his rent beyond what he was prepared to pay. He counterclaimed for a declaration that the notices of increase were void. The local authority applied to strike out the defence and counterclaim as an abuse of the process of the court. The House of Lords refused to do so on the ground that it was perfectly proper for the defendant, by way of reply to an action brought against him by a public authority, to claim that his existing *private law* rights (arising under a contract with the authority) had been infringed by a decision of the authority. He was not obliged in such circumstances to mount his challenge to the decision by judicial review.

Where, however, the defendant's private law rights have not been violated he will have no defence to the authority's possession proceedings. If he wishes to challenge the proceedings on the basis of an alleged infringement of some *public law* right (for example, that it was irrational for the authority to apply for possession) he cannot do so in the authority's proceedings, but must himself initiate a separate application for permission to apply for judicial review (*Avon County Council* v *Buscott* [1988] 1 All ER 841, CA, distinguishing *Wandsworth London Borough Council* v *Winder*, HL, above, and approving the decision of Scott J in *Waverley Borough Council* v *Hilden* [1988] 1 All ER 807).

In *Doyle* v *Northumbria Probation Committee* [1991] 4 All ER 294, the claimants were probation officers aggrieved at the decision of their employers to discontinue the payment of a daily home-to-office car mileage allowance as provided for in their contracts of employment. Just before the expiry of the six-year limitation period for bringing an action, they commenced private law proceedings claiming damages for breach of contract. The employers applied to the court to dismiss the action as an abuse of the process of the court on the ground that it raised issues of public law which should have been brought before the court by way of judicial review proceedings. The public law element referred to was the employers' assertion by way of defence that they had no statutory power to pay the allowance.

Applying the reasoning in *Wandsworth London Borough Council v Winder* [1984] 3 All ER 976, HL, above, Henry J held that the action would not be dismissed since it had been rightly commenced as a private law claim. The claimants were seeking a *private law* remedy in a genuine *private law* action; they were not alleging any breach of their *public law* rights. Moreover, to dismiss their action and require them to proceed by way of judicial review would effectively deprive them of their private law rights because, in view of the delay in seeking relief, it would be most unlikely that they would obtain permission to apply for judicial review.

In *R v East Berkshire Health Authority (ex parte Walsh)* [1984] 3 All ER 425, CA, a preliminary point was raised whether the purported dismissal of a senior nursing officer could be questioned in judicial review proceedings, or whether the appropriate procedure was to make a private law complaint to an employment tribunal. The Court of Appeal held that public law proceedings were inappropriate since a breach of an ordinary contract of employment with a public authority gives rise to private law remedies only.

It might be otherwise if there is a special statutory code of discipline governing the employment and dismissal of the employee. The existence of such a code might import a sufficient public law element into the purported dismissal to allow the employee to seek judicial review, as in *R v Secretary of State for the Home Department (ex parte Benwell)* [1984] 3 All ER 854, in which the *Walsh* case was distinguished.

It is interesting to note that, unlike Mr Walsh, Mr Benwell, as a prison officer, was precluded from making a complaint about dismissal to an employment tribunal. Accordingly, unless he was able to have his unfair treatment remedied in judicial review proceedings he would be without a remedy at all, having exhausted the internal appeal procedures. The judge referred to this predicament in his judgment (see [1984] 3 All ER 854 at p. 866) but did not, of course, treat it as a ground for holding that he had jurisdiction in the circumstances to grant a declaration and a quashing order against the Home Secretary.

It has been held that whether or not the relationship between general medical practitioners and family practitioner committees (since abolished) could be described as 'contractual', a doctor's claim for unpaid practice allowances rested in private law and was rightly pursued by private law action rather than judicial review, notwithstanding the fact that it involved a challenge to a public law decision (*Roy v Kensington and Chelsea and Westminster Family Practitioner Committee* [1992] 1 All ER 705, HL, per Lords Bridge and Lowry at pp. 709 and 725, respectively; see also *Lonrho plc v Tebbit* [1992] 4 All ER 280, CA, in which *Roy* was applied).

Again, in a case brought by way of private law action by a prison officer who refused to work on a new shift system, and in which he claimed a declaration that he was still employed on the old shift system, it was said to be at least arguable that the relationship between a prison officer and the Home Office is contractual, and that, in any event, the issues raised were private law ones. Accordingly, the claim was held to have been properly brought by way of private law action rather than by application for judicial review (*McClaren* v *Home Office* [1990] ICR 824, CA).

In *R* v *Derbyshire County Council (ex parte Noble)* [1990] ICR 808, CA, a deputy police surgeon's application for judicial review of the authority's decision to terminate his employment was dismissed on the ground that judicial review was inappropriate since the applicant's claim rested in the private law of contract and involved no sufficient public law element.

Similarly, the dismissal of a barrister employed by the Crown Prosecution Service was held not to be a matter of public law in *R* v *Crown Prosecution Service (ex parte Hogg)* (1994) *The Times*, 14 April, CA, in which *McClaren* v *Home Office*, above, was followed.

In *R* v *Secretary of State for the Home Department (ex parte Moore)* [1994] COD 67, DC, the *Walsh*, *Noble*, and *McClaren* cases were preferred to *Benwell*, and it was held that the refusal of the Permanent Secretary at the Home Office to accept a recommendation of the Civil Service Appeal Board that the applicant, who had been unfairly dismissed from his employment, be reinstated as a prison officer was not a matter of public law and could not, therefore, be challenged in judicial review proceedings.

On the other hand, in *R* v *Legal Aid Board (ex parte Donn & Co.)* [1996] 3 All ER 1, it was held that a decision—taken by a sub-committee of the Legal Aid Board—to award a contract for the management of the Gulf War Syndrome litigation to two particular firms of solicitors contained a sufficient public law element to justify allowing the decision to be challenged in judicial review proceedings.

In *Gillick* v *West Norfolk and Wisbech Area Health Authority* [1985] 3 All ER 402, HL, Lord Scarman (with whom Lord Fraser expressly agreed) said that a parent who wishes to proceed against a public authority in respect of the threatened infringement of parental rights is entitled to proceed by way of a private law action rather than by way of judicial review. Such a claim has a sufficient private law content in it not to make a private law action an abuse of the process of the court. Lord Scarman was also of the opinion that, in any event, Mrs Gillick's case fell within both exceptions mentioned in Lord Diplock's speech in *O'Reilly* v *Mackman* [1982] 3 All ER 1124, at p. 1134, namely that her attack on the validity of the contraceptive advice issued by the Department of Health and Social Security was collateral to her claim for infringement of her private law rights, and none of the parties to the case had objected to the use of the private law procedure (*Gillick*, above, at pp. 415–16).

O'Reilly v *Mackman*, above, was considered by the House of Lords in *Mercury Communications Ltd* v *Director General of Telecommunications* [1996] 1 All ER 575, HL. Here, the importance of avoiding an over-rigid demarcation between procedures at the expense of maintaining flexibility of choice was again emphasised. It was held that the overriding consideration in deciding whether private law or public law proceedings are the more appropriate is whether the proceedings constitute an abuse of the process of the court. Mercury's challenge to a determination of the Director General of Telecommunications was held to have been properly brought in a private law action for a declaration. The argument of the defence that the use of private law proceedings was an abuse of the process of the court since any determination by the Director General was governed solely by public law, and could only be challenged by seeking judicial review in public law proceedings, was unsuccessful.

The House of Lords took the view that the Director General was not precluded under all circumstances from entering the realms of private law merely because he held a statutory office and performed public duties. Their Lordships cited with apparent approval *Wandsworth London Borough Council* v *Winder* [1984] 3 All ER 976, HL, *Gillick* v *West Norfolk and Wisbech Area Health Authority* [1985] 3 All ER 402, HL, and *Roy* v *Kensington and Chelsea and Westminster Family Practitioner Committee* [1992] 1 All ER 705, HL, as examples of claims which arose in a public law context but which were rightly dealt with in private law proceedings.

The problem of over-technicality arose again in *Trustees of the Dennis Rye Pension Fund* v *Sheffield City Council* [1997] 4 All ER 747, CA. Here, the claimants' applications for improvement grants were approved by the Council under its statutory powers. When the Council later refused to pay the grants, the claimants brought a private law action claiming the sums (approaching £100,000) due under the grants. The Council argued that private law proceedings were an abuse of the process of the court since the appropriate remedy for the claimants was to apply for judicial review. The Court of Appeal disagreed, holding that in the circumstances an ordinary private law action was the more appropriate and convenient procedure. It was decided that, although in general the Council's function in relation to the making of grants is a public one, once an application for a grant has been approved the Council's duty to pay it is enforceable in an ordinary private law action. In addition, the claimants were seeking the payment of a sum of money—a remedy which is not available in judicial review proceedings.

In his judgment in the *Dennis Rye* case [1997] 4 All ER 747, CA, Lord Woolf MR (at p. 754) referred to the 'constant unprofitable litigation over the divide between public and private law proceedings'. He suggested that it was time to go back to first principles and the guidance given by Lord Diplock in *O'Reilly* v *Mackman* [1982] 3 All ER 1124. Lord Woolf emphasised the importance of recognising, first, that, in the public interest, judicial review provides protection for public bodies (in the form of the requirement of permission to apply for it and the protection against delay) which is not available in a private law action, and, secondly, that the *general* rule is that it is against public policy and, as such, an abuse of the process of the court, to circumvent these protective measures by bringing a *private law* action in order to enforce *public law* rights. He also said (at p. 755) that if it is not clear which is the correct procedure, it is 'safer' to use judicial review since the applicant cannot then be accused of evading the protection afforded thereby to public bodies.

In considering whether judicial review or a private law action is the more appropriate procedure, a highly relevant factor is that judicial review 'is not a fact-finding exercise' (*R* v *Chief Constable of the Warwickshire Constabulary (ex parte Fitzpatrick)* [1998] 1 All ER 65, DC, *per* Jowitt J at p. 80). Judicial review is concerned with the *legality* of decisions, not with their *merits*. Any disputed questions of fact involved in a case are usually more appropriately resolved in private law proceedings, where there can be a full-scale trial with no permission required, no special time-limit, no special restrictions on the discovery of documents or on cross-examination, and where the relief granted is not necessarily discretionary.

The presence of disputed questions of fact was another reason for holding in the *Dennis Rye* case [1997] 4 All ER 747, CA, above, that a private law action was a more appropriate and convenient procedure than judicial review, and in *R* v *Chief Constable of the Warwickshire Constabulary (ex parte Fitzpatrick)* [1998] 1 All ER 65, above, it was said that as a means for deciding whether there has been an unlawful seizure of material under a search warrant, judicial review in most cases 'has only disadvantages and no advantages when compared with the private law remedy' (*per* Jowitt J at p. 80).

In *Clark* v *University of Lincolnshire and Humberside* [2000] 3 All ER 752, CA, the claimant sued the University for breach of contract three years after the event, alleging that she had been wrongly treated over an examination failure. The University of Lincolnshire and Humberside was a 'new' university known as a 'higher education corporation'. It had no charter and no visitor. The University argued that since it was a public body the claimant should have proceeded against it by way of a claim for judicial review, and that her claim in the private law of contract was an abuse of process designed to avoid the three-month time limit for judicial review. It was held that the claim would not be struck out just because private law, rather than public law, proceedings had been used, or because the claim had been made so long after the event.

The Court of Appeal emphasised that there had been considerable changes in procedure since 1982 when *O'Reilly* v *Mackman* [1982] 3 All ER 1124, HL, was decided. In particular, it was pointed out that, under the Civil Procedure Rules 1998, the court now has a tighter control over private law claims.

10.8.2.4 Public v private functions

An application for judicial review, claiming any of the prerogative remedies or an injunction or declaration, is available to impugn only the decisions of a body which is performing a *public* function.

It follows, for example, that judicial review is not available to challenge the decisions of a voluntary (i.e., non-statutory) domestic tribunal which is performing what are essentially *private* functions (*Law* v *National Greyhound Racing Club Ltd* [1983] 3 All ER 300, CA (company limited by guarantee established to enforce rules of greyhound racing held not susceptible to judicial review), approving *R* v *British Broadcasting Corporation (ex parte Lavelle)* [1983] 1 All ER 241 (decision of disciplinary tribunal established by BBC for private internal purposes held not susceptible to judicial review); see also *R* v *Lord Chancellor's Department (ex parte Nangle)* [1992] 1 All ER 897, DC (internal disciplinary proceedings in government department held not amenable to judicial review); *R* v *Lord Chancellor (ex parte Hibbit and Saunders)* [1993] COD 326, DC (Lord Chancellor treated applicants unfairly in manner in which a contract for court reporting services was awarded, but his decision was held not to be reviewable because it lacked a sufficient public law element); *R* v *Fernhill Manor School (ex parte Brown)* (1992) *The Independent*, 25 June (decision of independent school to expel pupil not amenable to judicial review)).

The appropriate remedy in cases like these against a voluntary domestic tribunal exercising purely private functions is to seek declaratory or injunctive relief in private law proceedings.

In *R v Panel on Takeovers and Mergers (ex parte Datafin plc)* [1987] 1 All ER 564, CA, the Court of Appeal held that decisions of the Panel on Takeovers and Mergers (the 'City Takeover Panel') *are* subject to challenge by means of judicial review. (See also *R v Panel on Takeovers and Mergers (ex parte Guinness plc)* [1989] 1 All ER 509, CA, and note that in both of these cases judicial relief was refused on the facts.) The City Takeover Panel is a non-statutory unincorporated association with no statutory, prerogative, or common law powers. It was established privately by financial institutions for the purpose, *inter alia*, of enforcing a City code of practice. Nevertheless, it performs a *public* function and makes decisions which may indirectly affect the rights of ordinary citizens.

Similarly, the Advertising Standards Authority performs a *public* function (which, if the Authority did not exist, would be performed by the Office of Fair Trading). Accordingly, the Authority's decisions are susceptible to challenge by judicial review even though it has no statutory, prerogative, or common law powers (*R v Advertising Standards Authority Ltd (ex parte Insurance Services plc)* [1990] COD 42, DC, in which a decision of the Authority was quashed by a quashing order where it had been reached without regard to a material consideration). Decisions of the Committee of Advertising Practice are likewise susceptible to judicial review (*R v Committee of Advertising Practice (ex parte Bradford Exchange)* [1991] COD 43, where, on the facts, it was held that the decision-making process had not been unfair).

The question whether the decisions of particular non-statutory bodies are amenable to judicial review is a troublesome one which continues to occupy the time of the courts while the law is developed on a case-by-case basis. The position of the Jockey Club, for example, was considered by the Queen's Bench Divisional Court in two somewhat inconclusive cases decided in 1989 and 1990.

In the first, *R v Jockey Club (ex parte Massingberd-Mundy)* (1989) [1993] 2 All ER 207, DC, it was held by a two-man court consisting of Neill LJ and Roch J (with some reluctance on the part of the latter) that, applying *Law v National Greyhound Racing Club Ltd* [1983] 3 All ER 300, CA, above, the decisions of the Jockey Club are *not* susceptible to judicial review.

The second case, *R v Jockey Club (ex parte RAM Racecourses Ltd)* (1990) [1993] 2 All ER 225, DC, was also decided in favour of the Jockey Club—not on the jurisdiction point but on the doctrine of legitimate expectation—by a differently constituted two-man court.

On the question of jurisdiction, Stuart-Smith LJ said, *obiter*, that it was clear that the decision in *Massingberd-Mundy*, above, was not given *per incuriam*, and he felt bound by *Law v National Greyhound Racing Club Ltd* [1983] 3 All ER 300, CA, above, to hold that decisions of the Jockey Club are not reviewable. In addition, neither he nor Simon Brown J was convinced that *Massingberd-Mundy* was wrongly decided (which, as noted in para 5.3.2.5, is an additional test, laid down in *R v Greater Manchester Coroner (ex parte Tal)* [1984] 3 All ER 240, DC, for determining whether the Queen's Bench Divisional Court can refuse to follow one of its own earlier decisions). Simon Brown J, however, expressed the opinion that *Law v National Greyhound Racing Club Ltd*, above, was distinguishable and that, in the light of

R v *Panel on Takeovers and Mergers (ex parte Datafin plc)* [1987] 1 All ER 564, CA, above, *some* decisions of the Jockey Club *are* reviewable.

Some, at least, of the doubt surrounding the position of the Jockey Club was dispelled by the Court of Appeal in *R* v *Disciplinary Committee of the Jockey Club (ex parte Aga Khan)* [1993] 2 All ER 853, CA. Here it was held that the Jockey Club's decision to disqualify the applicant's horse, which had won a major race, following the detection of a prohibited substance (camphor) in its urine was not susceptible to judicial review. Applying *Law* v *National Greyhound Racing Club Ltd* [1983] 3 All ER 300, CA, above, and distinguishing *R* v *Panel on Takeovers and Mergers (ex parte Datafin plc)* [1987] 1 All ER 564, CA, above, the Court of Appeal decided that, although in its regulation of horse racing it exercises powers which affect the public, the Jockey Club is not a public body and its powers are not governmental. It is a private body whose powers emanate from the agreement of those who subscribe to its rules. This agreement creates private rights for breach of which judicial review is not, but private law remedies are, appropriate.

The question as to whether *all* decisions of the Jockey Club are similarly immune from judicial review was left open in *R* v *Disciplinary Committee of the Jockey Club (ex parte Aga Khan)*, CA, above. Sir Thomas Bingham MR declined (at p. 867) to speculate on the matter. Farquharson LJ (at p. 873) refused to dismiss the possibility that some decisions might be reviewable, as where the persons affected by them had no contractual relationship with the Jockey Club and their only remedy thus lay in public law. On the other hand, Hoffmann LJ (at pp. 875–6) was cautious about adopting an 'improvisatory' approach which pretends that domestic bodies are organs of government in order to give access to judicial review in public law as a means of supplementing inadequate or non-existent private law remedies.

Decisions of the following non-statutory bodies have been held to be susceptible to judicial review: the Professional Conduct Committee of the Bar Council (*R* v *General Council of the Bar (ex parte Percival)* [1990] 3 All ER 137, DC, although, on the facts, it was decided that the Committee had acted properly); the Code of Practice Committee of the British Pharmaceutical Industry Association, a voluntary self-regulating body but one which performs a public duty in the control of advertising and the promotion of medicines (*R* v *British Pharmaceutical Industry Association Code of Practice Committee (ex parte Professional Counselling Aids Ltd)* (1990) *The Independent*, 1 November; on the facts, relief was refused); the visitor of a university if he acts outside his jurisdiction, or abuses his powers, or violates the principles of natural justice (*Page* v *Hull University Visitor* [1993] 1 All ER 97, HL, where it was held that, on the facts, the visitor's decision was not reviewable); subject to the same conditions as apply to a university visitor, High Court judges acting as visitors to the Inns of Court in matters relating to the fitness of persons to become or remain barristers (*R* v *Visitors to the Inns of Court (ex parte Calder)* [1993] 2 All ER 876, CA).

On the other hand, decisions of the Chief Rabbi, of the Imam, and of a Beth Din have been held *not* to be subject to judicial review (*R* v *Chief Rabbi (ex parte Wachmann)* (1991) [1993] 2 All ER 249; *R* v *Imam of Bury Park Mosque, Luton (ex parte Sulaiman Ali)* (1993) *The Times*, 20 May, CA; *R* v *London Beth Din (Court of the Chief Rabbi) (ex parte Bloom)* [1998] COD 131). Their functions are essentially religious and spiritual ones lacking any public law character. Furthermore, a secular court would not feel competent to adjudicate on matters of religious law, customs, and traditions.

It seems that the public function which a body must be exercising in order that the High Court's supervisory jurisdiction can be invoked against it must involve more than the mere fact that the decisions of that body might have consequences for the public at large. There must, according to some of the cases, be a potential *governmental* interest in the particular decision-making body in the sense that if the body did not exist the government might have to consider asking Parliament to impose a statutory regime of control. Some of the financial services cases, and the advertising cases, fall into this category, but the cases of the Chief Rabbi, the Imam, and the Beth Din clearly do not (see also *R v Football Association Ltd (ex parte Football League Ltd)* (1991) [1993] 2 All ER 833, where it was held that the Football Association is a body whose decisions are not subject to judicial review, and note that this case was heavily relied on by the Court of Appeal in *R v Disciplinary Committee of the Jockey Club (ex parte Aga Khan)* [1993] 2 All ER 853, CA, above).

It has been held that, although the powers exercised by Lloyd's derive from a (private) Act of Parliament, the relationship between Lloyd's and the 'names' who join it in its insurance business is governed by the contract existing between the parties so that the relationship is a matter of private law and does not contain any public law element such as to make Lloyd's amenable to judicial review at the suit of 'names' (*R v Corporation of Lloyd's (ex parte Briggs)* (1992) *The Times*, 30 July, DC).

Decisions of the Parliamentary Commissioner for Administration (the Ombudsman) are reviewable (*R v Parliamentary Commissioner for Administration (ex parte Dyer)* [1994] 1 All ER 375, DC), but not those of the Parliamentary Commissioner for Standards (*R v Parliamentary Commissioner for Standards (ex parte Al Fayed)* [1998] 1 All ER 93, CA).

This distinction is based on a desire by the courts to leave Parliament to regulate its own internal affairs and not to be seen to be interfering with proceedings in Parliament. Both Parliamentary Commissioners have duties of a public nature, are subject to the supervision of standing committees of Parliament, and make reports to Parliament. However, while the Parliamentary Commissioner for Administration is concerned with the activities of the public service *outside* Parliament, the functions of the Parliamentary Commissioner for Standards relate directly to what happens *inside* Parliament, and, in the performance of those functions, Parliament has placed him under the supervision of the Select Committee on Standards and Privileges of the House of Commons.

In view of all this, the Court of Appeal decided in the *Al Fayed* case, above, that it would be inappropriate for the activities of the Parliamentary Commissioner for Standards to be reviewable by the courts. Accordingly, Mr Al Fayed's application for judicial review of a report of the Parliamentary Commissioner for Standards, which had concluded that there was no basis for Mr Al Fayed's allegation that Mr Michael Howard (a member of Parliament and a government minister) had received a corrupt payment, was dismissed.

Decisions of the Criminal Cases Review Commission created by the Criminal Appeal Act 1995 are susceptible to judicial review (*R v Criminal Cases Review*

Commission (ex parte Pearson) [1999] 3 All ER 498, DC, where, however, the challenge failed on the facts).

10.8.3 **Permission to apply for judicial review**

The discretionary nature of prerogative relief is emphasised by the rule that the claimant must obtain permission of a judge of the High Court in order to apply for judicial review (CPR, r. 54.4; Supreme Court Act 1981, s. 31(3)). Permission is required so that the court may act as a sieve to exclude unmeritorious claims at an early stage. It is thus possible to 'prevent the time of the court being wasted by busybodies with misguided or trivial complaints of administrative error' (*Commissioners of Inland Revenue* v *National Federation of Self-Employed & Small Businesses Ltd* [1981] 2 All ER 93, HL, *per* Lord Diplock at p. 105).

It has already been noted that permission will be required if it is proposed to apply by way of judicial review for a declaration or an injunction, whereas permission is *not* required if these remedies are sought as private law remedies (para 10.8.2.1 above).

The application for permission is the first, or 'threshold', stage in a two-stage judicial review procedure. The two stages are:

(a) the application for permission to claim judicial review; and

(b) if permission is granted, the hearing of the substantive claim itself.

A claim for judicial review is begun when the court issues a claim form at the request of the claimant. The claim form must contain, or be accompanied by, the documents required by the *Practice Direction* to CPR, Part 54. These include a detailed statement of the claimant's grounds for bringing the claim; a statement of the facts relied on; any written evidence in support of the claim; a copy of any order that the claimant seeks to have quashed; copies of any relevant statutory material; and a list of essential documents for advance reading by the court.

The claim form must be served on the defendant and, unless the court directs otherwise, any person the claimant considers to be an interested party (CPR, r. 54.7).

At the first stage of the procedure, the application for permission to claim judicial review can be determined by the judge without a hearing (CPR, r. 54.12). The claimant cannot *appeal* against a decision to refuse permission, or a decision to give permission subject to conditions or on certain grounds only, but he can request the decision to be *reconsidered* at a hearing (ibid.). Neither the defendant, nor any other person served with the claim form, can apply to set aside an order giving permission (ibid., r. 54.13).

If the judge grants permission to apply for judicial review, he can give directions, which can include a stay of proceedings to which the claim relates (ibid., r. 54.10). He can also grant interim relief, in the form of an interim injunction or an interim declaration, pending the hearing of the substantive claim on the merits. The power to grant an interim injunction includes the power to give a form of interim relief which is rarely granted, namely, an interim *mandatory* injunction. Such an injunction was ordered in the unusual circumstances of *R* v *Kensington and Chelsea Royal London Borough Council (ex parte Hammell)* [1989] 1 All ER 1202, CA.

At the second stage of the procedure, the substantive claim can be decided without a hearing where all the parties agree (ibid., r. 54.18). If the claim succeeds and the court makes a *quashing order* in respect of the decision under review, it can send the matter back to the decision-maker with a direction to reconsider it and reach a decision in accordance with the court's judgment (Supreme Court Act 1981, s. 31(5), as amended; CPR, r. 54.19). Alternatively, where the court considers that there is no purpose to be served in remitting the matter to the decision-maker, it may, subject to any statutory provision, take the decision itself (CPR, r. 54.19).

Claims for judicial review are dealt with in the Administrative Court, which took over from the 'Crown Office List', and was established as part of the Queen's Bench Division of the High Court, on 2 October 2000 (*Practice Direction (QBD: Administrative Court: Establishment)* [2000] 1 WLR 1654). On the same date, the citation of judicial review cases was changed (ibid.). A claim by Smith for judicial review of a decision of the Home Secretary decided before that date is cited as '*R v Secretary of State for the Home Department (ex parte Smith)*'. A claim by Smith decided on or after that date has the citation, '*R (Smith) v Secretary of State for the Home Department*'.

10.8.4 *Locus standi*

10.8.4.1 Introduction

Permission to apply for judicial review will not be granted unless the court considers that the applicant has a 'sufficient interest in the matter to which the application relates' (Supreme Court Act 1981, s. 31(3)). The requirement of *locus standi* affords protection against misuse of the legal process, enabling the court to 'prevent abuse by busybodies, cranks and other mischief makers' (*Commissioners of Inland Revenue v National Federation of Self-Employed & Small Businesses Ltd* [1981] 2 All ER 93, HL, *per* Lord Scarman at p. 113).

Locus standi is not the same as *capacity*. A body may have capacity to apply for judicial review, but it may later turn out that it has no *locus standi* in the proceedings. On the other hand, if capacity is found (at the permission stage) to be lacking, the proceedings will go no further (*R v Darlington Borough Council (ex parte Association of Darlington Taxi Owners)* [1994] COD 424, where it was held that an unincorporated association has no capacity to apply for judicial review).

The need to show *locus standi* arises initially at the first stage of the procedure; i.e., when applying for permission to apply for judicial review. Nevertheless, it is not usually possible to consider the question of whether the applicant has a 'sufficient interest' in total isolation from the merits of his claim. The court must first identify 'the matter' to which the application relates before deciding whether the applicant has 'a sufficient interest' in it (*Commissioners of Inland Revenue v National Federation of Self-Employed & Small Businesses Ltd*, above, *per* Lords Wilberforce, Diplock, Fraser, Scarman, and Roskill at pp. 96–7, 106, 107, 113, and 115, respectively). This will involve an examination of the legal powers or duties which are called in question, their alleged breach, and the position of the applicant in relation to those powers and duties (ibid., *per* Lord Wilberforce at p. 96).

When applying for permission to apply for judicial review, the applicant must make out a prima facie case of illegality, irrationality, or procedural impropriety. If he fails to do so, permission to apply will be refused. If he succeeds in making out such a prima facie case, the application may be adjourned pending a later hearing *inter partes* or permission to apply may be granted. If the evidence given later, either at the resumed hearing or at the hearing on the merits, shows that the applicant has no sufficient interest in the matter to which the application relates, the application for permission to apply, or the application for judicial review itself, will be dismissed.

Commissioners of Inland Revenue v *National Federation of Self-Employed & Small Businesses Ltd* [1981] 2 All ER 93, HL, was concerned with a challenge by the applicant Federation, which claimed to represent a body of 50,000 taxpayers, to an extra-statutory income tax concession made by the Revenue to a group of casual employees of national newspapers. The House of Lords held unanimously that at earlier hearings of the case both the Divisional Court and the Court of Appeal were wrong to have treated the question of *locus standi* as a preliminary issue divorced from a consideration of the duty alleged to have been broken or not performed by the Revenue.

Their Lordships said there were two reasons why the Federation had no sufficient interest to apply for judicial review. First, the tax legislation sought to achieve complete confidentiality of assessments, and no right on the part of one taxpayer to inquire about the tax affairs of another could be implied. Secondly, the evidence showed that the Revenue had not acted *ultra vires*, or otherwise unlawfully, in granting the extra-statutory concession. They had acted properly in the interests of the good management of taxes under the statutory powers conferred on them.

10.8.4.2 'Sufficient interest'

Whether an applicant has a sufficient interest is not a question to be determined solely as a matter of discretion. It is a mixed question of law and fact (*Commissioners of Inland Revenue* v *National Federation of Self-Employed & Small Businesses Ltd* [1981] 2 All ER 93 *per* Lords Wilberforce, Scarman, and Roskill at pp. 97, 113, and 117, respectively).

Every applicant for judicial review, no matter whether he is seeking a mandatory order, a prohibiting order, a quashing order, a declaration, or an injunction, must show the same *locus standi*. He must have a sufficient interest in the matter to which the application relates. But the fact that the same words are used to cover all the remedies available on an application for judicial review does not necessarily mean that the test is the same in all cases.

Thus, in an application for a mandatory order, where a person is seeking to compel a public authority to perform a public duty, the test of 'sufficient interest' may well be stricter than in an application for a quashing order, where a person is claiming that some court or tribunal has wronged him by exceeding its powers or by disregarding the requirements of natural justice.

For example, in *R* v *Felixstowe Justices (ex parte Leigh)* [1987] 1 All ER 551, DC, a newspaper reporter claimed a declaration that the policy of a local bench of magistrates to withhold the names of justices from the public and the press during and after the hearing of cases was unlawful. He also claimed a mandatory order to compel the disclosure of the names of those magistrates who had tried a particular case. He was granted a declaration, but a mandatory order was refused. As a public-spirited citizen and a representative of the press, he had a sufficient interest in the matter to which the application for a *declaration* related because the preservation of open justice in magistrates' courts was a subject of national importance and of vital concern in the administration of justice. It was held by the court, however, evidently applying a stricter test, that he did not have *locus standi* to apply for a *mandatory order* because he had not been present during the particular trial, and the identity of the justices was not essential, or even material, to the newspaper article he intended to write.

What constitutes a 'sufficient interest' will be decided on a case-by-case basis. It seems likely that the draftsman of s. 31(3) of the Supreme Court Act 1981, by choosing ordinary English words, was attempting to escape from the old technical rules about *locus standi*. He avoided using phrases such as 'a person aggrieved', 'a particular grievance', or 'a specific legal right'. All of these expressions have at some time been used in decided cases on mandatory, quashing, or prohibiting orders, though there was evidence from more modern cases that they were being discarded in favour of a more liberal test—that of 'sufficient interest'.

In *Commissioners of Inland Revenue* v *National Federation of Self-Employed & Small Businesses Ltd* [1981] 2 All ER 93, Lord Diplock (at pp. 103, 104, and 107) in particular welcomed the move away from the former technicalities. He regarded it as fundamental that there should be no reversion to the technical restrictions on *locus standi* which had allowed flagrant breaches of the law by public bodies to go unchecked. It is important that there should be as few restrictions as possible on the freedom of the ordinary private citizen to seek judicial review. The Attorney-General cannot always be relied on to act for the protection of the public interest. As Lord Diplock pointed out (at p. 107), the Attorney-General in practice never seeks prerogative relief against central government departments. He will only proceed, if at all, against public authorities which are not part of central government.

Commissioners of Inland Revenue v *National Federation of Self-Employed & Small Businesses Ltd*, above, was applied in *R* v *Secretary of State for the Environment (ex parte Rose Theatre Trust Co.)* [1990] 1 All ER 754, in which the judgment of Schiemann J (at pp. 766–8) contains a number of important observations on the requirement of *locus standi*. The case was a consequence of the Secretary of State's decision that the recently discovered remains of the Elizabethan Rose Theatre, acknowledged by him to be of national importance, should not be scheduled as a monument under the Ancient Monuments and Archaeological Areas Act 1979, as amended.

To the dismay of environmentalists, it was held, first, that the Secretary of State had not acted unlawfully in so deciding since he had not taken into account irrelevant factors when exercising his discretion, and, secondly, that, in any case, the applicant Company had no *locus standi* to apply for judicial review.

The Company was said to lack standing because the decision not to schedule the theatre site was the type of governmental decision in which ordinary citizens (even distinguished archaeologists, actors, and writers) did not have a 'sufficient interest' for the purposes of s. 31(3) of the Supreme Court Act 1981, and they could not acquire a 'sufficient interest', either by the expedient of incorporating themselves as a company with express power to press for scheduling, or by making representations to, and receiving a considered reply from, the Secretary of State about scheduling.

In *R* v *Inspectorate of Pollution (ex parte Greenpeace Ltd) (No. 2)* [1994] 4 All ER 329, Otton J gave cursory treatment to, and declined to follow, the *Rose Theatre* decision, above, in holding that Greenpeace had *locus standi* to challenge authorisations granted by the Pollution Inspectorate allowing British Nuclear Fuels Ltd to discharge liquid and gaseous radioactive waste from its premises. Among the factors taken into account were the nature of Greenpeace, the extent of its interest in the issues raised, and the nature of the remedy it sought.

The *Rose Theatre* case, above, was not even cited, but *Greenpeace Ltd (No. 2)*, above, was applied, in *R* v *Secretary of State for Foreign Affairs (ex parte World Development Movement Ltd)* [1995] 1 All ER 611, DC. Here, the World Development Movement, a non-partisan pressure group concerned with the misuse of overseas aid money, wished to challenge the Foreign Secretary's decision to enter into an agreement with the Malaysian government to provide aid and trade support for the Pergau dam scheme. The Foreign Secretary argued that the World Development Movement had no *locus standi*.

The court, however, held that the Movement did have *locus standi* in view of the importance of the issue raised and of safeguarding the rule of law, the fact that there was probably no other responsible challenger, and the Movement's prominent international role in promoting and protecting aid to underdeveloped nations. The Pergau dam scheme was uneconomic and the Foreign Secretary's decision to support it was declared unlawful because it was not concerned with promoting economically sound development as required by s. 1 of what was then the Overseas Development and Co-operation Act 1980.

In *Equal Opportunities Commission* v *Secretary of State for Employment* [1994] 1 All ER 910, HL, it was held that the EOC has *locus standi* to bring proceedings for the purpose of determining whether the Secretary of State has acted in breach of European Community law. The EOC was created by the Sex Discrimination Act 1975. Its duties include working towards the elimination of discrimination and harassment, promoting equality of opportunity between men and women generally, and keeping under review the working of (and making proposals for amendments to) the 1975 Act and the Equal Pay Act 1970.

The Court of Appeal had decided that the EOC did not have *locus standi*, but the House of Lords held that, looked at in the light of previous cases in which it had simply been assumed on all sides that the EOC had the necessary standing, it would be a retrograde step to deny it *locus standi* 'to agitate in judicial review proceedings questions related to sex discrimination which are of public importance and affect a large section of the population' ([1994] 1 All ER 910 *per* Lord Keith at p. 919).

The Law Society has been held to have *locus standi* to challenge proposed cuts in legal aid provision (*R* v *The Lord Chancellor (ex parte The Law Society)* (1993) *The Times*, 25 June, DC, where, however, the challenge was unsuccessful; see also *R* v *The Lord Chancellor (ex parte The Law Society)* (1993) *The Times*, 11 August, CA, where an unsuccessful attempt was made to prevent the introduction of a standard fees scheme for solicitors doing criminal work in the magistrates' courts). It also has *locus standi* to appeal to the Master of the Rolls under the Solicitors Act 1974 against the restoration of a name to the roll of solicitors (*R* v *Master of the Rolls (ex parte McKinnell)* [1993] 1 All ER 193, DC).

It is well established that a council taxpayer has the necessary *locus standi* to take action, for example, to prevent his local council from overspending (*Prescott v Birmingham Corporation* [1955] Ch 210, CA, para 10.2.2.1 above), or to challenge (what are now) council tax assessments made on other taxpayers (*Arsenal Football Club Ltd v Ende* [1979] AC 1, HL).

In *R v Bassetlaw District Council (ex parte Oxby)* (1997) *The Times,* 18 December, CA, it was held that the leader of a local council has *locus standi* to apply for judicial review of planning decisions of his own council which turn out to have been tainted by actual or apparent bias on the part of the councillors who made them. Since it is not appropriate for the council itself to be both applicant and respondent in the proceedings, according *locus standi* to the council leader is a convenient way of avoiding this difficulty. (See further on the *Bassetlaw* case, para 10.10 below.)

10.8.4.3 *Locus standi* and legitimate expectation

It is clear that a person has a sufficient interest to enable him to apply for judicial review where his *private law* rights have been infringed, or are threatened, and there is an adequate public law element in the case (*Council of Civil Service Unions v Minister for the Civil Service* [1984] 3 All ER 935, HL, *per* Lord Diplock at p. 949; *Gillick v West Norfolk and Wisbech Area Health Authority* [1985] 3 All ER 402, HL, *per* Lord Scarman at p. 416).

What is equally important is that judicial review may be available in cases where legal rights arising under private law are not involved at all. A person has a sufficient interest to enable him to apply for judicial review of an administrative decision where he has a *legitimate expectation* that the public authority will act fairly towards him.

Thus, in *O'Reilly v Mackman* [1982] 3 All ER 1124, HL, the prisoners who complained about unfair loss of remission could not point to the violation of any *private law* right because, under the Prison Rules, remission of sentence is not a legal right but a matter of indulgence on the part of the prison authorities. Nevertheless, the prisoners did have a legitimate expectation of remission which, in *public law*, would give them a sufficient interest for the purpose of challenging the validity of the board of visitors' decision in judicial review proceedings.

In *Council of Civil Service Unions v Minister for the Civil Service* [1984] 3 All ER 935, HL (the '*GCHQ* case'), the civil servants whose right to trade union membership was withdrawn by the government without consultation had no legal right in *private law* to prior consultation. But the House of Lords held, applying *O'Reilly v Mackman,* above, that they did have a legitimate expectation of consultation in *public law* based on the regular practice of consultation between management and unions at GCHQ about changes in employment terms. This gave the civil servants a sufficient interest to challenge the government's decision, although, on the facts, the challenge was unsuccessful.

10.8.5 Delay in making an application

The claim form required to begin judicial review proceedings must be filed promptly and in any event *within three months* from the date when grounds for the claim first arose (CPR, r. 54.5). The court has a discretion to extend this period if there is good reason for doing so (ibid.). 'Good reason' would include difficulty in obtaining publicly-funded legal assistance which was not the fault of the applicant (*R v Stratford-on-Avon District Council (ex parte Jackson)* [1985] 3 All ER 769, CA).

If a *quashing order* is claimed, the date when grounds for the claim first arose will be taken to be the date of the judgment, order, conviction, or other proceedings which it is sought to quash (*Practice Direction* to CPR, Part 54).

If the court considers that there has been undue delay in making an application for judicial review (which includes both an application for permission to apply and an application for substantive relief), the court has a discretion to refuse permission to apply, or to refuse a remedy if permission to apply has already been granted, where the granting of relief 'would be likely to cause substantial hardship to, or substantially prejudice the rights of, any person or would be detrimental to good administration' (Supreme Court Act 1981, s. 31(6); *Caswell v Dairy Produce Quota Tribunal for England and Wales* [1990] 2 All ER 434, HL, below).

A failure to apply promptly or within three months, as required by r. 54.5, constitutes 'undue delay' for the purposes of s. 31(6) even though the court is satisfied that there is 'good reason' for the delay. Accordingly, the court can still refuse to grant relief, on the grounds mentioned in s. 31(6), at the hearing of the substantive application even though there was good reason for failing to apply promptly (*Caswell*, above, approving *R v Stratford-on-Avon District Council (ex parte Jackson)*, CA, above; *R v Secretary of State for Health (ex parte Furneaux)* [1994] 2 All ER 652, CA).

If, at the permission stage, the court grants permission to apply for judicial review notwithstanding the undue delay in applying for it, this has the effect of extending the application period. It follows that the court which later deals with the substantive application on the merits will not refuse to grant relief solely on the basis of the original delay. By that time, the question of delay has already been resolved in the applicant's favour, and the court's powers are limited to refusing relief on the grounds contained in s. 31(6), namely hardship, prejudice, or detriment to good administration.

In *R v Criminal Injuries Compensation Board (ex parte A)* [1999] 2 WLR 974, HL, where permission to apply for judicial review had been granted by a judge despite the applicant's delay, the House of Lords held that the decision of the different judge who later dismissed the substantive application—solely on the ground of delay—was wrong. It is, moreover, probably the case that the court which later deals with the substantive application has no power to refuse permission to apply for judicial review on the basis of hardship, prejudice, or detriment to good administration. Once permission to apply has been granted (and has not been set aside), it is too late to 'refuse' it, although the court at the substantive hearing would still have power under s. 31(6) to refuse to grant relief (ibid., *per* Lord Slynn at p. 979).

In *Caswell v Dairy Produce Quota Tribunal for England and Wales* [1990] 2 All ER 434, HL, the Dairy Produce Quota Tribunal in 1985 had fixed a farmer's quota on the basis of the expected production in 1985 from 70 cows, and had indicated that the farmer could reapply to have his quota increased when, as anticipated, he expanded his herd to 150. The farmer later discovered that the quota set in 1985 could not be changed. Well over two years after the Tribunal's decision, he applied for, and was granted, permission to apply for judicial review of it.

At the hearing of the substantive application, it was held that the Tribunal had misconstrued the relevant statutory regulations and should have set the quota on the basis of milk production from 150 cows. Notwithstanding the fact that the Tribunal's decision was wrong in law, the judge exercised his discretion under s. 31(6) to refuse relief because there had been undue delay in seeking permission to apply for judicial review, and the granting of relief would be detrimental to good administration.

This decision was affirmed by the Court of Appeal and by the House of Lords. Their Lordships held, first, that there had clearly been undue delay in applying for permission, and, secondly, that granting relief would be detrimental to good administration within the meaning of s. 31(6) since it would be likely to lead to applications for permission to challenge the Tribunal's decisions by other disappointed applicants and to the reopening of the milk quota for previous years going back to 1984. This, according to Lord Goff (at pp. 440–1), delivering the leading opinion, was 'precisely the type of situation which Parliament was minded to exclude by the provision in s. 31(6) relating to detriment to good administration'.

The essence of CPR, r. 54.5, is that the application for permission must be made 'promptly'. The court has power to refuse permission to apply for judicial review under r. 54.5 even though the application for permission has been made within the three-month period (*R v Stratford-on-Avon District Council (ex parte Jackson)* [1985] 3 All ER 769, CA, *per* Ackner LJ at p. 772).

Again, in *R v Herrod (ex parte Leeds City District Council)* [1976] 1 All ER 273, CA, which was decided when the time-limit was *six* months, the Court of Appeal held that, on the facts, a delay of five and a half months was fatal to an application for permission to apply for a quashing order to quash a decision of the Crown Court. The Court of Appeal said that the time-limit was not an entitlement, but a maximum rarely to be exceeded.

10.8.6 Non-prerogative relief in judicial review proceedings

10.8.6.1 Declarations and injunctions

On an application for judicial review, it is now possible to obtain non-prerogative relief, although it is, of course, only granted at the discretion of the court. The court can grant a declaration or an injunction, provided that the remedy has been claimed, if it would be 'just and convenient' to do so (Supreme Court Act 1981, s. 31(2)).

In deciding on the 'justice' and 'convenience' of the matter, the court must apply the same rules relating to availability and scope as are applied in the case of applications for mandatory, prohibiting, and quashing orders (ibid.). In addition, the court must take into account all the circumstances of the case.

A declaration or an injunction can be claimed as a public law remedy by way of an application for judicial review without the applicant having to show that some legal right of his has been, or is being, infringed. He need only show that he has a 'sufficient interest' in the matter at issue.

Section 31(2) of the Supreme Court Act 1981 does not say that in judicial review proceedings a declaration can only be granted instead of a prerogative remedy. Accordingly, it seems that a declaration can be granted to an applicant who has *locus standi* even though, on the facts, the court could not grant a prerogative remedy—for example, because there is no 'decision' to be quashed (*Factortame Ltd* v *Secretary of State for Transport (No. 2)* [1991] 1 All ER 70, HL; *Equal Opportunities Commission* v *Secretary of State for Employment* [1994] 1 All ER 910, HL, *per* Lord Browne-Wilkinson at pp. 926–8, supported by Lords Jauncey and Lowry at p. 925).

10.8.6.2 Damages

It is also possible to obtain damages (or restitution or the recovery of a sum due) on an application for judicial review (Supreme Court Act 1981, s. 31(4), as substituted by the Civil Procedure (Modification of Supreme Court Act 1981) Order 2004, SI 2004, No. 1033). This will save the expense and inconvenience of bringing a separate private law action. However, damages (or restitution or the recovery of a sum due) can only be awarded if three conditions are satisfied.

First, they must be claimed in the claim form. The court has no power to award damages of its own motion on an application for judicial review. Secondly, some other remedy must be claimed at the same time. It is not possible to claim damages on their own (CPR, r. 54.3). Thirdly, the court must be satisfied that the applicant's claim is such that he would have been awarded damages if he had begun a private law action instead of making an application for judicial review. This rule ensures that damages are not awarded on judicial review for causes of action which are unknown to the law, and that the same principles (for example, as to causation and remoteness of loss) are applied in applications for judicial review as are applied in private law actions for damages.

10.8.6.3 Allowing proceedings to continue as if private law proceedings

If judicial review is claimed but it transpires for whatever reason that such a claim is inappropriate, the court can, instead of dismissing the claim, order it to continue as if it had not been started under CPR, Part 54 (r. 54.20). When the court does this, it can give directions about the future management of the claim as a claim in private law (ibid.). Transfers to and from the Administrative Court are made possible by CPR, Part 30.

Rule 54.20 is a sensible provision which makes it unnecessary to condemn the applicant to the costly procedure of having to start his action all over again before a different court.

The court, however, has no power to take the opposite course; i.e., there is no jurisdiction to permit a private law action to continue as though it were an application for judicial review.

In *R* v *South Glamorgan Health Authority (ex parte Phillips)* (1986) *The Independent*, 25 November, it was decided, for reasons similar to those in *R* v *East Berkshire Health Authority (ex parte Walsh)* [1984] 3 All ER 425, CA (para 10.8.2.3 above), that the court had no jurisdiction to deal with the case by way of judicial review. However, the judge exercised his discretion under what is now r. 54.20 to allow the case to proceed as if begun as private law proceedings. A declaration was granted.

In the *Walsh* case itself, above, Sir John Donaldson MR pointed out (at p. 432) that what is now r. 54.20 is an 'anti-technicality' rule designed to preserve the position of an applicant who finds that the basis for *the relief he has claimed* is private law rather than public law.

Rule 54.20 only operates where the applicant for judicial review has initially alleged a breach of his *public law* rights. It cannot be used to order the proceedings to continue as if they were private law proceedings where the applicant's complaint is essentially about a breach of his *private law* rights and does not contain a *public law* element.

An example is provided by the case of *R v Secretary of State for the Home Department (ex parte Dew)* [1987] 2 All ER 1049. Here, the applicant had received a bullet wound in the arm during the course of being arrested. While in prison on remand a bone graft was recommended by surgeons on two separate occasions, but by the time of his trial (some 16 months after his arrest and at which he was sentenced to 18 years' imprisonment) he had still not received proper treatment. Thereafter he applied for judicial review. He claimed a mandatory order and/or an injunction, and a quashing order, against the respondents (the Home Secretary and the governor and medical officer of Wandsworth Prison) to compel the provision of appropriate treatment for his arm. He also claimed damages for pain and suffering caused by the delay in treatment.

Before his application was heard the applicant was given proper treatment so that his claim for a mandatory order and/or an injunction, and a quashing order, was no longer relevant. He persisted with his claim for damages and sought an order that the proceedings should continue as if they had been begun as a private law claim.

McNeill J held that the application for judicial review would be struck out as an abuse of the process of the court, and that, accordingly, no order could be made under what is now r. 54.20. The application for judicial review was an abuse of process because the applicant's case did not allege any breach of his *public law* rights but was essentially a *private law* action for damages for the tort of negligence arising out of a failure to provide proper medical treatment. As such, it should have been commenced as a private law claim and not as a claim for judicial review in public law.

10.8.6.4 Injunctions against the Crown in public law proceedings

Non-prerogative relief in the form of an interim injunction is not generally available against the Crown or its ministers in *private law* proceedings (see para 10.8.6.5 below). However, in *R v Licensing Authority (ex parte Smith Kline & French Laboratories Ltd) (No. 2)* [1989] 2 All ER 113, CA, a majority of the Court of Appeal held, approving and applying the decision of Hodgson J in *R v Secretary of State for the Home Department (ex parte Herbage)* [1986] 3 All ER 209, that such relief was available in *public law* proceedings for judicial review. On the facts, however, an interim injunction was not granted in either case.

The reasoning behind these two revolutionary decisions was, first, that s. 21 of the Crown Proceedings Act 1947, by which the court is prevented from granting an injunction against the Crown in civil proceedings, did not apply because, by s. 38 of the same Act, the expression 'civil proceedings' does not for this purpose include judicial review proceedings, and, secondly, that s. 31(2) of the Supreme Court Act 1981 had, by implication, extended the jurisdiction of the High Court so as to empower it to grant an interim injunction against the Crown in an application for judicial review.

Both the *Smith Kline & French* case and the *Herbage* case were overruled by a unanimous House of Lords in *Factortame Ltd* v *Secretary of State for Transport (No. 1)* [1989] 2 All ER 692, HL. Their Lordships decided that the courts below were erroneous in their interpretation of the relevant provisions of the Crown Proceedings Act 1947 and the Supreme Court Act 1981. The House held that the position at common law had always been that injunctions could not be granted against the Crown in what are now called judicial review proceedings, and that the absence from the Crown Proceedings Act 1947 of any express prohibition against the grant of injunctions against the Crown in judicial review proceedings was of no significance in view of the common law prohibition, which it would have been superfluous to repeat in the Act. In effect, according to *Factortame (No. 1)*, the Act had preserved the common law rule (see [1989] 2 All ER 692 *per* Lord Bridge at p. 706). It was further held that s. 31(2) of the Supreme Court Act 1981 did *not* confer on the court a new jurisdiction to grant an interim injunction against the Crown because, among other reasons, if Parliament had intended to confer such a new and radical power it would have done so in express terms rather than leaving it to be created by mere implication (ibid., *per* Lord Bridge at p. 708).

In *M* v *Home Office* [1993] 3 All ER 537, HL, the whole question of the availability of injunctive relief against the Crown in judicial review proceedings was reopened. (For the facts of this case, see para 8.3.) The House of Lords retreated from the position it had adopted in *Factortame (No. 1)*, above, by holding that both final and interim injunctions are available against ministers acting in their official capacity as representatives of the Crown.

The leading judgment was delivered by Lord Woolf, with whose speech the other Law Lords expressed agreement. Lord Woolf pointed out that in *Factortame (No. 1)* the House had not been primarily concerned with the question as to whether injunctions were available against the Crown and its officers, and, accordingly, had not had the benefit of as full an argument on the history of both civil and prerogative proceedings involving the Crown as had been presented in the instant case (see [1993] 3 All ER 537 at pp. 549 and 561). In his view, the approach to the interpretation of s. 31 of the Supreme Court Act 1981 taken by Lord Bridge in *Factortame (No. 1)* (see above) was too narrow. In the opinion of Lord Woolf, the unqualified language of s. 31(2) has succeeded in conferring on the court jurisdiction to grant injunctions, including interim injunctions, against ministers and other officers of the Crown in judicial review proceedings (see [1993] 3 All ER 537 at pp. 562 and 564).

In the light of the decision of the House of Lords in *M* v *Home Office*, the views on the effect of s. 31 expressed by Hodgson J and the majority of the Court of Appeal in, respectively, *R* v *Secretary of State for the Home Department (ex parte Herbage)* [1986] 3 All ER 209, and *R* v *Licensing Authority (ex parte Smith Kline & French Laboratories Ltd (No. 2))* [1989] 2 All ER 113, CA, above, must be taken to have been rehabilitated.

Lord Woolf in *M* v *Home Office* [1993] 3 All ER 537, HL, thought that the newly-discovered power to grant injunctions against the Crown in judicial review proceedings would be exercised only in 'the most limited circumstances', and he foresaw no change in the usual practice whereby *final* relief in the form of a declaration (which the Crown invariably respects) is granted as the appropriate remedy in judicial review proceedings involving Crown servants. With regard to *interim* relief against Crown servants, he thought that an interim declaration might have advantages over an interim injunction. The court now has power to grant an interim declaration by virtue of Part 25 of the Civil Procedure Rules.

The decision of the House of Lords in *M v Home Office* has the further effect of achieving a desirable harmonisation in the way rights are protected by means of injunctive relief. The same protection is now afforded to rights not derived from European Community law as is enjoyed by rights which are so derived.

Although before 1993 interim injunctions were not generally available against the Crown in judicial review proceedings, an exception had been established in *Factortame Ltd v Secretary of State for Transport (No. 2)* [1991] 1 All ER 70, CJEC and HL, to cover the situation where the case was governed by Community law and interim relief was required in order to protect a person's Community law rights pending the final outcome of the litigation. In *Factortame (No. 2)*, above, it was held by the European Court of Justice that a rule of national law must be set aside by a national court if it is the sole impediment to granting interim relief.

In consequence, the House of Lords granted an interim injunction against the Secretary of State for Transport which had the effect of temporarily disapplying certain statutory regulations. Their Lordships held, first, that whether an interim injunction should be granted in this limited class of case will be determined according to the balance of convenience after taking into account the public interest in seeing that the law is upheld, and, secondly, that an apparently authentic law should not be disapplied unless the challenge to its validity is prima facie so firmly based as to justify the exceptional course of setting it aside by means of an interim injunction (see *Factortame (No. 2)* [1991] 1 All ER 70 *per* Lord Goff at pp. 119–20).

This approach was applied in *R v HM Treasury (ex parte British Telecommunications plc)* (1993) *The Times*, 2 December, CA, albeit with a different result. The Court of Appeal refused an interim injunction to disapply some statutory regulations so as to relieve British Telecom from certain obligations pending a preliminary ruling from the European Court on a reference made to it under art. 234 of the EC Treaty. It was said that the court should not adopt a formulaic approach to its assessment of the balance of convenience, but should instead attach varying degrees of weight to all the relevant matters thrown up by the facts of the case. Thus, the court might be more reluctant to disapply a major piece of primary legislation than a minor piece of subordinate legislation. One of the relevant matters which weighed with the Court of Appeal in the *British Telecom* case, above, was that, unlike the Spanish fishermen in *Factortame (No. 2)*, the survival of British Telecom was not put at risk by a denial of interim relief.

10.8.6.5 Injunctions against the Crown in private law proceedings

M v Home Office [1993] 3 All ER 537, above, does not affect the availability or otherwise of injunctions against the Crown in *private law* proceedings. This matter continues to be governed by s. 21 of the Crown Proceedings Act 1947, which (unlike s. 31 of the Supreme Court Act 1981) was not designed to extend the court's power to grant injunctions against the Crown (see *M v Home Office*, above, *per* Lord Woolf at p. 556).

By s. 21(1)(a) of the Crown Proceedings Act 1947, the court is forbidden to grant an injunction against the Crown in any civil proceedings (defined in s. 38(2) in a way which excludes what are now called judicial review proceedings).

Moreover, s. 21(2) forbids the court to grant an injunction against an *officer* of the

Crown in any civil proceedings if its effect would be to circumvent s. 21(1)(a). Thus, if an officer of the Crown is sued in civil proceedings in a *representative* capacity, no injunction can be granted against him because that would be tantamount to granting it against the Crown. If, however, the officer is sued in his *personal* capacity for some wrongdoing, an injunction can be granted against him because to do so does not affect the Crown.

10.9 Distinction between judicial review and an appeal

10.9.1 Introduction

In essence, the distinction is that an appeal is concerned with the *merits* of the decision under appeal while judicial review is only concerned with the *legality* of the decision or act under review. The determination of an appeal is always a statutory function whereas judicial review is based on an inherent common law power. It is implicit in the nature of judicial review that, so long as a public authority has acted *intra vires* and lawfully, the court will not interfere merely because it considers that the public authority came to a wrong conclusion. In the case of a successful appeal, however, the appellate body is called on to substitute its own discretion for that of some inferior court or tribunal.

In *Chief Constable of the North Wales Police* v *Evans* [1982] 3 All ER 141, HL (para 10.2.2.4 above), the Court of Appeal had not only held that natural justice demanded a fair hearing but had also stated that the decision under review must itself be 'fair and reasonable'. This confusion between the purposes of judicial review on the one hand, and an appeal on the other, was seized on by the House of Lords. Their Lordships made it quite clear that judicial review is not an *appeal* from a decision, but a *review* of the manner in which the decision was made. It follows that, on judicial review, the court is not entitled to consider whether the decision itself was fair and reasonable for that would be to usurp the power of the decision-making body under the pretence of preventing the abuse of power (see especially *per* Lord Hailsham LC at pp. 143–4 and Lord Brightman at pp. 154–5).

The difference in nature and purpose between an appeal and judicial review is re-emphasised by the fact that judicial review of a decision is still possible notwithstanding the existence of statutory words by which any right of appeal against that decision is successfully excluded (as in *R* v *Medical Appeal Tribunal (ex parte Gilmore)* [1957] 1 QB 574, CA, para 10.5.3.3 above, and see *R (Sivasubramaniam)* v *Wandsworth County Court* [2003] 1 WLR 475, CA, para 10.11.1 below). This matter will be examined more closely in para 10.11.

10.9.2 Overlap between judicial review and appeals

The fact that a right of appeal exists against a decision does not necessarily prevent that decision being quashed by a quashing order, following a judicial review, on the traditional public law grounds of excess of power, violation of natural justice, or error of law. It is quite common for both an appeal and an application for judicial review to be available at the same time, and there is often a considerable overlap between the two procedures. The courts occasionally give guidance about which procedure is the more suitable in particular circumstances.

For example, both sides have a right of appeal by way of case stated to the Queen's Bench Division of the High Court from the decision of a magistrates' court (Magistrates' Courts Act 1980, s. 111) and of the Crown Court in a non-indictable matter, such as a decision on appeal from the justices (Supreme Court Act 1981, s. 28(1)). The grounds on which the appeal may be taken are that the decision was wrong in law or that it was made in excess of jurisdiction. Alternatively, either side may challenge the decision by way of an application for judicial review on slightly wider, but substantially the same, grounds, namely, error of law, excess of jurisdiction, or violation of the principles of natural justice.

In *R v Crown Court at Ipswich (ex parte Baldwin)* [1981] 1 All ER 596, it was said that the more convenient procedure where the facts of a case are complicated is an appeal by way of case stated. That procedure enables the High Court to get at the facts found by the justices more easily than in an application for judicial review. However, in *R v Hereford Magistrates' Court (ex parte Rowlands)* [1997] 2 WLR 854, DC, it was held that judicial review is a more appropriate procedure than appealing to the Crown Court where the complaint against the justices is one of procedural irregularity or bias (unless, it might be added, judicial review is applied for with the ulterior motive of causing delay which may thereby lead to the dropping of the prosecution before the defendant's appeal to the Crown Court is heard: *R v Peterborough Magistrates' Court (ex parte Dowler)* [1997] 2 WLR 843, DC).

It was said in the *Rowlands* case, above, that it was important to retain the High Court's supervisory jurisdiction over magistrates' courts in order to ensure the maintenance of high standards of fairness and procedural propriety, having regard to the central role of those courts in the administration of the criminal justice system and to the fact that the Crown Court has no power to supervise their proceedings.

A challenge to a sentence imposed by a magistrates' court has been held to be better made by appealing to the Crown Court than either stating a case for the opinion of the High Court (*Tucker v Director of Public Prosecutions* [1992] 4 All ER 901, DC) or applying for judicial review (*R v Ealing Justices (ex parte Scrafield)* (1993) *The Times*, 29 March, DC, applying *Tucker*, above).

Similarly, it has been held that taxation issues, including the conduct of a hearing before General or Special Commissioners of income tax, are more properly dealt with by an appeal, by way of case stated to the Chancery Division of the High Court, than by an application for judicial review (*R v Commissioner for the Special Purposes of the Income Tax Acts (ex parte Napier)* [1988] 3 All ER 166, CA). Judicial review might, however, be the appropriate course where it is alleged that the commissioner's conduct of the hearing raised matters extraneous to taxation issues (ibid., *per* Purchas LJ at p. 171).

In *R v Birmingham City Council (ex parte Ferrero Ltd)* (1991) [1993] 1 All ER 530, CA, it was held that the better way of challenging a suspension notice served by a local authority under the Consumer Protection Act 1987 (prohibiting the supply of goods on the ground that a safety provision has been contravened) is by exercising the statutory right of appeal to a magistrates' court provided by the Act rather than by taking judicial review proceedings, even where there are alleged defects in the local authority's decision-making process. In judicial review proceedings at first instance, the judge had quashed the local authority's issue of a suspension notice which prohibited Ferrero Ltd from supplying 'Kinder Surprise' chocolate eggs containing kits for making Pink Panther toys. The notice had been issued following the death of a three-year-old child by asphyxiation after swallowing a toy foot which had come loose.

The Court of Appeal allowed the local authority's appeal, holding that the applicants had adopted the wrong procedure in their challenge to the notice. The court held that the statutory appeal procedure is the more appropriate in the circumstances since it does not require permission of the court, it is quicker and more suitable for getting at the facts, and there is provision in the 1987 Act for the payment of compensation to the trader if it transpires that no safety provision has been contravened.

On the other hand, it has been said that the proper way to challenge the validity of regulations and administrative directions made under the authority of a statute is by way of an application for judicial review rather than by the exercise of a right of appeal (*Moss of London* v *Commissioners of Customs & Excise* [1981] 2 All ER 86, CA, *per* Lord Denning MR at p. 90).

An application for judicial review will be more effective than a special case stated under the provisions of the Mental Health Act 1983 as a means of challenging the reasons given by a mental health review tribunal for refusing to order the release of a mental patient (*Bone* v *Mental Health Review Tribunal* [1985] 3 All ER 330 *per* Nolan J at p. 334). This is because judicial review proceedings allow a broader consideration of the issues involved and provide more and varied remedies. For example, the High Court can quash the decision of a mental health review tribunal in judicial review proceedings but cannot do so under the special case-stated procedure.

10.10 Judicial review and alternative remedies

The granting of permission to claim judicial review is discretionary. Until fairly recently it could be said with some confidence that, with the exception of the remedy of a mandatory order (as to which see para 10.7.4 above), it did not matter as a general rule that an aggrieved citizen had not exhausted his rights of appeal, or whatever administrative procedures might be open to him, before seeking judicial review.

Now, however, the general rule is that judicial review is to be regarded as a remedy 'of last resort' (*R* v *Inland Revenue Commissioners (ex parte Opman International UK)* [1986] 1 All ER 328 *per* Woolf J at p. 330), and permission to apply for it will not normally be granted where there exists an alternative remedy which has not been used (*Preston* v *Inland Revenue Commissioners* [1985] 2 All ER 327, HL, *per* Lords Scarman and Templeman at pp. 330 and 337, respectively; *R* v *Epping and Harlow General Commissioners (ex parte Goldstraw)* [1983] 3 All ER 257, CA).

Only in exceptional circumstances will judicial review be allowed where the alternative remedy has not been pursued or exhausted. What circumstances are 'exceptional' for this purpose will be developed on a case-by-case basis. In *R v Secretary of State for the Home Department (ex parte Swati)* [1986] 1 All ER 717, CA, Sir John Donaldson MR said (at p. 724) that '[b]y definition, exceptional circumstances defy definition'. In the same case, Parker LJ said (at p. 728) that each case will depend on its own facts and that it would be impossible and legally wrong to attempt to define what are 'exceptional circumstances'.

It is highly probable that no single circumstance by itself is sufficiently 'exceptional' to persuade the court to grant permission to apply for judicial review where there is an unused alternative remedy. Nevertheless, it is possible to distil from the cases some of the factors which may influence the court in the exercise of its discretion. Such factors include whether the alternative remedy would be quicker or slower than judicial review, and whether the case involves some technical knowledge which the alternative appellate body is more likely to have than the High Court (*R v Hallstrom (ex parte W)* [1985] 3 All ER 775, CA, *per* Glidewell LJ at pp. 789–90), and whether the alternative remedy is as convenient and effective as judicial review.

In *R v Chief Constable of the Merseyside Police (ex parte Calveley)* [1986] 1 All ER 257, CA, Glidewell LJ (at p. 267) stood by his remarks in the *Hallstrom* case, above, and they received the tacit approval of Sir John Donaldson MR (at p. 262). May LJ, however, was of the view (at p. 267) that the slowness of the alternative remedy should only be regarded as exceptional where it amounts to an abuse of process. He also doubted (at pp. 264–5) whether the fact that the alternative remedy is not as convenient and effective is in itself exceptional enough to justify giving permission to apply for judicial review.

On the facts of the *Calveley* case, above, the Court of Appeal was unanimous that permission to apply should be granted and that the Chief Constable's decision should be quashed. In proceedings under the statutory Police (Discipline) Regulations the applicants had been found guilty of disciplinary offences. Two of them were dismissed by the Chief Constable and the other three were required to resign. The applicants gave notice of appeal to the Home Secretary under the statutory disciplinary procedure. However, before the hearing of the appeal they applied for a quashing order to quash the Chief Constable's decision on the ground that the formal written notification, required by the regulations to be made 'as soon as is practicable', informing them of the allegations or complaints against them had been served so late as to prejudice their right to a fair hearing.

The Court of Appeal held that the circumstances were exceptional. There had been such a serious departure from the disciplinary procedure that the court should intervene even though the internal appeal machinery had not been exhausted.

In *R v Westminster City Council (ex parte Hilditch)* [1990] COD 434, CA, a ratepayer sought permission to apply for judicial review of a local authority's policy on the sale of council flats. The relief claimed was a declaration. Permission to apply was refused because the applicant had already invoked an alternative statutory procedure by way of complaint to the auditor under the Local Government Finance Act 1982. The alternative remedy had not been exhausted, and the court said that there were no exceptional circumstances to justify granting permission to apply for judicial review.

The fact that only a declaration was sought meant that there was no urgency requiring judicial review proceedings to be given priority, and, in any event, the alternative procedure in question was more appropriate for resolving disputed issues of fact.

In *R v Director of Public Prosecutions (ex parte Camelot Group plc)* (1997) *The Independent*, 22 April, DC, Camelot, the organisers of the National Lottery, sought judicial review of the decision of the Crown Prosecution Service not to prosecute the promoters of another lottery scheme for running an illegal lottery. It was held that Camelot should make use of the alternative remedy of a private prosecution, which would be equally as effective as judicial review and would avoid the undesirable situation of inviting the court to declare in a civil case that a third party's actions are criminal.

R v Bassetlaw District Council (ex parte Oxby) (1997) *The Times*, 18 December, CA, concerned planning decisions which were tainted by bias. The District Council was held entitled to use judicial review proceedings to have the decisions declared illegal and void as an alternative to revoking them itself and paying compensation under the statutory procedure provided by the Town and Country Planning Act 1990. (See further on this case, para 10.8.4.2 above.)

In *Scott v National Trust for Places of Historic Interest or Natural Beauty* [1998] 2 All ER 705, permission to apply for judicial review was refused on the ground that decisions of the National Trust, although it is a public body amenable to judicial review, are best challenged by means of the alternative remedy provided by Parliament ('charity proceedings' under the Charities Act 1993).

In *R (Sivasubramaniam) v Wandsworth County Court* [2003] 1 WLR 475, CA, it was held that the new régime for civil appeals provided by the Access to Justice Act 1999 has not impliedly abolished the High Court's judicial review jurisdiction over the decisions of county court judges. However, the Court of Appeal further held that the coherent statutory scheme governing appeals from county court decisions laid down by the Act is a suitable alternative remedy, and should not be by-passed by means of an application for judicial review unless the circumstances are exceptional.

In *R (Davies) v Financial Services Authority* [2004] 1 WLR 185, CA, the Court of Appeal emphasised, *obiter*, that only in the most exceptional cases should judicial review be used as a means of challenging the actions and decisions of the FSA. This is because the Financial Services and Markets Act 2000 provides a special procedure for such challenges in the form of a reference to the Financial Services and Markets Tribunal and by appeal therefrom to the Court of Appeal on a point of law.

The new reluctance to make available the discretionary remedy of judicial review where an alternative remedy exists but has not been used can sometimes verge on the oppressive. In *R v Secretary of State for the Home Department (ex parte Swati)* [1986] 1 All ER 717, CA, the Court of Appeal refused permission to apply for judicial review in an immigration case on the ground that the applicant ought first to exhaust the statutory right of appeal provided by the immigration legislation.

The court did not regard it as an 'exceptional' circumstance that, under the legislation, a person who is refused permission to enter the United Kingdom must first leave the United Kingdom before exercising his right of appeal. It might be otherwise if the applicant's country of origin is hostile, as in *R v Chief Immigration Officer, Gatwick*

Airport (ex parte Kharrazi) [1980] 3 All ER 373, CA, where permission to apply for judicial review was granted in the case of a 13 year-old Iranian boy. If he had been forced to return to Iran in order to pursue his right of appeal, he would not have been allowed to leave that country again until he was much older.

With a view to reducing expense and delay, guidance issued by the Court of Appeal emphasises that, before resorting to litigation, a claimant for judicial review may be expected to make use of any statutory or informal complaints procedure which exists in the circumstances of the case, or some other form of alternative dispute resolution, in order to resolve or reduce the issues in dispute. This is so even though the alternative remedy does not cover exactly the same ground as judicial review.

Indeed, the guidance goes so far as to say that 'the courts should not permit, except for good reason, proceedings for judicial review to proceed if a significant part of the issues between the parties could be resolved outside the litigation process' (*R(Cowl)* v *Plymouth City Council (Practice Note)* [2002] 1 WLR 803, CA, *per* Lord Woolf CJ at [14]).

It seems that the remedy of a mandatory order has always been subject to the rule that the discretion to grant it will not be exercised if there exists an adequate alternative remedy. This is dealt with in para 10.7.4 above.

10.11 Statutory exclusion of appeal and judicial review

10.11.1 'Finality' or 'ouster' clauses

It is sometimes provided by a particular statute that a decision taken under it 'shall be final', or 'shall be final and conclusive', or 'shall not be appealable', or 'shall not be questioned in any legal proceedings whatsoever'. Such expressions are known as 'finality' or 'ouster' clauses because they make the original decision final by attempting to oust the jurisdiction of the courts.

Traditionally, the courts have interpreted these expressions narrowly so as to mean that, although there is to be no further *appeal*, the decision is still subject to *judicial review*, which, as noted in para 10.9.1 above, is not an appeal. In other words, the decision under attack may be final on the facts, but it is not to be regarded as final on the law.

For example, a finality or ouster clause will not necessarily prevent the quashing of a decision by means of a quashing order on the grounds of excess or abuse of power, breach of the requirements of natural justice, or error of law. Nor will it necessarily preclude the making of a declaration. In *R* v *Medical Appeal Tribunal (ex parte Gilmore)* [1957] 1 QB 574, CA (para 10.5.3.3 above), the Tribunal's decision was quashed for error of law even though the statute in question said that the Tribunal's decision 'shall be final'. Denning LJ said (at p. 583):

I find it very well settled that the remedy by certiorari is never to be taken away by any statute except by the most clear and explicit words.

In *South East Asia Fire Bricks Sdn Bhd* v *Non-Metallic Mineral Products Manufacturing Employees Union* [1981] AC 363, PC, it was held that words in a Malaysian statute that 'an award . . . shall be final and conclusive, and no award shall be challenged, appealed against, reviewed, *quashed* or *called into question in any court of law*' (italics supplied) were clear and explicit enough to exclude the remedy of a quashing order for error of law. But, applying the *Anisminic* case [1969] 2 AC 147, HL (para 10.11.3 below), it was made clear that those words would not effectively exclude a quashing order if the award was *ultra vires* or was a nullity because of violation of the principles of natural justice.

In *R* v *Hallstrom (ex parte W)* [1985] 3 All ER 775, CA, the court had to construe s. 139(1) of the Mental Health Act 1983, which provides that 'no person shall be liable, whether on the ground of want of jurisdiction or on any other ground, to any civil . . . proceedings . . . in respect of any act purporting to be done in pursuance of this Act . . . unless the act was done in bad faith or without reasonable care'. It was held that those words were not sufficiently wide or clear to exclude judicial review by way of a quashing order or a declaration.

In *R (Sivasubramaniam)* v *Wandsworth County Court* [2003] 1 WLR 475, CA, it was argued that s. 54(4) of the Access to Justice Act 1999, which provides that 'no appeal may be made against a decision of a court under this section to give or refuse permission' to appeal, had impliedly ousted the jurisdiction of the High Court to carry out a judicial review of a county court decision to give or refuse permission to appeal. The Court of Appeal, applying Lord Denning's dictum in the *Gilmore* case [1957] 1 QB 574, CA, above, that 'the remedy by certiorari is never to be taken away by any statute except by the most clear and explicit words', rejected the argument.

It will, however, be rare that permission to apply for judicial review of a county court decision giving or refusing permission to appeal will be granted. Such cases will be limited to circumstances where the county court judge lacked jurisdiction to make the decision, or violated the fair hearing principle of natural justice, or acted in complete disregard of his duties (*Sivasubramaniam*, above, *per* Lord Phillips MR, delivering the judgment of the court, at [56]; *Gregory* v *Turner* [2003] 1 WLR 1149, CA, below).

Sivasubramaniam, above, was applied in *Gregory* v *Turner*, CA, above, in which a circuit judge sitting in a county court had refused permission to appeal from the decision of a district judge. By virtue of s. 54(4) of the Access to Justice Act 1999, there could be no *appeal* against the circuit judge's refusal of permission to appeal. That left *judicial review* as the only route by which to challenge the circuit judge's decision, but the claimants were some two years over the usual three month time-limit and had not applied for judicial review. The Court of Appeal pointed out that permission to apply for judicial review would, in any event, have been refused because there was no evidence of lack of jurisdiction, or unfairness, or disregard of duties in the county court proceedings.

The reason for the restrictive approach towards any attempt to exclude judicial remedies is the fear of the ordinary courts of law that public authorities and tribunals might otherwise acquire arbitrary and uncontrollable power. The courts seek to prevent this by means of a presumption that Parliament always intends statutory powers to be exercised lawfully.

Desirable though this approach is, it must be remembered that it can only be adopted against public authorities and tribunals. It probably cannot be used against an inferior *court of law* because Parliament must be taken to have assumed that a court of law is competent to decide questions of law as well as questions of fact. It cannot be used against a superior court for the same reason (*Re Racal Communications Ltd* [1980] 2 All ER 634, HL, *per* Lord Diplock at pp. 639–40). Moreover, the decisions of a superior court of record are not subject to judicial review, with the exception of decisions of the Crown Court on non-indictable matters (para 10.5.2.3 above).

It must also be remembered that the understandable judicial aspiration to prevent the acquisition and exercise of arbitrary and uncontrollable power by an unaccountable executive must yield to any express parliamentary instruction to the contrary. The presumption that powers conferred by Parliament are intended to be exercised lawfully can be rebutted by express statutory words which make the exercise of particular powers 'judge-proof'.

For example, the Regulation of Investigatory Powers Act 2000 created the Investigatory Powers Tribunal with the function of investigating and, if appropriate, remedying, complaints about the unlawful interception of communications and complaints about the activities of the intelligence services (the Security Service, the Secret Intelligence Service, and GCHQ). The Act (replacing earlier provisions contained in the Interception of Communications Act 1985 and the Security Service Act 1989) provides that

... determinations, awards, orders and other decisions of the Tribunal (*including decisions as to whether they have jurisdiction*) shall not be subject to appeal or be liable to be questioned in any court (s. 67(8); italics supplied).

The use of the word 'jurisdiction' would appear to constitute a successful attempt to exclude even the effect of the *Anisminic* case (para 10.11.3 below) from the interpretation of the statute. The result is that the Tribunal's actions, whether they are wrong or a complete nullity, are not only unappealable but also immune from challenge by way of judicial review and beyond the reach of a quashing order. (This result was acknowledged in relation to the Interception of Communications Act 1985 by Taylor J, *obiter*, in *R v Secretary of State for the Home Department (ex parte Ruddock)* [1987] 2 All ER 518 at pp. 527–8.)

The ouster clause in the Regulation of Investigatory Powers Act 2000, above, would have been eclipsed by the even more explicit one initially contained in the Asylum and Immigration (Treatment of Claimants, etc.) Bill of 2004. Clause 14 of the Bill was intended to insert into the Nationality, Immigration and Asylum Act 2002 a new section making the jurisdiction of the Asylum and Immigration Tribunal exclusive and final. The new section would have provided that:

(1) No court shall have any supervisory or other jurisdiction (whether statutory or inherent) in relation to the Tribunal.

(2) No court may entertain proceedings for questioning (whether by way of appeal or otherwise) ... any determination, decision or other action of the Tribunal ...

(3) Subsections (1) and (2) . . . prevent a court, in particular, from entertaining proceedings to determine whether a purported determination, decision or action of the Tribunal was a nullity by reason of—

    (i)   lack of jurisdiction,

    (ii) irregularity,

    (iii) error of law,

    (iv) breach of natural justice, or

    (v) any other matter . . .

The clause was dropped from the Bill in March 2004 in the face of mounting criticism from judges and others.

10.11.2 *Re Racal Communications Ltd*

In *Re Racal Communications Ltd* [1980] 2 All ER 634, HL, a High Court judge had refused to make an order under what was then s. 441 of the Companies Act 1948 authorising the inspection of a company's books or papers. It was provided by s. 441(3) of the 1948 Act that the decision of the High Court judge 'is not appealable'. Notwithstanding these clear words, the Director of Public Prosecutions, who had applied for the order, appealed. The Court of Appeal assumed jurisdiction, reversed the judge's decision, and made the order sought. The reason given was that the judge had erred in law and the words, 'is not appealable', only excluded an appeal on the facts and not on the law.

On a further appeal, the House of Lords unanimously reversed the decision of the Court of Appeal. Their Lordships held that the words, 'is not appealable', mean exactly what they say. There was no right of appeal under s. 441 from the High Court judge's decision either on the facts or on the law. Therefore, the Court of Appeal had no jurisdiction under s. 441 to hear the appeal.

In any case, the jurisdiction of the Court of Appeal is entirely statutory, and the relevant statute (now s. 18(1)(c) of the Supreme Court Act 1981) makes it clear that no appeal lies to the Court of Appeal from any decision of the High Court which by virtue of any statutory provision is final (see *Re Racal Communications Ltd*, above, *per* Lord Diplock at p. 638).

Nor would the Court of Appeal have any jurisdiction to reconsider a High Court judge's decision by way of judicial review. In the first place, the High Court is a superior court of record so that its decisions are not subject to judicial review. And, secondly, the jurisdiction of the Court of Appeal is entirely appellate. It has no original jurisdiction and cannot, therefore, hear original applications for judicial review but is limited to hearing appeals from decisions of the High Court made on such applications.

In *Re Racal Communications Ltd* [1980] 2 All ER 634, HL, above, the earlier decision in *Pearlman* v *Keepers & Governors of Harrow School* [1979] QB 56, CA, in which it was held that the words 'shall be final and conclusive' did not preclude judicial review where a county court judge has exceeded his jurisdiction, was disapproved, as was Lord Denning's dictum in that case (at p. 71) that the operative words would exclude appeals from the county court to the Court of Appeal on questions of fact but not on questions of law.

One of the effects of *Re Racal Communications Ltd*, above, is to make it clear that mistakes of law made by High Court judges can only be corrected by an appeal to an appellate court, and that they cannot be corrected at all if the statute provides that the judge's decision shall not be appealable (see [1980] 2 All ER 634 *per* Lord Diplock at p. 640).

10.11.3 *Anisminic Ltd v Foreign Compensation Commission*

In *Re Racal Communications Ltd*, above, their Lordships were simply explaining and applying what had been laid down by the House in the earlier case of *Anisminic Ltd v Foreign Compensation Commission* [1969] 2 AC 147, HL.

The Foreign Compensation Commission was set up by the Foreign Compensation Act 1950 to decide claims by British subjects to participate in compensation funds paid to Her Majesty's Government by a foreign government. The Commission decided that the claimant company had failed to establish a claim for compensation for the loss of its Egyptian assets following the Suez crisis of 1956. This determination was based on a ground which the Commission had no right to take into account. The 1950 Act provided that the Commission's determination of an application 'shall not be called in question in any court of law'.

The Company applied for a declaration (under what is now r. 40.20 of the CPR) that the Commission's determination was a nullity and that the Company was entitled to participate in the compensation fund. The House of Lords held that it was only a real, and not a purported, 'determination' which could not be called in question. The Commission's 'determination' was *ultra vires* and a nullity. It was no determination at all and could, therefore, be called in question in the courts of law. In the circumstances, the declaration claimed by the company was held to be the appropriate remedy.

The importance of the *Anisminic* case is twofold. First, it is another example of the firm resolution of the courts to prevent the acquisition by public authorities and tribunals of arbitrary and uncontrollable power. Secondly, it provides an extended definition of 'jurisdiction' within which tribunals must confine themselves.

The opinion of Lord Reid is particularly illuminating on this point. He said that the decision of a tribunal could be *ultra vires* and a nullity on many more grounds than simply lack of jurisdiction to consider the matter at issue in the first place. Lord Reid ([1969] 2 AC 147 at p. 171) listed the following additional grounds:

(a) where its decision was given in bad faith;

(b) where it has made a decision which it had no power to make;

(c) where it has failed to comply with the requirements of natural justice;

(d) where, in perfect good faith, it has misconstrued the provisions giving it power to act so that it failed to deal with the question remitted to it and, instead, decided some question which was not remitted to it;

(e) where it has refused to take into account something which it was required to take into account;

(f) where it has taken into account something which it had no right to take into account.

It is clear from the language used in the *Anisminic* case that their Lordships were confining their comments to the decisions of administrative tribunals as opposed to the ordinary courts of law. In the light of the decision in *Re Racal Communications Ltd* [1980] 2 All ER 634, HL, para 10.11.2 above, it is apparent that Anisminic Ltd would have had no remedy at all, either by way of appeal or judicial review, if the Foreign Compensation Act 1950 had provided for the 'determination' to be made by a High Court judge rather than by a statutory tribunal.

The *Anisminic* case was further explained by the House of Lords in *Page v Hull University Visitor* [1993] 1 All ER 97, HL (para 10.5.3.3 above).

10.11.4 Section 12 of the Tribunals and Inquiries Act 1992

As long ago as 1932, it was recommended that ouster clauses in statutes 'should be abandoned in all but the most exceptional cases' (*Report of the Committee on Ministers' Powers*, Cmd 4060, 1932, p. 65). Parliament did nothing in response to this proposal, and reliance continued to be placed on ouster clauses. In 1957, a more fundamental recommendation was made to the effect that the prerogative remedies should not be ousted by statute (*Report of the Committee on Tribunals and Enquiries*, Cmnd 218, 1957). Legislative provision of a limited nature to implement this proposal was contained in the Tribunals and Inquiries Act 1958. The legislation only applies to ouster clauses embodied in pre-1958 statutes, but it does at least make ouster clauses clearer targets for attack.

The rule is now to be found in s. 12 of the Tribunals and Inquiries Act 1992:

As respects England and Wales—

(a) any provision in an Act passed before 1st August 1958 that any order or determination shall not be called into question in any court, or

(b) any provision in such an Act which by similar words excludes any of the powers of the High Court,

shall not have effect so as to prevent the removal of the proceedings into the High Court by order of certiorari or to prejudice the powers of the High Court to make orders of mandamus.

Exceptional cases, where it is still possible for a pre-1958 statute to attempt to oust the jurisdiction of the courts, are mentioned in s. 12(3) of the 1992 Act:

Nothing in this section shall apply—

(a) to any order or determination of a court of law, or

(b) where an Act makes special provision for application to the High Court . . . within a time limited by the Act.

Section 12 of the Tribunals and Inquiries Act 1992 thus seeks to prevent the exclusion of the prerogative remedies of quashing and mandatory orders by pre-1958 statutes. Presumably, it is hoped that statutes passed *after* 1958 will not contain such ouster clauses in the first place, although there is nothing to stop Parliament resorting to them (as noted in para 10.11.1 above). If, however, an ouster clause in a statute passed after 1958 is simply a repetition of a pre-1958 clause, the courts will treat it as a pre-1958 clause and it will be caught by s. 12 (*R v Preston Supplementary Benefits Appeal Tribunal (ex parte Moore)* [1975] 2 All ER 807, CA).

The fact that the decisions of even inferior courts of law are expressly excluded indicates that s. 12 is intended to apply only to the decisions of public authorities and statutory tribunals.

The most significant exception to the operation of s. 12 of the Tribunals and Inquiries Act 1992 is the one relating to statutes which allow an application to be made to the High Court within a stated time-limit, usually six weeks. Provisions of this type are contained in, for example, the Highways Act 1980, the Acquisition of Land Act 1981, the New Towns Act 1981, the Wildlife and Countryside Act 1981, the Housing Act 1985, and the Town and Country Planning Act 1990.

By way of illustration, under the time-limited procedure in the Acquisition of Land Act 1981 (formerly contained in the Acquisition of Land (Authorisation Procedure) Act 1946), any person aggrieved by a compulsory purchase order who wishes to question the validity of the order on the ground that it was *ultra vires* the Act may apply to the High Court. But an application must be made within six weeks from the date on which notice of the confirmation or making of the order was first published. If an application is not made within the six-week period, then the order 'shall not . . . be questioned in any legal proceedings whatsoever'.

There is some justification for an ouster clause of this sort, although the time-limit of six weeks has been criticised as 'pitifully inadequate' (*Smith v East Elloe Rural District Council* [1956] AC 736, HL, *per* Lord Radcliffe at p. 769). Such a clause enables public authorities to proceed with confidence to commit public money to projects after the expiry of the time-limit, knowing that the order cannot be invalidated or their title to land called in question (*R v Secretary of State for the Environment (ex parte Ostler)* [1977] QB 122, CA, *per* Lord Denning MR at p. 136).

In *Smith v East Elloe Rural District Council*, above, the House of Lords, by a majority of three to two, held that once the six-week period has expired it is not possible to challenge the validity of the compulsory purchase order even on the ground that it had been obtained wrongfully and in bad faith. The argument that Parliament could not have intended to take away a remedy where powers had been exercised in bad faith was rejected.

The *East Elloe* case, above, was criticised by three of the Law Lords who later decided the *Anisminic* case [1969] 2 AC 147—Lords Reid, Pearce, and Wilberforce. But the Court of Appeal has since held that the *East Elloe* case was not overruled by the decision in *Anisminic*. This was decided in *R v Secretary of State for the Environment (ex parte Ostler)* [1977] QB 122, CA, in which it was held that a provision in the Highways Act 1959 (since repealed and replaced by a similar provision in the Highways Act 1980) that a compulsory purchase order is not to 'be questioned in any legal proceedings whatever' after the expiry of a six-week time-limit is absolute. This is so even where bad faith and violation of natural justice are alleged. The *East Elloe* case was applied. The *Anisminic* case was distinguished on a number of grounds, two of which may be mentioned here.

First, it was said that the *Anisminic* case only applies where there has been a complete ouster of the jurisdiction of the courts. It does not apply where the ouster operates only after the expiry of a time-limit. Secondly, a distinction was drawn between an administrative decision, which involves considerations of policy and the public interest, and a judicial decision, which does not. The decision of the Secretary of State in the *Ostler* case to confirm the order after an inquiry was an administrative one whereas the decision of the Foreign Compensation Commission in the *Anisminic* case was a judicial one which could be challenged on the ground that it was *ultra vires* and a nullity.

The *Ostler* case [1977] QB 122, CA, above, was followed in *R v Secretary of State for the Environment (ex parte Kent)* [1990] COD 78, CA, where it was held that an appeal outside the six-week time-limit imposed by the planning legislation could not be allowed to proceed (see also *R v Secretary of State for the Environment (ex parte Upton Brickworks)* [1992] COD 301).

In *R v Cornwall County Council (ex parte Huntington)* [1994] 1 All ER 694, CA, decided under the Wildlife and Countryside Act 1981, an attempt was made by counsel for the applicant to distinguish the *Ostler* case on the grounds that:

(a) it only applies to administrative decisions and not to judicial or quasi-judicial decisions; and

(b) it does not apply where the decision under attack is 'fundamentally invalid'.

The attempt failed. The distinction between administrative decisions and judicial or quasi-judicial decisions as a basis for ignoring a finality clause was rejected (despite what was said on this matter in the *Ostler* case (see above)), as was the notion that there are different degrees or grounds of invalidity.

According to the Court of Appeal in *R v Cornwall County Council (ex parte Huntington)*, above, the true position, applying the *Ostler* case, is that:

(a) an *Anisminic*-type statutory clause does not exclude challenges based on invalidity;

(b) an *Ostler*-type clause only allows questions of invalidity to be raised on the specified grounds, within the prescribed time and in the prescribed manner, but otherwise the clause successfully excludes the jurisdiction of the court in the interests of certainty.

In making use of the *Ostler*-type clause, it is taken to be the intention of Parliament that the statutory application to the High Court should be the exclusive remedy available to those affected by the impugned decision. The usual *Ostler*-type clause will, additionally, preclude any legal challenge whatsoever to a decision or order before it has had a chance to take effect or be confirmed (*R v Cornwall County Council (ex parte Huntington)*, CA, above).

Index